OXFORD MASTER SERIES IN CONDENSED MATTER PHY

OXFORD MASTER SERIES IN PHYSICS

The Oxford Master Series is designed for final-year undergraduate and beginning graduate students in physics and related disciplines. It has been driven by a perceived gap in the literature today. While basic undergraduate physics texts often show little or no connection with the huge explosion of research over the last two decades, more advanced and specialized texts tend to be rather daunting for students. In this series, all topics and their consequences are treated at a simple level, while pointers to recent developments are provided at various stages. The emphasis in on clear physical principles like symmetry, quantum mechanics, and electromagnetism which underlie the whole of physics. At the same time, the subjects are related to real measurements and to the experimental techniques and devices currently used by physicists in academe and industry. Books in this series are written as course books, and include ample tutorial material, examples, illustrations, revision points, and problem sets. They can likewise be used as preparation for students starting a doctorate in physics and related fields, or for recent graduates starting research in one of these fields in industry.

CONDENSED MATTER PHYSICS

1. M. T. Dove: *Structure and dynamics: an atomic view of materials*
2. J. Singleton: *Band theory and electronic properties of solids*
3. A. M. Fox: *Optical properties of solids*
4. S. J. Blundell: *Magnetism in condensed matter*
5. J. F. Annett: *Superconductivity, superfluids, and condensates*
6. R. A. L. Jones: *Soft condensed matter*
17. S. Tautz: *Surfaces of condensed matter*
18. H. Bruus: *Theoretical microfluidics*

ATOMIC, OPTICAL, AND LASER PHYSICS

7. C. J. Foot: *Atomic physics*
8. G. A. Brooker: *Modern classical optics*
9. S. M. Hooker, C. E. Webb: *Laser physics*
15. A. M. Fox: *Quantum optics: an introduction*
16. S. M. Barnett: *An introduction to quantum information*

PARTICLE PHYSICS, ASTROPHYSICS, AND COSMOLOGY

10. D. H. Perkins: *Particle astrophysics*
11. T. P. Cheng: *Relativity, gravitation and cosmology*

STATISTICAL, COMPUTATIONAL, AND THEORETICAL PHYSICS

12. M. Maggiore: *A modern introduction to quantum field theory*
13. W. Krauth: *Statistical mechanics: algorithms and computations*
14. J. P. Sethna: *Statistical mechanics: entropy, order parameters, and complexity*

Theoretical Microfluidics

HENRIK BRUUS

Department of Micro- and Nanotechnology
Technical University of Denmark

OXFORD
UNIVERSITY PRESS

Great Clarendon Street, Oxford OX2 6DP

Oxford University Press is a department of the University of Oxford.
It furthers the University's objective of excellence in research, scholarship,
and education by publishing worldwide in

Oxford New York

Auckland Cape Town Dar es Salaam Hong Kong Karachi
Kuala Lumpur Madrid Melbourne Mexico City Nairobi
New Delhi Shanghai Taipei Toronto

With offices in

Argentina Austria Brazil Chile Czech Republic France Greece
Guatemala Hungary Italy Japan Poland Portugal Singapore
South Korea Switzerland Thailand Turkey Ukraine Vietnam

Oxford is a registered trade mark of Oxford University Press
in the UK and in certain other countries

Published in the United States
by Oxford University Press Inc., New York

First published 2008

Reprinted with corrections 2009, 2010, 2011

British Library Cataloguing in Publication Data

Data available

Library of Congress Cataloging in Publication Data

Data available

Typeset by Newgen Imaging Systems (P) Ltd., Chennai, India
Printed in Great Britain
on acid-free paper by
CPI Antony Rowe, Chippenham, Wilts.

ISBN 978–0–19–923508–7 (Hbk)
ISBN 978–0–19–923509–4 (Pbk)

10 9 8 7 6 5 4

Preface

This book on theoretical microfluidics has grown out from a set of lecture notes that I began writing in the summer of 2004. Much of the material has been tested in my teaching at the Technical University of Denmark at BSc-, MSc- and PhD-level lecture courses. The courses have been followed by students of both experimental and of theoretical inclination, and it is my experience that both groups of students have benefitted from the lecture notes. The more than 200 students I have been in contact with during the past three years have helped me shape the presentation of the material in a way that appears useful for them in their studies.

Microfluidics is a vast and rapidly evolving research field, and this textbook is in no way meant to be an exhaustive review. Instead, my ambition has been to write a final-year undergraduate textbook, which in a self-contained manner presents the basic theoretical concepts and methods used in the cross-disciplinary field of microfluidics, thus closing the gap between a number of basic physics textbooks and contemporary research in microfluidics. It is my hope that the presentation of basic theory, many worked-through examples and exercises with solutions, will get the advanced undergraduate students or first-year graduate students to understand the foundation of the theory, to be able to use the theory as a practical tool, and to be able to read research papers about microfluidics and lab-on-a-chip systems. Moreover, I have tried to write the text so that the students in principle can read the book as a self-study. The gaps in the logical and mathematical progression are deliberately made relatively small, thus making it possible for a student to fill them in by herself.

To write a textbook is hard work, but I have been so fortunate that many students and colleagues have helped me on the way with inspiring discussions and in many cases direct comments on the book as it evolved. In particular, I would like to thank my talented PhD students for many interesting joint research projects: Goran Goranović, Christian Mikkelsen, Anders Brask, Mads Jakob Jensen, Lennart Bitsch, Laurits Højgaard Olesen, Martin Heller and Misha Gregersen, as well as PhD students S. Melker Hagsäter and Kristian Smistrup, whom I have co-supervised. Also, my latest MSc students, Thomas Eilkær Hansen, Thomas Glasdam Jensen and Peder Skafte-Pedersen, have been very helpful. Among my local theory colleagues I have in particular enjoyed the feedback from Niels Asger Mortensen and Fridolin Okkels in connection with the courses on microfluidics that we have taught together. Several of my experimental colleagues have contributed with pictures and measurement results, and they have been thanked at each individual instance throughout the book.

In a broader context, I have received much inspiration from my colleagues at the Center for Fluid Dynamics at the Technical University of Denmark (Fluid•DTU) Hassan Aref, Tomas Bohr, Morten Brøns, Ole Hassager, Jens Nørkær Sørensen, Jens H. Walther, and

Kirstine Berg-Sørensen, and the same can be said about my international colleagues Armand Ajdari, Daniel Attinger, Martin Bazant, Steffen Hardt, Ralph Lindken, Howard Stone and Patrick Tabeling.

Finally, I am grateful to Pernille and Christian for letting me use their beach house at Kandestederne as an author's refuge in two absolutely critical phases during my writing of this book: a couple of weeks in August 2004, when I wrote what is now the first couple of chapters in the book, and again several weeks in March, April and May 2007, when I wrote the last chapters and finished the book.

Professor Henrik Bruus
Department of Micro- and Nanotechnology
Technical University of Denmark
May 2007

www.nanotech.dtu.dk/bruus

Contents

List of symbols

Symbol	Meaning	Definition
$\equiv$	Equal to by definition	Eq. (1.4a)
$\approx$	Approximately equal to	Eq. (1.1)
$\propto$	Proportional to	Eq. (1.1)
$\ll$, $\gg$	Much smaller than, much bigger than	Section 1.5
$\cdot$	Scalar product	Eq. (1.11)
$:$	Tensor contraction, trace	Eq. (1.14)
$\times$	Cross-product or multiplication sign	Eqs. (1.24) and (1.7)
$\partial_j = \frac{\partial}{\partial x_j}$	Partial derivative after jth co-ordinate	Eq. (1.15)
$\boldsymbol{\nabla}$	Nabla or gradient operator	Eq. (1.17)
$\boldsymbol{\nabla}\cdot$	Divergence operator	Eq. (1.20)
$\boldsymbol{\nabla}\times$	Rotation operator	Eq. (1.25)
∇^2	Laplace operator	Eq. (1.18)
$\langle\bullet\|$, $\|\bullet\rangle$, $\langle\bullet\|\bullet\rangle$	Dirac bra-ket notion for functions	Eq. (1.42)
$\mathcal{A}$, $\mathcal{A}_n$	Area and effective area	Eqs. (3.25) and (4.15a)
$\mathcal{A}$	Absorbance	Eq. (16.22)
a	Radius of a cylinder or a sphere	Figs. 3.7 and 3.12
$\mathbf{B}$	Magnetic induction	Eqs. (11.1) and (16.1)
b	Stokes mobility	Eq. (11.15b)
Bo	Bond number	Eq. (7.30)
C_α	Dimensionless concentration, mass fraction	Eq. (5.16)
Ca	Capillary number	Eq. (7.31)
C_{hyd}	Compliance	Section 4.6
$\mathcal{C}$	Compactness	Eq. (4.19)
c, c_α	Concentration, molecules per volume	Eq. (5.24)
c_0	Speed of light in vacuum	Eq. (16.7)
c_{a}	Speed of sound	Eq. (15.12)
c_{p}	Specific heat, constant pressure	Eq. (2.65)
$\mathcal{D}$	Arbitrary differential operator	Eq. (15.12)
$\mathbf{D}$	Electric displacement	Eqs. (8.2) and (8.12)
D	Molecular diffusivity	Eqs. (5.9) and (5.21)
D_t	Material time derivative	Eq. (2.34)
D_{th}	Thermal diffusivity	Eq. (2.67)
$\mathrm{d}_t = \frac{\mathrm{d}}{\mathrm{d}t}$	Total time derivative	Eq. (1.16)
$\mathbf{d}$	Distance vector in polarization	Fig. 8.1(a)

Symbol	Meaning	Definition
$\mathbf{E}$	Electric field	Eq. (8.2)
e	The exponential constant exp(1)	Eq. (1.29)
e	The elementary charge	Eq. (8.13)
F, F_1	Helmholtz free energy	Eqs. (D.2) and (D.4)
$\mathbf{F}_{\mathrm{DEP}}$	Dielectrophoretic force	Eqs. (10.23) and (10.43)
$\mathbf{F}_{\mathrm{drag}}$	Stokes' drag force	Section 3.7
$\mathbf{F}_{\mathrm{el}}$	Electric force	Eqs. (8.4) and (8.13)
f	Frequency	Eqs. (16.8b) and (E.5b)
$\mathbf{f}$	Force density	Section 2.2.6
G	Gibbs free energy	Eqs. (7.1) and (D.5)
$\mathbf{g}$, g	Gravitational acceleration	Section 3.1
$\mathbf{H}$	Magnetic field	Eqs. (11.5) and (16.1)
h	Channel height	Fig. 3.6
h	Planck's constant	Eq. (16.21)
i	The imaginary unit $\sqrt{-1}$	Eq. (1.29)
I_{eo}	Electro-osmotic current	Eq. (9.18)
I_{th}	Heat current	Eq. (12.58)
$\mathbf{J}$	Flux density, current density	Eq. (2.3)
$\mathbf{J}_{\mathrm{heat}}$	Heat-flux density	Eq. (2.55)
$K(\epsilon_1, \epsilon_2)$	Clausius–Mossotti factor	Eq. (10.19)
k_{B}	Boltzmann's constant	Eq. (6.50)
L	Length of a channel	Fig. 3.5
$\mathbf{M}$	Magnetization	Eq. (11.3)
N_{St}	Stokes number	Eq. (7.32)
$\mathcal{O}(x^n)$	Terms of order x^n and higher powers	Eq. (1.36)
$P(s)$	Normal distribution	Eq. (5.30)
$\mathbf{P}$	Polarization	Eqs. (8.8) and (10.1)
$\mathbf{p}$	Electric dipole moment	Eqs. (8.5b) and (10.18)
$\mathcal{P}$	Perimeter	Eq. (4.15b)
p^*	Standard pressure	Eq. (3.3)
$p(\mathbf{r}, t)$	Pressure field	Sections 1.3.3 and 2.2.2
p_{eo}	Electro-osmotic pressure	Eq. (9.35b)
$P\acute{e}$	Péclet number	Eq. (5.53)
$P\acute{e}_{\mathrm{th}}$	Thermal Péclet number	Eq. (12.45)
Q, Q_{mass}	Volumetric and mass flow rates	Eqs. (3.21a) and (3.21b)
$\mathbf{q}$	2D mass flux vector	Eq. (14.34)
Q_{eo}	Electro-osmotic flow rate	Eqs. (9.15) and (9.35a)
$\mathcal{R}$	Half the hydraulic diameter	Eq. (3.26)
R_{hyd}	Hydraulic resistance	Chapter 4
$\mathbf{r}$	Position vector	Eq. (1.9)
Re	Reynolds number	Section 2.2.7
s	Entropy per unit mass	Section 2.3
T	Temperature	Eqs. (2.50) and (12.2)
$\mathcal{T}$	Transmittance	Eq. (16.22)
Tr	Trace of a tensor	Eq. (1.14)
$\mathcal{V}$, $\Delta\mathcal{V}$	Volume	Eq. (1.4a)
$\mathbf{v}(\mathbf{r}, t)$	Eulerian velocity field	Section 1.3.3
$\mathbf{v}(\mathbf{r}(t), t)$	Lagrangian velocity field	Section 1.3.3

Symbol	Meaning	Definition
$V_{\mathrm{LJ}}(\mathbf{r})$	Lennard-Jones potential	Eqs. (1.54) and (17.36)
v_{eo}	Electro-osmotic velocity	Eq. (9.11)
Z	Integer valence number	Eq. (8.13)
Z	Acoustic impedance	Eq. (15.61)
Z_1, Z_N	Single- and N-particle partition function	Section D.1
α	Perturbation parameter	Section 1.5
α_{p}	Permeability of a porous medium	Eq. (14.39)
β	Dimensionless viscosity ratio	Eq. (2.19c)
Γ	Radius ratio	Eq. (10.30)
γ	Surface tension	Eq. (7.2)
γ	Dimensionless damping coefficient	Eqs. (15.38) and (16.14)
$\dot{\gamma}_{ij}$	Shear rate tensor	Eq. (2.46a)
δ	Dimensionless extinction coefficient	Eq. (6.42a)
δ_{ij}	Kronecker delta	Eq. (1.22)
$\delta(x)$	the Dirac delta function	Section 5.3.1
ε	Energy per unit mass	Section 2.3.1
ϵ,ϵ_0	Electric permittivity	Eqs. (8.12) and (16.2a)
$\epsilon(\omega)$	Complex dielectric function	Eq. (10.42)
ϵ_{ijk}	Levi–Civita symbol	Eq. (1.23)
ζ	Dimensionless z co-ordinate	Eq. (3.60)
ζ	Second viscosity	Eq. (2.19b)
ζ	Zeta-potential	Eq. (8.21)
η	Dimensionless y co-ordinate	Eq. (3.60)
η	Dynamic viscosity	Eq. (2.19a)
η^*	Standard dynamic viscosity	Eq. (3.61)
Θ	Dimensionless temperature difference	Eq. (12.2)
θ	Polar angle	Section C.3
κ	Thermal conductivity	Eq. (2.55)
$\kappa(\mathbf{r})$, $\kappa(s)$	Curvature of surfaces and curves	Eqs. (7.10) and (7.11)
λ	Wavelength	Appendix E
λ^*	Size of a fluid particle	Eq. (1.2)
$\lambda(\xi)$	Shape perturbation function	Eqs. (3.69) and (14.4)
λ_{D}	Debye length	Eq. (8.26)
λ_{s}	Navier slip length	Eq. (17.1)
μ, μ_{r}	Magnetic and relative magnetic permeability	Eqs. (11.8b) and (16.2b)
μ_0	Magnetic permeability of vacuum	Eq. (16.9)
μ_{eo}	Electro-osmotic mobility	Eq. (9.12)
μ_{ion}	Ionic mobility	Eq. (8.15)
ν	Kinematic viscosity	Eq. (5.44)
ξ	Dimensionless x co-ordinate	Eqs. (3.60) and (14.4)
Π_{ij}	Momentum flux density tensor	Eq. (2.15)
$\rho(\mathbf{r},t)$	Density field	Sections 1.3.3 and 2.1
ρ^*	Standard density	Eq. (3.6)
ρ_{el}	Electric charge density	Eqs. (8.3b) and (8.23)
σ_{ij}	Full stress tensor	Eq. (2.26)
σ'_{ij}	Viscous stress tensor	Eq. (2.18)
σ_{el}	Electric conductivity	Eqs. (8.2d) and (16.4)
σ_{ion}	Ionic conductivity	Eq. (8.18)

Symbol	Meaning	Definition
τ	Characteristic time scale	Eq. (6.11), (7.35),(12.16b)
τ_{th}	Thermal diffusion time	Eq. (12.16a)
ϕ	Azimuthal angle	Sections C.2 and C.3
ϕ, ϕ_i	Velocity potential	Eq. (13.6), (15.3), (15.15)
$\phi_{\mathrm{eq}}, \phi_{\mathrm{ext}}$	Equilibrium and external electric potential	Eq. (9.2)
ϕ_{dip}	Potential from a point dipole	Eq. (10.6)
χ	Electric susceptibility	Eqs. (8.12) and (10.1)
χ	Magnetic susceptibility	Eq. (11.7)
ψ	Stream function	Eq. (6.32)
Ω	Region of interest in 2D or 3D	Eq. (2.1)
$\partial\Omega$	Surface of region Ω	Eq. (2.1)
$\boldsymbol{\omega}$	Vorticity	Eq. (2.43)
ω	Angular frequency	Eq. (E.5b)
ω_{c}	Critical dielectric frequency	Eq. (10.45)
ω_{D}	Debye frequency	Eqs. (8.45) and (14.52c)
ᚨ	Imprint speed ratio	Exercise 17.6

1 Basic concepts in microfluidics

Theoretical microfluidics deals with the theory of flow of fluids and of suspensions in submillimeter-sized systems influenced by external forces. Although an old discipline in hydrodynamics, the scientific and technological interest in and development of microfluidics has been particularly significant during the past decade and a half following the emerging and rapidly evolving field of lab-on-a-chip systems. This field is mainly driven by technological applications, the vision being to develop entire bio/chemical laboratories on the surface of silicon or polymer chips. Many of the amazing techniques developed over the past fifty years in connection with the silicon-based microelectronics industry are now used to fabricate lab-on-chip systems. In recent years, also polymer-based lab-on-a-chip systems have emerged, and these systems promise cheaper and faster production cycles. As microfluidic technology advances the demand for better theoretical insight grows, and this has been one of the motivating factors behind this book.

1.1 Lab-on-a-chip technology

There are several advantages of scaling down standard laboratory setups by a factor of 1000 or more from the decimeter scale to the 100 µm scale. One obvious advantage is the dramatic reduction in the amount of required sample. A linear reduction by a factor of 10^3 amounts to a volume reduction by a factor of 10^9, so instead of handling 1 L or 1 mL a lab-on-a-chip system could easily deal with as little as 1 nL or 1 pL. Such small volumes allow for very fast analysis, efficient detection schemes, and analysis, even when large amounts of sample are unavailable. Moreover, the small volumes makes it possible to develop compact and portable systems that might ease the use of bio/chemical handling and analysis systems tremendously. Finally, as has been the case with microelectronics, it is the hope by mass production to manufacture very cheap lab-on-a-chip systems.

Lab-on-a-chip (LOC) systems can be thought of as the natural generalization of the existing electronic integrated circuits and microelectromechanical systems (MEMS). Why confine the systems to contain only electric and mechanical parts? Indeed, a lab-on-chip system can really be thought of as the shrinking of an entire laboratory to a chip. Two examples of a systems evolving in that direction are shown in Fig. 1.1: In panel (a) is shown the CalTech microfluidics large-scale integration chip containing 256 subnanoliter reaction chambers controlled by 2056 on-chip microvalves, while in panel (b) is shown the DTU Nanotech integrated optochemical lab-on-a-chip system containing optical (lasers and waveguides), chemical (channels and mixers), and electronic (photodiodes) components. Perhaps, only our imagination sets the limits of what could be integrated in a lab-on-a-chip system. It is

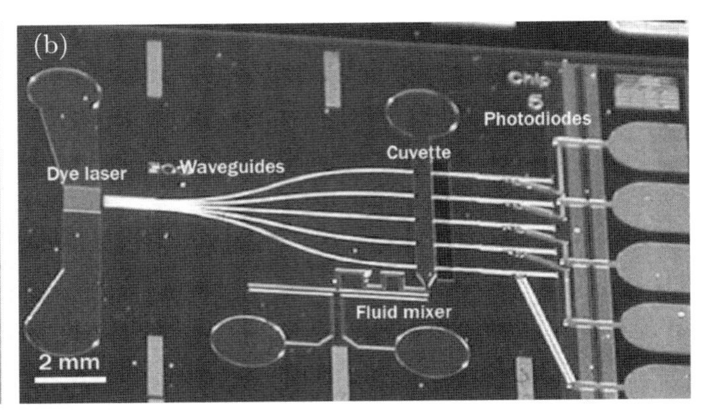

Fig. 1.1 (a) An optical micrograph of a 27 mm by 27 mm polydimethylsiloxane-based (PDMS) large-scale integrated microfluidic comparator containing 256 subnanoliter reaction chambers and 2056 microvalves fabricated at CalTech by Thorsen *et al.*, Science **298**, 580-584 (2002), reprinted with permission from AAAS. (b) An optical micrograph of a 15 mm by 20 mm integrated opto-chemical lab-on-a-chip system for optical analysis of chemical reactions fabricated at DTU Nanotech by Balslev *et al.*, Lab Chip **6**, 213-217 (2006), reproduced by permission of The Royal Society of Chemistry. The system is a hybrid polymer/silicon device made on a silicon substrate containing the integrated photodiodes, while the laser, waveguides, mixer and cuvette are made in a SU-8 polymer film on top of the substrate.

expected that lab-on-a-chip systems will have great impact in biotechnology, pharmacology, medical diagnostics, forensics, environmental monitoring and basic research.

The fundamental laws of Nature underlying our understanding of the operation of lab-on-a-chip systems are all well known. Throughout the book we shall draw on our knowledge from mechanics, fluid dynamics, acoustics, electromagnetism, thermodynamics and physical chemistry. What is new, however, is the interplay between many different forces and the change of the relative importance of these forces as we pass from the m- and mm-sized macrosystems to µm- and nm-sized micro- and nanosystems.

1.2 Scaling laws in microfluidics

When analyzing the physical properties of microsystems, it is helpful to introduce the concept of scaling laws. A scaling law expresses the variation of physical quantities with the size ℓ of the given system or object, while keeping other quantities such as time, pressure, temperature, *etc.* constant. As an example, consider volume forces, such as gravity and inertia, and surface forces, such as surface tension gradients and shear stresses. The basic scaling law for the ratio of these two classes of forces can generally be expressed by

$$\frac{\text{surface forces}}{\text{volume forces}} \propto \frac{\ell^2}{\ell^3} = \ell^{-1} \underset{\ell\to 0}{\longrightarrow} \infty. \tag{1.1}$$

This scaling law implies that when scaling down to the microscale in lab-on-a-chip systems, the volume forces, which are very prominent in our daily life, become largely unimportant. Instead, the surface forces become dominant, and as a consequence, we must rebuild our intuition and be prepared for some surprises on the way.

Table 1.1 The scaling laws as a function of a typical length scale, object size or distance ℓ for a number of physical quantities studied in this book.

Area	ℓ^2	Eq. (1.1)	Time	ℓ^0	Section 1.2
Volume	ℓ^3	Eq. (1.1)	Velocity	ℓ^1	Section 1.2
Rel. fluctuations	$\ell^{-\frac{1}{2}}$	Exercise 1.3	Hydrostatic pressure	ℓ^1	Eq. (3.3)
Reynolds number *Re*	ℓ^2	Eq. (2.39)	Hydraulic resistance	ℓ^{-4}	Table 4.1
Péclet number	ℓ^2	Eq. (5.53)	Stokes drag	ℓ^1	Eq. (3.128)
Diffusion time	ℓ^2	Eq. (5.10)	Particle diffusion const.	ℓ^{-1}	Eq. (6.49)
Fluid acceleration time	ℓ^2	Eq. (6.25)	Taylor dispersion time	ℓ^{-2}	Eq. (5.71)
Young–Laplace pressure	ℓ^{-1}	Eq. (7.8)	Contact angle	ℓ^0	Eq. (7.14)
Bond number *Bo*	ℓ^2	Eq. (7.30)	Capillary rise height	ℓ^{-1}	Eq. (7.21)
Marangoni force	ℓ^{-1}	Eq. (7.39)	Capillary speed	ℓ^1	Eq. (7.36)
Electric field	ℓ^{-1}	Eq. (8.3a)	EO velocity	ℓ^{-1}	Eq. (9.11)
Ionic mobility	ℓ^{-1}	Eq. (8.15)	EO mobility	ℓ^0	Eq. (9.12)
Debye length	ℓ^0	Eq. (8.26)	EO flow rate	ℓ^1	Eq. (9.35a)
Debye frequency	ℓ^{-1}	Eq. (8.45)	EO pressure	ℓ^{-2}	Eq. (9.35b)
DEP force, particle	ℓ^3	Eq. (10.23)	MAP force, particle	ℓ^3	Eq. (11.13)
DEP force, system	ℓ	Eq. (10.31)	MAP force, system	ℓ	Eq. (11.13)
Thermal diffusion time	ℓ^2	Eq. (12.16a)	Acoustic impedance	ℓ^0	Eq. (15.61)
Thermal resistance	ℓ^{-1}	Eq. (12.59)	Acoustic radiation force	ℓ^3	Eq. (15.81)
Thermal capacitance	ℓ^3	Eq. (12.61)	Optical absorbance	ℓ^1	Eq. (16.25b)
Thermal RC-time	ℓ^2	Eq. (12.63)	Optical damping coeff.	ℓ^0	Eq. (16.14)

In Table 1.1 are listed the scaling laws for a number of physical quantities studied in this book. Depending on the context the length scale ℓ is either controlling all lengths in the system while maintaining constant aspect ratios or it represents a single object-size or -distance, which is being downscaled while maintaining all other lengths of the system. The table gives a first impression of the intricate interplay between the many physical forces present in microfluidic systems.

1.3 Fluids and fields

The main purpose of a lab-on-a-chip system is to handle fluids. A fluid, i.e. either a liquid or a gas, is characterized by the property that it will deform continuously and with ease under the action of external forces. The shape of a fluid is determined by the vessel containing it, and different parts of the fluid may be rearranged freely without affecting the macroscopic properties of it. In a fluid the presence of shear forces, however small in magnitude, will result in large changes in the relative positions of the fluid elements. In contrast, the changes in the relative positions of the atoms in a solid remain small under the action of any small external force. When applied external forces cease to act on a fluid, it will not necessarily retract to its initial shape. This property is also in contrast to a solid, which relaxes to its initial shape when no longer influenced by (small) external forces.

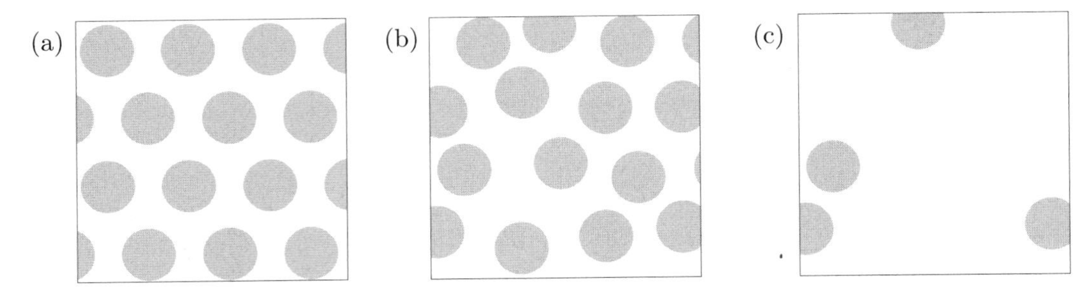

Fig. 1.2 (a) A sketch of a typical solid with 0.1 nm wide atoms (or molecules) and a lattice constant of 0.3 nm. The molecules oscillate around the indicated equilibrium points forming a regular lattice. (b) A sketch of a liquid with the same molecules and same average intermolecular distance 0.3 nm as in panel (a). The molecules move around in a thermally induced irregular pattern. (c) A sketch of a gas with the same molecules as in panel (a). The average intermolecular distance is 3 nm, and the motion is free between the frequent intermolecular collisions.

1.3.1 Fluids: liquids and gases

The two main classes of fluids, the liquids and the gases, differ primarily by the densities and by the degree of interaction between the constituent molecules as sketched in Fig. 1.2. The density $\rho_{\mathrm{gas}} \approx 1\ \mathrm{kg\,m^{-3}}$ of an ideal gas is so low, at least a factor of 10^3 smaller than that of a solid, that the molecules move largely as free particles that only interact by direct collisions at atomic distances, ≈ 0.1 nm. The relatively large distance between the gas molecules, ≈ 3 nm, makes the gas compressible. The density $\rho_{\mathrm{liq}} \approx 10^3\ \mathrm{kg\,m^{-3}}$ of a liquid is comparable to that of a solid, i.e. the molecules are packed as densely as possible with a typical average intermolecular distance of 0.3 nm, and a liquid can, for many practical purposes, be considered incompressible.

The intermolecular forces in a liquid are of quite an intricate quantum and electric nature since each molecule is always surrounded by a number of molecules within atomic distances. In model calculations of simple liquids many features can be reproduced by assuming the basic Lennard-Jones pair-interaction potential, $V_{\mathrm{LJ}}(r) = 4\varepsilon\big[(\sigma/r)^{12} - (\sigma/r)^{6}\big]$, between any pair of molecules. Here, r is the distance between the molecules, while the maximal energy of attraction ε and the collision diameter σ are material parameters typically of the order $100\ \mathrm{K}\times k_{\mathrm{B}}$ and 0.3 nm, respectively. The corresponding intermolecular force is given by the derivative $F_{\mathrm{LJ}}(r) = -\mathrm{d}V_{\mathrm{LJ}}/\mathrm{d}r$. The Lennard-Jones potential is shown in Fig. 1.3(a) and discussed further in Exercise 1.2.

At short time intervals and up to a few molecular diameters the molecules in a liquid are ordered almost as in a solid. However, whereas the ordering in solids remains fixed in time and space,[1] the ordering in liquids fluctuates. In some sense the thermal fluctuations are strong enough to overcome the tendency to order, and this is the origin of the ability of liquids to flow.

[1]The molecules in a solid execute only small, thermal oscillations around equilibrium points well described by a regular lattice.

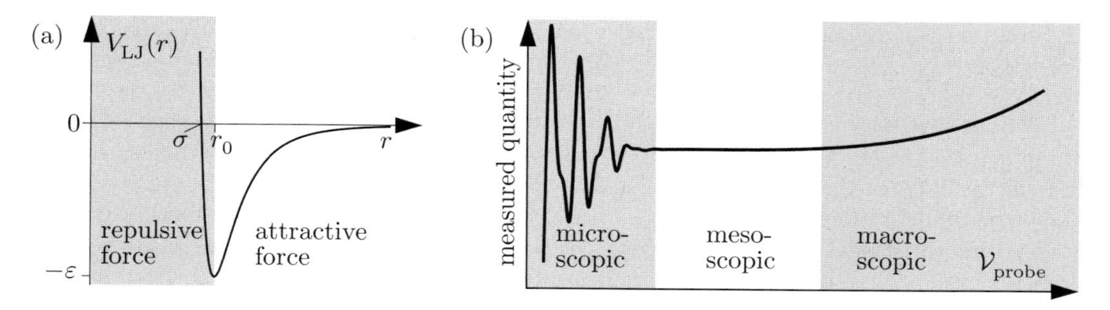

Fig. 1.3 (a) The Lennard-Jones pair-potential $V_{\rm LJ}(r)$ often used to describe the interaction potential between two molecules at distance r, see also Exercise 1.2. For small distances, $r < r_0 \approx 0.3$ nm the interaction forces are strongly repulsive (gray region), while for large distances, $r > r_0$, they are weakly attractive. (b) A sketch adopted from Batchelor (2000) of some measured physical quantity of a liquid as a function of the volume $\mathcal{V}_{\rm probe}$ probed by some instrument. For microscopic probe volumes (left gray region) large molecular fluctuations will be observed. For mesoscopic probe volumes (white region) a well-defined local value of the property can be measured. For macroscopic probe volumes (right gray region) gentle variations in the fluid due to external forces can be observed.

1.3.2 The continuum hypothesis and fluid particles

Although fluids are quantized on the length scale of intermolecular distances (of the order 0.3 nm for liquids and 3 nm for gases), they appear continuous in most lab-on-a-chip applications, since these typically are defined on macroscopic length scales of the order 10 µm or more. In this book we shall therefore assume the validity of the continuum hypothesis, which states that the macroscopic properties of a fluid are the same if the fluid were perfectly continuous in structure instead of, as in reality, consisting of molecules. Physical quantities such as the mass, momentum and energy associated with a small volume of fluid containing a sufficiently large number of molecules are to be taken as the sum of the corresponding quantities for the molecules in the volume.

The continuum hypothesis leads to the concept of fluid particles, the basic constituents in the theory of fluids. In contrast to an ideal point particle in ordinary mechanics, a fluid particle in fluid mechanics has a finite size. But how big is it? Well, the answer to this question is not straightforward. Imagine, as illustrated in Fig. 1.3(b), that we probe a given physical quantity of a fluid with some probe sampling a volume $\mathcal{V}_{\rm probe}$ of the fluid at each measurement. Let $\mathcal{V}_{\rm probe}$ change from (sub-)atomic to macroscopic dimensions. At the atomic scale (using, say, a modern AFM or STM tool) we would encounter large fluctuations due to the molecular structure of the fluid, but as the probe volume increases we soon enter a size where steady and reproducible measurements are obtained. This happens once the probe volume is big enough to contain a sufficiently large number of molecules, such that well-defined average values with small statistical fluctuations are obtained. As studied in Exercise 1.3 a typical possible side length λ^* in a cubic fluid particle in a liquid is

$$\lambda^* \approx 10 \text{ nm}, \quad \text{(for a liquid).} \tag{1.2}$$

Such a liquid particle contains approximately 4×10^4 molecules and exhibits number fluctuations of the order 0.5%. For a fluid particle in a gas λ^* is roughly ten times larger. If the size of the fluid particle is taken too big the probe volume could begin to sample regions of the fluid with variations in the physical properties due to external forces. In that case we are beyond the concept of a constituent particle and enter the regime we actually would like to study, namely, how do the fluid particles behave in the presence of external forces.

A fluid particle must thus be ascribed a size λ^* in the mesoscopic range. It must be larger than microscopic lengths ($\simeq 0.3$ nm) to contain a sufficiently large number of molecules, and it must be smaller than macroscopic lengths ($\simeq 10$ µm) over which external forces change the property of the fluid. Of course, this does not define an exact size, and in fluid mechanics it is therefore natural to work with physical properties per volume, such as mass density, energy density, force density and momentum density. In such considerations the volume is taken to the limit of a small, but finite, fluid-particle volume, and not to the limit of an infinitesimal volume.

The continuum hypothesis breaks down when the system under consideration approaches the molecular scale. This happens in nanofluidics, e.g. in liquid transport through nanopores in cell membranes or in artificially made nanochannels.

1.3.3 The velocity, pressure and density field

Once the concept of fluid particles in a continuous fluid has been established we can move on and describe the physical properties of the fluid in terms of fields. This can basically be done in two ways, as illustrated in Fig. 1.4 for the case of the velocity field. In this book we shall use the Eulerian description, Fig. 1.4(a), where one focuses on fixed points $\mathbf{r}$ in space and observes how the fields evolve in time at these points, i.e. the position $\mathbf{r}$ and the time t are independent variables. The alternative is the Lagrangian description, Fig. 1.4(b), where one follows the history of individual fluid particles as they move through the system, i.e. the co-ordinate $\mathbf{r}_a(t)$ of particle a depends on time.

In the Eulerian description the value of any field variable $F(\mathbf{r},t)$ is defined as the average value of the corresponding molecular quantity $F_\mathrm{mol}(\mathbf{r}',t)$ for all the molecules contained in some liquid particle of volume $\Delta\mathcal{V}(\mathbf{r})$ positioned at $\mathbf{r}$ at time t,

$$F(\mathbf{r},t) = \big\langle F_\mathrm{mol}(\mathbf{r}',t)\big\rangle_{\mathbf{r}'\in\Delta\mathcal{V}(\mathbf{r})}. \tag{1.3}$$

If we for brevity let m_i and $\mathbf{v}_i$ be the mass and the velocity of molecule i, respectively, and

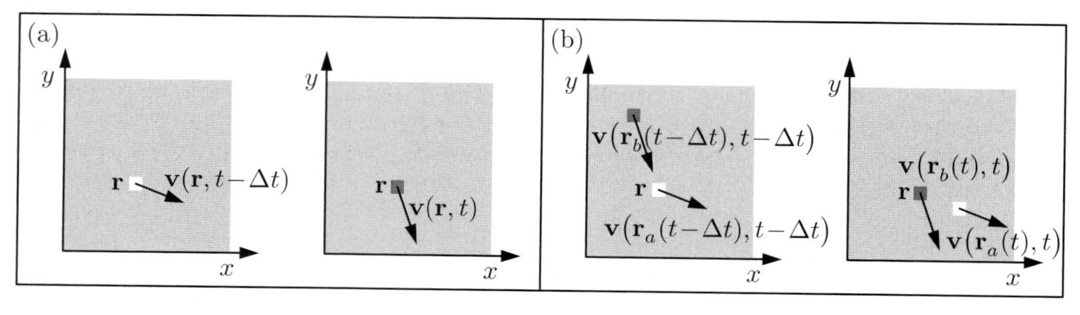

Fig. 1.4 (a) The velocity field $\mathbf{v}(\mathbf{r},t)$ in the Eulerian description at the point $\mathbf{r}$ at the two times $t-\Delta t$ and t. The spatial co-ordinates $\mathbf{r}$ are independent of the temporal co-ordinate t. (b) The Lagrangian velocity fields $\mathbf{v}(\mathbf{r}_a(t),t)$ and $\mathbf{v}(\mathbf{r}_b(t),t)$ of fluid particles a (white) and b (dark gray). The particles pass the point $\mathbf{r}$ at time $t-\Delta t$ and t, respectively. The particle co-ordinates $\mathbf{r}_{a,b}(t)$ depend on t. Note that $\mathbf{r}_a(t-\Delta t) = \mathbf{r}$ and $\mathbf{r}_b(t) = \mathbf{r}$.

furthermore let $i \in \Delta\mathcal{V}$ stand for all molecules i present inside the volume $\Delta\mathcal{V}(\mathbf{r})$ at time t, then the definition of the density $\rho(\mathbf{r},t)$ and the velocity field $\mathbf{v}(\mathbf{r},t)$ can be written as

$$\rho(\mathbf{r},t) \equiv \frac{1}{\Delta\mathcal{V}} \sum_{i\in\Delta\mathcal{V}} m_i, \tag{1.4a}$$

$$\mathbf{v}(\mathbf{r},t) \equiv \frac{1}{\rho(\mathbf{r},t)\Delta\mathcal{V}} \sum_{i\in\Delta\mathcal{V}} m_i\mathbf{v}_i. \tag{1.4b}$$

Here, we have introduced the "equal-to-by-definition sign" $\equiv$. Notice how the velocity is defined through the more fundamental concept of momentum. Through the technique of micro particle-image velocimetry, micro-PIV, see Santiago *et al.* (1998), it is possible to measure the velocity field in a transparent microfluidic device. Microparticles with diameters of the order 1 µm are suspended in the flow to be measured. Their positions are recorded as gray-scale values in a CCD camera through an optical microscope using either transmitted or reflected light. Two pictures are recorded by sending two light pulses in quick succession of the order milliseconds apart at time t_1 and t_2, and the corresponding light intensities in each CCD camera pixel positioned at $\mathbf{r}$ are denoted $I_1(\mathbf{r})$ and $I_2(\mathbf{r})$. The CCD pixel array is divided into a number of interrogation areas n, and for each of those a cross-correlation function $R_n(\Delta\mathbf{r})$ is defined as the average over all pixel co-ordinates in the given interrogation area n as $R_n(\Delta\mathbf{r}) \equiv \langle I_1(\mathbf{r})I_2(\mathbf{r}+\Delta\mathbf{r})\rangle_n$, where $\Delta\mathbf{r}$ is some pixel displacement vector. The value $\Delta\mathbf{r}_n$ of $\Delta\mathbf{r}$ that maximizes $R_n(\Delta\mathbf{r})$ is a statistical measure of the overall displacement of the fluid inside the given interrogation area n. Thus the average flow velocity $\mathbf{v}_n$ of that area is given by

$$\mathbf{v}_n \equiv \frac{\Delta\mathbf{r}_n}{t_2 - t_1}. \tag{1.5}$$

Examples of micro-PIV measurements of velocity fields can be seen in Figs. 15.4 and 15.5.

In general, the field variables in microfluidics can be scalars (such as density ρ, viscosity η, pressure p, temperature T, and free energy $\mathcal{F}$), vectors (such as velocity $\mathbf{v}$, current density $\mathbf{J}$, pressure gradient $\boldsymbol{\nabla}p$, force densities $\mathbf{f}$, and electric fields $\mathbf{E}$) and tensors (such as stress tensor σ and velocity gradient $\boldsymbol{\nabla}\mathbf{v}$).

To obtain a complete description of the state of a moving fluid it is necessary to know the three components of the velocity field $\mathbf{v}(\mathbf{r},t)$ and any two of the thermodynamical variables of the fluid, e.g. the pressure field $p(\mathbf{r},t)$ and the density field $\rho(\mathbf{r},t)$. All other thermodynamical quantities can be derived from these fields together with the equation of state of the fluid.

1.4 SI units and mathematical notation

Notation is an important part in communicating scientific and technical material. Especially in fluid mechanics a precise mathematical notation is important due to the elaborate many-variable differential calculus on the scalar, vector and tensor fields mentioned in the previous section. Instead of regarding units and notation as an annoying burden the student should instead regard it as part of the trade that needs to be mastered by the true professional. Learn the basic rules, and stick to them thereafter.

1.4.1 SI units

Throughout this book we shall use the SI units. If not truly familiar with this system, the name and spelling of the units, or the current best values of the fundamental physical constants of Nature, the reader is urged to consult the websites of the Bureau International des Poids et Mesures (BIPM) or the National Institute of Standards and Technology (NIST) for constants, units, and uncertainty at

http://www.bipm.fr/en/si/ , (1.6a)

http://physics.nist.gov/cuu/ . (1.6b)

A scalar physical variable is given by a number of significant digits, a power of ten and a proper SI unit. The power of ten can be moved to the unit using prefixes such as giga, kilo, micro, atto, *etc.* The SI unit can be written in terms of the seven fundamental units or suitable derived units. As an example the viscosity η of water at 20 ℃ is written as

$$\eta = 1.002 \times 10^{-3}\ \mathrm{kg\,m^{-1}s^{-1}} = 1.002\ \mathrm{mPa\,s}. \tag{1.7}$$

Note the multiplication sign before the power of ten and the space after it, and note that the SI units are written in roman and *not in italics*. Unfortunately, most typesetting systems will automatically use italics for letters written in equations. Note also the space inserted between the units. Be aware that even though many units are capitalized, as are the names of the physicists that gave rise to them, e.g. Pa and Pascal, the unit itself is never capitalized when written in full, e.g. pascal. Also, the unit is written pascal without plural form whether there is one, five or 3.14 of them.

There will be two exceptions from the strict use of SI units. Sometimes, just as above, temperatures will be given in °C, so be careful when inserting values for temperature in formulae. Normally, a temperature T in an expression calls for values in kelvin. The other exception from SI units is the atomic unit of energy, electronvolt (eV),

$$1\ \mathrm{eV} = 1.602 \times 10^{-19}\ \mathrm{J} = 0.1602\ \mathrm{aJ}. \tag{1.8}$$

Note that it would be possible to use attojoule instead of electronvolt, but this is rarely done.

1.4.2 Vectors, derivatives and the index notation

The mathematical treatment of microfluidic problems is complicated due to the presence of several scalar, vector and tensor fields and the non-linear partial differential equations that govern them. To facilitate the treatment some simplifying notation is called for.

First, a suitable co-ordinate system must be chosen. We shall encounter three, as summarized in Appendix C: Cartesian co-ordinates (x, y, z) with corresponding basis vectors $\mathbf{e}_x$, $\mathbf{e}_y$, and $\mathbf{e}_z$; cylindrical co-ordinates (r, ϕ, z) with corresponding basis vectors $\mathbf{e}_r$, $\mathbf{e}_\phi$, and $\mathbf{e}_z$; and spherical co-ordinates (r, θ, ϕ) with corresponding basis vectors $\mathbf{e}_r$, $\mathbf{e}_\theta$, and $\mathbf{e}_\phi$. All sets of basis vectors are orthonormal, which means that the vectors involved have unity length and are mutually orthogonal, but the Cartesian basis vectors are special since they are constant in space, whereas all other sets of basis vectors depend on position in space. For simplicity, we postpone the usage of the curvilinear co-ordinates to later chapters and use only Cartesian co-ordinates in the following.

The position vector $\mathbf{r} = (r_x, r_y, r_z) = (x, y, z)$ can be written as

$$\mathbf{r} = r_x\,\mathbf{e}_x + r_y\,\mathbf{e}_y + r_z\,\mathbf{e}_z = x\,\mathbf{e}_x + y\,\mathbf{e}_y + z\,\mathbf{e}_z. \tag{1.9}$$

In fact, any vector $\mathbf{v}$ can be written in terms of its components v_i (where for Cartesian co-ordinates $i = x, y, z$) as

$$\mathbf{v} = \sum_{i=x,y,z} v_i\,\mathbf{e}_i \equiv v_i\,\mathbf{e}_i. \tag{1.10}$$

In the last equality we have introduced the Einstein summation convention: by definition a repeated index always implies a summation over that index. Other examples of this handy notation, the so-called index notation, is the scalar product,

$$\mathbf{v}\cdot\mathbf{u} = v_i u_i, \tag{1.11}$$

the length v of a vector $\mathbf{v}$,

$$v = |\mathbf{v}| = \sqrt{\mathbf{v}^2} = \sqrt{\mathbf{v}\cdot\mathbf{v}} = \sqrt{v_i v_i}, \tag{1.12}$$

and the ith component of the vector-matrix equation $\mathbf{u} = M\mathbf{v}$,

$$u_i = M_{ij}\,v_j. \tag{1.13}$$

Likewise, the full contraction or double-dot product of two tensors T and S, which in fact is the trace $\mathrm{Tr}\,(TS)$, can be written as

$$T{:}S \equiv \sum_{i,j} T_{ij}S_{ji} = T_{ij}S_{ji} = \mathrm{Tr}\,(TS). \tag{1.14}$$

Further studies of the index notation can be found in Exercise 1.4.

For the partial derivatives of some function $F(\mathbf{r}, t)$ we use the symbols ∂_i, with $i = x, y, z$, and ∂_t,

$$\partial_x F \equiv \frac{\partial F}{\partial x}, \quad \text{and} \quad \partial_t F \equiv \frac{\partial F}{\partial t}, \tag{1.15}$$

while for the total time derivative, as, e.g. in the case of the Lagrangian description of some variable $F\big(\mathbf{r}(t), t\big)$ following the fluid particles, see Fig. 1.4(b), we use the symbol d_t,

$$\mathrm{d}_t F \equiv \frac{\mathrm{d}F}{\mathrm{d}t} = \partial_t F + \big(\partial_t r_i\big)\partial_i F = \partial_t F + v_i \partial_i F. \tag{1.16}$$

The nabla operator $\boldsymbol{\nabla}$ containing the spatial derivatives plays an important role in differential calculus. In Cartesian co-ordinates it is given by

$$\boldsymbol{\nabla} \equiv \mathbf{e}_x\partial_x + \mathbf{e}_y\partial_y + \mathbf{e}_z\partial_z = \mathbf{e}_i\partial_i. \tag{1.17}$$

Note that we have written the differential operators to the right of the unit vectors. While not important for Cartesian co-ordinates it is crucial when working with curvilinear co-ordinates. The Laplace operator, which appears in numerous partial differential equations in theoretical physics, is just the square of the nabla operator,

$$\boldsymbol{\nabla}^2 \equiv \nabla^2 \equiv \partial_i \partial_i. \tag{1.18}$$

In terms of the nabla operator the total time derivative in Eq. (1.16) can be written as

$$\mathrm{d}_t F\big(\mathbf{r}(t), t\big) = \partial_t F + (\mathbf{v} \cdot \boldsymbol{\nabla}) F. \tag{1.19}$$

Since $\boldsymbol{\nabla}$ is a differential operator, the order of the factors does matter in a scalar product containing it. So, whereas $\mathbf{v} \cdot \boldsymbol{\nabla}$ in the previous equation is a differential operator, the product $\boldsymbol{\nabla} \cdot \mathbf{v}$ with the reversed order of the factors is a scalar quantity. It appears so often in mathematical physics that it has acquired its own name, namely the divergence of the vector field,

$$\boldsymbol{\nabla} \cdot \mathbf{v} \equiv \partial_x v_x + \partial_y v_y + \partial_z v_z = \partial_i v_i. \tag{1.20}$$

Concerning integrals, we denote the 3D integral measure by $\mathrm{d}\mathbf{r}$, so that in Cartesian co-ordinates we have $\mathrm{d}\mathbf{r} = \mathrm{d}x\, \mathrm{d}y\, \mathrm{d}z$, in cylindrical co-ordinates $\mathrm{d}\mathbf{r} = r \mathrm{d}r\, \mathrm{d}\phi\, \mathrm{d}z$, and in spherical co-ordinates $\mathrm{d}\mathbf{r} = r^2 \mathrm{d}r\, \sin\theta\, \mathrm{d}\theta\, \mathrm{d}\phi$. We also consider definite integrals as operators acting on integrands, thus we keep the integral sign and the associated integral measure together to the left of the integrand. As an example, the integral over a spherical body with radius a of the scalar function $S(\mathbf{r})$ is written as

$$\int_{\text{sphere}} S(x, y, z)\, \mathrm{d}x\, \mathrm{d}y\, \mathrm{d}z = \int_{\text{sphere}} \mathrm{d}\mathbf{r}\, S(\mathbf{r}) = \int_0^a r^2 \mathrm{d}r \int_0^\pi \sin\theta \mathrm{d}\theta \int_0^{2\pi} \mathrm{d}\phi\, S(r, \theta, \phi). \tag{1.21}$$

When working with vectors and tensors it is advantageous to use the following two special symbols: the Kronecker delta δ_{ij},

$$\delta_{ij} = \begin{cases} 1, & \text{for } i = j, \\ 0, & \text{for } i \neq j, \end{cases} \tag{1.22}$$

and the Levi–Civita symbol ϵ_{ijk},

$$\epsilon_{ijk} = \begin{cases} +1, & \text{if } (ijk) \text{ is an even permutation of } (123) \text{ or } (xyz), \\ -1, & \text{if } (ijk) \text{ is an odd permutation of } (123) \text{ or } (xyz), \\ 0, & \text{otherwise.} \end{cases} \tag{1.23}$$

In the index notation, the Levi–Cevita symbol appears directly in the definition of the cross-product of two vectors $\mathbf{u}$ and $\mathbf{v}$,

$$(\mathbf{u} \times \mathbf{v})_i \equiv \epsilon_{ijk}\, u_j v_k, \tag{1.24}$$

and in the definition of the rotation $\boldsymbol{\nabla} \times \mathbf{v}$ of a vector $\mathbf{v}$. The expression for the ith component of the rotation is:

$$(\boldsymbol{\nabla} \times \mathbf{v})_i \equiv \epsilon_{ijk}\, \partial_j v_k. \tag{1.25}$$

To calculate in the index notation the rotation of a rotation, such as $\boldsymbol{\nabla} \times (\boldsymbol{\nabla} \times \mathbf{v})$, or the rotation of a cross-product it is very helpful to know the following expression for the product of two Levi–Civita symbols with one pair of repeated indices (here k):

$$\epsilon_{ijk}\, \epsilon_{lmk} = \delta_{il}\delta_{jm} - \delta_{im}\delta_{jl}. \tag{1.26}$$

Note the plus sign when pairing index 1 with 1 and 2 with 2 (direct pairing), while a minus sign appears when pairing index 1 with 2 and 2 with 1 (exchange pairing).

Let us end this short introduction to the index notation by an explicit example, namely the proof of the expression for the rotation of a rotation of a vector field $\mathbf{v}$,

$$\boldsymbol{\nabla} \times (\boldsymbol{\nabla} \times \mathbf{v}) = \boldsymbol{\nabla}(\boldsymbol{\nabla}\cdot\mathbf{v}) - \nabla^2\mathbf{v}, \tag{1.27a}$$

or the equivalent expression for the gradient of the divergence,

$$\boldsymbol{\nabla}(\boldsymbol{\nabla}\cdot\mathbf{v}) = \nabla^2\mathbf{v} + \boldsymbol{\nabla} \times (\boldsymbol{\nabla} \times \mathbf{v}). \tag{1.27b}$$

These expressions are used in hydrodynamics, acoustics and electromagnetism. First we write out the ith component of the left-hand side of Eq. (1.27a) using the Levi–Civita symbol for each cross-product, one at a time,

$$\big[\boldsymbol{\nabla} \times (\boldsymbol{\nabla} \times \mathbf{v})\big]_i = \epsilon_{ijk}\partial_j(\boldsymbol{\nabla} \times \mathbf{v})_k = \epsilon_{ijk}\partial_j(\epsilon_{klm}\partial_l v_m) = \epsilon_{ijk}\epsilon_{klm}\,\partial_j\partial_l v_m. \tag{1.28a}$$

Then we permute the indices in the second Levy–Civita symbol and apply Eq. (1.26),

$$\big[\boldsymbol{\nabla}\times(\boldsymbol{\nabla}\times\mathbf{v})\big]_i = \epsilon_{ijk}\epsilon_{lmk}\,\partial_j\partial_l v_m = (\delta_{il}\delta_{jm}-\delta_{im}\delta_{jl})\,\partial_j\partial_l v_m = \delta_{il}\delta_{jm}\,\partial_j\partial_l v_m - \delta_{im}\delta_{jl}\,\partial_j\partial_l v_m. \tag{1.28b}$$

Finally, we perform the sum over the indices appearing in the Kronecker deltas and get

$$\big[\boldsymbol{\nabla}\times(\boldsymbol{\nabla}\times\mathbf{v})\big]_i = \partial_j\partial_i v_j - \partial_j\partial_j v_i = \partial_i(\partial_j v_j) - (\partial_j\partial_j)v_i = \partial_i(\boldsymbol{\nabla}\cdot\mathbf{v}) - \nabla^2 v_i = \big[\boldsymbol{\nabla}(\boldsymbol{\nabla}\cdot\mathbf{v}) - \nabla^2\mathbf{v}\big]_i, \tag{1.28c}$$

which indeed proves Eq. (1.27a) and therefore also Eq. (1.27b).

Finally, we use upright letters for the two mathematical constants

$$\mathrm{e} \equiv \text{the exponential constant } \exp(1), \quad \text{and} \quad \mathrm{i} \equiv \text{the imaginary unit } \sqrt{-1}. \tag{1.29}$$

1.5 Perturbation theory

As we shall see shortly, the governing differential equations of microfluidics can only be solved analytically in a few idealized cases. Given sufficiently powerful computers, it is of course possible to solve almost any problem numerically, however, it is often of great value also to find analytical approximations to these solutions. One general applicable approximation scheme is the so-called perturbation theory, which we shall se examples of throughout the book. Here, we give a short introduction to the method.

Imagine that a given problem can be formulated in terms of some partial differential operator $\mathcal{D}$ acting on the field $f(\mathbf{r}, t)$ as

$$\mathcal{D}f = 0. \tag{1.30}$$

The goal is to determine f. Now, assume that the differential operator can be written as a series expansion

$$\mathcal{D} = \mathcal{D}_0 + \alpha\mathcal{D}_1 + \alpha^2\mathcal{D}_2 + \cdots, \tag{1.31}$$

where $\mathcal{D}_0$ is a differential operator, which represents a simpler problem that we can solve, where α is a small dimensionless parameter, known as the perturbation parameter, which describes how far the actual problem deviates from the simpler solvable problem, and where $\mathcal{D}_i$ for $i > 0$ are known differential operators. The simpler problem is also denoted the unperturbed problem, while terms proportional to α and higher powers in α are denoted the perturbation terms. For $\alpha = 0$ the actual problem is identical to the simpler problem, and can thus be solved, while for $0 < |\alpha| \ll 1$ the actual problem deviates slightly from the simpler problem. The idea is to calculate the field f by successively finding the higher-order terms in the expansion

$$f = f_0 + \alpha f_1 + \alpha^2 f_2 + \cdots . \tag{1.32}$$

Inserting the perturbation series for $\mathcal{D}$ and f results, under the assumption of proper convergence, in the following expression for $\mathcal{D}f$,

$$\begin{aligned} \mathcal{D}f &= (\mathcal{D}_0 + \alpha\mathcal{D}_1 + \alpha^2\mathcal{D}_2 + \cdots)(f_0 + \alpha f_1 + \alpha^2 f_2 + \cdots) \\ &= \mathcal{D}_0 f_0 + \alpha(\mathcal{D}_1 f_0 + \mathcal{D}_0 f_1) + \alpha^2(\mathcal{D}_2 f_0 + \mathcal{D}_1 f_1 + \mathcal{D}_0 f_2) + \cdots . \end{aligned} \tag{1.33}$$

The original problem Eq. (1.30), $\mathcal{D}f = 0$ is consequently reformulated as

$$\mathcal{D}_0 f_0 + \alpha(\mathcal{D}_1 f_0 + \mathcal{D}_0 f_1) + \alpha^2(\mathcal{D}_2 f_0 + \mathcal{D}_1 f_1 + \mathcal{D}_0 f_2) + \cdots = 0. \tag{1.34}$$

For this to be true for any value of α each term must be zero, and we get the following infinite system of equations to solve,

$$\mathcal{D}_0 f_0 = 0, \qquad \text{order } \alpha^0 \text{ terms}, \tag{1.35a}$$

$$\mathcal{D}_0 f_1 = -\mathcal{D}_1 f_0, \qquad \text{order } \alpha^1 \text{ terms}, \tag{1.35b}$$

$$\mathcal{D}_0 f_2 = -\mathcal{D}_2 f_0 - \mathcal{D}_1 f_1, \qquad \text{order } \alpha^2 \text{ terms}, \tag{1.35c}$$

$$\vdots \qquad\qquad \vdots .$$

By assumption, the homogeneous zero-order equation (1.35a) is the unperturbed, solvable problem, and f_0 can therefore be found. This implies that the first-order equation (1.35b) becomes an inhomogeneous differential equation for the first-order contribution f_1 with a known right-hand side, so in principle we can now find f_1. This in turn means that the second-order equation (1.35c) has become an inhomogeneous differential equation for f_2 with a known right-hand side determined by the, at this point known, lower-order fields f_0 and f_1, and in principle f_2 can be found. In this way the perturbation scheme allows for consecutive determination of f_n once all lower-order contributions f_i, for $i < n$, have been found.

In practice, it is only possible or worthwhile to calculate a few of the terms, rarely going beyond the second order contribution. The series is therefore truncated by neglecting all terms $\mathcal{O}(\alpha^3)$ containing α^3 or higher powers, and an approximate result is found,

$$f = f_0 + \alpha f_1 + \alpha^2 f_2 + \mathcal{O}(\alpha^3) \approx f_0 + \alpha f_1 + \alpha^2 f_2. \tag{1.36}$$

In some cases it may be advantageous to absorb the perturbation parameter α into the operators and functions as $\alpha^n\mathcal{D}_n \to \mathcal{D}_n$ and $\alpha^n f_n \to f_n$, so that only the index indicates the order. The perturbation expansions Eqs. (1.31) and (1.32) then become

$$\mathcal{D} = \mathcal{D}_0 + \mathcal{D}_1 + \mathcal{D}_2 + \cdots , \tag{1.37a}$$

$$f = f_0 + f_1 + f_2 + \cdots . \tag{1.37b}$$

In the differential equation $\mathcal{D}f = 0$, terms of order n are those where the sum of indices is n, e.g. $\mathcal{D}_k f_{n-k}$, so the expanded differential equation (1.34) is now written as

$$(\mathcal{D}_0 f_0) + (\mathcal{D}_1 f_0 + \mathcal{D}_0 f_1) + (\mathcal{D}_2 f_0 + \mathcal{D}_1 f_1 + \mathcal{D}_0 f_2) + \cdots = 0, \tag{1.38}$$

where the parentheses contain the zero-order term, the first-order terms, the second-order terms, *etc.* The infinite system of equations (1.35) is unaffected by making the perturbation parameter α implicit.

1.6 Eigenfunction expansion

A second general approach involves expansion of the hydrodynamic fields in certain basis functions or eigenfunctions $\phi_n(\mathbf{r})$, which are found as the eigenfunctions to simpler differential equation eigenvalue problems related to the problem to be solved. Often, the basis functions $\phi_n(\mathbf{r})$ are determined from the Helmholtz equation with Dirichlet boundary conditions involving the Laplace operator with eigenvalues k_n^2 in a domain Ω with the boundary $\partial\Omega$,

$$\nabla^2\phi_n(\mathbf{r}) = -k_n^2\,\phi_n(\mathbf{r}), \quad \text{for } \mathbf{r}\in\Omega \text{ and } n = 1,2,3,\ldots, \tag{1.39a}$$

$$\phi_n(\mathbf{r}) = 0, \qquad \text{for } \mathbf{r}\in\partial\Omega. \tag{1.39b}$$

For a 1D domain the expansion is the standard Fourier expansion in sine functions, while for a 2D circular domain it is the Fourier–Bessel expansion in Bessel functions. Any field $f(\mathbf{r})$ can be expanded in the basis- or eigenfunctions as

$$f(\mathbf{r}) = \sum_{n=1}^{\infty} a_n\phi_n(\mathbf{r}), \tag{1.40}$$

and the problem is solved once the coefficients a_n are determined.

The basis- or eigenfunctions are mutually orthogonal in the sense that the integral over the domain Ω of a product of two different eigenfunctions is zero. Furthermore, by proper normalization the integral over the square of a single eigenfunction is unity,

$$\int_\Omega d\mathbf{r}\; \phi_n(\mathbf{r})\phi_m(\mathbf{r}) = \delta_{nm}, \tag{1.41}$$

where we have used the Kronecker delta Eq. (1.22). This special orthonormality property of the eigenfunctions forms the basis for the so-called Hilbert space theory of functions, where functions can be thought of as abstract vectors in a linear vector space. In physics, Hilbert spaces are used particularly in quantum physics, and the reader unfamiliar with the concept is referred to any basic textbook on quantum theory for further reading. We shall briefly use some of the concepts in fluid dynamics mainly as a convenient shorthand notation. First, we introduce the Dirac bra-ket notation for real-valued functions and integrals,

$$\langle f| \equiv f(\mathbf{r}), \qquad \text{the bra of } f, \tag{1.42a}$$

$$|g\rangle \equiv g(\mathbf{r}), \qquad \text{the ket of } g, \tag{1.42b}$$

$$\langle f|g\rangle \equiv \int_\Omega d\mathbf{r}\; f(\mathbf{r})\,g(\mathbf{r}), \text{ the bra(c)ket of } f\,g. \tag{1.42c}$$

For complex-valued functions the function f associated with the bra-vector is to be complex conjugated. The bra(c)ket $\langle f|g\rangle$ is also known as the inner product of f and g, and the bra-ket notation makes it possible to interpret the complicated operation of "*multiplying by a function f followed by integration over the domain*" as a simple "*multiply from the left by the bra $\langle f|$*". From the linearity of integrals it is easy to show, see Exercise 1.6, that

$$\langle a_1 f_1 + a_2 f_2 | b_1 g_1 + b_2 g_2 \rangle = a_1 b_1 \langle f_1 | g_1 \rangle + a_1 b_2 \langle f_1 | g_2 \rangle + a_2 b_1 \langle f_2 | g_1 \rangle + a_2 b_2 \langle f_2 | g_2 \rangle. \tag{1.43}$$

In the Dirac notation the expansion Eq. (1.40) becomes

$$|f\rangle = \sum_{n=1}^{\infty} a_n |\phi_n\rangle, \tag{1.44}$$

and the othonormality property Eq. (1.41) appears as

$$\langle \phi_n | \phi_m \rangle = \delta_{nm}. \tag{1.45}$$

If we assume that the function $f(\mathbf{r})$ is known, then we can determine the expansion coefficients a_n by multiplying Eq. (1.44) from the left by $\langle \phi_m|$ and utilizing the linearity of the inner product as well as the othonormality of the eigenfunctions,

$$\langle \phi_m | f \rangle = \langle \phi_m | \left[\sum_{n=1}^{\infty} a_n |\phi_n\rangle \right] \rangle = \sum_{n=1}^{\infty} a_n \langle \phi_m | \phi_n \rangle = a_m. \tag{1.46}$$

This simple result is the generalization of the familiar method of determining Fourier coefficients. We note that by combining Eqs. (1.44) and (1.46) we obtain

$$|f\rangle = \sum_{n=1}^{\infty} \langle \phi_n | f \rangle \, |\phi_n\rangle = \sum_{n=1}^{\infty} |\phi_n\rangle \langle \phi_n | f \rangle = \left[\sum_{n=1}^{\infty} |\phi_n\rangle \langle \phi_n | \right] |f\rangle, \tag{1.47}$$

and consequently the parenthesis on the right-hand side must be unity, and we have derived the so-called completeness condition

$$\sum_{n=1}^{\infty} |\phi_n\rangle \langle \phi_n | \equiv 1. \tag{1.48}$$

As a simple example of how to use the eigenfunction expansion to solve problems, let us consider the Poisson equation for the unknown field $f(\mathbf{r})$ with a known source term $g(\mathbf{r})$,

$$\nabla^2 f(\mathbf{r}) = -g(\mathbf{r}). \tag{1.49}$$

Inserting the expansions $|f\rangle = \sum_{n=1}^{\infty} a_n |\phi_n\rangle$ and $|g\rangle = \sum_{n=1}^{\infty} \langle \phi_n | g \rangle |\phi_n\rangle$ into the Poisson equation leads to

$$\sum_{n=1}^{\infty} a_n \nabla^2 |\phi_n\rangle = -\sum_{n=1}^{\infty} \langle \phi_n | g \rangle \, |\phi_n\rangle. \tag{1.50}$$

Employing the fundamental relation Eq. (1.39a) and multiplying from the left by $\langle \phi_m|$ leads, as shown in Exercise 1.6 to the determination of the coefficients a_m, and the solution can be written as

$$|f\rangle = \sum_{n=1}^{\infty} \frac{\langle \phi_n | g \rangle}{k_n^2} \, |\phi_n\rangle. \tag{1.51}$$

This simple example points to the usefulness of the eigenfunction-expansion approach. The eigenvalues k_n^2 are increasing rapidly for increasing values of n, so for many practical purposes it suffices to truncate the infinite sum and include only the first few terms.

In Section 3.4.1 we shall use eigenfunction expansion to study liquid flow through straight channels of constant but arbitrarily shaped cross-sections. In this context the domain Ω is the 2D cross-section, and issues related to the area and areal coverage become important. In the Dirac notation the area $\mathcal{A}$ of the domain Ω is written as

$$\mathcal{A} = \int_\Omega \mathrm{d}\mathbf{r}\ 1 = \int_\Omega \mathrm{d}\mathbf{r}\ 1^2 = \langle 1|1\rangle. \tag{1.52}$$

Likewise, when integrating over an eigenfunction $\phi_n(\mathbf{r})$ over the domain we get a measure of how much area this eigenfunction effectively covers. However, since $\langle 1|1\rangle = \mathcal{A}$ and $\langle \phi_n|\phi_n\rangle = 1$ the dimension of $\langle 1|\phi_n\rangle$ is seen to be length. Therefore the so-called effective area $\mathcal{A}_n$ of eigenfunction $\phi_n(\mathbf{r})$ is defined as the square of the area integral,

$$\mathcal{A}_n = \left[\int_\Omega \mathrm{d}\mathbf{r}\ \phi_n(\mathbf{r})\right]^2 = \left[\int_\Omega \mathrm{d}\mathbf{r}\ \phi_n(\mathbf{r})\ \phi_n(\mathbf{r})\right]^2 = \left[\langle 1|\phi_n\rangle\right]^2 = \langle \phi_n|1\rangle\langle 1|\phi_n\rangle. \tag{1.53}$$

1.7 Further reading

Two classic textbooks on the fundamentals of fluid dynamics are Landau and Lifshitz (1993) and Batchelor (2000), and more are listed in Section 2.4. Books focusing in particular on microfluidics or lab-on-a-chip systems are Karniadakis and Beskok (2002), Geschke, Klank and Telleman (2004), Tabeling (2005), and Berthier and Silberzan (2006); these titles are well supplemented by the review papers on microfluidics by Stone, Stroock and Ajdari (2004) and Squires and Quake (2005). For perturbation theory in fluid mechanics see Van Dyke (1975), while eigenfunction expansion in the Dirac notation is treated in quantum mechanics textbooks such as Merzbacher (1998) and Bruus and Flensberg (2004).

1.8 Exercises

Exercise 1.1
The intermolecular distance in air
Assume that air at room temperature and a pressure of 1000 hPa is an ideal gas. Estimate the average intermolecular distance. Compare the result with that of liquids.

Exercise 1.2
The Lennard-Jones potential for intermolecular pair-interaction
An approximative but quite useful expression for intermolecular pair-interactions is the so-called Lennard-Jones potential,

$$V_\mathrm{LJ}(r) = 4\varepsilon\left[\left(\frac{\sigma}{r}\right)^{12} - \left(\frac{\sigma}{r}\right)^{6}\right]. \tag{1.54}$$

Let r_0 be the distance at which the pair of molecules experience the smallest possible interaction energy.

(a) Determine r_0 in units of the collision diameter σ and calculate the corresponding interaction energy $V(r_0)$ in units of the maximum attraction energy ε.

(b) Calculate $V_\mathrm{LJ}(3\sigma)$ and use the result to discuss the applicability of the ideal-gas model to air, given that for nitrogen $\sigma_{\mathrm{N}_2} = 0.3667$ nm and $\varepsilon_{\mathrm{N}_2}/k_\mathrm{B} = 99.8$ K.

Exercise 1.3
The size of the fundamental fluid particle in a liquid
Consider a small cube of side length λ^* in the middle of some liquid. The typical average intermolecular distance in the liquid is the one discussed in Section 1.3.1. Due to random thermal fluctuations the molecules inside the cube are continuously exchanged with the surrounding liquid, but on average there are N molecules inside the cube. For sufficiently small fluctuations the cube can play the role of a fundamental fluid particle.

(a) Use the standard result from basic statistics that the standard deviation of the counting number of uncorrelated random events (here, the number N of molecules inside the cube) is given by $\sqrt{N}$ to estimate the side length λ, such that the relative uncertainty $\sqrt{N}/N$ of the number of molecules is 1%.

(b) Determine λ^* such that the relative uncertainty of the number of molecules is 0.1%.

Exercise 1.4
The index notation
To become familiar with the index notation try to work out the following problems.

(a) Use the index notation to prove that $\partial_k(p\,\delta_{ik}) = (\boldsymbol{\nabla}p)_i$.

(b) Use the index notation to prove that $\boldsymbol{\nabla}\cdot(\rho\mathbf{v}) = (\boldsymbol{\nabla}\rho)\cdot\mathbf{v} + \rho\boldsymbol{\nabla}\cdot\mathbf{v}$.

(c) Prove that Eq. (1.25) for the rotation of a vector is correct.

(d) Use Eqs. (1.24) and (1.26) to prove that $\mathbf{a}\times(\mathbf{b}\times\mathbf{c}) = (\mathbf{a}\cdot\mathbf{c})\mathbf{b} - (\mathbf{a}\cdot\mathbf{b})\mathbf{c}$.

Exercise 1.5
First-order perturbation of the damped, harmonic oscillator
Consider the 1D, damped, harmonic oscillator of mass m, force constant k, damping coefficient γ and position co-ordinate $x(t)$ described by the equation of motion

$$m\,\partial_t^2 x = -kx - m\gamma\,\partial_t x. \tag{1.55}$$

The initial condition is given by $x(0) = \ell$ and $\partial_t x(0) = 0$. We study the solution of this problem using perturbation theory with the damping as the perturbation. The unperturbed oscillator has $\gamma = 0$ and the solution $x_0(t) = \ell\,\cos(\omega_0 t)$ with $\omega_0 \equiv \sqrt{k/m}$.

(a) Introduce the following dimensionless variable $\tilde{x}$ and $\tilde{t}$ by the definitions $x \equiv \ell\,\tilde{x}$ and $t \equiv \tilde{t}/\omega_0$, and let $\alpha = \gamma/\omega_0$ be the dimensionless perturbation parameter. Calculate the first-order perturbation result $\tilde{x} = \tilde{x}_0 + \alpha\,\tilde{x}_1$.

(b) Find the exact solution using a trial solution of the complex form $\tilde{x} = \exp(\mathrm{i}\beta\tilde{t})$, and compare a first-order expansion in α of the result with the first-order perturbation result.

Exercise 1.6
The Dirac bra-ket notation
The Dirac bra-ket notation is a compact notation that makes it possible to maintain the overview in a complex calculation without getting swamped by details.

(a) Prove the linearity relation for the inner product given in Eq. (1.43):
$\langle a_1 f_1 + a_2 f_2 | b_1 g_1 + b_2 g_2\rangle = a_1 b_1 \langle f_1|g_1\rangle + a_1 b_2 \langle f_1|g_2\rangle + a_2 b_1 \langle f_2|g_1\rangle + a_2 b_2 \langle f_2|g_2\rangle$.

(b) Use the Dirac notation expansions $|f\rangle = \sum_{n=1}^{\infty} a_n|\phi_n\rangle$ and $|g\rangle = \sum_{n=1}^{\infty}\langle\phi_n|g\rangle|\phi_n\rangle$ to solve the Poisson equation for the unknown field $f(\mathbf{r})$ with a known source term $g(\mathbf{r})$.

1.9 Solutions

Solution 1.1
The intermolecular distance in air
A single air molecule occupies the volume $\lambda^3 = \mathcal{V}/N$, where $\mathcal{V}$ is the volume of air containing N molecules. The length scale λ thus represents the average intermolecular distance. Using $p\mathcal{V} = Nk_\mathrm{B}T$, with $p = 10^5$ Pa and $T = 300$ K, we find

$$\lambda = \left(\frac{\mathcal{V}}{N}\right)^{\frac{1}{3}} = \left(\frac{k_\mathrm{B}T}{p}\right)^{\frac{1}{3}} = 3.5 \text{ nm}. \tag{1.56}$$

Thus the intermolecular distance in air is roughly one order of magnitude larger than the intermolecular distance in a typical liquid.

Solution 1.2
The Lennard-Jones potential for intermolecular pair interaction

(a) The minimum is found by solving $\partial_r V_\mathrm{LJ}(r) = 0$, which yields $r_0 = 2^{\frac{1}{6}}\sigma \approx 1.12\sigma$, and a corresponding interaction energy of $V_\mathrm{LJ}(r_0) = -\varepsilon$.

(b) $V_\mathrm{LJ}(3\sigma) = -0.0055\varepsilon$. For nitrogen this means that in the distance $3\sigma_{\mathrm{N}_2} = 1.1$ nm the interaction energy in kelvin is $V_\mathrm{LJ}(3\sigma_{\mathrm{N}_2})/k_\mathrm{B} = -0.5$ K. The average intermolecular distance is 3.5 nm, while the average kinetic translation energy in kelvin is $\frac{3}{2}T = 450$ K. Thus, the interaction effects are minute and can be neglected.

Solution 1.3
The size of the fundamental fluid particle in a liquid
Consider a cube of liquid with side length λ^* in which $\alpha = \sqrt{N}/N$ is a given relative uncertainty in the number of molecules inside the cube. Each molecule occupies the volume λ^3, where $\lambda = 0.3$ nm is a typical value of the intermolecular distance in a liquid. Clearly $(\lambda^*)^3 = N\lambda^3$ and $N = \alpha^{-2}$ and thus $\lambda^*(\alpha) = \alpha^{-\frac{2}{3}}\lambda$.

(a) With $\alpha = 10^{-2}$ we find $\lambda^* = 6.5$ nm.

(b) With $\alpha = 10^{-3}$ we find $\lambda^* = 30$ nm. For a gas λ^* is roughly ten times larger.

Solution 1.4
The index notation

(a) Since δ_{ij} is a constant we have $\partial_k\delta_{ij} \equiv 0$ for any value of i, j and k.
We thus find $\partial_k(p\delta_{ik}) = (\partial_k p)\delta_{ik} + p(\partial_k\delta_{ik}) = \partial_i p + 0 = (\boldsymbol{\nabla}p)_i$.

(b) For the divergence of the current density we get
$\boldsymbol{\nabla}\cdot(\rho\mathbf{v}) = \partial_j(\rho v_j) = (\partial_j\rho)v_j + \rho(\partial_j v_j) = (\boldsymbol{\nabla}\rho)\cdot\mathbf{v} + \rho\boldsymbol{\nabla}\cdot\mathbf{v}$

(c) Let us consider the z component of the rotation. By definition we have $(\boldsymbol{\nabla}\times\mathbf{v})_z = \partial_x v_y - \partial_y v_x$. Using index notation we obtain $(\boldsymbol{\nabla}\times\mathbf{v})_z = \epsilon_{zjk}\partial_j v_k$. The only non-zero terms are carrying the indices $(j,k) = (x,y)$ or $(j,k) = (y,x)$, and since $\epsilon_{zxy} = +1$ and $\epsilon_{zyx} = -1$ we get the desired result: $(\boldsymbol{\nabla}\times\mathbf{v})_z = \epsilon_{zjk}\partial_j v_k = \partial_x v_y - \partial_y v_x$. Likewise, for the x and y components of the rotation.

(d) For the double cross-product identity we get $\left[\mathbf{a}\times(\mathbf{b}\times\mathbf{c})\right]_i = \epsilon_{ijk}a_j(\mathbf{b}\times\mathbf{c})_k = \epsilon_{ijk}a_j(\epsilon_{klm}b_l c_m) = \epsilon_{ijk}\epsilon_{lmk}\, a_j b_l c_m$, where in the last equation we have made an even permutation of the indices in the second Levi–Civita symbol, $\epsilon_{klm} = \epsilon_{lmk}$. Finally, we use Eq. (1.26) to express the product of the two Levi–Civita symbols as a linear combination of Kronecker deltas, $\epsilon_{ijk}\epsilon_{lmk}\, a_j b_l c_m = (\delta_{il}\delta_{jm} - \delta_{im}\delta_{jl})\, a_j b_l c_m = a_j c_j b_i - a_j b_j c_i = (\mathbf{a}\cdot\mathbf{c})b_i - (\mathbf{a}\cdot\mathbf{b})c_i = \left[(\mathbf{a}\cdot\mathbf{c})\mathbf{b} - (\mathbf{a}\cdot\mathbf{b})\mathbf{c}\right]_i$, which proves the relation.

Solution 1.5
First-order perturbation of the damped, harmonic oscillator
Introducing the dimensionless variables yields the following equation of motion,

$$\partial_{\tilde{t}}^2 \tilde{x} = -\tilde{x} - \alpha\, \partial_{\tilde{t}} \tilde{x}. \tag{1.57}$$

(a) With $\tilde{x} = \tilde{x}_0 + \alpha\tilde{x}_1$ the zero-order equation becomes $\partial_{\tilde{t}}^2 \tilde{x}_0 = -\tilde{x}_0$ with the solution $\tilde{x}_0(\tilde{t}) = \cos(\tilde{t})$. The first-order equation becomes $\partial_{\tilde{t}}^2 \tilde{x}_1 = -\tilde{x}_1 - \partial_{\tilde{t}} \tilde{x}_0 = -\tilde{x}_1 + \sin(\tilde{t})$ with the solution $\tilde{x}_1(\tilde{t}) = -\frac{1}{2}\, \tilde{t}\, \cos(\tilde{t})$. So the complete first-order perturbation result is

$$\tilde{x}(\tilde{t}) = \cos(\tilde{t}) \Big[1 - \tfrac{1}{2}\alpha\tilde{t}\Big] + \mathcal{O}(\alpha^2). \tag{1.58}$$

(b) Insertion of the trial function $\tilde{x} = \exp(\mathrm{i}\beta\tilde{t})$ into Eq. (1.57) leads to the simple algebraic equation $-\beta^2 = -1 - \mathrm{i}\,\alpha\beta$ with the solution $\beta = \sqrt{1-(\alpha/2)^2} + \mathrm{i}\,\alpha/2 \approx 1 + \mathrm{i}\,\alpha/2$. Hence

$$\mathrm{Re}\Big[\mathrm{e}^{\mathrm{i}\beta\tilde{t}}\Big] \approx \mathrm{Re}\Big[\mathrm{e}^{\mathrm{i}\tilde{t}}\ \mathrm{e}^{-\alpha\tilde{t}/2}\Big] = \cos(\tilde{t})\, \mathrm{e}^{-\alpha\tilde{t}/2} \approx \cos(\tilde{t})\, \Big[1 - \tfrac{1}{2}\alpha\tilde{t}\Big] \tag{1.59}$$

in agreement with the first-order result Eq. (1.58).

Solution 1.6
The Dirac bra-ket notation
(a) Use the basic definition Eq. (1.42c) to obtain

$$\begin{aligned}
&\langle a_1 f_1 + a_2 f_2 | b_1 g_1 + b_2 g_2 \rangle \\
&\quad = \int_\Omega \mathrm{d}\mathbf{r}\ \big[a_1 f_1(\mathbf{r}) + a_2 f_2(\mathbf{r})\big]\ \big[b_1 g_1(\mathbf{r}) + b_2 g_2(\mathbf{r})\big] \\
&\quad = \int_\Omega \mathrm{d}\mathbf{r}\ \big[a_1 b_1 f_1(\mathbf{r}) g_1(\mathbf{r}) + a_1 b_2 f_1(\mathbf{r}) g_2(\mathbf{r}) + a_2 b_1 f_2(\mathbf{r}) g_1(\mathbf{r}) + a_2 b_2 f_2(\mathbf{r}) g_2(\mathbf{r})\big] \\
&\quad = a_1 b_1 \langle f_1 | g_1 \rangle + a_1 b_2 \langle f_1 | g_2 \rangle + a_2 b_1 \langle f_2 | g_1 \rangle + a_2 b_2 \langle f_2 | g_2 \rangle.
\end{aligned} \tag{1.60}$$

(b) With the expansion $|f\rangle = \sum_{n=1}^{\infty} a_n |\phi_n\rangle$ we get

$$\nabla^2 |f\rangle = \sum_{n=1}^{\infty} a_n \nabla^2 |\phi_n\rangle = \sum_{n=1}^{\infty} a_n (-k_n^2) |\phi_n\rangle, \tag{1.61}$$

and with $|g\rangle = \sum_{n=1}^{\infty} \langle \phi_n | g \rangle |\phi_n\rangle$ the Poisson equation $\nabla^2 f(\mathbf{r}) = -g(\mathbf{r})$ therefore becomes

$$\sum_{n=1}^{\infty} a_n k_n^2 |\phi_n\rangle = \sum_{n=1}^{\infty} \langle \phi_n | g \rangle\ |\phi_n\rangle. \tag{1.62}$$

When multiplied by $\langle \phi_m |$ and using the orthonormality property Eq. (1.45) this reduces to

$$a_m k_m^2 = \langle \phi_m | g \rangle. \tag{1.63}$$

Division by k_m^2 leads to an expression for a_m and thus a determination of the solution $|f\rangle$,

$$|f\rangle = \sum_{m=1}^{\infty} a_m |\phi_m\rangle = \sum_{m=1}^{\infty} \frac{\langle \phi_n | g \rangle}{k_m^2}\ |\phi_m\rangle. \tag{1.64}$$

2
Governing equations

The governing equations for hydrodynamics in general and microfluidics in particular are derived from the fundamental equations describing the rate of change of the flux densities of mass, momentum and energy. In the following, we shall repeatedly make use of Gauss's theorem, which for a given vector field $\mathbf{V}(\mathbf{r})$ relates the volume integral in a given region Ω of the divergence $\boldsymbol{\nabla}\cdot\mathbf{V}$ to the integral over the surface $\partial\Omega$ of the flux $\mathbf{V}\cdot\mathbf{n}\,\mathrm{d}a$ through an area element $\mathrm{d}a$ with the surface normal $\mathbf{n}$,

$$\int_{\Omega} \mathrm{d}\mathbf{r}\; \boldsymbol{\nabla}\cdot\mathbf{V} = \int_{\partial\Omega} \mathrm{d}a\; \mathbf{n}\cdot\mathbf{V} \qquad \text{or} \qquad \int_{\Omega} \mathrm{d}\mathbf{r}\; \partial_j V_j = \int_{\partial\Omega} \mathrm{d}a\; n_j V_j. \tag{2.1}$$

By definition, the surface normal $\mathbf{n}$ of a closed surface is an outward-pointing unit vector perpendicular to the surface, and we have used the notation of Eq. (1.21) for the integral.

2.1 Mass flux, conservation of mass, and the continuity equation

The first governing equation of fluid dynamics to be derived is the continuity equation. This equation expresses the conservation of mass in classical mechanics.

2.1.1 The continuity equation for compressible fluids

We begin by considering the general case of a compressible fluid, i.e. a fluid where the density ρ may vary as function of space and time. Consider an arbitrarily shaped, but fixed, region Ω in the fluid as sketched in Fig. 2.1. The total mass $M(\Omega,t)$ inside Ω can be expressed as a volume integral over the density ρ,

$$M(\Omega,t) = \int_{\Omega} \mathrm{d}\mathbf{r}\; \rho(\mathbf{r},t). \tag{2.2}$$

Since mass can neither appear nor disappear spontaneously in non-relativistic mechanics, $M(\Omega,t)$ can only vary due to a mass flux through the surface $\partial\Omega$ of the region Ω. The mass current density $\mathbf{J}$ is defined as the mass density ρ times the convection velocity $\mathbf{v}$, or the mass flow per oriented unit area per unit time (hence the unit $\mathrm{kg\,m^{-2}\,s^{-1}}$):

$$\mathbf{J}(\mathbf{r},t) = \rho(\mathbf{r},t)\,\mathbf{v}(\mathbf{r},t), \tag{2.3}$$

where $\mathbf{v}$ is the Eulerian velocity field.

Since the region Ω is fixed the time derivative of the mass $M(\Omega,t)$ can be calculated either by differentiating the volume integral Eq. (2.2),

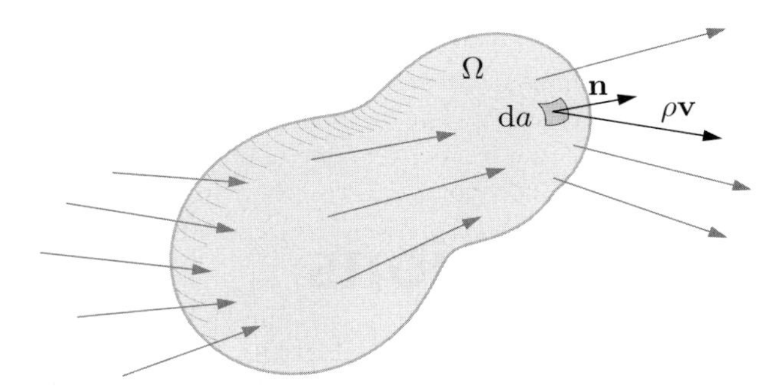

Fig. 2.1 A sketch of the mass current density field $\rho\mathbf{v}$ (long arrows) flowing through an arbitrarily shaped region Ω (gray). Any infinitesimal area da (dark gray) is associated with an outward-pointing unit vector $\mathbf{n}$ (short arrow) perpendicular to the local surface. The current through the area da is given by da times the projection $\rho\mathbf{v}\cdot\mathbf{n}$ of the current density on the surface unit vector.

$$\partial_t M(\Omega,t) = \partial_t \int_\Omega d\mathbf{r}\,\rho(\mathbf{r},t) = \int_\Omega d\mathbf{r}\,\partial_t\rho(\mathbf{r},t), \tag{2.4}$$

or as a surface integral over $\partial\Omega$ of the mass current density using Eq. (2.3) and Fig. 2.1,

$$\partial_t M(\Omega,t) = -\int_{\partial\Omega} da\,\mathbf{n}\cdot\Big(\rho(\mathbf{r},t)\mathbf{v}(\mathbf{r},t)\Big) = -\int_\Omega d\mathbf{r}\,\boldsymbol{\nabla}\cdot\Big(\rho(\mathbf{r},t)\mathbf{v}(\mathbf{r},t)\Big). \tag{2.5}$$

The last expression is obtained by applying Gauss's theorem Eq. (2.1) to the vector field $\mathbf{V}\equiv\rho\mathbf{v}$. The minus sign appears since the mass inside Ω diminishes if $\rho\mathbf{v}$ is parallel to the outward pointing surface vector $\mathbf{n}$. It follows immediately from Eqs. (2.4) and (2.5) that

$$\int_\Omega d\mathbf{r}\,\partial_t\rho(\mathbf{r},t) = -\int_\Omega d\mathbf{r}\,\boldsymbol{\nabla}\cdot\Big(\rho(\mathbf{r},t)\mathbf{v}(\mathbf{r},t)\Big). \tag{2.6}$$

This result is true for any choice of region Ω. However, this is only possible if the integrands are identical. Thus we have derived the continuity equation,

$$\partial_t\rho = -\boldsymbol{\nabla}\cdot(\rho\mathbf{v}) \quad \text{or} \quad \partial_t\rho = -\boldsymbol{\nabla}\cdot\mathbf{J}, \tag{2.7}$$

which in the index notation reads

$$\partial_t\rho = -\partial_j(\rho v_j). \tag{2.8}$$

Note that since electric charge is also a conserved quantity, the argument holds if ρ is substituted by the charge density ρ_{el}, and Eq. (2.7) can be read as the continuity equation for charge instead of for mass.

2.1.2 The continuity equation for incompressible fluids

In many cases, especially in microfluidics, where the flow velocities are much smaller than the velocity of pressure waves (sound) in the liquid, the fluid can be treated as being incompressible. This means that ρ is constant in space and time, and the continuity equation (2.7) is simplified to the following form,

$$\boldsymbol{\nabla}\cdot\mathbf{v} = 0 \quad \text{or} \quad \partial_i v_i = 0, \tag{2.9}$$

a result we shall use extensively throughout the book, except for Section 3.6 and Chapter 15.

2.2 Momentum flux, force densities, and the equation of motion

To derive the second governing equation, the so-called equation of motion for the Eulerian velocity field, we now turn from the mass density ρ of the fluid to its momentum density $\rho\mathbf{v}$ using an approach similar to that which led us to the continuity equation. We consider the ith component $P_i(\Omega,t)$ of the total momentum of the fluid inside an arbitrarily shaped, but fixed, region Ω. In analogy with the mass equation (2.4) the rate of change of the momentum is given by

$$\partial_t P_i(\Omega,t) = \partial_t \int_\Omega \mathrm{d}\mathbf{r}\,\rho(\mathbf{r},t)v_i(\mathbf{r},t) = \int_\Omega \mathrm{d}\mathbf{r}\,\Big[(\partial_t\rho)v_i + \rho\partial_t v_i\Big]. \tag{2.10}$$

In contrast to the mass inside Ω, which according to Eq. (2.5) can only change by convection through the surface $\partial\Omega$, the momentum $P_i(\Omega,t)$ can change both by convection and by the action of forces given by Newton's second law. The forces can be divided into contact forces that act on the surface $\partial\Omega$ of Ω, e.g. pressure and viscosity forces, and body forces that act on the interior of Ω, e.g. gravitational and electrical forces. Thus, the rate of change of the ith component of the momentum can be written as

$$\partial_t P_i(\Omega,t) = \partial_t P_i^{\mathrm{conv}}(\Omega,t) + \partial_t P_i^{\mathrm{pres}}(\Omega,t) + \partial_t P_i^{\mathrm{visc}}(\Omega,t) + \partial_t P_i^{\mathrm{body}}(\Omega,t). \tag{2.11}$$

2.2.1 Rate of change in momentum due to convection of momentum

Just as convection of the mass density scalar ρ in the velocity field $\mathbf{v}$ is described by the vector $\rho\mathbf{v}$, convection of the momentum vector $\rho\mathbf{v}$ is described by a tensor Π' as

$$\Pi' \equiv \rho\mathbf{v}\mathbf{v}, \qquad \text{or} \qquad \Pi'_{ij} \equiv \rho v_i v_j. \tag{2.12}$$

The tensor Π' is known as the momentum flux density tensor, and the prime indicates that we have not taken the pressure p into account. The amount of ith component of the momentum that enters or leaves Ω through the infinitesimal area $\mathbf{n}\mathrm{d}a$ is given by $(\rho v_i)\mathbf{v}\cdot\mathbf{n}\mathrm{d}a$, and thus the total change $\partial_t P_i^{\mathrm{conv}}(\Omega,t)$ of momentum in Ω due to convection is given by

$$\partial_t P_i^{\mathrm{conv}}(\Omega,t) = -\int_{\partial\Omega} \mathrm{d}a\,\mathbf{n}\cdot(\rho v_i\,\mathbf{v}) = -\int_{\partial\Omega} \mathrm{d}a\,n_j\,\rho v_i v_j. \tag{2.13}$$

The sign is due to the previously mentioned convention that $\mathbf{n}$ points outwards.

2.2.2 Rate of change in momentum due to pressure forces

At each infinitesimal area $\mathrm{d}a$ on the surface of $\partial\Omega$ the surroundings act with the pressure force $-p\mathbf{n}\mathrm{d}a$ onto Ω. As a result, the ith component of the momentum will change due to the force $(-p\mathbf{n}\mathrm{d}a)\cdot\mathbf{e}_i = -n_i p\mathrm{d}a$, where $\mathbf{e}_i$ is the unit vector corresponding to the ith component. Hence, we obtain

$$\partial_t P_i^{\mathrm{pres}}(\Omega,t) = -\int_{\partial\Omega} \mathrm{d}a\,\mathbf{n}\cdot(p\mathbf{e}_i) = -\int_{\partial\Omega} \mathrm{d}a\,n_j\,p\delta_{ij}. \tag{2.14}$$

In the last equation we use that $\mathbf{n}\cdot\mathbf{e}_i = n_j\delta_{ij}$, whereby $\mathbf{n}$ can be ascribed the same free index j different from the momentum component index i as in Eq. (2.13). From Eqs. (2.13) and (2.14) it is seen that it is quite natural to extend the definition of the momentum flux density tensor to include the pressure,

$$\Pi_{ij} \equiv p\delta_{ij} + \Pi'_{ij} = p\delta_{ij} + \rho v_i v_j. \tag{2.15}$$

This expression is the common definition of the momentum flux density tensor.

2.2.3 Rate of change in momentum due to viscous forces

Due to the viscous nature of the fluid, the region Ω will be subject to frictional forces on its surface $\partial\Omega$ from the flow of the surrounding liquid. The frictional force $\mathrm{d}\mathbf{F}$ on a surface element da with the normal vector $\mathbf{n}$ must be characterized by a tensor rank of two since two vectors are needed to determine it: the force and the surface normal. This tensor is denoted as the viscous stress tensor σ'_{ij}, and it expresses the ith component of the friction force per area acting on a surface element oriented with its surface normal parallel to the jth unit vector $\mathbf{e}_j$. Thus

$$\mathrm{d}F_i = \sigma'_{ij} n_j \,\mathrm{d}a. \tag{2.16}$$

This expression leads immediately to the change in the momentum of Ω due to the viscous forces at the surface $\partial\Omega$,

$$\partial_t P_i^{\mathrm{visc}}(\Omega, t) = \int_{\partial\Omega} \mathrm{d}a \; n_j \, \sigma'_{ij}. \tag{2.17}$$

The internal friction is only non-zero when fluid particles move relative to each other, hence the viscous stress tensor σ' depends only on the spatial derivatives of the velocity. For the small velocity gradients encountered in microfluidics we can safely assume that only first-order derivatives enter the expression for σ', thus σ'_{ij} must depend linearly on the velocity gradients $\partial_i v_j$. We can pinpoint the expression for σ'_{ij} further by noticing that it must vanish when the liquid is rotating as a whole, i.e. when the velocity field has the form $\mathbf{v} = \boldsymbol{\omega} \times \mathbf{r}$, where $\boldsymbol{\omega}$ is an angular velocity vector. As shown in Exercise 2.4, this velocity field obeys the antisymmetric relation $\partial_j v_i = -\partial_i v_j$, so σ' can only vanish for this particular velocity field, as it should, if it solely contains the symmetric combinations $\partial_j v_i + \partial_i v_j$ and $\partial_k v_k$ of the first-order derivatives. The most general tensor of rank two satisfying these conditions is

$$\sigma'_{ij} = \eta\left(\partial_j v_i + \partial_i v_j - \tfrac{2}{3}\delta_{ij}\partial_k v_k\right) + \zeta\,\delta_{ij}\partial_k v_k \tag{2.18a}$$

$$= \eta\left(\partial_j v_i + \partial_i v_j\right) + \left(\zeta - \tfrac{2}{3}\eta\right)\left(\partial_k v_k\right)\delta_{ij} \tag{2.18b}$$

$$= \eta\left(\partial_j v_i + \partial_i v_j\right) + (\beta - 1)\eta\left(\partial_k v_k\right)\delta_{ij}, \tag{2.18c}$$

where the following material parameters have been introduced:

$$\eta, \text{ dynamic viscosity (internal friction due to shear stress)}, \tag{2.19a}$$

$$\zeta, \text{ second viscosity (internal friction due to compression)}, \tag{2.19b}$$

$$\beta \equiv \frac{\zeta}{\eta} + \frac{1}{3}, \text{ dimensionless viscosity ratio.} \tag{2.19c}$$

Note, that the viscous stress tensor in Eq. (2.18a) has been normalized such that the term with the dynamic viscosity η has zero trace, i.e. the sum of the diagonal elements is not proportional to the compression-related quantity $\partial_k v_k$, while the term with the second viscosity is proportional to exactly that quantity. The viscosity ratio β is introduced to simplify expressions involving the divergence of the stress tensor, see Chapter 15.

For an incompressible fluid $\partial_k v_k = 0$ and σ'_{ij} becomes

$$\sigma'_{ij} = \eta\left(\partial_i v_j + \partial_j v_i\right), \quad \text{(incompressible fluids).} \tag{2.20}$$

To determine the values of the viscosity coefficients one must go beyond the symmetry consideration presented here, and either measure them experimentally or calculate them by

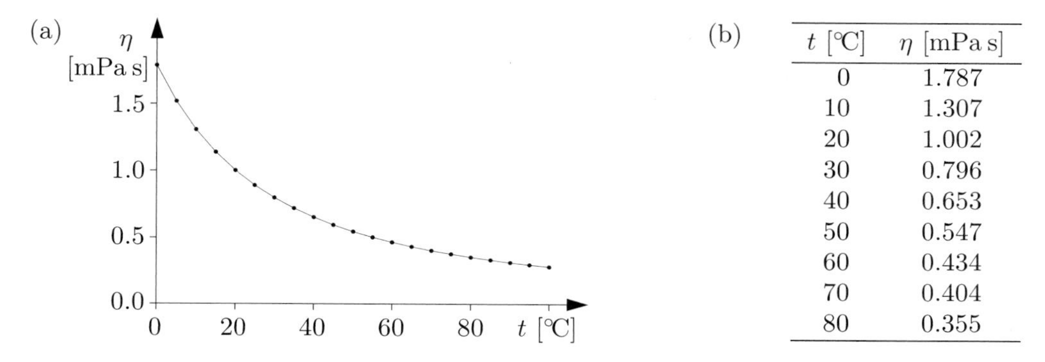

t [℃]	η [mPa s]
0	1.787
10	1.307
20	1.002
30	0.796
40	0.653
50	0.547
60	0.434
70	0.404
80	0.355

Fig. 2.2 (a) Graph of the viscosity η of water as a function of temperature. (b) A table of some of the data points. Data are taken from *CRC Handbook of Chemistry and Physics*.

some microscopic model of the liquid. We shall take the phenomenological approach and simply employ the experimental values, however, in Exercise 5.8 we study one example of a simple theoretical model leading to an estimate of the viscosity η. The SI unit for viscosity is Pa s, and the value for pure water at 20 ℃ is

$$\eta_{\mathrm{water}}(20\ ^\circ\mathrm{C}) = 1.002 \times 10^{-3}\ \mathrm{Pa\,s} = 1.002\ \mathrm{mPa\,s}. \tag{2.21}$$

The viscosity of water has a strong dependence on temperature as seen in Fig. 2.2. A table of viscosity values for other fluids is given in Appendix A. The second viscosity ζ only plays a role for compressible fluids (acoustics), and for many simple fluids the Stokes surmise

$$\zeta = \left(\beta - \tfrac{1}{3}\right)\eta \approx \tfrac{4}{3}\,\eta, \text{ or } \beta \approx \tfrac{5}{3}, \tag{2.22}$$

is a good approximation.

2.2.4 Rate of change in momentum due to body forces

The body forces are external forces that act throughout the entire body of the fluid. We shall, in particular, work with the gravitational force (in terms of the density ρ and the acceleration of gravity $\mathbf{g}$) and the electrical force (in terms of the charge density ρ_{el} of the fluid and the external electric field $\mathbf{E}$). The resulting change in the momentum of Ω due to these two body forces is

$$\partial_t P_i^{\mathrm{body}}(\Omega, t) = \int_\Omega \mathrm{d}\mathbf{r}\ (\rho\mathbf{g} + \rho_{\mathrm{el}}\mathbf{E})_i = \int_\Omega \mathrm{d}\mathbf{r}\ (\rho g_i + \rho_{\mathrm{el}} E_i). \tag{2.23}$$

2.2.5 The equation of motion and the Navier–Stokes equation

The general equation of motion for a viscous fluid can now be found from Eq. (2.11) by collecting the results from the previous subsections. In integral form we obtain

$$\int_\Omega \mathrm{d}\mathbf{r}\ \Big[(\partial_t\rho)v_i + \rho\partial_t v_i\Big] = \int_{\partial\Omega} \mathrm{d}a\, n_j \Big[-\rho v_i v_j - p\delta_{ij} + \sigma'_{ij}\Big] + \int_\Omega \mathrm{d}\mathbf{r}\ (\rho g_i + \rho_{\mathrm{el}} E_i). \tag{2.24}$$

Utilizing Gauss's theorem the surface integral involving n_j can be rewritten as a volume integral involving ∂_j. Since the resulting volume integral equation is valid for any region Ω the integrands must be identical, hence we arrive at the partial differential equation

$$(\partial_t\rho)v_i + \rho\partial_t v_i = -\partial_j(\rho v_i v_j) - \partial_j(p\delta_{ij}) + \partial_j\sigma'_{ij} + \rho g_i + \rho_{\text{el}}E_i. \tag{2.25}$$

This expression can be simplified by noting that $-\partial_j(\rho v_j v_i) = -\partial_j(\rho v_j)v_i - \rho v_j\partial_j v_i$, of which according to the continuity equation (2.8) the first term cancels out with the term $(\partial_t\rho)v_i$ on the left-hand side in Eq. (2.25). Moreover, it is customary to combine the pressure p and the viscous stress tensor σ' into the full stress tensor σ,

$$\sigma_{ij} \equiv -p\,\delta_{ij} + \sigma'_{ij}, \tag{2.26}$$

so we end up with the general equation of motion for the Eulerian velocity field of a viscous fluid,

$$\rho\partial_t v_i + \rho v_j\partial_j v_i = \partial_j\sigma_{ij} + \rho g_i + \rho_{\text{el}}E_i. \tag{2.27}$$

The left-hand side can be interpreted as inertial force densities, while the right-hand side is intrinsic or applied force densities.

Normally, for the so-called Newtonian fluids, the viscosity coefficients η and ζ in the viscous stress tensor vary only a little, and they can therefore be taken as constants. In that case, as shown in Exercise 2.7, the force density $\partial_j\sigma_{ij}$ due to pressure and viscosity becomes

$$\partial_j\sigma_{ij} = \partial_j(-p\delta_{ij}) + \partial_j\sigma'_{ij} = -\partial_i p + \eta\,\partial_j\partial_j v_i + \beta\eta\,\partial_i(\partial_j v_j), \tag{2.28}$$

and the equation of motion becomes the Navier–Stokes equation, which has the following form in the index and the vector notation, respectively, including the body forces

$$\rho\big[\partial_t v_i + v_j\partial_j v_i\big] = -\partial_i p + \eta\,\partial_j^2 v_i + \beta\eta\,\partial_i(\partial_j v_j) + \rho\,g_i + \rho_{\text{el}}E_i, \tag{2.29a}$$

$$\rho\big[\partial_t \mathbf{v} + (\mathbf{v}\cdot\boldsymbol{\nabla})\mathbf{v}\big] = -\boldsymbol{\nabla}p + \eta\nabla^2\mathbf{v} + \beta\eta\boldsymbol{\nabla}(\boldsymbol{\nabla}\cdot\mathbf{v}) + \rho\,\mathbf{g} + \rho_{\text{el}}\mathbf{E}. \tag{2.29b}$$

Due to the low fluid velocities realized in microfluidic systems, much lower than the respective sound velocities, the fluids can, to a very good accuracy, be treated as incompressible. In that case, the Navier–Stokes equation including the body forces becomes

$$\rho\big[\partial_t v_i + v_j\partial_j v_i\big] = -\partial_i p + \eta\,\partial_j^2 v_i + \rho\,g_i + \rho_{\text{el}}E_i, \quad \text{(incompressible fluids)} \tag{2.30a}$$

$$\rho\big[\partial_t \mathbf{v} + (\mathbf{v}\cdot\boldsymbol{\nabla})\mathbf{v}\big] = -\boldsymbol{\nabla}p + \eta\nabla^2\mathbf{v} + \rho\,\mathbf{g} + \rho_{\text{el}}\mathbf{E}, \quad \text{(incompressible fluids).} \tag{2.30b}$$

2.2.6 Newton's second law and the material time derivative

In the previous subsections the equation of motion of the Eulerian velocity field was derived using the concept of momentum flux. However, as we shall see in the following, it can also be derived directly from Newton's second law for fluid particles using the concept of material time derivatives.

For a particle of mass m influenced by external forces $\sum_j \mathbf{F}_j$ Newton's second law reads

$$m\,\mathrm{d}_t\mathbf{v} = \sum_j \mathbf{F}_j. \tag{2.31}$$

In fluid dynamics, as discussed in Section 1.3.3, we divide by the volume of the fluid particle and thus work with the density ρ and the force densities $\mathbf{f}_j$. Moreover, we must be careful with the time derivative of the velocity field $\mathbf{v}$. As illustrated in Fig. 1.4 the Eulerian velocity field $\mathbf{v}(\mathbf{r}, t)$ is not the velocity of any particular fluid particle, as it should be in Newton's second law Eq. (2.31). To obtain a physically correct equation of motion a special time derivative, the so-called material time derivative D_t defined in the following, is introduced for Eulerian velocity fields. Our first version of the equation of motion thus takes the form

$$\rho\, D_t\mathbf{v} = \sum_j \mathbf{f}_j. \tag{2.32}$$

The material (or substantial) time derivative is the time derivative obtained when following the flow of a particle as in ordinary Newtonian particle mechanics, i.e. when adopting a Lagrangian description. We have already in Eq. (1.19) found the appropriate expression, so using that on the velocity field $\mathbf{v}$ we get

$$\rho\, D_t\mathbf{v}(\mathbf{r},t) \equiv \rho\, \mathrm{d}_t\mathbf{v}\big(\mathbf{r}(t),t\big) = \rho\, \big[\partial_t\mathbf{v}(\mathbf{r},t) + (\mathbf{v}\cdot\boldsymbol{\nabla})\mathbf{v}(\mathbf{r},t)\big]. \tag{2.33}$$

Note the use of the Eulerian velocity field in the first and third expression, and the Lagrangian velocity field in the second expression.

We can derive the same result by first noting that the total differential of the Eulerian velocity field in general is given by $\mathrm{d}\mathbf{v} = \mathrm{d}t\, \partial_t\mathbf{v} + (\mathrm{d}\mathbf{r}\cdot\boldsymbol{\nabla})\mathbf{v}$. Secondly, if we insist on calculating the change due to the flow of a particular fluid particle we must have $\mathrm{d}\mathbf{r} = \mathbf{v}\mathrm{d}t$. Combining these two expression leads to Eq. (2.33). The same analysis applies for any flow variable, and we can conclude that the material time derivative D_t is given by

$$D_t = \partial_t + (\mathbf{v}\cdot\boldsymbol{\nabla}). \tag{2.34}$$

The equation of motion now takes the form

$$\rho\big[\partial_t\mathbf{v} + (\mathbf{v}\cdot\boldsymbol{\nabla})\mathbf{v}\big] = \sum_j \mathbf{f}_j, \tag{2.35}$$

and we see that this is exactly the expression given in Eq. (2.27). The force densities are simply the pressure, viscosity and body-force densities. Instead of involving the momentum flux, the derivation was based on the material time derivative.

2.2.7 The dimensionless Reynolds number and Stokes flow

Mathematically the richness and beauty of hydrodynamic phenomena is spawned by the non-linear term $\rho(\mathbf{v}\cdot\boldsymbol{\nabla})\mathbf{v}$ in the Navier–Stokes equation. On the other hand, the non-linear term is also responsible for making the mathematical treatment of the equation more complex and difficult; the solutions of the equation have never been fully characterized. However, as we shall see in the following, in the limit of low flow velocities, a limit highly relevant for microfluidic systems, the non-linear term can be neglected. We enter the regime of the so-called Stokes flow or creeping flow, where analytical solutions to a number of flow problems can be found.

The proper way to analyze when the non-linear term is negligible, is to make the Navier–Stokes equation dimensionless. This means that we express all physical variables, such as length and velocity, in units of the characteristic scales, e.g. L_0 for length and V_0 for velocity. If the system under consideration is characterized by only one length scale L_0 and one velocity scale V_0, the expression of co-ordinates and velocity in terms of dimensionless co-ordinates and velocity is

$$\mathbf{r} = L_0\,\tilde{\mathbf{r}}, \tag{2.36a}$$

$$\mathbf{v} = V_0\,\tilde{\mathbf{v}}, \tag{2.36b}$$

where the tilde on top of a symbol indicates that the symbol is a quantity without physical dimension, i.e. pure numbers. Once the length and velocity scales L_0 and V_0 have been fixed the scales T_0 and P_0 for time and pressure follow,

$$t = \frac{L_0}{V_0}\,\tilde{t} = T_0\,\tilde{t}, \tag{2.36c}$$

$$p = \frac{\eta V_0}{L_0}\,\tilde{p} = P_0\,\tilde{p}. \tag{2.36d}$$

Note that a quantity often can be made dimensionless in more than one way. Regarding the pressure, it is, for example, possible to choose P_0 either as $\eta V_0/L_0$ or as $\rho\,{V_0}^2$. The former gives the scale of pressure in the case of small velocities where viscosity dominates, such as in microfluidics, whereas the latter is used at high velocities.

By insertion of Eq. (2.36) into the Navier–Stokes equation for incompressible fluids, Eq. (2.30b) excluding the body-forces, and using the straightforward scaling of the derivatives, $\partial_t = (1/T_0)\,\tilde{\partial}_t$ and $\boldsymbol{\nabla} = (1/L_0)\,\tilde{\boldsymbol{\nabla}}$, we get

$$\rho\left[\frac{V_0}{T_0}\,\tilde{\partial}_t\tilde{\mathbf{v}} + \frac{{V_0}^2}{L_0}\,\big(\tilde{\mathbf{v}}\cdot\tilde{\boldsymbol{\nabla}}\big)\tilde{\mathbf{v}}\right] = -\frac{P_0}{L_0}\,\tilde{\boldsymbol{\nabla}}\tilde{p} + \frac{\eta V_0}{{L_0}^2}\,\tilde{\boldsymbol{\nabla}}^2\tilde{\mathbf{v}}, \tag{2.37}$$

which after reduction becomes

$$Re\left[\tilde{\partial}_t\tilde{\mathbf{v}} + \big(\tilde{\mathbf{v}}\cdot\tilde{\boldsymbol{\nabla}}\big)\tilde{\mathbf{v}}\right] = -\tilde{\boldsymbol{\nabla}}\tilde{p} + \tilde{\boldsymbol{\nabla}}^2\tilde{\mathbf{v}}. \tag{2.38}$$

Here, we have introduced the dimensionless number Re, the so-called Reynolds number,

$$Re \equiv \frac{\rho V_0 L_0}{\eta}. \tag{2.39}$$

We clearly see from Eq. (2.38) that for $Re \ll 1$ the viscous term $\tilde{\boldsymbol{\nabla}}^2\tilde{\mathbf{v}}$ dominates, whereas in steady state for $Re \gg 1$ the inertia term $\big(\tilde{\mathbf{v}}\cdot\tilde{\boldsymbol{\nabla}}\big)\tilde{\mathbf{v}}$ is the most important term.

The corresponding dimensionless form of the incompressibility condition $\partial_i v_i = 0$ is quite simple since $\partial_i = (1/L_0)\tilde{\partial}_i$ and $v_i = V_0\,\tilde{v}_i$,

$$\tilde{\partial}_i\tilde{v}_i = 0. \tag{2.40}$$

Returning to physical variables in the limit of low Reynolds number, the non-linear Navier–Stokes equation is reduced to the linear Stokes equation,

$$\mathbf{0} = -\boldsymbol{\nabla} p + \eta\nabla^2\mathbf{v}. \tag{2.41}$$

In deriving this approximation we assumed that the time derivative $\partial_t\mathbf{v}$ was controlled by the intrinsic time scale $T_0 = L_0/V_0$. If, however, the time dependence is controlled by some external time scale different from T_0, say the oscillation period of an oscillating boundary, the time derivative is not necessarily negligible, and we must employ the time-dependent, linear Stokes equation,

$$\rho\, \partial_t \mathbf{v} = -\boldsymbol{\nabla} p + \eta \nabla^2 \mathbf{v}. \tag{2.42}$$

Due to the specific structure of the Stokes equation for incompressible fluids, it is often useful to solve directly for the pressure p and for the so-called vorticity $\boldsymbol{\omega}$ defined as

$$\boldsymbol{\omega} \equiv \boldsymbol{\nabla} \times \mathbf{v}. \tag{2.43}$$

Utilizing that $\boldsymbol{\nabla} \times \boldsymbol{\nabla} = \mathbf{0}$ when taking the rotation $\boldsymbol{\nabla}\times$ of the Stokes equation, we get a simple equation of motion for $\boldsymbol{\omega}$,

$$\partial_t \boldsymbol{\omega} = \frac{\eta}{\rho}\, \nabla^2 \boldsymbol{\omega}. \tag{2.44}$$

Similary, utilizing that $\boldsymbol{\nabla}\cdot\mathbf{v} = 0$ when taking the divergence $\boldsymbol{\nabla}\cdot$ of the Stokes equation, we get a simple Laplace equation for p,

$$\nabla^2 p = 0. \tag{2.45}$$

The linearity of the Stokes equation and the simplicity of the latter two equations makes it possible in several cases to derive analytical solutions to the creeping flow problem.

2.2.8 Non-Newtonian fluids

For a wide class of fluids and solutions, the so-called non-Newtonian fluids, the assumption of constant viscosity does not apply, see e.g. Bird, Armstrong and Hassager (1987), Probstein (1994) or Morrison (2001). If a liquid, such as polymer solutions or blood, contains large deformable molecules or particles, these can be stretched out at an increased shear stress, which then can lead to a decrease in viscosity. In other fluids containing small, strongly interacting particles, the interparticle interaction can impede the flow at an increased shear stress, which then can lead to an increased viscosity. These two opposite effects are denoted shear thinning and shear thickening, respectively, and they form a central topic in the field of rheology, which comprises the study of deformation and flow of matter that does not obey Newton's and Hooke's classical laws for viscous fluids and elastic materials.

A detailed molecular description of shear thinning and thickening is difficult to obtain, but empirical constitutive equations for the viscosity η can be found and expressed in terms of the velocity gradient through scalar invariants of the shear rate tensor $\dot{\gamma}_{ij}$,

$$\dot{\gamma}_{ij} \equiv \partial_i v_j + \partial_j v_i, \tag{2.46a}$$

such as its trace $\mathrm{Tr}(\dot{\gamma})$ or its magnitude $|\dot{\gamma}|$ defined as

$$\mathrm{Tr}(\dot{\gamma}) \equiv \dot{\gamma}_{ii} = 2\, \boldsymbol{\nabla}\cdot\mathbf{v}, \tag{2.46b}$$

$$|\dot{\gamma}| \equiv \sqrt{\tfrac{1}{2}\mathrm{Tr}(\gamma^2)} = \sqrt{\tfrac{1}{2}\, \dot{\gamma}_{ij}\dot{\gamma}_{ji}}. \tag{2.46c}$$

For incompressible fluids we have $\mathrm{Tr}(\dot{\gamma}) = 2\,\boldsymbol{\nabla}\cdot\mathbf{v} = 0$, so for these the viscosity is taken to be a function of $|\dot{\gamma}|$ only, and the viscous stress tensor, Eq. (2.20), is generalized to be

$$\sigma'_{ij} = \eta\big(|\dot{\gamma}|\big)\, \dot{\gamma}_{ij}, \quad \text{(incompressible non-Newtonian fluids).} \tag{2.47}$$

One elementary constitutive equation is the Carreau–Yasuda model, which successfully has been applied to a number of non-Newtonian fluids,

$$\eta(|\dot{\gamma}|) = \eta_\infty + (\eta_0 - \eta_\infty)\left[1 + (\lambda|\dot{\gamma}|)^a\right]^{\frac{n-1}{a}}. \tag{2.48}$$

This five-parameter model describes well the experimentally observed transition from the zero-shear-stress viscosity η_0 to the infinite-shear-stress viscosity η_∞, which happens around the time scale λ. The curvature at the transition-point in a $\log(\eta) - \log(|\dot{\gamma}|)$ plot is controlled by a, while the slope of the curve in the transition region of the plot is determined by n.

A simpler model, which focuses on the transition region where shear thinning or thickening happens, and which makes it possible to obtain analytical results of various flow problems, is the two-parameter Ostwald–de Waele power-law model,

$$\eta(|\dot{\gamma}|) = m\,|\dot{\gamma}|^{n-1}, \quad \begin{cases} n < 1, & \text{non-Newtonian shear thinning,} \\ n = 1, & \text{Newtonian fluid with } \eta = m, \\ n > 1, & \text{non-Newtonian shear thickening.} \end{cases} \tag{2.49}$$

We note that the power-law model is identical to the Carreau–Yasuda model in the large-shear-stress limit $(\lambda|\dot{\gamma}|)^a \gg 1$ when taking $\eta_\infty = 0$ and $\eta_0\lambda^{n-1} = m$.

Except for Exercises 2.9 and 3.8 we shall not study the otherwise very fascinating topic of non-Newtonian fluids further in this book.

2.3 Energy flux and the heat-transfer equation

The third and last governing equation to be established is the heat-transfer equation of the fluid relating the rate of change of the energy density to the energy density flux. When working with thermodynamics of fluids it is natural to work with the thermodynamic quantities per unit mass, which are directly related to the molecules present in the fluid. Thus, we will work with the internal energy ε per unit mass, the entropy s per unit mass, the enthalpy h per unit mass and the volume $1/\rho$ per unit mass instead of the energy E, the entropy S, the enthalpy H and the volume $\mathcal{V}$ of the fluid. The first law of thermodynamics relates internal energy $\mathrm{d}\epsilon$, heat $T\mathrm{d}s$, and pressure work $-p\mathrm{d}(1/\rho)$. When it is expressed per unit mass, it takes the form

$$\mathrm{d}\varepsilon = T\,\mathrm{d}s - p\,\mathrm{d}\big(\tfrac{1}{\rho}\big) = T\,\mathrm{d}s + \frac{p}{\rho^2}\,\mathrm{d}\rho. \tag{2.50}$$

The densities of the quantities involved are obtained by multiplying them by the mass density ρ, e.g. the energy density is written as $\rho\varepsilon$.

In analogy with the study of the momentum in the previous section, we consider the rate of change $\partial_t E(\Omega, t)$ of the energy, i.e. the power conversion, of the fluid inside some fixed region Ω. As the energy density is given by the sum of the kinetic energy density $\frac{1}{2}\rho v^2$ and the internal energy density $\rho\varepsilon$, the rate of change is given by

$$\partial_t E(\Omega, t) = \partial_t \int_\Omega \mathrm{d}\mathbf{r}\,\left[\tfrac{1}{2}\rho v^2 + \rho\varepsilon\right] = \int_\Omega \mathrm{d}\mathbf{r}\,\partial_t\left[\tfrac{1}{2}\rho v^2 + \rho\varepsilon\right]. \tag{2.51}$$

In analogy with the momentum changes Eq. (2.11), the energy of the fluid inside the region Ω can change by convection through the surface $\partial\Omega$, work done by pressure and friction forces from the surroundings acting on the surface $\partial\Omega$ of Ω, and by heat conduction due to thermal gradients at the surface. For simplicity, we disregard heat sources and sinks that in principle could be present inside Ω. Thus, the rate of change of the energy can be written as

$$\partial_t E(\Omega,t) = \partial_t E^{\mathrm{conv}}(\Omega,t) + \partial_t E^{\mathrm{pres}}(\Omega,t) + \partial_t E^{\mathrm{visc}}(\Omega,t) + \partial_t E^{\mathrm{cond}}(\Omega,t). \tag{2.52}$$

2.3.1 Energy convection and work by stress forces

The convection of energy out of the region is easily expressed in terms of the energy flux density $\mathbf{J}_\varepsilon = (\frac{1}{2}\rho v^2 + \rho\varepsilon)\mathbf{v}$, which is constructed in complete analogy with the mass current density $\mathbf{J}$ of Eq. (2.3),

$$\partial_t E^{\mathrm{conv}}(\Omega,t) = -\int_{\partial\Omega} \mathrm{d}a\,\mathbf{n}\cdot\mathbf{J}_\varepsilon = -\int_{\partial\Omega} \mathrm{d}a\, n_j v_j \big[\tfrac{1}{2}\rho v^2 + \rho\varepsilon\big]. \tag{2.53}$$

As before the minus sign is due to the convention that the surface normal points outwards.

The power transferred into the region Ω through the work done by the stress forces due to pressure and viscosity at the surface is given by the product $\mathbf{v}\cdot(\sigma\mathbf{n}\mathrm{d}a)$ of the velocity of the fluid and the stress force vector,

$$\partial_t E^{\mathrm{pres}}(\Omega,t) + \partial_t E^{\mathrm{visc}}(\Omega,t) = \int_{\partial\Omega} \mathrm{d}a\, v_k \sigma_{kj} n_j = \int_{\partial\Omega} \mathrm{d}a\, n_j \big[-p\delta_{jk} + \sigma'_{jk}\big] v_k. \tag{2.54}$$

In the last equation we have utilized the symmetry $\sigma'_{kj} = \sigma'_{jk}$, see Eq. (2.18).

2.3.2 Thermal conduction and Fourier's law

Thermal conduction occurs in any medium given a spatially varying temperature field $T(\mathbf{r})$. The heat flux density $\mathbf{J}_{\mathrm{heat}}$, which is the heat-transfer per area per time given in $\mathrm{J\,m^{-2}\,s^{-1}}$ or $\mathrm{W\,m^{-2}}$, can therefore be expanded in derivatives of the temperature. Given only small temperature variations this expansion can be approximated by taking only the first derivative $\boldsymbol{\nabla}T$ into account, and we arrive at Fourier's law of heat conduction for an isotropic medium,

$$\mathbf{J}_{\mathrm{heat}} = -\kappa\,\boldsymbol{\nabla}T, \tag{2.55}$$

where the coefficient κ, which has the unit $\mathrm{W\,m^{-1}\,K^{-1}}$, is called the thermal conductivity of the fluid. The value of κ for water at 20 ℃ is

$$\kappa_{\mathrm{water}}(20\ ℃) = 0.597\ \mathrm{W\,m^{-1}\,K^{-1}}. \tag{2.56}$$

The rate of change of energy due to conduction is readily found through the heat flux density and by applying Fourier's law,

$$\partial_t E^{\mathrm{cond}}(\Omega,t) = -\int_{\partial\Omega} \mathrm{d}a\,\mathbf{n}\cdot\mathbf{J}_{\mathrm{heat}} = \int_{\partial\Omega} \mathrm{d}a\, n_j\,(\kappa\partial_j T). \tag{2.57}$$

Note that the double sign change results in a contribution with a plus sign.

2.3.3 The general equation of heat transfer and Fourier's equation

One version of the energy or heat-transfer equation can now be found by combining Eq. (2.52) with the results for the various rates of energy change found in the previous subsections.

Moreover, in analogy with the momentum equation, all surface integrals involving n_j are by use of Gauss's theorem rewritten into volume integrals involving ∂_j. Finally, since the resulting integral equation is valid for any region Ω, the integrands have to be identical, and we arrive at the third governing equation, the heat-transfer equation, here written both in the index and in the vector notation,

$$\partial_t\big[\tfrac{1}{2}\rho v^2+\rho\varepsilon\big] = -\partial_j\Big(\big[\tfrac{1}{2}\rho v^2+\rho\varepsilon+p\big]v_j-\sigma'_{jk}v_k-\kappa\partial_j T\Big), \tag{2.58a}$$

$$\partial_t\big[\tfrac{1}{2}\rho v^2+\rho\varepsilon\big] = -\boldsymbol{\nabla}\cdot\Big(\big[\tfrac{1}{2}\rho v^2+\rho\varepsilon+p\big]\mathbf{v}-\sigma'\cdot\mathbf{v}-\kappa\boldsymbol{\nabla}T\Big). \tag{2.58b}$$

The explicit minus sign in front of the divergence allows us to interpret the equation as a continuity equation for the energy density in analogy with Eq. (2.7) for the mass density. The total energy flux density $\mathbf{J}_{\mathrm{erg}}$ can therefore be identified with the vector

$$\mathbf{J}_{\mathrm{erg}} \equiv \big[\tfrac{1}{2}\rho v^2+\rho\varepsilon+p\big]\mathbf{v}-\sigma'\cdot\mathbf{v}-\kappa\boldsymbol{\nabla}T. \tag{2.59}$$

To make it more explicit that Eq. (2.58) is indeed a heat-transfer equation, it is customary to rewrite it in terms of the entropy s per unit mass times ρT. The first step is to perform the time derivative on the left-hand side,

$$\partial_t\big[\tfrac{1}{2}\rho v^2+\rho\varepsilon\big] = \big(\tfrac{1}{2}v^2+\varepsilon\big)\partial_t\rho+\rho v_j\partial_t v_j+\rho\partial_t\varepsilon \tag{2.60a}$$

$$= -\big(\tfrac{1}{2}v^2+\varepsilon\big)\partial_j(\rho v_j)-\rho v_k\partial_k\big(\tfrac{1}{2}v^2\big)-v_j\partial_j p+v_j\partial_k\sigma'_{jk}+\rho\partial_t\varepsilon, \tag{2.60b}$$

where the continuity equation (2.8) and the equation of motion Eq. (2.27) have been used to rewrite $\partial_t\rho$ and $\partial_t v_j$, respectively. The last term $\rho\partial_t\varepsilon$ can be rewritten by using the first law of thermodynamics, Eq. (2.50), and thereby bringing the entropy s into play

$$\rho\partial_t\varepsilon = \rho T\partial_t s+\frac{p}{\rho}\,\partial_t\rho = \rho T\partial_t s-\frac{p}{\rho}\,\partial_j(\rho v_j). \tag{2.61}$$

Likewise, the third term containing $v_j\partial_j p$ can also be rewritten by use of the first law, $\mathrm{d}(\varepsilon+p/\rho) = [T\mathrm{d}s-p\mathrm{d}(1/\rho)]+[p\mathrm{d}(1/\rho)+(1/\rho)\mathrm{d}p] = T\mathrm{d}s+(1/\rho)\mathrm{d}p$, from which follows

$$-v_j\partial_j p = -\rho v_j\partial_j\big(\varepsilon+\frac{p}{\rho}\big)+\rho T v_j\partial_j s. \tag{2.62}$$

Substituting Eqs. (2.61) and (2.62) into Eq. (2.60b) leads to

$$\partial_t\Big[\tfrac{1}{2}\rho v^2+\rho\varepsilon\Big] = -\Big[\tfrac{1}{2}v^2+\varepsilon+\frac{p}{\rho}\Big]\partial_j(\rho v_j)-\rho v_j\partial_j\Big[\tfrac{1}{2}v^2+\varepsilon+\frac{p}{\rho}\Big]+v_j\partial_k\sigma'_{jk}+\rho T\big[\partial_t s+v_j\partial_j s\big]$$

$$= -\partial_j\Big(\big[\tfrac{1}{2}\rho v^2+\rho\varepsilon+p\big]v_j\Big)+v_j\partial_k\sigma'_{jk}+\rho T\big[\partial_t s+v_j\partial_j s\big]. \tag{2.63}$$

Since the right-hand sides of Eqs. (2.58a) and (2.63) must be identical, we can deduce that

$$\rho T\big[\partial_t s+v_j\partial_j s\big] = \sigma'_{jk}\partial_k v_j+\partial_j\big[\kappa\partial_j T\big], \tag{2.64a}$$

or in vector notation

$$\rho T\big[\partial_t s+(\mathbf{v}\cdot\boldsymbol{\nabla})s\big] = \sigma':\boldsymbol{\nabla}v+\boldsymbol{\nabla}\cdot(\kappa\boldsymbol{\nabla}T), \tag{2.64b}$$

where we use the double-dot product defined in Eq. (1.14). This equation is the second version of the third governing equation of the fluid dynamic system, and it is also called the general heat-transfer equation. Its left-hand side is ρT times the total time derivative of the entropy per unit mass, hence it expresses the total gain in heat density per unit time. The right-hand side represents the sources for heat gain: viscous friction and thermal conduction. Without these sources we have an ideal fluid with conservation of entropy.

In microfluidics, the fluid velocities are generally much smaller than the speed of sound in the fluid. As a consequence, the pressure deviations due to the motion of the fluid are minute and can be neglected. It is therefore natural to work with thermodynamic quantities as a function of temperature and pressure and keeping the latter constant. For the entropy we get $s = s(T,p)$ with constant p and the derivatives encountered in the heat equation (2.64b) become $\partial_t s = (\partial_T s)_p\, \partial_t T$ and $\boldsymbol{\nabla} s = (\partial_T s)_p\, \boldsymbol{\nabla} T$. It is well known from thermodynamics that $T(\partial_T s)_p = c_\mathrm{p}$, where c_p is the specific heat at constant pressure. We can conclude that in microfluidics it often suffices to use the following simpler form of the heat-transfer equation,

$$\rho c_\mathrm{p}\big[\partial_t T + (\mathbf{v}\cdot\boldsymbol{\nabla})T\big] = \boldsymbol{\nabla}\cdot(\kappa\boldsymbol{\nabla}T) + \sigma' : \boldsymbol{\nabla} v. \tag{2.65}$$

In the limit of small temperature differences in the microfluidic system the heat-transfer equation can be simplified even further, because in that case we can neglect the temperature dependence of ρ, η, κ and c_p and take them to be constant. In Exercise 2.10 it is shown that Eq. (2.65) reduces to

$$\partial_t T + (\mathbf{v}\cdot\boldsymbol{\nabla})T = D_\mathrm{th}\nabla^2 T + \frac{\eta}{2\rho c_\mathrm{p}}\,(\partial_k v_i + \partial_i v_k)^2, \tag{2.66}$$

where we have introduced the thermal diffusivity or the thermal diffusion constant D_th given by

$$D_\mathrm{th} \equiv \frac{\kappa}{\rho c_\mathrm{p}}. \tag{2.67}$$

For an incompressible fluid at rest we obtain the simplest version of the heat-transfer equation, namely the equation of thermal conduction also known as Fourier's equation,

$$\partial_t T = D_\mathrm{th}\nabla^2 T. \tag{2.68}$$

The value of the thermal diffusivity of water is

$$D_\mathrm{th} = \frac{0.597\ \mathrm{W\,m^{-1}K^{-1}}}{998\ \mathrm{kg\,m^{-3}} \times 4182\ \mathrm{J\,kg^{-1}\,K^{-1}}} = 1.43\times 10^{-7}\ \mathrm{m^2\,s^{-1}},\ \text{for water at 20 ℃}. \tag{2.69}$$

2.4 Further reading

There exists a large number of textbooks on general aspects of fluid dynamics. Besides the already mentioned books by Landau and Lifshitz (1993) and Batchelor (2000), the reader can find more elementary treatments in Tritton (1977) and Faber (1995). Although old-fashioned in style due to its appearance in six editions between 1879 and 1932, Lamb (1997) contains analyses of many, by now classic, flow problems. Of particular relevance for microfluidics are the textbooks that focuses on viscous flow in the low Reynolds number regime. Among these are Happel and Brenner (1983), Sherman (1990), and White (1991). For further reading on rheology and non-Newtonian fluids, the reader is referred to Bird, Armstrong and Hassager (1987), Probstein (1994) or Morrison (2001). Finally, besides what is presented in the above mentioned books, the reader can find more material on the heat-transfer equation in Bird, Stewart, and Lightfoot (2002).

2.5 Exercises

Exercise 2.1
The mass current density J
Argue why it is correct as stated in Eq. (2.3) that $\mathbf{J} = \rho\mathbf{v}$ indeed is the mass current density. Determine the SI unit of $\mathbf{J}$.

Exercise 2.2
A heuristic derivation of the continuity equation

The continuity equation (2.7) can be derived heuristically by considering the rate of change, $\partial_t(\rho\Delta x\Delta y\Delta z)$, of the mass inside the small cube (shown to the right) due to the flow of mass through the walls. Show that $\partial_t\rho = -\partial_x J_x$ if only the x component J_x of the current density is non-zero, and obtain the full continuity equation by generalization.

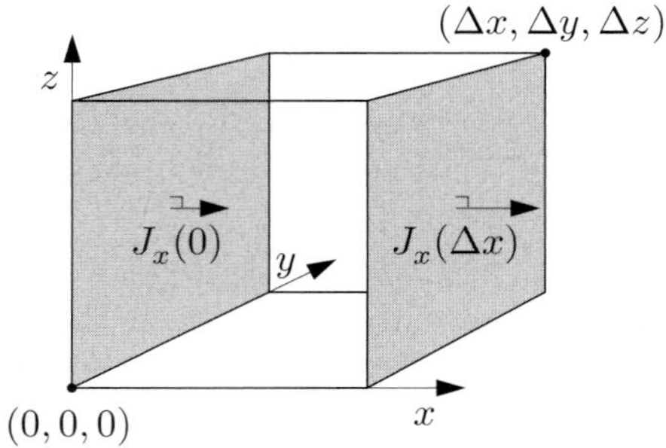

Fig. 2.3 Sketch of an elementary fluid volume.

Exercise 2.3
A heuristic derivation of the force densities from pressure and viscosity
Use the figure below to give a heuristic derivation of the pressure gradient and viscous force density on the right-hand-side of the incompressible Navier–Stokes equation (2.30b).

Hints: Consider the three pairs of opposite sides in the cubic fluid element defined by the corners $(0,0,0)$ and $(\Delta x, \Delta y, \Delta z)$. For the pressure, see panel (a), use the fact that the force $\Delta\mathbf{F}$ on an area $\Delta\mathcal{A}$ with surface normal $\mathbf{n}$ is given by $p\,\Delta\mathcal{A}\,(-\mathbf{n})$. For the viscosity, see panel (b), use the fact that the ith force component $(\Delta\mathbf{F})_i$ on an area $\Delta\mathcal{A}$ with a surface normal $\mathbf{n}$ is given by $\sigma'_{ik}n_k\,\Delta\mathcal{A} = \eta(\partial_i v_k + \partial_k v_i)n_k\,\Delta\mathcal{A}$. Let Δx, Δy, and Δz go to zero at the end.

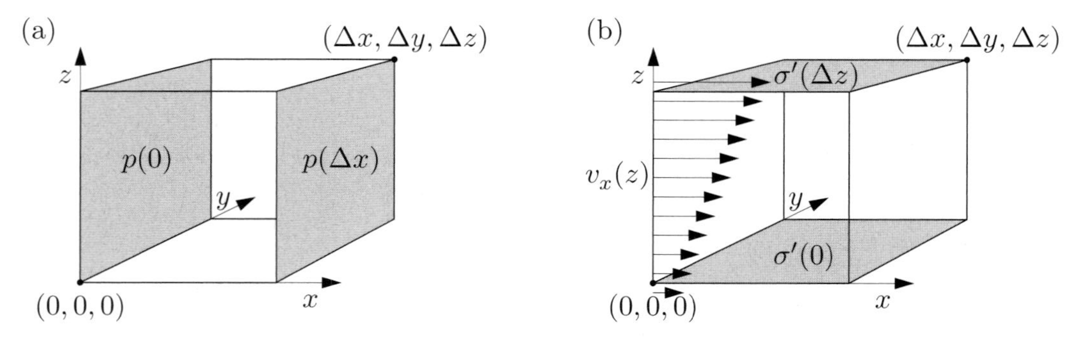

Fig. 2.4 (a) A sketch of a control volume of size $\Delta x\Delta y\Delta z$ with pressures $p(0)$ and $p(\Delta x)$ on the left and right side, respectively. (b) A sketch of a control volume of size $\Delta x\Delta y\Delta z$ in a velocity field $v_x(z)$ with shear stresses $\sigma'(0)$ and $\sigma'(\Delta z)$ on the bottom and top faces, respectively.

Exercise 2.4
Velocity gradients in a uniformly rotating fluid
The velocity field in a uniformly rotating fluid has the form $\mathbf{v} = \boldsymbol{\omega} \times \mathbf{r}$, where $\boldsymbol{\omega}$ is an angular velocity vector. Use the index notation Eq. (1.24) for the vector cross-product to show that this velocity field obeys the antisymmetric relation $\partial_j v_i = -\partial_i v_j$.

Exercise 2.5
The equation of motion for the Eulerian velocity field
Show that Eq. (2.25) leads to the equation of motion Eq. (2.27).

Exercise 2.6
Viscosity of water: measured temperature dependence
Based on Table A.2, in which the viscosity of water is listed as a function of temperature, make a so-called Arrhenius plot, of $\ln(\eta)$ versus $1/T$, where T is the temperature in kelvin. Discuss the temperature dependence of the viscosity of water.

Exercise 2.7
The divergence of the stress tensor
Use the index notation and Eqs. (2.18) and (2.26) to prove that for constant η and ζ Eq. (2.28), i.e. $\partial_j \sigma_{ij} = -\partial_i p + \eta \partial_j \partial_j v_i + \left(\frac{1}{3}\eta + \zeta\right)\partial_i(\partial_j v_j)$.

Exercise 2.8
The Navier–Stokes equation for a fluid in a magnetic field
The Navier–Stokes equation is given in Eq. (2.30b) for an incompressible fluid in a gravitational and electric field. Modify this equation to take into account the presence of an external magnetic field $\mathbf{B}$.

Exercise 2.9
The non-Newtonian shear rate tensor
Consider the shear rate tensor $\dot{\gamma}_{ij}$ and its magnitude $|\dot{\gamma}|$ defined in Eq. (2.46).

(a) Assume that the velocity field is given by $\mathbf{v} = v_x(z)\, \mathbf{e}_x$ consistent with a pressure-driven flow in a parallel-plate channel along the x axis. Determine $\dot{\gamma}_{ij}$ and $|\dot{\gamma}|$.

(b) Assume that the velocity field is given by $\mathbf{v} = v_x(r)\, \mathbf{e}_x$ consistent with a pressure-driven flow in a cylindrical channel along the x axis. Determine $\dot{\gamma}_{ij}$ and $|\dot{\gamma}|$.

Exercise 2.10
The heat-transfer equation with constant coefficients
In general, the coefficients of the heat-transfer equation (2.65) depend on the spatial coordinates. Show that in the special case, where they are constant the heat equation reduces to Eq. (2.66).

2.6 Solutions

Solution 2.1
The mass current density J
Consider a fluid of mass density ρ occupying the volume $\Delta\mathcal{V} = \Delta x\, \Delta y\, \Delta z$. Let the volume move along the x axis with speed v_x so that the entire volume has passed through the cross-sectional area $\Delta\mathcal{A} = \Delta y\, \Delta z$ in the time $\Delta t = \Delta x / v_x$. The mass current density is thus $J_x = \text{mass/area/time} = (\rho\Delta\mathcal{V})/(\Delta\mathcal{A})/\Delta t = \rho\Delta x/\Delta t = \rho v_x$, as was to be shown. The SI unit of $\mathbf{J}$ is $[\mathbf{J}] = [\rho\mathbf{v}] = (\mathrm{kg\, m^{-3}})(\mathrm{m\, s^{-1}}) = \mathrm{kg\, m^{-2}\, s^{-1}}$.

Solution 2.2
A heuristic derivation of the continuity equation
The mass ΔM inside the small, fixed volume $\Delta\mathcal{V} = \Delta x\, \Delta y\, \Delta z$ can change in time if and only if the density ρ changes in time:

$$\partial_t(\Delta M) = (\partial_t\rho)\, \Delta x\, \Delta y\, \Delta z. \tag{2.70}$$

However, due to mass conservation, this change in mass can only occur if the mass current density $\mathbf{J} = J_x\mathbf{e}_x$ causes different amounts of mass to enter the volume at the left side and to leave it at the right side. Therefore, the rate of change in mass can also be written as

$$\partial_t(\Delta M) = +J_x(0)\Delta y\Delta z - J_x(\Delta x)\Delta y\Delta z = -\big[J_x(\Delta x) - J_x(0)\big]\Delta y\Delta z. \tag{2.71}$$

Equating the two right-hand sides of Eqs. (2.70) and (2.71) and dividing by $\Delta\mathcal{V}$ leads to

$$\partial_t\rho = -\frac{J_x(\Delta x) - J_x(0)}{\Delta x} \underset{\Delta x\to 0}{\longrightarrow} -\partial_x J_x. \tag{2.72}$$

If the current density also has non-zero components along the y and the z directions, the corresponding two terms appear on the right-hand side of Eq. (2.72) leading to

$$\partial_t\rho = -\frac{J_x(\Delta x) - J_x(0)}{\Delta x} - \frac{J_y(\Delta y) - J_y(0)}{\Delta y} - \frac{J_z(\Delta z) - J_z(0)}{\Delta z} \underset{\Delta\mathcal{V}\to 0}{\longrightarrow} -\boldsymbol{\nabla}\cdot\mathbf{J}. \tag{2.73}$$

Solution 2.3
A heuristic derivation of the force densities from pressure and viscosity
The total force from the given pressure of surroundings on the cube is directed along the x axis and is given by $F_x = p(0)\, \Delta y\Delta z - p(\Delta x)\, \Delta y\Delta z$. This results in a force density

$$f_x = \frac{p(0)\, \Delta y\Delta z - p(\Delta x)\, \Delta y\Delta z}{\Delta x\Delta y\Delta z} = -\frac{p(\Delta x) - p(0)}{\Delta x} \underset{\Delta x\to 0}{\longrightarrow} -\partial_x p, \tag{2.74}$$

which is the x component of $\mathbf{f} = -\boldsymbol{\nabla} p$. The other two components are derived similarly.

The total force due to the viscous friction by the given velocity field from the surroundings on the cube is directed along the x axis and given by $F_x = \eta\partial_z v_x(\Delta z)\Delta x\Delta y - \eta\partial_z v_x(0)\Delta x\Delta y$. This results in a force density

$$f_x = \eta\frac{\partial_z v_x(\Delta z)\, \Delta x\Delta y - \partial_z v_x(0)\, \Delta x\Delta y}{\Delta x\Delta y\Delta z} = \eta\frac{\partial_z v_x(\Delta z) - \partial_z v_x(0)}{\Delta z} \underset{\Delta z\to 0}{\longrightarrow} \eta\partial_z^2 v_x, \tag{2.75}$$

which is the z-dependent part of the x component of the viscous force density $\mathbf{f} = \eta\nabla^2\mathbf{v}$ in the Navier–Stokes equation (2.30b).

Solution 2.4
Velocity gradients in a uniformly rotating fluid
From Eq. (1.24) we get $v_j = \epsilon_{jkl}\omega_k r_l$, while differentiation of the position vector $\mathbf{r}$ yields $\partial_i r_l = \delta_{il}$. Hence $\partial_i v_j = \epsilon_{jkl}\omega_k\partial_i r_l = \epsilon_{jkl}\omega_k\delta_{il} = \epsilon_{jki}\omega_k$, and likewise $\partial_j v_i = \epsilon_{ikj}\omega_k$. By definition the Levi–Civita symbol is antisymmetric, $\epsilon_{jki} = -\epsilon_{ikj}$, so we can conclude $\partial_i v_j = -\partial_j v_i$.

Solution 2.5
The equation of motion for the Eulerian velocity field
On the right-hand side of Eq. (2.25) we note that $-\partial_j(p\delta_{ij}) + \partial_j\sigma'_{ij} = \partial_j(-p\delta_{ij} + \sigma'_{ij})$ and, by use of the continuity equation (2.8), that $-\partial_j(\rho v_j v_i) = -\partial_j(\rho v_j)v_i - \rho v_j\partial_j v_i = (\partial_t\rho)v_i - \rho v_j\partial_j v_i$. With these two results substituted into Eq. (2.25) we obtain after cancellation of the terms $(\partial_t\rho)v_i$ the result, $\rho\partial_t v_i = -\rho v_j\partial_j v_i + \partial_j\sigma_{ij} + \rho g_i + \rho_{\mathrm{el}}E_i$.

Solution 2.6
Viscosity of water: measured temperature dependence
Let η_{373} denote the viscosity of water at the boiling point $T = 373$ K. Based on Table A.2 we then plot $\ln\left[\frac{\eta}{\eta_{373}}\right]$ versus $1/T$ as shown to the right. It is seen that to a good approximation we can write $\eta(T) \propto \exp(\Delta E/k_{\mathrm{B}}T)$, where ΔE is some kind of activation energy for molecular motion giving rise to viscosity. See also the molecular model of viscosity treated in Exercise 5.8.

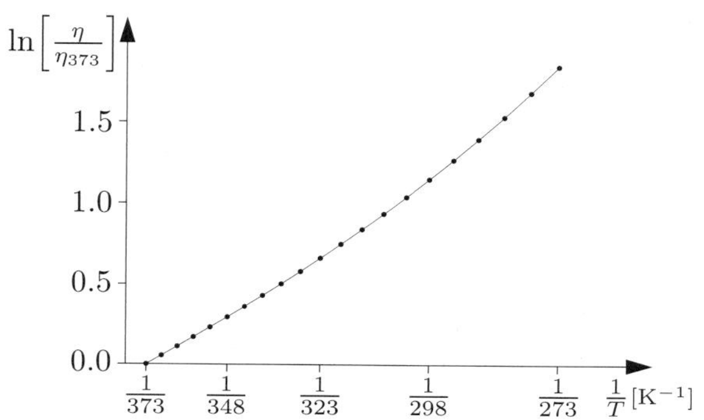

Fig. 2.5 An Arrhenius plot of the viscosity of water.

Solution 2.7
The divergence of the stress tensor
Constant η and ζ in Eq. (2.18) as well as $\partial_i\partial_j = \partial_j\partial_i$ and $\partial_j v_j = \partial_k v_k$ give $\partial_j\sigma_{ij} = \partial_j(-p\delta_{ij}) + \eta\partial_j\partial_j v_i + \eta\partial_j\partial_i v_j + \left(\zeta - \frac{2}{3}\eta\right)\delta_{ij}\partial_j(\partial_k v_k) = -\partial_i p + \eta\partial_j\partial_j v_i + \left(\zeta + \frac{1}{3}\eta\right)\partial_i(\partial_j v_j)$.

Solution 2.8
The Navier–Stokes equation for a fluid in a magnetic field
The force on a single charge q moving with velocity $\mathbf{v}$ in an external magnetic field $\mathbf{B}$ is given by the Lorentz force $\mathbf{F} = q\mathbf{v}\times\mathbf{B}$. The Navier–Stokes equation involves force densities instead of forces, so the charge q must be replaced by the electrical charge density ρ_{el}, leading to the magnetic force density $\rho_{\mathrm{el}}\mathbf{v}\times\mathbf{B}$. The Navier–Stokes equation is thus extended to

$$\rho\big[\partial_t v_i + v_j\partial_j v_i\big] = -\partial_i p + \eta\partial_j^2 v_i + \rho g_i + \rho_{\mathrm{el}}E_i + \epsilon_{ijk}\rho_{\mathrm{el}}v_j B_k, \tag{2.76a}$$

$$\rho\big[\partial_t\mathbf{v} + (\mathbf{v}\cdot\boldsymbol{\nabla})\mathbf{v}\big] = -\boldsymbol{\nabla}p + \eta\nabla^2\mathbf{v} + \rho\,\mathbf{g} + \rho_{\mathrm{el}}\mathbf{E} + \rho_{\mathrm{el}}\mathbf{v}\times\mathbf{B}. \tag{2.76b}$$

Solution 2.9
The non-Newtonian shear rate tensor

(a) The only non-zero components in $\dot{\gamma}_{ij}$ are $\dot{\gamma}_{xz} = \dot{\gamma}_{zx} = \partial_z v_x$, so the only non-zero components in $\dot{\gamma}^2$ are $(\dot{\gamma}^2)_{xx} = (\dot{\gamma}^2)_{zz} = (\partial_z v_x)^2$. Consequently $|\dot{\gamma}| = \sqrt{\frac{1}{2}\mathrm{Tr}(\dot{\gamma}^2)} = |\partial_z v_x|$.

(b) The only non-zero components in $\dot{\gamma}_{ij}$ are $\dot{\gamma}_{rx} = \dot{\gamma}_{xr} = \partial_r v_x$, so the only non-zero components in $\dot{\gamma}^2$ are $(\dot{\gamma}^2)_{xx} = (\dot{\gamma}^2)_{rr} = (\partial_r v_x)^2$. Consequently $|\dot{\gamma}| = \sqrt{\frac{1}{2}\mathrm{Tr}(\dot{\gamma}^2)} = |\partial_r v_x|$.

Solution 2.10
The heat-transfer equation with constant coefficients
Being constant, all the coefficients can be moved outside the respective differential operators, and the main problem is really to rewrite the viscous term.

For an incompressible fluid Eq. (2.20) states $\sigma'_{ij} = \eta\,(\partial_i v_j + \partial_j v_i)$, which is symmetric in the indices, $\sigma'_{ij} = \sigma'_{ji}$. From this and the definition in Eq. (1.14) of the double-dot product we get $\sigma' : \boldsymbol{\nabla} v = \sigma'_{ij}\partial_i v_j = \frac{1}{2}\big[\sigma'_{ij} + \sigma'_{ji}\big]\partial_i v_j = \frac{1}{2}\big[\sigma'_{ij}\partial_i v_j + \sigma'_{ji}\partial_i v_j\big]$. Being summed over, we can interchange the i and j indeces in the second product, $\sigma' : \boldsymbol{\nabla} v = \frac{1}{2}\big[\sigma'_{ij}\partial_i v_j + \sigma'_{ij}\partial_j v_i\big] = \frac{1}{2}\sigma'_{ij}\big[\partial_i v_j + \partial_j v_i\big]$, and we end with $\sigma' : \boldsymbol{\nabla} v = \frac{1}{2}\eta\,(\partial_i v_j + \partial_j v_i)(\partial_i v_j + \partial_j v_i)$, from which Eq. (2.66) follows readily.

3

Basic flow solutions

The Navier–Stokes equation is notoriously difficult to solve analytically because it is a non-linear differential equation. Analytical solutions can, however, be found in a few, but very important cases. Some of these solutions are the topic of this chapter. In particular, we shall solve a number of steady-state problems, among them Poiseuille-flow problems, i.e. pressure induced steady-state fluid flow in infinitely long, translation-invariant channels. It is important to study such idealized flows, since they provide us with a basic understanding of the behavior of liquids flowing in the microchannels of lab-on-a-chip systems.

Before analyzing the Poiseuille problem we treat three even simpler flow problems: fluids in mechanical equilibrium, the gravity-driven motion of a thin liquid film on an inclined plane, and the motion of a fluid between two parallel plates driven by the relative motion of theses plates (Couette flow).

In all cases we shall employ the so-called no-slip boundary condition for the velocity field at the part $\partial\Omega$ of the boundary that is a solid wall moving with velocity $\mathbf{v}_{\text{wall}}$,

$$\mathbf{v}(\mathbf{r}) = \mathbf{v}_{\text{wall}}, \quad \text{for } \mathbf{r} \in \partial\Omega \ \text{(no-slip)}. \tag{3.1}$$

The microscopic origin of this condition is the assumption of complete momentum relaxation between the molecules of the wall, which are at rest, and the outermost molecules of the fluid that collide with the wall. The momentum is relaxed on a length scale of the order of the molecular mean free path in the fluid, which for liquids and high-density fluids means one intermolecular distance ($\simeq 0.3$ nm). Only for rarified gases or narrow channels, where the mean free path of the gas molecules is comparable with the channel dimensions, is it necessary to abandon the no-slip boundary condition. Dealing mainly with liquids we shall adopt the no-slip boundary in this book, however, in Section 17.1 the reader can find a short discussion on recent experimental investigations of its validity.

3.1 Fluids in mechanical equilibrium

A fluid in mechanical equilibrium must be at rest relative to the walls of the vessel containing it, because otherwise it would continuously lose kinetic energy by heat conversion due to internal friction originating from viscous forces inside the fluid. The velocity field is therefore trivially zero everywhere, a special case of steady state defined by $\partial_t \mathbf{v} \equiv \mathbf{0}$. If we let gravity, described by the gravitational acceleration $\mathbf{g} = -g\mathbf{e}_z$ in the negative z direction, be the only external force, the Navier–Stokes equation reduces to

$$\mathbf{v}(\mathbf{r}) = \mathbf{0}, \tag{3.2a}$$

$$\mathbf{0} = -\boldsymbol{\nabla} p - \rho g \mathbf{e}_z. \tag{3.2b}$$

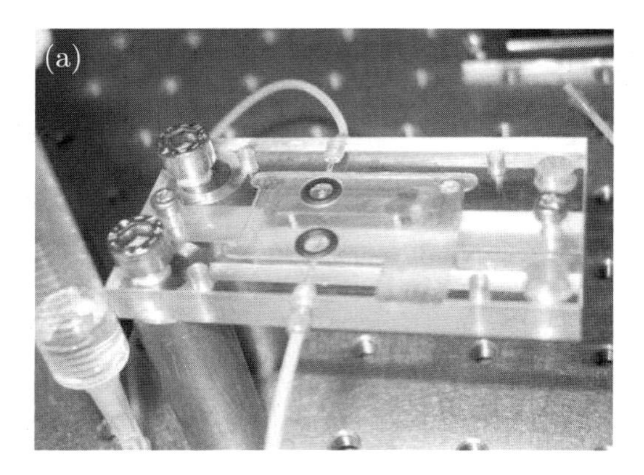

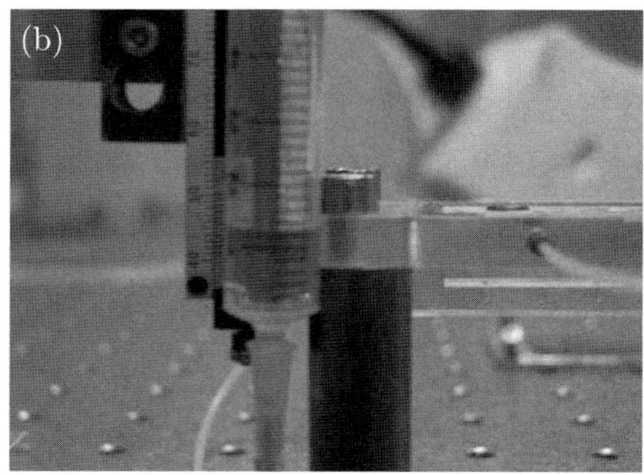

Fig. 3.1 (a) A water column (the syringe to the left) used in a lab-on-a-chip testing system (the polymer chip holder with two black o-rings in the center) to establish the pressure needed for sending water through the silicone tube into the microchannels of the chip. (b) Adjusting the water level in the syringe to level with the chip. Pictures courtesy of A.N. Hansen, M.B.L. Mikkelsen and A. Kristensen, DTU Nanotech.

For an incompressible fluid, say water, Eq. (3.2b) is easily integrated to give

$$p(z) = p^* - \rho g z, \tag{3.3}$$

where p^* is the pressure at the arbitrarily defined zero level $z = 0$. The z-dependent contribution to the pressure is denoted the hydrostatic pressure p_{hs},

$$p_{\mathrm{hs}}(z) \equiv -\rho g z. \tag{3.4}$$

In many microfluidic applications this is the only manifestation of gravity. It is therefore customary to write the total pressure p_{tot} as

$$p_{\mathrm{tot}} = p + p_{\mathrm{hs}}, \tag{3.5}$$

such that in the Navier–Stokes equation the gravitational body force is cancelled by the gradient of hydrostatic pressure. The resulting Navier–Stokes equation thus contains the auxiliary pressure p and no gravitational body force. We shall use this point of view frequently in the book.

The form of the hydrostatic pressure p_{hs} in Eq. (3.4) points to an easy way of generating pressure differences in liquids: the pressure at the bottom of a liquid column of height H is $\rho g H$ higher than the pressure at height H. Liquids with different densities, such as mercury with $\rho_{\mathrm{Hg}} = 1.36 \times 10^4~\mathrm{kg~m^{-3}}$ and water with $\rho_{\mathrm{H_2O}} = 1.00 \times 10^3~\mathrm{kg~m^{-3}}$ can be used to generate different pressures for given heights. The use of this technique in lab-on-a-chip systems is illustrated in Fig. 3.1 and in Exercise 3.1.

The hydrostatic pressure is less straightforward to compute for a compressible fluid. Let us consider an ideal gas under isothermal conditions. In this case, the density ρ and the pressure p are related by

$$\rho = \frac{\rho^*}{p^*}\, p, \tag{3.6}$$

where ρ^* and p^* is the density and pressure, respectively, for one particular state of the gas. With this equation of state Eqs. (3.2a) and (3.2b) are changed into

$$\mathbf{v}(\mathbf{r}) = \mathbf{0}, \tag{3.7a}$$

$$\mathbf{0} = -\boldsymbol{\nabla} p - \frac{\rho^*}{p^*}\, p g \mathbf{e}_z. \tag{3.7b}$$

Integration of Eq. (3.7b) yields

$$p(z) = p^* \exp\Big(-\frac{\rho^* g}{p^*}\, z\Big). \tag{3.8}$$

Inserting the parameter values for air at the surface of the Earth (hardly a microfluidic system), the thickness of the atmosphere is readily estimated to be $\ell = p^*/(\rho^* g) \approx 10$ km, see Exercise 3.2.

3.2 Liquid film flow on an inclined plane

The first example of a non-trivial velocity field is that of a liquid film flowing down along an infinitely long and infinitely wide inclined plane. Consider the geometry defined in Fig. 3.2. The component g_z of the gravitational acceleration normal to the inclined plane is balanced by the normal forces. The component g_x parallel to the plane accelerates the film down along the inclined plane until the velocity of the film is so large that the associated viscous friction forces in the film compensate g_x. When this happens the motion of the film has reached steady state, a situation we analyze in the following.

The translation invariance of the setup along the x and y directions dictates that the velocity field can only depend on z. Moreover, since the driving force points along the x direction only the x component of the velocity field is non-zero. Finally, no pressure gradients play any role in this free-flow problem, so the steady-state Navier–Stokes equation (i.e. $\partial_t \mathbf{v} = 0$) becomes

$$\mathbf{v}(\mathbf{r}) = v_x(z)\, \mathbf{e}_x, \tag{3.9a}$$

$$\rho(\mathbf{v}\cdot\boldsymbol{\nabla})\mathbf{v} = \eta\, \partial_z^2 \mathbf{v} + \rho g \sin\alpha\, \mathbf{e}_x. \tag{3.9b}$$

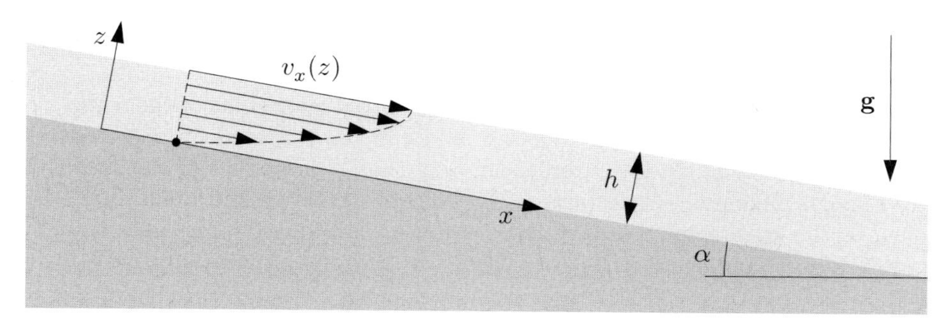

Fig. 3.2 A liquid film (light gray) of uniform thickness h flowing down along an inclined plane (dark gray). The plane has the inclination angle α and is assumed to be infinitely long and infinitely wide. The x and z axes are chosen parallel and normal to the plane, respectively. The gravitational acceleration is thus given by $\mathbf{g} = g\sin\alpha\, \mathbf{e}_x - g\cos\alpha\, \mathbf{e}_z$. In the steady state the resulting velocity profile of the liquid film is parabolic as shown.

The special symmetry of the velocity field given in Eq. (3.9a) implies an enormous simplification of the flow problem. Straightforward differentiation shows namely that the non-linear term in the Navier–Stokes equation (3.9b) vanishes,

$$(\mathbf{v}\cdot\boldsymbol{\nabla})\mathbf{v} = v_x(z)\partial_x\big[v_x(z)\big] = 0. \tag{3.10}$$

We thus only need to solve a linear second-order ordinary differential equation. This demands two boundary conditions, which are given by demanding no-slip of $\mathbf{v}$ at the plane $z = 0$ and no viscous stress on the free surface, i.e. σ'_{xz} from Eq. (2.20) is zero at $z = h$. We arrive at

$$\eta\,\partial_z^2 v_x(z) = -\rho g \sin\alpha, \tag{3.11a}$$

$$v_x(0) = 0, \quad \text{(no-slip)} \tag{3.11b}$$

$$\eta\,\partial_z v_x(h) = 0, \quad \text{(no stress)}. \tag{3.11c}$$

The solution is seen to be the well-known half-parabola

$$v_x(z) = \sin\alpha\,\frac{\rho g}{2\eta}\,(2h - z)z = \sin\alpha\,\frac{\rho g h^2}{2\eta}\left(2 - \frac{z}{h}\right)\frac{z}{h}. \tag{3.12}$$

As studied in Exercise 3.3 a typical speed for a 100 µm thick film of water is 1 cm/s.

3.3 Couette flow

Couette flow is a flow generated in a liquid by moving one or more of the walls of the vessel containing the fluid relative to the other walls. An important and very useful example is the Couette flow set up in a fluid held in the space between two concentric cylinders rotating axisymmetrically relative to each other. This setup is used extensively in rheology because it is possible to determine the viscosity η of the fluid very accurately by measuring the torque necessary to sustain a given constant speed of relative rotation.

Here, we study the simpler case of planar Couette flow as illustrated in Fig. 3.3. A liquid is placed between two infinite planar plates. The plates are oriented horizontally in the xy-plane perpendicular to the gravitational acceleration $\mathbf{g}$. The bottom plate at $z = 0$ is kept fixed in the laboratory, while the top plate at $z = h$ is moved in the x direction with the constant speed v_0.

As in the previous example the presence of translation invariance of the setup along the x and y directions implies that the velocity field can only depend on z. Moreover, since the driving force points along the x direction, only the x component of the velocity field is non-zero. Finally, the only pressure present is the hydrostatic pressure, see Eq. (3.4), which is cancelled by the gravitational body force. As in the previous example the symmetry again implies $(\mathbf{v}\cdot\boldsymbol{\nabla})\mathbf{v} = 0$, and the steady-state Navier–Stokes equation then reads

$$\mathbf{v}(\mathbf{r}) = v_x(z)\,\mathbf{e}_x, \tag{3.13a}$$

$$\eta\,\partial_z^2\mathbf{v} = \mathbf{0}. \tag{3.13b}$$

The boundary conditions on $\mathbf{v}$ is no-slip at the top and bottom plane at $z = 0$ and $z = h$, respectively, so we arrive at the following second-order ordinary differential equation with two boundary conditions:

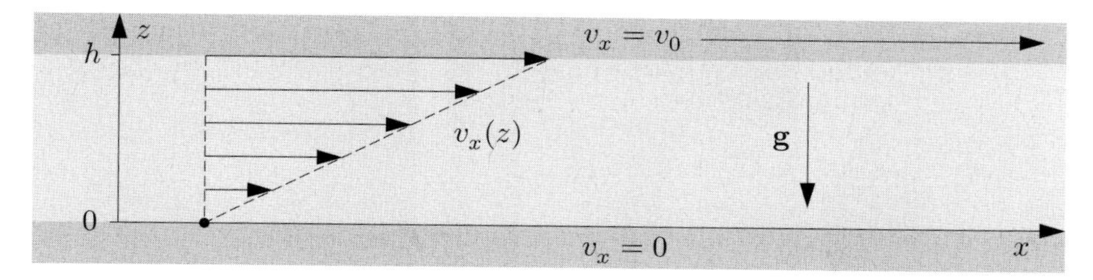

Fig. 3.3 An example of Couette flow. A fluid is occupying the space of height h between two horizontally placed, parallel, infinite planar plates. The top plate is moved with the constant speed v_0 relative to the bottom plate. The no-slip boundary condition at both plates forces the liquid into motion, resulting in the linear velocity profile shown.

$$\eta\, \partial_z^2 v_x(z) = 0, \tag{3.14a}$$

$$v_x(0) = 0, \quad \text{(no-slip)} \tag{3.14b}$$

$$v_x(h) = v_0, \quad \text{(no-slip)}. \tag{3.14c}$$

The solution is seen to be the well-known linear profile

$$v_x(z) = v_0\, \frac{z}{h}. \tag{3.15}$$

Assuming this expression to be valid for large, but finite, plates with area $\mathcal{A}$ we can, by use of the viscous stress tensor σ', determine the horizontal external force $\mathbf{F} = F_x \mathbf{e}_x$ necessary to apply to the top plate to pull it along with fixed speed v_0,

$$F_x = \sigma'_{xz}\, \mathcal{A} = \eta\, \frac{v_0 \mathcal{A}}{h}. \tag{3.16}$$

This expression allows for a simple experimental determination of the viscosity η.

3.4 Poiseuille flow

We now turn to the final class of analytical solutions to the Navier–Stokes equation: the pressure-driven, steady state flows in channels, also known as Poiseuille flows or Hagen–Poiseuille flows[1]. This class is of major importance for the basic understanding of liquid handling in lab-on-a-chip systems.

In a Poiseuille flow the fluid is driven through a long, straight, and rigid channel by imposing a pressure difference between the two ends of the channel. Originally, Hagen and Poiseuille studied channels with circular cross-sections, as such channels are straightforward to produce. However, especially in microfluidics, one frequently encounters other shapes. One example, shown in Fig. 3.4, is the Gaussian-like profile that results from producing microchannels by laser ablation in the surface of a piece of the polymer polymethylmethacrylate (PMMA). The heat from the laser beam cracks the PMMA into methylmethacrylate (MMA), which by evaporation leaves the substrate. A whole network of microchannels can then be created by sweeping the laser beam across the substrate in a well-defined pattern. The channels are sealed by placing and bonding a polymer lid on top of the structure.

[1] Whereas the pronunciation "Har-gen" with a hard "g" of the German name is straightforward for English speakers, the French name Poiseuille is often a minor stumbling block. Its pronunciation lies between "Pwa-soy" and "Pwa-say", but with the second vowel closer to the sound of "i" in the English word "Sir".

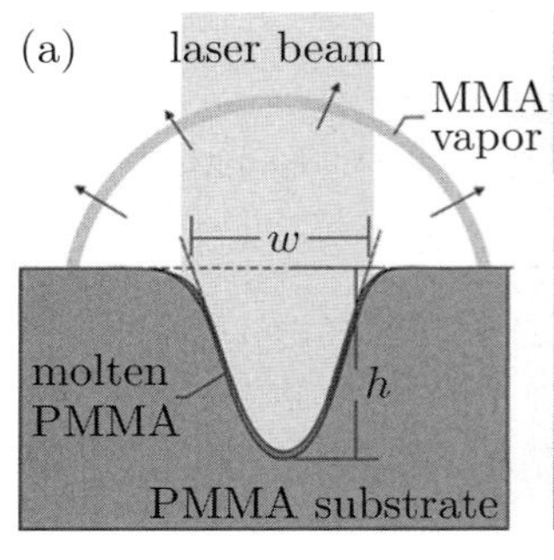

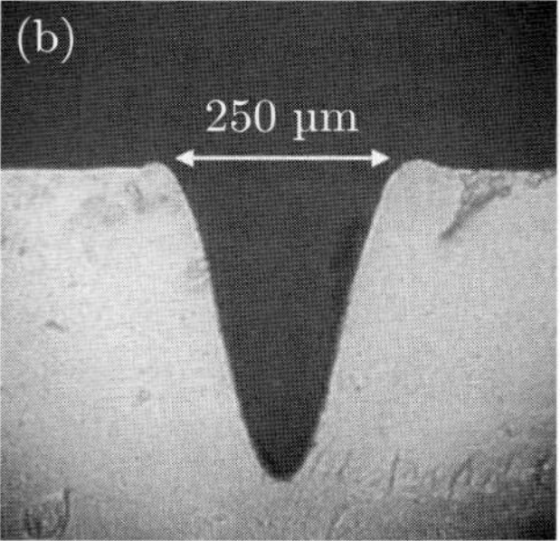

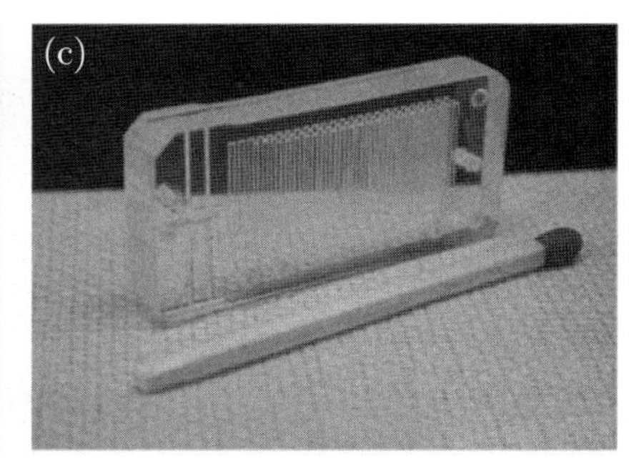

Fig. 3.4 Fabrication of microchannels by laser ablation in the surface a substrate made of the polymer PMMA. (a) Schematic diagram of the laser beam, the laser-ablated groove, including the molten PMMA and the vaporized hemispherical cloud of MMA leaving the cut zone. (b) Scanning electron microscope (SEM) micrograph of the cross-section of an actual microchannel showing the resulting Gaussian-like profile. (c) A three-layered PMMA-microfluidic system for the detection of ammonia in aqueous samples sent through the meandering microchannel fabricated by laser ablation. Courtesy of Oliver Geschke, DTU Nanotech.

3.4.1 Arbitrary cross-sectional shape

We first study the steady state Poiseuille-flow problem with an arbitrary cross-sectional shape as illustrated in Fig. 3.5. Although not analytically solvable, this example nevertheless provides us with the structural form of the solution for the velocity field.

The channel is parallel to the x axis, and it is assumed to be translationally invariant in that direction. The constant cross-section in the yz-plane is denoted $\mathcal{C}$ with boundary $\partial\mathcal{C}$, respectively. A constant pressure difference Δp is maintained over a segment of length L of the channel, i.e. $p(0) = p^* + \Delta p$ and $p(L) = p^*$. Here, p is the auxiliary pressure, which is left after cancellation of the trivial hydrostatic pressure p_hs by the gravitational body force, see Eq. (3.5). The translation invariance of the channel in the x direction combined with the vanishing of forces in the yz-plane implies the existence of a velocity field independent of x, while only its x component can be non-zero, $\mathbf{v}(\mathbf{r}) = v_x(y,z)\,\mathbf{e}_x$.[2] Consequently, $(\mathbf{v}\cdot\boldsymbol{\nabla})\mathbf{v} = 0$ and the steady-state Navier–Stokes equation becomes

$$\mathbf{v}(\mathbf{r}) = v_x(y,z)\,\mathbf{e}_x, \tag{3.17a}$$

$$\mathbf{0} = \eta\nabla^2\big[v_x(y,z)\,\mathbf{e}_x\big] - \boldsymbol{\nabla}p. \tag{3.17b}$$

Since the y and z components of the velocity field are zero, it follows that $\partial_y p = 0$ and $\partial_z p = 0$, and consequently that the pressure field only depends on x, $p(\mathbf{r}) = p(x)$. Using this result, the x component of the Navier–Stokes equation (3.17b) becomes

$$\eta\big[\partial_y^2 + \partial_z^2\big]v_x(y,z) = \partial_x p(x). \tag{3.18}$$

Here, it is seen that the left-hand side is a function of y and z, while the right-hand side is a function of x. The only possible solution is therefore that the two sides of the Navier–Stokes

[2]Although a valid mathematical solution at any flow speed, the translation-invariant velocity field is only stable at low velocities. The translation-invariance symmetry is spontaneously broken as the flow speed is increased, and eventually an unsteady turbulent flow appears as the physical solution having the smallest possible entropy production rate.

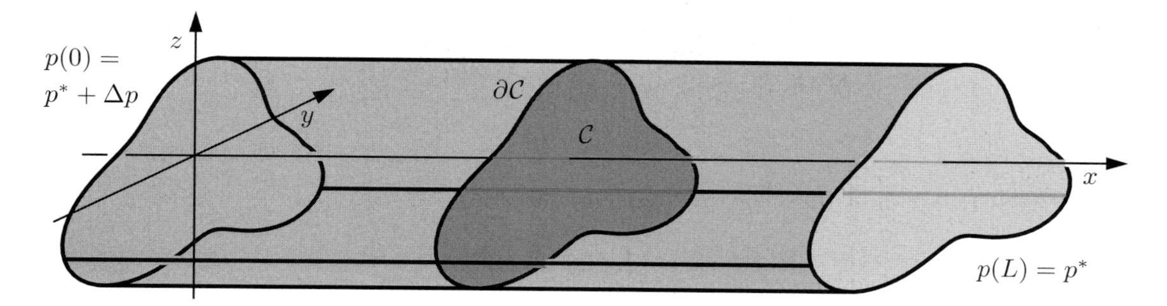

Fig. 3.5 The Poiseuille-flow problem in a channel, which is translationally invariant in the x direction, and that has an arbitrarily shaped cross-section $\mathcal{C}$ in the yz-plane. The boundary of $\mathcal{C}$ is denoted $\partial\mathcal{C}$. The pressure at the left end, $x = 0$, is an amount Δp higher than at the right end, $x = L$.

equation equal the same constant. However, a constant pressure gradient $\partial_x p(x)$ implies that the pressure must be a linear function of x, and using the boundary conditions for the pressure we obtain

$$p(\mathbf{r}) = \frac{\Delta p}{L}\,(L - x) + p^*. \tag{3.19}$$

With this we finally arrive at the second-order partial differential equation that $v_x(y,z)$ must fulfil in the domain $\mathcal{C}$ given the usual no-slip boundary conditions at the solid walls of the channel described by $\partial\mathcal{C}$,

$$\big[\partial_y^2 + \partial_z^2\big]\, v_x(y,z) = -\frac{\Delta p}{\eta L}, \quad \text{for } (y,z) \in \mathcal{C} \tag{3.20a}$$

$$v_x(y,z) = 0, \qquad \text{for } (y,z) \in \partial\mathcal{C}. \tag{3.20b}$$

Once the velocity field is determined it is possible to calculate the so-called volumetric flow rate Q, which is defined as the fluid volume discharged by the channel per unit time. For compressible fluids it becomes important to distinguish between the volumetric flow rate Q and the mass flow rate Q_{mass} defined as the discharged mass per unit time. In the case of the geometry of Fig. 3.5 we have

$$Q \equiv \int_{\mathcal{C}} \mathrm{d}y\,\mathrm{d}z\; v_x(y,z), \tag{3.21a}$$

$$Q_{\text{mass}} \equiv \int_{\mathcal{C}} \mathrm{d}y\,\mathrm{d}z\; \rho\, v_x(y,z). \tag{3.21b}$$

We can obtain a formal expression for the relation between the flow rate Q and the applied pressure difference Δp by employing the eigenfunction expansion technique of Section 1.6. Since the velocity field v_x is given by the Poisson-type equation (3.20a), we can write a formal solution using Eq. (1.51). In the present context the source function $g(\mathbf{r})$ is the constant $(\Delta p/\eta L) \times 1$, while the inner product is the 2D integral over the cross-section,

$$\langle f|g\rangle \equiv \int_{\mathcal{C}} \mathrm{d}y\,\mathrm{d}z\; f(y,z)\, g(y,z). \tag{3.22}$$

The velocity field $v_x(y,z)$ is therefore, by Eq. (1.51), given in the Dirac notation as

$$|v_x\rangle = \frac{\Delta p}{\eta L} \sum_{n=1}^{\infty} \frac{\langle \phi_n|1\rangle}{k_n^2}\, |\phi_n\rangle. \tag{3.23}$$

Consequently, the flow rate Q is given by

$$Q = \int_{\mathcal{C}} \mathrm{d}y\,\mathrm{d}z\; 1\, v_x(y,z) = \langle 1|v_x\rangle = \frac{\Delta p}{\eta L} \sum_{n=1}^{\infty} \frac{1}{k_n^2}\, \langle \phi_n|1\rangle\langle 1|\phi_n\rangle. \tag{3.24}$$

It is convenient to make the factors inside the sum dimensionless. The eigenvalue k_n^2 has the dimension of inverse length squared, we therefore multiply this term by some length $\mathcal{R}$. By Eq. (1.53) we further recognize the appearance of the effective area $\mathcal{A}_n = \langle \phi_n|1\rangle\langle 1|\phi_n\rangle$, and we therefore divide this term by the cross-sectional area $\mathcal{A}$. The expression for the flow rate can thus be written as

$$Q = \langle 1|v_x\rangle = \frac{\Delta p}{\eta L}\, \mathcal{R}^2 \mathcal{A} \sum_{n=1}^{\infty} \frac{1}{(k_n \mathcal{R})^2}\, \frac{\mathcal{A}_n}{\mathcal{A}}. \tag{3.25}$$

The proper length scale $\mathcal{R}$ turns out to be half the so-called hydrodynamics diameter, and in terms of the perimeter $\mathcal{P}$ and the area $\mathcal{A}$ of the cross-section Ω it is given by

$$\mathcal{R} \equiv \frac{2\mathcal{A}}{\mathcal{P}}. \tag{3.26}$$

For a circular cross-section of radius a, where $\mathcal{A} = \pi a^2$ and $\mathcal{P} = 2\pi a$, it is seen that $\mathcal{R} = a$. Remarkably, the introduction of the perimeter $\mathcal{P}$ and the area $\mathcal{A}$ captures most of the geometry dependence of the problem. Numerical calculations have revealed that the infinite sum is very close to $1/8$ for a wide range of shapes and exactly equal to this number for the circular cross-section. A factor of 8 is therefore introduced and the resulting dimensionless sum is denoted γ^{-1}. The final expression for the flow rate is

$$Q = \frac{1}{\gamma}\, \frac{\Delta p}{2\eta L}\, \frac{\mathcal{A}^3}{\mathcal{P}^2}, \tag{3.27a}$$

$$\frac{1}{\gamma} \equiv \sum_{n=1}^{\infty} \frac{8}{(k_n \mathcal{R})^2}\, \frac{\mathcal{A}_n}{\mathcal{A}} \approx 1. \tag{3.27b}$$

This is how far we can get theoretically without specifying the actual shape of the channel.

3.4.2 Infinite parallel-plate channel

In microfluidics the aspect ratio of a rectangular channel can often be so large that the channel is well approximated by an infinite parallel-plate configuration. This is our first and most simple example. The geometry shown in Fig. 3.6 is much like the one shown for the Couette flow in Fig. 3.3, but now both plates are kept fixed and a pressure difference Δp is

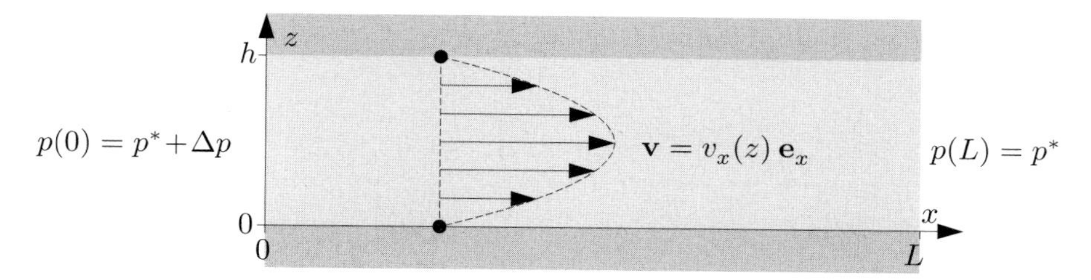

Fig. 3.6 A sketch in the xz-plane of an infinite, parallel-plate channel of height h. The system is translation invariant in the y direction and fluid is flowing in the x direction due to a pressure drop Δp over the section of length L.

applied. Due to the symmetry the y co-ordinate drops out and we end with the following ordinary differential equation,

$$\partial_z^2 v_x(z) = -\frac{\Delta p}{\eta L}, \tag{3.28a}$$

$$v_x(0) = 0, \quad \text{(no-slip)} \tag{3.28b}$$

$$v_x(h) = 0, \quad \text{(no-slip)}. \tag{3.28c}$$

The solution is a simple parabola

$$v_x(z) = \frac{\Delta p}{2\eta L}\,(h-z)z, \tag{3.29}$$

and the flow rate Q through a section of width w is found as

$$Q = \int_0^w \mathrm{d}y \int_0^h \mathrm{d}z\, \frac{\Delta p}{2\eta L}\,(h-z)z = \frac{h^3 w}{12\eta L}\,\Delta p. \tag{3.30}$$

This approximate expression for the flow rate in flat rectangular channels can be used instead of the more accurate expression Eq. (3.58) to obtain good order-of-magnitude estimates. However, it should be noted that the error is as much as 23% for an aspect ratio of one third, $h = w/3$, but falls to 7% for an aspect ratio of one tenth, $h = w/10$.

3.4.3 Elliptic cross-section

Our next example is the elliptic cross-section. We let the center of the ellipse be at $(y,z) = (0,0)$. The major axis of length a and the minor axis of length b are parallel to the y axis and z axis, respectively, as shown in Fig. 3.7(a). The boundary $\partial\mathcal{C}$ of the ellipse is given by the expression

$$\partial\mathcal{C}: \quad 1 - \frac{y^2}{a^2} - \frac{z^2}{b^2} = 0. \tag{3.31}$$

If we therefore, as a trial solution, choose

$$v_x(y,z) = v_0\left(1 - \frac{y^2}{a^2} - \frac{z^2}{b^2}\right), \tag{3.32}$$

we are guaranteed that $v_x(y,z)$ satisfies the no-slip boundary condition Eq. (3.20b). Insertion of the trial solution into the left-hand side of the Navier–Stokes equation (3.20a) yields

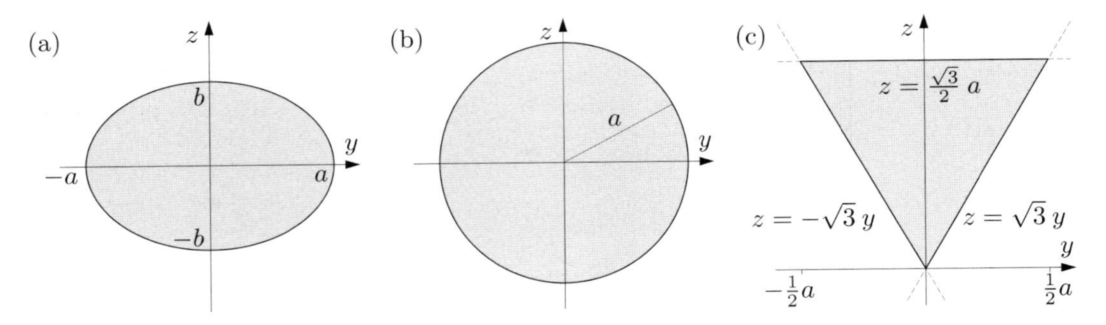

Fig. 3.7 The definition of three specific cross-sectional shapes for the Poiseuille-flow problem in long, straight channels. (a) The ellipse with major axis a and minor axis b, (b) the circle with radius a, and (c) the equilateral triangle with side length a.

$$\big[\partial_y^2 + \partial_z^2\big] v_x(y,z) = -2v_0 \left(\frac{1}{a^2} + \frac{1}{b^2}\right). \tag{3.33}$$

Thus, the Navier–Stokes equation (3.20a) will be satisfied by choosing the constant v_0 as

$$v_0 = \frac{\Delta p}{2\eta L} \frac{a^2 b^2}{a^2 + b^2}\,. \tag{3.34}$$

To calculate the flow rate Q for the elliptic channel we need to evaluate a 2D integral in an elliptically shaped integration region. This we handle by the following co-ordinate transformation. Let (ρ,ϕ) be the polar co-ordinates of the unit disk, i.e. the radial and azimuthal co-ordinates obey $0 \le \rho \le 1$ and $0 \le \phi \le 2\pi$, respectively. Our physical co-ordinates (y,z) and the velocity field v_x can then be expressed as functions of (ρ,ϕ):

$$y(\rho,\phi) = a\rho\, \cos\phi, \tag{3.35a}$$

$$z(\rho,\phi) = b\rho\, \sin\phi, \tag{3.35b}$$

$$v_x(\rho,\phi) = v_0\big(1-\rho^2\big). \tag{3.35c}$$

The advantage is that now the boundary $\partial\mathcal{C}$ can be expressed in terms of just one co-ordinate instead of two,

$$\partial\mathcal{C}: \quad \rho = 1. \tag{3.36}$$

The (y,z) surface integral in Eq. (3.21a) is transformed into (ρ,ϕ) co-ordinates by use of the Jacobian determinant $|\partial_{(\rho,\phi)}(y,z)|$,

$$\begin{aligned}\int_{\mathcal{C}} \mathrm{d}y\,\mathrm{d}z &= \int_{\mathcal{C}} \mathrm{d}\rho\,\mathrm{d}\phi \left|\frac{\partial(y,z)}{\partial(\rho,\phi)}\right| = \int_{\mathcal{C}} \mathrm{d}\rho\,\mathrm{d}\phi \begin{vmatrix} \partial_\rho y & \partial_\rho z \\ \partial_\phi y & \partial_\phi z \end{vmatrix} \\ &= \int_0^1 \mathrm{d}\rho \int_0^{2\pi} \mathrm{d}\phi \begin{vmatrix} +a\cos\phi & +b\sin\phi \\ -a\rho\sin\phi & +b\rho\cos\phi \end{vmatrix} = ab \int_0^{2\pi} \mathrm{d}\phi \int_0^1 \mathrm{d}\rho\, \rho. \end{aligned} \tag{3.37}$$

The flow rate Q for the elliptic channel is now easily calculated as

$$Q = \int_{\mathcal{C}} \mathrm{d}y\,\mathrm{d}z\, v_x(y,z) = ab \int_0^{2\pi} \mathrm{d}\phi \int_0^1 \mathrm{d}\rho\, \rho\, v_x(\rho,\phi) = \frac{\pi}{4}\, \frac{1}{\eta L}\, \frac{a^3 b^3}{a^2+b^2}\, \Delta p. \tag{3.38}$$

3.4.4 Circular cross-section

Since the circle Fig. 3.7(b) is just the special case $a = b$ of the ellipse, we can immediately write down the result for the velocity field and flow rate for the Poiseuille-flow problem in a circular channel. From Eqs. (3.32), (3.34), and (3.38) using $a = b$ it follows that

$$v_x(y,z) = \frac{\Delta p}{4\eta L}\Big(a^2 - y^2 - z^2\Big), \tag{3.39a}$$

$$Q = \frac{\pi a^4}{8\eta L}\,\Delta p. \tag{3.39b}$$

However, the same result can also be obtained by direct calculation using cylindrical co-ordinates (x, r, ϕ) thereby avoiding the trial solution Eq. (3.32). For cylindrical co-ordinates, see Appendix C.2, with the x axis chosen as the cylinder axis we have

$$(x,\ y,\ z) = (x,\ r\cos\phi,\ r\sin\phi), \tag{3.40a}$$

$$\mathbf{e}_x = \mathbf{e}_x, \tag{3.40b}$$

$$\mathbf{e}_r = +\cos\phi\,\mathbf{e}_y + \sin\phi\,\mathbf{e}_z, \tag{3.40c}$$

$$\mathbf{e}_\phi = -\sin\phi\,\mathbf{e}_y + \cos\phi\,\mathbf{e}_z, \tag{3.40d}$$

$$\nabla^2 = \partial_x^2 + \partial_r^2 + \frac{1}{r}\partial_r + \frac{1}{r^2}\partial_\phi^2. \tag{3.40e}$$

The symmetry considerations reduce the velocity field to $\mathbf{v} = v_x(r)\,\mathbf{e}_x$, so that the Navier–Stokes equation (3.20a) becomes an ordinary differential equation of second order,

$$\Big[\partial_r^2 + \frac{1}{r}\,\partial_r\Big]v_x(r) = -\frac{\Delta p}{\eta L}. \tag{3.41}$$

The solution to this inhomogeneous equation is the sum of a general solution to the homogeneous equation, $v_x'' + v_x'/r = 0$, and one particular solution to the inhomogeneous equation. It is easy to see that the general homogeneous solution has the linear form $v_x(r) = A + B\ln r$, while a particular inhomogeneous solution is $v_x(r) = -(\Delta p/4\eta L)\,r^2$. Given the boundary conditions $v_x(a) = 0$ and $v_x'(0) = 0$ we arrive at

$$v_x(r,\phi) = \frac{\Delta p}{4\eta L}\,(a^2 - r^2) \tag{3.42a}$$

$$Q = \int_0^{2\pi} \mathrm{d}\phi \int_0^a \mathrm{d}r\, r\,\frac{\Delta p}{4\eta L}\,(a^2 - r^2) = \frac{\pi}{8}\,\frac{a^4}{\eta L}\,\Delta p. \tag{3.42b}$$

3.4.5 Equilateral triangular cross-section

There exists no analytical solution to the Poiseuille-flow problem with a general triangular cross-section. In fact, it is only for the equilateral triangle defined in Fig. 3.7(c) that an analytical result is known.

The domain $\mathcal{C}$ in the yz-plane of the equilateral triangular channel cross-section can be thought of as the union of the three half-planes $(\sqrt{3}/2)a \geq z$, $z \geq \sqrt{3}y$, and $z \geq -\sqrt{3}y$. Inspired by our success with the trial solution of the elliptic channel, we now form a trial solution by multiplying together the expression for the three straight lines defining the boundaries of the equilateral triangle,

$$v_x(y,z) = \frac{v_0}{a^3}\left(\frac{\sqrt{3}}{2}a - z\right)\left(z - \sqrt{3}y\right)\left(z + \sqrt{3}y\right) = \frac{v_0}{a^3}\left(\frac{\sqrt{3}}{2}a - z\right)\left(z^2 - 3y^2\right). \tag{3.43}$$

By construction this trial solution satisfies the no-slip boundary condition on $\partial\mathcal{C}$. Luckily, it turns out that the Laplacian acting on the trial solution yields a constant,

$$\left[\partial_y^2 + \partial_z^2\right] v_x(y,z) = -2\sqrt{3}\,\frac{v_0}{a^2}. \tag{3.44}$$

Thus the Navier–Stokes equation will be satisfied by choosing the constant v_0 as

$$v_0 = \frac{1}{2\sqrt{3}}\,\frac{\Delta p}{\eta L}\,a^2. \tag{3.45}$$

The flow rate Q is most easily found by first integrating over y and then over z,

$$\begin{aligned} Q &= 2\int_0^{\frac{\sqrt{3}}{2}a} \mathrm{d}z \int_0^{\frac{1}{\sqrt{3}}z} \mathrm{d}y\, v_x(y,z) = \frac{4v_0}{3\sqrt{3}\,a^3}\int_0^{\frac{\sqrt{3}}{2}a} \mathrm{d}z \left(\frac{\sqrt{3}}{2}\,a - z\right) z^3 \\ &= \frac{3}{160}\,v_0\,a^2 = \frac{\sqrt{3}}{320}\,\frac{a^4}{\eta L}\,\Delta p. \end{aligned} \tag{3.46}$$

3.4.6 Rectangular cross-section

For lab-on-a-chip systems many fabrication methods lead to microchannels having a rectangular cross-section. One example is the microreactor shown in panels (a) and (b) of Fig. 3.8. This device is made in the epoxy-based polymer SU-8 by hot embossing, i.e. the SU-8 is heated to slightly above its glass-transition temperature, where it gets soft, and then a hard stamp containing the negative of the desired pattern is pressed into the polymer. The stamp is removed and later a polymer lid is placed on top of the structure and bonded to make a leakage-free channel.

It is perhaps a surprising fact that no analytical solution is known to the Poiseuille-flow problem with a rectangular cross-section. In spite of the high symmetry of the boundary the best we can do analytically is to find a Fourier sum representing the solution.

In the following, as illustrated in Fig. 3.8(c), we always take the width to be larger than the height, $w > h$. By rotation this situation can always be realized. The Navier–Stokes equation and associated boundary conditions are

$$\left[\partial_y^2 + \partial_z^2\right] v_x(y,z) = -\frac{\Delta p}{\eta L}, \quad \text{for} \quad -\frac{1}{2}w < y < \frac{1}{2}w,\ \ 0 < z < h, \tag{3.47a}$$

$$v_x(y,z) = 0, \qquad \text{for}\ \ y = \pm\frac{1}{2}w,\ \ z = 0,\ \ z = h. \tag{3.47b}$$

We begin by expanding all functions in the problem as Fourier series along the short vertical z direction. To ensure the fulfilment of the boundary condition $v_x(y,0) = v_x(y,h) = 0$ we

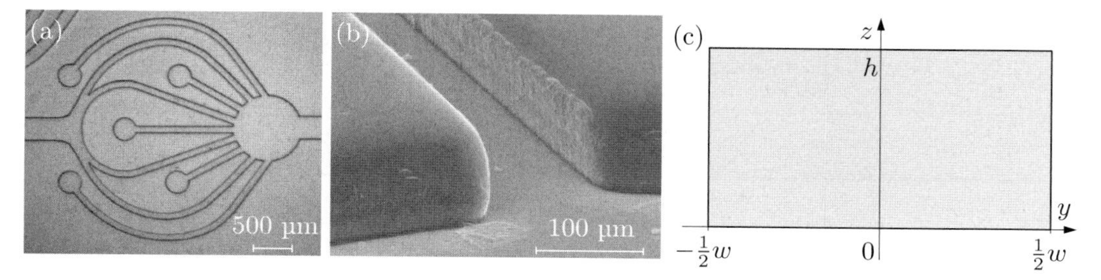

Fig. 3.8 (a) A top-view picture of a microreactor with nine inlet microchannels made by hot embossing in the polymer SU-8 before bonding on a polymer lid. (b) A zoom-in on one of the inlet channels having a near perfect rectangular shape of height $h = 50$ µm and width $w = 100$ µm. Courtesy of Oliver Geschke, DTU Nanotech. (c) The definition of the rectangular channel cross-section of height h and width w, which is analyzed in the text.

use only terms proportional to $\sin(n\pi z/h)$, where n is a positive integer. A Fourier expansion of the constant on the right-hand side in Eq. (3.47a) yields,

$$-\frac{\Delta p}{\eta L} = -\frac{\Delta p}{\eta L}\,\frac{4}{\pi}\sum_{n,\mathrm{odd}}^{\infty}\frac{1}{n}\,\sin\Big(n\pi\frac{z}{h}\Big), \tag{3.48}$$

a series containing only odd integers n. The coefficients $f_n(y)$ of the Fourier expansion in the z co-ordinate of the velocity are constants in z, but functions of y,

$$v_x(y,z) \equiv \sum_{n=1}^{\infty} f_n(y)\sin\Big(n\pi\frac{z}{h}\Big). \tag{3.49}$$

Inserting this series in the left-hand side of Eq. (3.47a) leads to

$$\big[\partial_y^2+\partial_z^2\big]v_x(y,z) = \sum_{n=1}^{\infty}\Big[f_n''(y) - \frac{n^2\pi^2}{h^2}\,f_n(y)\Big]\,\sin\Big(n\pi\frac{z}{h}\Big). \tag{3.50}$$

A solution to the problem must satisfy that for all values of n the nth coefficient in the pressure term Eq. (3.48) must equal the nth coefficient in the velocity term Eq. (3.50). The functions $f_n(y)$ are therefore given by

$$f_n(y) = 0, \qquad \text{for } n \text{ even}, \tag{3.51a}$$

$$f_n''(y) - \frac{n^2\pi^2}{h^2}\,f_n(y) = -\frac{\Delta p}{\eta L}\,\frac{4}{\pi}\,\frac{1}{n}, \qquad \text{for } n \text{ odd}. \tag{3.51b}$$

To determine $f_n(y)$, for n being odd, we need to solve the inhomogeneous second-order differential equation (3.51b). A general solution can be written as

$$f_n(y) = f_n^{\mathrm{inhom}}(y) + f_n^{\mathrm{homog}}(y), \tag{3.52}$$

where $f_n^{\mathrm{inhom}}(y)$ is a particular solution to the inhomogeneous equation (3.51b) and $f_n^{\mathrm{homog}}(y)$ is a general solution to the homogeneous equation, where the right-hand side of Eq. (3.51b)

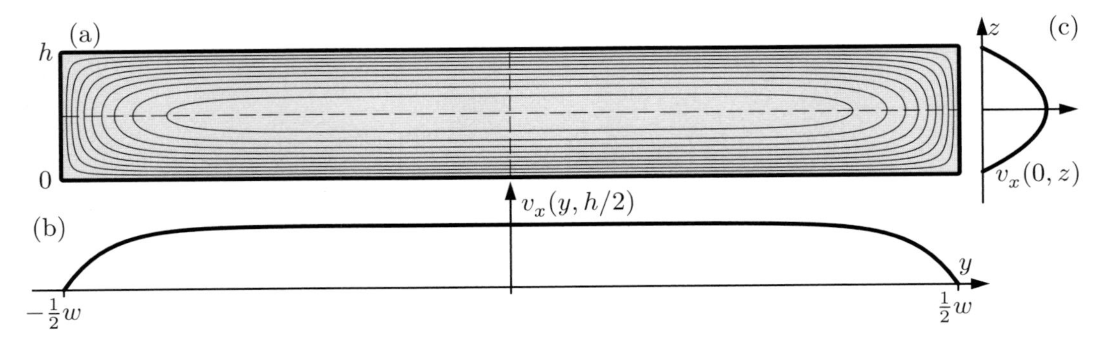

Fig. 3.9 (a) Contour lines for the velocity field $v_x(y,z)$ for the Poiseuille-flow problem in a rectangular channel. The contour lines are shown in steps of 10% of the maximal value $v_x(0,h/2)$. (b) A plot of $v_x(y,h/2)$ along the long centerline parallel to $\mathbf{e}_y$. (c) A plot of $v_x(0,z)$ along the short centerline parallel to $\mathbf{e}_z$.

is put equal to zero. It is easy to find one particular solution to Eq. (3.51b). One can simply insert the trial function $f_n^{\mathrm{inhom}}(y) = \mathrm{const}$ and solve the resulting algebraic equation,

$$f_n^{\mathrm{inhom}}(y) = \frac{4h^2\Delta p}{\pi^3\eta L}\,\frac{1}{n^3}, \quad \text{for } n \text{ odd.} \tag{3.53}$$

The general solution to the homogeneous equation, $f_n''(y) - (n^2\pi^2/h^2)\,f_n(y) = 0$ is the linear combination

$$f_n^{\mathrm{homog}}(y) = A\cosh\left(\frac{n\pi}{h}\,y\right) + B\sinh\left(\frac{n\pi}{h}\,y\right). \tag{3.54}$$

The solution $f_n(y)$ that satisfies the no-slip boundary conditions $f_n\big(\pm\frac{1}{2}w\big) = 0$ is

$$f_n(y) = \frac{4h^2\Delta p}{\pi^3\eta L}\,\frac{1}{n^3}\left[1 - \frac{\cosh\left(n\pi\frac{y}{h}\right)}{\cosh\left(n\pi\frac{w}{2h}\right)}\right], \quad \text{for } n \text{ odd,} \tag{3.55}$$

which leads to the velocity field for the Poiseuille flow in a rectangular channel,

$$v_x(y,z) = \frac{4h^2\Delta p}{\pi^3\eta L}\sum_{n,\mathrm{odd}}^{\infty}\frac{1}{n^3}\left[1 - \frac{\cosh\left(n\pi\frac{y}{h}\right)}{\cosh\left(n\pi\frac{w}{2h}\right)}\right]\sin\left(n\pi\frac{z}{h}\right). \tag{3.56}$$

In Fig. 3.9 are shown some plots of the contours of the velocity field and of the velocity field along the symmetry axes.

The flow rate Q is found by integration as follows,

$$Q = 2\int_0^{\frac{1}{2}w} \mathrm{d}y \int_0^h \mathrm{d}z\, v_x(y,z) \tag{3.57a}$$

$$= \frac{4h^2\Delta p}{\pi^3\eta L} \sum_{n,\mathrm{odd}}^{\infty} \frac{1}{n^3}\frac{2h}{n\pi}\left[w - \frac{2h}{n\pi}\tanh\left(n\pi\frac{w}{2h}\right)\right] \tag{3.57b}$$

$$= \frac{8h^3 w\Delta p}{\pi^4\eta L} \sum_{n,\mathrm{odd}}^{\infty} \left[\frac{1}{n^4} - \frac{2h}{\pi w}\frac{1}{n^5}\tanh\left(n\pi\frac{w}{2h}\right)\right] \tag{3.57c}$$

$$= \frac{h^3 w\Delta p}{12\eta L}\left[1 - \sum_{n,\mathrm{odd}}^{\infty} \frac{1}{n^5}\frac{192}{\pi^5}\frac{h}{w}\tanh\left(n\pi\frac{w}{2h}\right)\right], \tag{3.57d}$$

where we have used $\sum_{n,\mathrm{odd}}^{\infty} \frac{1}{n^4} = \frac{\pi^4}{96}$.

Very useful approximate results can be obtained in the limit $\frac{h}{w} \to 0$ of a flat and very wide channel, for which $\frac{h}{w}\tanh\left(n\pi\frac{w}{2h}\right) \to \frac{h}{w}\tanh(\infty) = \frac{h}{w}$, and Q becomes

$$Q \approx \frac{h^3 w\Delta p}{12\eta L}\left[1 - \frac{192}{\pi^5}\frac{h}{w}\sum_{n,\mathrm{odd}}^{\infty}\frac{1}{n^5}\right]$$

$$= \frac{h^3 w\Delta p}{12\eta L}\left[1 - \frac{192}{\pi^5}\frac{31}{32}\zeta(5)\frac{h}{w}\right]$$

$$\approx \frac{h^3 w\Delta p}{12\eta L}\left[1 - 0.630\,\frac{h}{w}\right], \quad \text{for } h < w. \tag{3.58}$$

Here, we have on the way used the Riemann zeta function, $\zeta(x) \equiv \sum_{n=1}^{\infty} 1/n^x$,

$$\sum_{n,\mathrm{odd}}^{\infty}\frac{1}{n^5} = \sum_{n=1}^{\infty}\frac{1}{n^5} - \sum_{n,\mathrm{even}}^{\infty}\frac{1}{n^5} = \zeta(5) - \sum_{k=1}^{\infty}\frac{1}{(2k)^5} = \zeta(5) - \frac{1}{32}\zeta(5) = \frac{31}{32}\zeta(5). \tag{3.59}$$

The approximative result Eq. (3.58) for Q is surprisingly good. For the worst case, the square with $h = w$, the error is just 13%, while already at an aspect ratio of a half, $h = w/2$, the error is down to 0.2%.

If we neglect the side walls completely we arrive at the case of an infinitely wide channel, which is studied in the following subsection.

3.5 Poiseuille flow in shape-perturbed channels

By use of shape-perturbation theory it is possible to extend the analytical results for Poiseuille flow beyond the few cases of regular geometries that we have treated above. In shape-perturbation theory, as for any other perturbation calculation, see Section 1.5, the starting point is an analytically solvable case, which then is deformed slightly, characterized by some small perturbation parameter α.

We shall study two examples of such shape-perturbed channels, and to simplify the mathematics it is useful to introduce dimensionless co-ordinates. Traditionally, these are denoted (ξ, η, ζ) in the x, y and z direction respectively, and given some characteristic length scales ℓ_x, ℓ_y and ℓ_z in the three directions, they are defined as

$$\xi \equiv \frac{x}{\ell_x}, \qquad \eta \equiv \frac{y}{\ell_y}, \qquad \zeta \equiv \frac{z}{\ell_z}. \tag{3.60}$$

To avoid a possible confusion between the dimensionless y co-ordinate and the dynamic viscosity, we supply the latter with an asterisk,

$$\eta^*, \text{ standard dynamic viscosity.} \tag{3.61}$$

3.5.1 Perturbation of a circular cross-section

The first example is a perturbation of a channel with circular cross-section. As illustrated in Fig. 3.10 the unperturbed shape is described by parametric co-ordinates (η, ζ) in Cartesian form or (ρ, θ) in polar form. The co-ordinates of the physical problem we would like to solve are (y, z) in Cartesian form and (r, ϕ) in polar form.

As an actual example we take the multipolar deformation of the circle defined by the transformation

$$\phi = \theta, \qquad 0 \le \theta \le 2\pi, \tag{3.62a}$$

$$r = a\,\rho\big[1 + \alpha \sin(k\theta)\big], \qquad 0 \le \rho \le 1, \tag{3.62b}$$

$$y(\rho, \theta) = a\,\rho\big[1 + \alpha \sin(k\theta)\big] \cos\theta, \tag{3.62c}$$

$$z(\rho, \theta) = a\,\rho\big[1 + \alpha \sin(k\theta)\big] \sin\theta, \tag{3.62d}$$

where a is length scale and k is an integer defining the order of the multipolar deformation. Note that for $\alpha = 0$ the shape is the unperturbed circle. The boundary of the perturbed shape is simply described by fixing the unperturbed co-ordinate $\rho = 1$ and sweeping in θ,

$$(y, z) = \big(y(1, \theta),\ z(1, \theta)\big), \quad \text{the perturbed boundary.} \tag{3.63}$$

It is therefore desirable to formulate the perturbed Poiseuille problem using the unperturbed co-ordinates. To obtain analytical results it is important to make the appearance of the perturbation parameter explicit. When performing a perturbation calculation to order m, see Section 1.5, all terms containing α^l with $l > m$ are discarded, while the remaining terms containing the same power of α are grouped together, and the equations are solved power by power. To carry out the perturbation calculation the velocity field $v_x(y, z)$ is written as

$$v_x(y, z) = v_x\big(y(\rho, \theta), z(\rho, \theta)\big) = v_x^{(0)}(\rho, \theta) + \alpha\, v_x^{(1)}(\rho, \theta) + \alpha^2\, v_x^{(2)}(\rho, \theta) + \cdots. \tag{3.64}$$

Likewise, the Laplacian operator in the Navier–Stokes equation must be expressed in terms of ρ, θ, and α. The starting point of this transformation is the transformation of the gradients

$$\partial_r = (\partial_r \rho)\, \partial_\rho + (\partial_r \theta)\, \partial_\theta, \tag{3.65a}$$

$$\partial_\phi = (\partial_\phi \rho)\, \partial_\rho + (\partial_\phi \theta)\, \partial_\theta. \tag{3.65b}$$

The derivatives $(\partial_r \rho)$, $(\partial_r \theta)$, $(\partial_\phi \rho)$, and $(\partial_\phi \theta)$ are obtained from the inverse transformation of Eqs. (3.62b) and (3.62a),

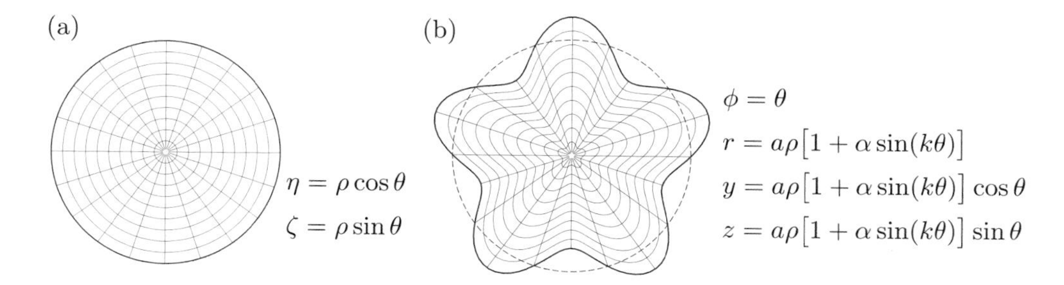

Fig. 3.10 (a) The geometry of the unperturbed and analytically solvable cross-section, the unit circle, described by co-ordinates (η, ζ) or (ρ, θ). (b) The geometry of the perturbed cross-section described by co-ordinates (y, z) or (r, ϕ) and the perturbation parameter α. Here, $a = 1$, $k = 5$ and $\alpha = 0.2$.

$$\rho(r, \phi) = \frac{1}{1 + \alpha\sin(k\phi)}\,\frac{r}{a}, \tag{3.66a}$$

$$\theta(r, \phi) = \phi. \tag{3.66b}$$

The expansion Eq. (3.64) can now be inserted into the Navier–Stokes equation and by use of the derivatives, Eqs. (3.65a) and (3.65b), we can carry out the perturbation scheme. The calculation is straightforward but tedious. We shall here just quote the first-order perturbation result for the velocity field:

$$v_x(\rho, \theta) = \Big[\big(1 - \rho^2\big) - 2\big(\rho^2 - \rho^k\big)\sin(k\theta)\,\alpha\Big]\,\frac{a^2\Delta p}{4\eta L} + \mathcal{O}\big(\alpha^2\big). \tag{3.67}$$

This example may appear rather artificial. However, almost any shape deformation of the circle can by analyzed based on this example. An arbitrarily shaped boundary can be written as a Fourier series involving a sum over infinitely many multipole deformations like the kth one studied in this section.

3.5.2 Perturbation of a flat cross-section

The second example is Poiseuille flow in a flat channel with an arbitrarily shaped cross-section in the yz-plane. We orient the cross-section such that it is cut by the y axis at its maximal width w, see Fig. 3.11. The origin of the yz co-ordinate system is placed at the left-most edge, such that the right-most edge has the yz co-ordinate $(w, 0)$. The cross-section is flipped around the y axis to ensure that the point on the perimeter furthest away from the y axis, a distance denoted h_0, is in the upper half-plane. The top edge $h_+(y)$ and the bottom edge $h_-(y)$ are described by the dimensionless shape functions λ_+ and λ_- as

$$h_+(y) \equiv h_0\,\lambda_+(y/w), \qquad h_-(y) \equiv h_0\,\lambda_-(y/w). \tag{3.68}$$

To facilitate the following calculations, we make the Navier–Stokes equation (3.20) dimensionless by introducing the following dimensionless variables and parameters,

$$\eta \equiv \frac{y}{w}, \qquad \zeta \equiv \frac{z}{h_0}, \qquad v(\eta, \zeta) \equiv \frac{\eta^* L}{h_0^2\Delta p}\,v_x(w\eta, h_0\zeta), \qquad \alpha \equiv \frac{h_0^2}{w^2}, \tag{3.69}$$

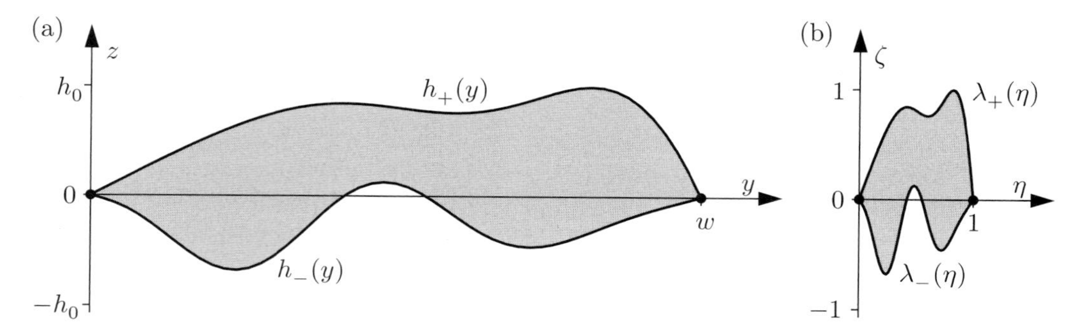

Fig. 3.11 (a) A flat channel with translation invariance along the x axis and with an arbitrary yz cross-section (dark gray) of width w and height h_0. (b) The rescaled cross-section of unity width and height in the dimensionless $\eta\zeta$-plane, where the curves defining the top and the bottom of the channel are denoted $\lambda_+(\eta)$ and $\lambda_-(\eta)$, respectively.

and obtain the following differential equation for $v(\eta,\zeta)$ with corresponding no-slip boundary conditions,

$$\Big[\alpha\,\partial_\eta^2+\partial_\zeta^2\Big]v(\eta,\zeta)=-1,\qquad v\big(\eta,\lambda_+(\eta)\big)=0,\qquad v\big(\eta,\lambda_-(\eta)\big)=0. \tag{3.70}$$

Note that $0\le\eta\le1$ and $-1\le\zeta\le1$.

For flat channels, where $h_0\ll w$, the dimensionless flatness parameter is much less than unity, $\alpha\equiv h_0^2/w^2\ll1$, and it can therefore be used as a perturbation parameter. Consequently, we write

$$v\equiv v_0+\alpha\,v_1+\alpha^2\,v_2+\cdots, \tag{3.71}$$

which upon insertion into Eq. (3.70) and expansion in α, as in Eq. (1.34), yields

$$\partial_\zeta^2v_0+\alpha\big(\partial_\zeta^2v_1+\partial_\eta^2v_0\big)+\alpha^2\big(\partial_\zeta v_2+\partial_\eta^2v_1\big)=-1. \tag{3.72}$$

Following the perturbation scheme of Eq. (1.35) we then obtain the zero- and first-order equations with corresponding no-slip boundary conditions,

$$\partial_\zeta^2v_0=-1,\qquad v_0\big(\eta,\lambda_+(\eta)\big)=0,\quad v_0\big(\eta,\lambda_-(\eta)\big)=0, \tag{3.73a}$$

$$\partial_\zeta^2v_1=-\partial_\eta^2v_0,\quad v_1\big(\eta,\lambda_+(\eta)\big)=0,\quad v_1\big(\eta,\lambda_-(\eta)\big)=0. \tag{3.73b}$$

The zero-order problem Eq. (3.73a) is equivalent to the infinite parallel-plate problem Eq. (3.28), and hence the solution for v is equivalent to Eq. (3.29),

$$v_0(\eta,\zeta)=\frac{1}{2}\big[\zeta-\lambda_-(\eta)\big]\big[\lambda_+(\eta)-\zeta\big]. \tag{3.74}$$

Equipped with the solution for v_0 we can determine $-\partial_\eta^2v_0$, which enters on the right-hand side in Eq. (3.73b) for v_1. The latter can then be found by straightforward integration over ζ, and with the shorthand notation $\lambda''_\pm\equiv\partial_\eta^2\lambda_\pm$ we get

$$v_1(\eta,\zeta)=\frac{\lambda''_+(\eta)-\lambda''_-(\eta)}{12}\Big[\lambda_-(\eta)-\zeta\Big]\Big[\lambda_+(\eta)-\zeta\Big]\Big[3\frac{\partial_\eta^2\big[\lambda_+(\eta)\lambda_-(\eta)\big]}{\lambda''_+(\eta)+\lambda''_-(\eta)}-\lambda_+(\eta)-\lambda_-(\eta)-\zeta\Big]. \tag{3.75}$$

Often, microfluidic fabrication methods involve a flat substrate bottom onto which is bonded a chip containing open channels. If the xy-plane is made to coincide with the flat bottom of the resulting closed channels, the above expressions are simplified since $\lambda_-(\eta) \equiv 0$. To first order in α the resulting velocity field in this case becomes

$$v(\eta,\zeta) = \frac{1}{2}\,\zeta\Big[\lambda_+(\eta) - \zeta\Big] + \alpha\,\frac{\lambda''_+(\eta)}{12}\,\zeta\Big[\lambda^2_+(\eta) - \zeta^2\Big]. \tag{3.76}$$

In fluid dynamics the above perturbation expansion in the flatness parameter $\alpha = h_0^2/w^2$ is denoted lubrication theory. The name stems from studies of lubrication by liquids in the narrow region between two closely spaced solid bodies. As can be seen from Eq. (3.75), lubrication theory is closely linked to systems with weakly varying boundaries, in our case expressed by the pre-factor $\alpha\big[\lambda''_+(\eta) - \lambda''_-(\eta)\big]$, which describes a very small curvature of the boundary curve. We shall return to lubrication theory in Section 14.1.

As an actual example of the theory we study the case of a flat, parabolic shape. Such a shape appears in microchannels fabricated by certain soft-lithography methods. The bottom shape function is put to zero, while the top shape function is a downward-pointing parabola of height h_0 and width w,

$$h_+(y) = \frac{4h_0}{w^2}\,y\,(w-y), \tag{3.77a}$$

$$h_-(y) = 0. \tag{3.77b}$$

In the dimensionless co-ordinates the shape functions are

$$\lambda_+(\eta) = 4\,\eta\,(1-\eta), \tag{3.78a}$$

$$\lambda_-(\eta) = 0. \tag{3.78b}$$

We note that $\lambda''_+(\eta) = -8$, and from Eq. (3.76) we obtain an expression for the flow rate Q,

$$\begin{aligned} Q &= \int_0^w \mathrm{d}y \int_0^{h(y)} \mathrm{d}z\; v_x(x,z) \\ &= \frac{h_0^3 w}{\eta^* L}\,\Delta p \int_0^1 \mathrm{d}\eta \int_0^{4\eta(1-\eta)} \mathrm{d}\zeta \left\{\frac{1}{2}\,\zeta\Big\{4\,\eta\,(1-\eta) - \zeta\Big\} - \alpha\,\frac{2}{3}\,\zeta\Big\{\big[4\,\eta\,(1-\eta)\big]^2 - \zeta^2\Big\}\right\}. \end{aligned} \tag{3.79}$$

The calculation of the integral is straightforward but tedious, see Exercise 3.7. The final result is

$$Q = \left[1 - \frac{16}{9}\,\alpha\right]\frac{4}{105}\,\frac{h_0^3 w}{\eta^* L}\,\Delta p. \tag{3.80}$$

3.6 Poiseuille flow for weakly compressible fluids

Relaxing the constraint of complete incompressibility, we now study Poiseuille flow of a weakly compressible fluid. The corrections to the well-known solution for incompressible fluids are calculated by first-order perturbation theory using the relative compressibility as the expansion parameter.

Consider a circular cylindrical channel of radius a and length L parallel to the x axis containing a compressible fluid of density $\rho(p)$ and constant viscosity η. Under the influence of a steady pressure drop Δp a steady-state flow profile $\mathbf{v}$ is established. Assuming perfect translation-invariance along the x direction and perfect hydrostatic pressure balance in the z direction the resulting velocity field contains only an x component,

$$\mathbf{v} = v(x, y, z)\,\mathbf{e}_x. \tag{3.81}$$

Further, assuming low Reynolds numbers, $Re \ll 1$, the steady state (Navier–)Stokes equation takes the form

$$(\nabla^2 v)\,\mathbf{e}_x = \frac{1}{\eta}\,\boldsymbol{\nabla} p, \tag{3.82}$$

and the vanishing of the y and z components implies that the pressure field depends only on x,

$$p = p(x) \tag{3.83a}$$

$$p(0) = p^* + \Delta p, \tag{3.83b}$$

$$p(L) = p^*. \tag{3.83c}$$

Finally, we model the compressibility of the fluid through the pressure-dependent density $\rho(p)$ as

$$\rho\big(p(x)\big) = \rho^* + \frac{1}{c_\mathrm{a}^2}\,(p - p^*), \tag{3.84}$$

where ρ^* and p^* are the density and pressure, respectively, at room temperature and standard pressure, while the compressibility constant of the fluid is given in terms of the speed of sound c_a in the fluid. Note that since the pressure depends only on x, so does ρ. Given the forms, Eqs. (3.81) and (3.84), of the velocity and density fields, the continuity equation becomes

$$0 = \boldsymbol{\nabla}\cdot\mathbf{v} = \partial_x(\rho v) = \rho\partial_x v + \frac{1}{c_\mathrm{a}^2}\,(\partial_x p)v. \tag{3.85}$$

This is how far we will get without further approximations. In the following we will introduce the small parameter α given by

$$\alpha \equiv \frac{\Delta p}{\rho^* c_\mathrm{a}^2}, \tag{3.86}$$

and solve the Stokes and continuity equation by a first-order perturbation expansion in this parameter.

The case of incompressible fluids is given by $\alpha = 0$ or an infinite speed of sound, $c = \infty$. The corresponding zero-order fields are well known and given by

$$v_0 = v_0(y, z) = \frac{\Delta p}{4\eta L}\left(a^2 - y^2 - z^2\right), \tag{3.87a}$$

$$p_0 = p_0(x) \quad = p^* + \left(1 - \frac{x}{L}\right)\Delta p, \tag{3.87b}$$

$$\rho_0 = \rho^*. \tag{3.87c}$$

Note especially that the unperturbed velocity field is independent of x, and that the unperturbed pressure gradient is a constant,

$$\partial_x p_0(x) = -\frac{\Delta p}{L}, \tag{3.88}$$

and that this zero-order solution satisfies the pressure boundary conditions

$$p_0(0) = p^* + \Delta p, \qquad p_0(L) = p^*. \tag{3.89}$$

For later use we also note the following two relations. The unperturbed Stokes equation reads

$$\nabla^2 v_0 = -\frac{\Delta p}{\eta L}, \tag{3.90}$$

while the unperturbed mass flow rate $Q_{\mathrm{mass},0}$ is given by

$$Q_{\mathrm{mass},0} = \int_{\mathrm{area}} \mathrm{d}y\, \mathrm{d}z\; \rho^* v_0(y,z) = \frac{\pi a^4 \rho^*}{8\eta L}\,\Delta p. \tag{3.91}$$

The fields of velocity, pressure and density are now written as the following first-order expansions, see Eq. (1.37b),

$$v = v_0(y,z) + v_1(x,y,z), \tag{3.92a}$$
$$p = p_0(x) + p_1(x), \tag{3.92b}$$
$$\rho = \rho^* + \rho_1(x), \tag{3.92c}$$

where we have made the variable dependencies explicit. Each of the first-order terms are assumed to be proportional to the expansion parameter α of Eq. (3.86), and consequently we are, in the following, going to neglect terms containing products of first-order terms.

The first-order perturbation ρ_1 to the density is found by considering the density $\rho\big(p(x)\big)$ given in Eq. (3.84). We note that the correction to ρ^* already contains the small pre-factor $1/c_{\mathrm{a}}^2$, hence it suffices to insert the unperturbed pressure p_0 from Eq. (3.87b), and we arrive at

$$\rho_1 = \frac{\Delta p}{c_{\mathrm{a}}^2}\left(1-\frac{x}{L}\right) = \alpha\left(1-\frac{x}{L}\right)\rho^*. \tag{3.93}$$

Similarly, to find the first-order expression in the continuity equation, we insert the expressions for v and ρ, Eqs. (3.92a) and (3.92c), into Eq. (3.85) and obtain

$$\begin{aligned} 0 &= (\rho_0+\rho_1)\partial_x(v_0+v_1) + \frac{1}{c_{\mathrm{a}}^2}\,[\partial_x(p_0+p_1)](v_0+v_1) \\ &= \rho_0\partial_x v_1 + \frac{1}{c_{\mathrm{a}}^2}[\partial_x p_0]v_0. \end{aligned} \tag{3.94a}$$

Here, we have utilized that $\partial_x v_0 = 0$ and that terms including the small pre-factor $1/c_{\mathrm{a}}^2$ need only contain zero-order fields. Isolating $\partial_x v_1$ we arrive at

$$\partial_x v_1 = \frac{1}{\rho^* c_{\mathrm{a}}^2}\,\frac{\Delta p}{L}\,v_0 = \alpha\,\frac{1}{L}\,v_0(y,z). \tag{3.95}$$

We can obtain the first-order velocity field v_1 by integrating the first-order continuity equation (3.95) with respect to x,

$$v_1(x,y,z) = \alpha\Big[\frac{x}{L}\,v_0(y,z) + u_1(y,z)\Big], \tag{3.96}$$

where the integration "constant" $u_1(y,z)$ is yet to be determined.

To find u_1 we consider the first-order Stokes equation found by combining Eqs. (3.82) and (3.92),

$$\nabla^2 v_1 = \frac{1}{\eta}\partial_x p_1(x). \tag{3.97}$$

Inserting expression (3.96) for v_1 into Eq. (3.97) leads to

$$\alpha\Big[\frac{x}{L}\,\nabla^2 v_0(y,z) + \nabla^2 u_1(y,z)\Big] = \frac{1}{\eta}\partial_x p_1(x), \tag{3.98}$$

where we have utilized that $\partial_x^2(x/L) = 0$ when operating with ∇^2. Expressing $\nabla^2 v_0(y,z)$ by the zero-order Stokes equation (3.90) we can separate the variables and obtain

$$\nabla^2 u_1(y,z) = \alpha\frac{\Delta p}{\eta L}\frac{x}{L} + \frac{1}{\eta}\partial_x p_1(x). \tag{3.99}$$

Clearly, the two sides of the equation must be equal to the same constant C,

$$\nabla^2 u_1(y,z) = C, \tag{3.100a}$$

$$\partial_x p_1(x) = \eta C - \alpha\frac{\Delta p}{L^2}\,x. \tag{3.100b}$$

The first-order pressure can be found by integrating Eq. (3.100b) using the boundary conditions

$$p_1(0) = 0, \qquad p_1(L) = 0, \tag{3.101}$$

which are valid since, according to Eq. (3.89) p_0 already satisfies the pressure boundary conditions, Eqs. (3.83b) and (3.83c). We obtain

$$p_1(x) = \frac{\alpha}{2}\Big[\frac{x}{L} - \frac{x^2}{L^2}\Big]\,\Delta p, \tag{3.102}$$

and

$$C = \frac{\alpha}{2}\,\frac{\Delta p}{\eta L}. \tag{3.103}$$

Finally, inserting this value of C into Eq. (3.100a) we note that u_1 satisfy a Stokes equation similar to Eq. (3.90) for v_0 with a pressure multiplied by $-\alpha/2$. Thus, $u_1 = -(\alpha/2)v_0$ and using Eq. (3.96) we can write the explicit expression for v_1,

$$v_1(x,y,z) = \alpha\Big[\frac{x}{L} - \frac{1}{2}\Big]\,v_0(y,z). \tag{3.104}$$

Equations (3.93), (3.102) and (3.104) constitute the first-order perturbation solution to the Poiseuille-flow problem for a weakly compressible fluid.

The mass flow rate Q at position x is defined by Eq. (3.21b), and using first-order perturbation theory, this becomes

$$Q_{\text{mass}} = Q_{\text{mass},0} + \int_{\text{area}} \mathrm{d}y\,\mathrm{d}z\,\big[\rho_1(x)v_0(y,z) + \rho^* v_1(x,y,z)\big]. \tag{3.105}$$

With the first-order results, Eqs. (3.93) and (3.104), for ρ_1 and v_1, respectively, we easily obtain

$$Q_{\text{mass}} = \Big[1 + \alpha\Big(1 - \frac{x}{L}\Big) + \alpha\Big(\frac{x}{L} - \frac{1}{2}\Big)\Big]\, Q_{\text{mass},0} = \Big[1 + \frac{\alpha}{2}\Big]\, Q_{\text{mass},0}. \tag{3.106}$$

As expected for a steady state flow, the mass flow rate is independent of x. Moreover, we see that for a given pressure drop Δp, the introduction of a small compressibility increases the mass flow rate as compared to the mass flow rate for incompressible fluids. This can be understood from the fact that increased pressure at the inlet compresses the fluid and lets more molecules through the channel.

If we want to interpret the Poiseuille flow in terms of a hydraulic resistance of the channel, the result in Eq. (3.106) for the mass flow rate forces us to introduce a pressure-dependent hydraulic resistance $R(\Delta p)$, e.g. by the definition

$$R(\Delta p) \equiv \frac{\Delta p}{Q_{\text{mass}}(\Delta p)}. \tag{3.107}$$

Inserting Eq. (3.106) in this expression and expanding to first order in α we obtain

$$R(\Delta p) = \Big(1 - \frac{\alpha}{2}\Big)\frac{\Delta p}{Q_{\text{mass},0}} = \Big(1 - \frac{\Delta p}{2\rho^* c_{\text{a}}^2}\Big)\frac{8\rho^*\eta L}{\pi a^4}. \tag{3.108}$$

In experimental microfluidics the most compressible fluid in use is air. We can estimate the limitations of the result in Eq. (3.108) for the pressure-dependent hydraulic resistance by demanding that the correction factor $\alpha/2$ should be less than, say, $0.1 = 10\%$, since this implies a second-order correction of the order $(0.1)^2 = 1\%$. For air at room temperature $c_{\text{a}} = 340$ m/s and $\rho^* = 1$ kg/m^3, which leads to the following estimate for the maximal overpressure Δp_{max} allowed,

$$\Delta p_{\text{max}} = 0.1(2\rho^* c_{\text{a}}^2) = 23\ \text{kPa} = 0.23p^*. \tag{3.109}$$

If the overpressure is created by compressing the air in a chamber attached to the inlet, then the first-order perturbation result can be trusted for relative, isothermal volume compressions up to about 25%.

The present result can also be applied to time-dependent situations in the quasi-static limit. It can be trusted if the characteristic time scale τ of the time dependence is much larger than the time it takes to establish any given pressure profile along the channel,

$$\tau \gg \frac{L}{c_{\text{a}}}. \tag{3.110}$$

For an air-filled channel of length 1 mm we get $\tau \gg 3$ µs.

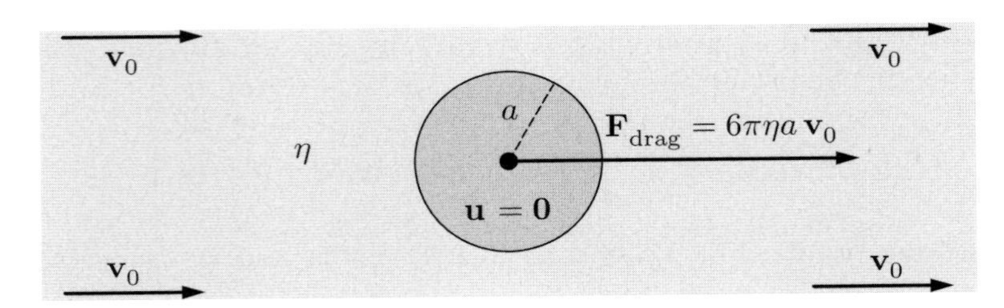

Fig. 3.12 The Stokes drag force $\mathbf{F}_{\mathrm{drag}}$ on a rigid sphere of radius a, when the sphere is at rest, $\mathbf{u} = \mathbf{0}$, and the fluid of viscosity η has the constant velocity $\mathbf{v}_0$ at infinity as indicated by the four vectors $\mathbf{v}_0$.

3.7 Stokes drag on a sphere moving in steady state

As the last example, we study the steady state motion of a rigid sphere in an incompressible fluid. This is relevant for many applications of lab-on-a-chip systems where small objects, such as magnetic beads, fluorescent markers, or biological cells, are moved around inside the microfluidic channels. We shall restrict our treatment to rigid spherical bodies.

As shown in Fig. 3.12, we choose a co-ordinate system where the sphere with radius a is at rest and the surrounding fluid moves past it. The goal is to calculate the velocity and pressure fields, and from these find the stress tensor at the surface of the sphere, which finally will give us the force acting on the sphere. At infinity the fluid is assumed to move with a constant velocity $\mathbf{v} = \mathbf{v}_0 = V_0\mathbf{e}_x$ along the x axis, while at the surface of the sphere we have the no-slip boundary condition, $\mathbf{v}(r = a) = \mathbf{0}$.

To simplify the problem we consider the low Reynolds number limit, $Re = \eta a V_0/\rho \ll 1$, where the steady-state Navier–Stokes equation reduces to the linear Stokes equation (2.41), $\nabla^2\mathbf{v} = \frac{1}{\eta}\boldsymbol{\nabla} p$. Due to the symmetry we choose to work with spherical co-ordinates (r, θ, ϕ), and we notice that only the radial co-ordinate r as well as the polar angle θ (the angle to the x axis) enters. The azimuthal angle ϕ is therefore suppressed in the following. In spherical co-ordinates the Navier–Stokes equation and the continuity equation become

$$\partial_r^2 v_r + \frac{2}{r}\,\partial_r v_r - \frac{2}{r^2}v_r + \frac{1}{r^2}\partial_\theta^2 v_r + \frac{\cot\theta}{r^2}\partial_\theta v_r - \frac{2}{r^2\sin\theta}\partial_\theta v_\theta - \frac{2\cot\theta}{r^2\sin\theta}v_\theta = \frac{1}{\eta}\,\partial_r p, \tag{3.111a}$$

$$\partial_r^2 v_\theta + \frac{2}{r}\,\partial_r v_\theta - \frac{1}{r^2\sin^2\theta}v_\theta + \frac{1}{r^2}\partial_\theta^2 v_\theta + \frac{\cot\theta}{r^2}\partial_\theta v_\theta + \frac{2}{r^2}v_r = \frac{1}{\eta}\,\frac{1}{r}\partial_\theta p, \tag{3.111b}$$

$$\partial_r v_r + \frac{2}{r}\,v_r + \frac{1}{r}\,\partial_\theta v_\theta + \frac{\cot\theta}{r}\,v_\theta = 0. \tag{3.111c}$$

Instead of solving this system of equations directly, we utilize the symmetry of the problem to find $\mathbf{v} - \mathbf{v}_0$ that vanishes at infinity. Since, by assumption, $\mathbf{v}$ represents an incompressible flow and $\mathbf{v}_0$ is a constant vector we have $\boldsymbol{\nabla}\cdot(\mathbf{v} - \mathbf{v}_0) = 0$, and $\mathbf{v} - \mathbf{v}_0$ can therefore be written as the rotation of some vector $\mathbf{A}$,

$$\mathbf{v} - \mathbf{v}_0 = \boldsymbol{\nabla}\times\mathbf{A}. \tag{3.112}$$

The velocity is a polar vector[3], so, due to the rotation operator, $\mathbf{A}$ must be an axial vector.

[3]Polar vectors are characterized by having a direction which is independent of the "handedness" of the co-ordinate system. In contrast, the direction of axial vectors is reversed when changing from a right- to a left-handed system, e.g. by $(\mathbf{e}_x, \mathbf{e}_y, \mathbf{e}_z) \to (-\mathbf{e}_x, \mathbf{e}_y, \mathbf{e}_z)$. The cross-product depends on the "handedness".

Moreover, as the Stokes equation is linear, we must require that $\mathbf{A}$ depends linearly on $\mathbf{v}_0$. Finally, due to the symmetry of the sphere, $\mathbf{A}$ cannot depend on other directions than that of $\mathbf{v}_0$. Combining these requirements on $\mathbf{A}$ we are led to $\mathbf{A} = \big[\boldsymbol{\nabla} f(r)\big] \times \mathbf{v}_0$, where $f(r)$ is an unknown function of r to be determined. Consequently,

$$\mathbf{v} - \mathbf{v}_0 = \boldsymbol{\nabla} \times \big[\boldsymbol{\nabla} f(r) \times \mathbf{v}_0\big] = \boldsymbol{\nabla} \times \boldsymbol{\nabla} \times \big[f(r)\mathbf{v}_0\big]. \tag{3.113}$$

To establish a differential equation for $f(r)$ we involve the rotation of the Stokes equation,

$$\nabla^2\big[\boldsymbol{\nabla} \times (\mathbf{v} - \mathbf{v}_0)\big] = \boldsymbol{\nabla} \times \big[\nabla^2(\mathbf{v} - \mathbf{v}_0)\big] = \boldsymbol{\nabla} \times \big[\nabla^2\mathbf{v}\big] = \frac{1}{\eta}\boldsymbol{\nabla} \times \boldsymbol{\nabla} p = 0. \tag{3.114a}$$

On the other hand, when acting with $\nabla^2\boldsymbol{\nabla}\times$ on Eq. (3.113) and using the operator identities $\boldsymbol{\nabla} \times (\boldsymbol{\nabla}\times) = \boldsymbol{\nabla}(\boldsymbol{\nabla}\cdot) - \nabla^2$, see Eq. (1.27a), and $\boldsymbol{\nabla}\cdot(\boldsymbol{\nabla}\times) = 0$, we obtain

$$\begin{aligned}
\nabla^2\big[\boldsymbol{\nabla} \times (\mathbf{v} - \mathbf{v}_0)\big] &= \nabla^2\Big[\boldsymbol{\nabla} \times \boldsymbol{\nabla} \times \Big(\boldsymbol{\nabla} \times \big[f(r)\mathbf{v}_0\big]\Big)\Big] \\
&= \nabla^2\Big[\boldsymbol{\nabla}\Big\{\boldsymbol{\nabla}\cdot\Big(\boldsymbol{\nabla} \times \big[f(r)\mathbf{v}_0\big]\Big)\Big\} - \nabla^2\Big(\boldsymbol{\nabla} \times \big[f(r)\mathbf{v}_0\big]\Big)\Big] \\
&= -\nabla^2\nabla^2\big[\boldsymbol{\nabla} f(r) \times \mathbf{v}_0\big] = -\boldsymbol{\nabla}\Big[\nabla^2\nabla^2 f(r)\Big] \times \mathbf{v}_0.
\end{aligned} \tag{3.114b}$$

From Eqs. (3.114a) and (3.114b) follows that the gradient of $\nabla^2\nabla^2 f(r)$ is zero. By integration we get that $\nabla^2\nabla^2 f(r)$ must be a constant, and given that $\mathbf{v} - \mathbf{v}_0$ vanishes at infinity, this constant must be zero,

$$\nabla^2\big[\nabla^2 f(r)\big] = 0. \tag{3.115}$$

Since $f(r)$ does not depend on the angular variables, the Laplacian reduces to $\nabla^2 = r^{-2}\partial_r(r^2\partial_r)$. By integration of the first Laplacian in Eq. (3.115) and utilizing that the derivatives of $f(r)$ must vanish at infinity to ensure that $\mathbf{v} - \mathbf{v}_0$ also vanishes there, we obtain

$$\nabla^2 f(r) = \frac{2c_1}{r}, \tag{3.116}$$

where c_1 is an integration constant, and where the factor of 2 is introduced for later convenience. By integration of Eq. (3.116) and again utilizing that the derivatives of $f(r)$ must vanish at infinity, we arrive at

$$f(r) = c_1\, r + \frac{c_2}{r}, \tag{3.117}$$

where c_2 is a second constant. To determine the integration constants c_1 and c_2 we express $\mathbf{v}$ in Eq. (3.113) in terms of $f(r)$ given by Eq. (3.117) and obtain

$$\mathbf{v}(r,\theta) = \mathbf{v}_0 + \boldsymbol{\nabla} \times \boldsymbol{\nabla} \times \big[f(r)\mathbf{v}_0\big] = \mathbf{v}_0 + \boldsymbol{\nabla}\Big(\boldsymbol{\nabla}\cdot\big[f(r)\mathbf{v}_0\big]\Big) - \nabla^2\big[f(r)\mathbf{v}_0\big] \tag{3.118a}$$

$$= \mathbf{v}_0 + \big[\boldsymbol{\nabla}\boldsymbol{\nabla} f(r)\big]\cdot\mathbf{v}_0 - \big[\nabla^2 f(r)\big]\mathbf{v}_0 \tag{3.118b}$$

$$= \mathbf{v}_0 - c_1\left[\frac{\mathbf{v}_0 + (\mathbf{v}_0\cdot\mathbf{e}_r)\,\mathbf{e}_r}{r}\right] + c_2\left[\frac{3(\mathbf{v}_0\cdot\mathbf{e}_r)\,\mathbf{e}_r - \mathbf{v}_0}{r^3}\right], \quad \text{for } r > a. \tag{3.118c}$$

The last equality is proven in Exercise 3.13.

The pressure gradient $\boldsymbol{\nabla} p$ is found by combining Stokes equation with Eq. (3.118b) and using Eqs. (3.115) and (3.116),

$$\boldsymbol{\nabla} p = \eta \nabla^2 \mathbf{v} = \eta \big[\boldsymbol{\nabla}\boldsymbol{\nabla}\nabla^2 f(r)\big]\cdot\mathbf{v}_0 - \big[\nabla^2\nabla^2 f(r)\big]\mathbf{v}_0 = \boldsymbol{\nabla}\left[\eta\mathbf{v}_0\cdot\boldsymbol{\nabla}\left(\frac{2c_1}{r}\right)\right]. \tag{3.119}$$

The pressure itself is found by integrating the gradient and then utilizing that $\boldsymbol{\nabla} = \mathbf{e}_r\partial_r$,

$$p(r,\theta) = p^* - \eta\,\frac{2c_1}{r^2}\,\mathbf{v}_0\cdot\mathbf{e}_r, \tag{3.120}$$

where the integration constant p^* is the ambient pressure.

The values of c_1 and c_2 are found from the no-slip boundary condition at the surface $r = a$ of the sphere,

$$\mathbf{0} = \mathbf{v}(a,\theta) = \left[1 - \frac{c_1}{a} - \frac{c_2}{a^3}\right]\mathbf{v}_0 + \left[\frac{3c_2}{a^3} - \frac{c_1}{a}\right](\mathbf{v}_0\cdot\mathbf{e}_r)\,\mathbf{e}_r. \tag{3.121}$$

The sum of the two velocity terms can only be zero everywhere on the sphere if the coefficients of each velocity are zero, and we find

$$c_1 = \frac{3}{4}\,a \qquad \text{and} \qquad c_2 = \frac{1}{4}\,a^3. \tag{3.122}$$

In spherical co-ordinates we have $\mathbf{v} = v_r\mathbf{e}_r + v_\theta\mathbf{e}_\theta$ and $\mathbf{v}_0 = V_0\cos\theta\,\mathbf{e}_r - V_0\sin\theta\,\mathbf{e}_\theta$. Inserting these expressions together with Eq. (3.122) in Eqs. (3.118c) and (3.120) gives us the solutions

$$v_r = +V_0\cos\theta\left[1 - \frac{3a}{2r} + \frac{a^3}{2r^3}\right], \tag{3.123a}$$

$$v_\theta = -V_0\sin\theta\left[1 - \frac{3a}{4r} - \frac{a^3}{4r^3}\right], \tag{3.123b}$$

$$p = p^* - \frac{3}{2}\,\frac{\eta V_0}{a}\,\cos\theta\,\frac{a^2}{r^2}. \tag{3.123c}$$

The total frictional force acting on the sphere in the x direction is derived from the stress tensor σ, see Eq. (2.26), as the integral of the x component of the surface force density $\mathbf{e}_x\cdot(\sigma\cdot\mathbf{n})$. In spherical co-ordinates $\mathbf{e}_x = \cos\theta\,\mathbf{e}_r - \sin\theta\,\mathbf{e}_\theta$, and for the sphere $\mathbf{n} = \mathbf{e}_r$, so we obtain the drag force

$$F_\text{drag} = \int_{\partial\Omega} da\,\mathbf{e}_x\cdot(\sigma\cdot\mathbf{e}_r) = \int_0^a r\,dr\int_{-1}^{1} d(\cos\theta)\,\Big(-p\,\cos\theta + \sigma'_{rr}\cos\theta - \sigma'_{\theta r}\sin\theta\Big). \tag{3.124}$$

The stress-tensor components in spherical co-ordinates are

$$\sigma'_{rr} = 2\eta\,\partial_r v_r, \qquad \text{and} \qquad \sigma'_{\theta r} = \eta\left(\frac{1}{r}\partial_\theta v_r + \partial_r v_\theta - \frac{1}{r}\,v_\theta\right), \tag{3.125}$$

so using our explicit results, Eq. (3.123), for the velocity and pressure fields we find at the surface $r = a$ of the sphere that

$$\sigma'_{rr} = 0, \qquad \sigma'_{\theta r} = -\frac{3\eta V_0}{2a}\sin\theta, \qquad p = p^* - \frac{3\eta V_0}{2a}\cos\theta. \tag{3.126}$$

Inserting this into Eq. (3.124) yields the famous formula for the Stokes drag

$$F_{\text{drag}} = 6\pi\eta\, aV_0. \tag{3.127}$$

Consider a particle moving with the velocity $\mathbf{u}$ at a position where the velocity of the fluid would have been $\mathbf{v}$ had the particle not been present. In this case, given that no walls or other obstacles are near by, the expression Eq. (3.127) can be generalized to

$$\mathbf{F}_{\text{drag}} = 6\pi\eta\, a\,(\mathbf{v} - \mathbf{u}). \tag{3.128}$$

A closer look at expressions (3.123a) and (3.123b) for the velocity components reveal that they cannot be correct at large distances from the sphere. The reason is that for $r \gg a$ we have $|\mathbf{v}| \approx V_0$ that is independent of a/r, while $|\boldsymbol{\nabla}\mathbf{v}| \approx V_0\, a/r^2$, and $|\nabla^2\mathbf{v}| \approx V_0\, a/r^3$. Hence, the basic assumption that $|\rho(\mathbf{v}\cdot\boldsymbol{\nabla})\mathbf{v}| \ll |\eta\nabla^2\mathbf{v}|$ becomes $\rho V_0^2\, a/r^2 \ll \eta V_0\, a/r^3$, or $r \ll \eta/(\rho V_0)$. A more careful analysis is therefore needed to obtain the correct velocity field and hence the correct drag force. It turns out that the drag force can be expressed as a series expansion in the Reynolds number $Re = \eta V_0 a/\rho$, and that Eq. (3.127) expresses the leading term. The first correction was found by Oseen in 1910,

$$F_{\text{drag}} = 6\pi\eta\, aV_0\left[1 + \frac{3}{8}\,Re\right], \text{ with } Re = \frac{\eta V_0 a}{\rho}. \tag{3.129}$$

3.8 Exercises

Exercise 3.1
Generation of hydrostatic pressure in microchannels

(a) Check that Eq. (3.3) is a solution to the static Navier–Stokes equation (3.2b).

(b) Consider the figure shown below illustrating a microchannel filled with water. Calculate the pressure generated by the water column of height $H = 10$ cm at the points A, B, and C inside the circular microchannel of radius $a = 100$ µm.

(c) Calculate the heights of mercury- and water-columns generating a pressure difference of 1 atm $= 1.013 \times 10^5$ Pa.

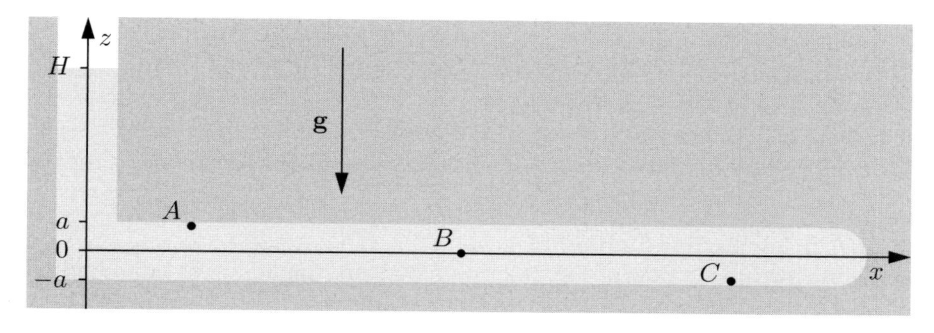

Fig. 3.13 Hydrostatic pressure in a liquid (light gray) inside a closed channel (dark gray).

Exercise 3.2
The thickness of the atmosphere of the Earth
Prove the validity of Eqs. (3.6) and (3.8) for an ideal gas under isothermal conditions, and estimate the thickness of the atmosphere. Discuss the result.

Exercise 3.3
The flow of a liquid film on an inclined plane
We study the flow defined in Fig. 3.2 of a liquid film of thickness h on an inclined plane with inclination angle α.

(a) Check the correctness of the form of the Navier–Stokes equation (3.11a) and of the solution for $v_x(z)$ given in Eq. (3.12).

(b) Let the liquid be water and calculate the speed $v_x(h)$ of the free surface of the film in the case of $h = 100$ µm and $\alpha = 30°$.

Exercise 3.4
Symmetry and the structure of the Poiseuille-flow solution
Prepare a blackboard presentation of the symmetry arguments leading from the Navier–Stokes equation in the general case to the special form of the solution, Eq. (3.20a), of the velocity field for the steady state Poiseuille flow.

Exercise 3.5
Poiseuille-flow profile in a circular channel
Sketch the flow profile $v_x(r, \phi)$ of Eq. (3.42a) valid for a circular channel.

Exercise 3.6
The physical origin of the correction term in the rectangular channel
Find a qualitative argument that explains the correction term in Eq. (3.58) for the Poiseuille flow rate Q in a flat, rectangular channel.

Exercise 3.7
The flow rate in a parabolic channel
Verify the expression given in Eq. (3.80) for the Poiseuille flow rate Q in a flat, parabolic channel.

Exercise 3.8
The flow rate of a non-Newtonian fluid in an infinite, parallel-plate channel
Consider a steady state flow of an incompressible, non-Newtonian fluid in a section of width w of the infinite, parallel-plate channel of length L shown in Fig. 3.6. Let the fluid be described by the Ostwald–de Waele power-law model given in Eq. (2.49), and assume a laminar velocity field of the form $\mathbf{v} = v_x(z)\,\mathbf{e}_x$.

(a) Calculate the x component $v_x(z)$ of the velocity field.

(b) Calculate the flow rate Q for a given applied pressure drop Δp along the channel.

Exercise 3.9
Couette flow in an inclined channel
Consider the inclined-plane flow of Fig. 3.2 but substitute the no-stress boundary condition $\partial_z v_x(h) = 0$ of Eq. (3.11c) with the Coeutte flow boundary $v_x(h) = v_0$ of Eq. (3.14c) similar to the one shown in Fig. 3.3. Determine the resulting velocity field $v_x(z)$.

Exercise 3.10
Poiseuille flow in an inclined channel
Extend the analysis presented in Section 3.4.1 for the Poiseuille flow through a channel with an arbitrary cross-section by inclining the channel at an angle α with respect to the yz-plane and taking the effect of gravity into account.

Exercise 3.11
Combined Poiseuille and Couette flow
Extend the analysis presented in Section 3.3 for the planar Couette flow by applying a pressure difference Δp over a section of length L in the x direction. Determine the velocity field $v_x(y,z)$ and the flow rate Q for this combined Poiseuille and Couette flow.

Exercise 3.12
Contour plots of the velocity fields
Use your favorite computer program to generate contour or surface plots illustrating the different Poiseuille velocity fields calculated in Section 3.4.

Exercise 3.13
The velocity field in the Stokes-drag problem
Use the explicit form $f(r) = c_1 r + c_2/r$ given in Eq. (3.117) to prove that Eq. (3.118c) follows from Eq. (3.118b).

Exercise 3.14
Stokes drag on a spherical particle in a microchannel
Consider the Stokes drag discussed in Section 3.7.

(a) Beginning from Eq. (3.127) prove the general expression, Eq. (3.128), for the Stokes drag force on a spherical, rigid particle.

(b) Discuss under which circumstances this expression can be applied to the motion of spherical particles inside microchannels.

3.9 Solutions

Solution 3.1
Generation of hydrostatic pressure in microchannels
We use $g = 9.82\ \mathrm{m/s^2}$, $\rho_{\mathrm{H_2O}} = 10^3\ \mathrm{kg/m^3}$, and $\rho_{\mathrm{Hg}} = 13.6\times10^3\ \mathrm{kg/m^3}$.

(a) $\boldsymbol{\nabla}p = \boldsymbol{\nabla}p^* - \boldsymbol{\nabla}(\rho g z) = 0 - \rho g\,(0,0,1) = -\rho g\mathbf{e}_z$.

(b) For height H we have $p(x,y,z) = p_H(z) = \rho g(H-z) = 9.82\times10^3\,\mathrm{Pa\,m^{-1}}\,(0.1\ \mathrm{m}-z)$, so $p_A = p_H(100\ \mathrm{\mu m}) = 981\ \mathrm{Pa}$, $p_B = p_H(0\ \mathrm{\mu m}) = 982\ \mathrm{Pa}$, $p_C = p_H(-100\ \mathrm{\mu m}) = 983\ \mathrm{Pa}$.

(c) $H(\rho) = p^*/(\rho g) = \frac{1}{\rho}\,9.82\times10^3\ \mathrm{kg/m^2}$ so $H_{\mathrm{H_2O}} = 10.3\ \mathrm{m}$ and $H_{\mathrm{Hg}} = 0.76\ \mathrm{m}$.

Solution 3.2
The thickness of the atmosphere of the Earth
Let $p^* = 10^5$ Pa and $\rho^* = 1\ \mathrm{kg\,m^{-3}}$ be the pressure and density of air at ground level. For an ideal isothermal gas $p\mathcal{V}$ or p/ρ is constant, so $\rho = (\rho^*/p^*)\,p$. Thus, Eq. (3.2b) becomes $0 = -\partial_z p - (\rho^* g/p^*)\,p$ or $\partial_z p = -(1/\lambda_{\mathrm{air}})\,p$, where $\lambda_{\mathrm{air}} = p^*/(\rho^* g) = 10^4$ m. This leads to $p(z) = p^*\exp(-z/\lambda_{\mathrm{air}})$.

In the isothermal model the density of the atmosphere decreases exponentially with a characteristic length of 10 km. This is in accordance with the fact that a 10 m high water column having $\rho = 1000\ \mathrm{kg\,m^{-3}}$ can be balanced by a 10 km high air column having

$\rho = 1$ kg m^{-3}. Also, commercial jet airliners fly at an altitude of 10 km, where there is enough air for the jet engines to work, but a lower density providing less air resistance.

Solution 3.3
The flow of a liquid film on an inclined plane
Check carefully the assumptions and calculations leading to Eqs. (3.9), (3.10), and (3.11).

(a) Upon insertion of Eq. (3.12) into the left-hand side of Eq. (3.11) we get $\eta\partial_z^2 v_x(z) = \eta\big[\sin\alpha\rho g/(2\eta)\big]\partial_z^2\big[2hz - z^2\big] = \sin\alpha(\rho g/2)[0-2] = -\rho g\sin\alpha$. Moreover, for the boundary conditions we find $v_x(0) = 0$ and $\partial_z v_x = \big[\sin\alpha\rho g/(2\eta)\big](2h - 2z) \Rightarrow \partial_z v_x(h) = 0$.

(b) $v_x(h\!=\!100\mu\mathrm{m}) = \big[\sin(30^\circ)(10^3\,\frac{\mathrm{kg}}{\mathrm{m}^3}\times 10\,\frac{\mathrm{m}}{\mathrm{s}^2})/(2\times10^{-3}\,\mathrm{Pa\,s})\big](10^{-4}\,\mathrm{m})^2 = 0.025\,\frac{\mathrm{m}}{\mathrm{s}}$.

Solution 3.4
Symmetry and the structure of the Poiseuille-flow solution
Distinguish clearly between physical and mathematical arguments. Begin by a clear formulation of the physical assumptions and arguments that leads to the special form of the velocity field. Insert this velocity field in the full Navier–Stokes equation and reduce it using mathematical arguments.

Solution 3.5
Poiseuille-flow profile in a circular channel
Utilizing the connection between Cartesian and cylindrical co-ordinates we get

$$v_x(y,z) = v_x(r\cos\phi, r\sin\phi) = v_0\Big[1 - \Big(\frac{r}{a}\Big)^2\Big],$$

which is plotted in the figure to the right.

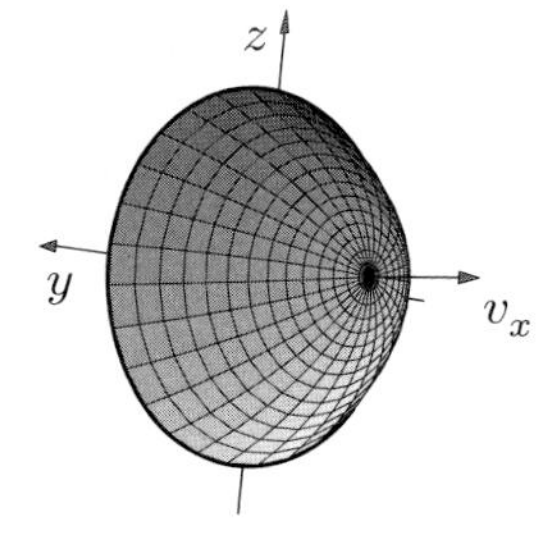

Fig. 3.14 The flow profile $v_x(y,z)$.

Solution 3.6
The physical origin of the correction term in the rectangular channel
The flow rate is $Q_{\parallel} = h^3 w\Delta p/(12\eta L)$ in a section of width w in the infinite parallel-plate channel. The flow rate $Q_{\square} = (1-\alpha)Q_{\parallel}$ in a rectangular channel of the same width must be smaller by a relative amount α, since the velocity at the side walls must be zero. The difference between the two channels is solely due to the two side regions closer than $h/2$ from the sides, since for any point outside these regions the two channels appear identical as the top and bottom walls are closer than the sides. In the parallel-plate channel the flow rate in one side-region of width $h/2$ is $Q_{\parallel}(h/2)/w$. Assuming a linear drop in local flow rate in the two side regions of the rectangular channel we can estimate the flow rate as $Q_{\square} = Q_{\parallel} - 2\times\frac{1}{2}\times Q_{\parallel}(h/2)/w = (1-0.5\,h/w)Q_{\parallel}$ not far from the more exact expression $Q_{\square} = (1-0.63\,h/w)Q_{\parallel}$.

Solution 3.7
The flow rate in a parabolic channel
With the notation $\lambda(\eta) = 4\eta(1-\eta)$ the integral in Eq. (3.80) for the Poiseuille flow rate Q in a flat, parabolic channel is calculated as follows.

$$\int_0^{\lambda} d\zeta\left[\frac{1}{2}(\zeta\lambda-\zeta^2) - \alpha\frac{2}{3}(\zeta\lambda^2-\zeta^3)\right] = \frac{1}{2}\left[\frac{1}{2}-\frac{1}{3}\right]\lambda^3 - \alpha\frac{2}{3}\left[\frac{1}{2}-\frac{1}{4}\right]\lambda^4 = \frac{1}{12}\lambda^3 - \frac{\alpha}{6}\lambda^4. \quad (3.130)$$

4 Hydraulic resistance and compliance

In Chapter 3 we studied the pressure-driven, steady-state flow of an incompressible Newtonian fluid through a straight channel, the Poiseuille flow. We found that a constant pressure drop Δp resulted in a constant flow rate Q. This result can be summarized in the Hagen–Poiseuille law

$$\Delta p = R_{\mathrm{hyd}}\, Q = \frac{1}{G_{\mathrm{hyd}}}\, Q, \tag{4.1}$$

where we have introduced the proportionality factors R_{hyd} and G_{hyd} known as the hydraulic resistance and conductance, respectively. The Hagen–Poiseuille law, Eq. (4.1), is completely analogous to Ohm's law, $\Delta V = R\, I$, relating the electrical current I through a wire with the electrical resistance R of the wire and the electrical potential drop ΔV along the wire. The SI units used in the Hagen–Poiseuille law are

$$[Q] = \frac{\mathrm{m}^3}{\mathrm{s}}, \qquad [\Delta p] = \mathrm{Pa} = \frac{\mathrm{N}}{\mathrm{m}^2} = \frac{\mathrm{kg}}{\mathrm{m\,s}^2}, \qquad [R_{\mathrm{hyd}}] = \frac{\mathrm{Pa\,s}}{\mathrm{m}^3} = \frac{\mathrm{kg}}{\mathrm{m}^4\,\mathrm{s}}. \tag{4.2}$$

The concept of hydraulic resistance is central in characterizing and designing microfluidic channels in lab-on-a-chip systems. In this chapter we study both fundamental and applied aspects of hydraulic resistance. We also introduce the concept of compliance, i.e. change in volume as a function of pressure, which is the hydraulic analogue of electrical capacitance.

4.1 Viscous dissipation of energy for incompressible fluids

Just as electrical resistance is intimately connected to dissipation of energy in the form of Joule heating, hydraulic resistance is due to viscous dissipation of mechanical energy into heat by internal friction in the fluid.

4.1.1 Viscous dissipation in time-dependent systems

To obtain an expression for the energy dissipation in terms of the viscosity and the velocity field, we study the thought experiment sketched in Fig. 4.1. Consider an incompressible fluid inside a channel performing an ideal steady-state Poiseuille flow at times $t < 0$. The constant velocity field $\mathbf{v}$ is maintained by a constant overpressure Δp applied to the left end of the channel. The overpressure Δp is suddenly removed at $t = 0$, but of course the fluid flow continues due to the inertia of the fluid. However, it is clear that the internal viscous friction of the fluid gradually will slow down the motion of the fluid, and eventually in the

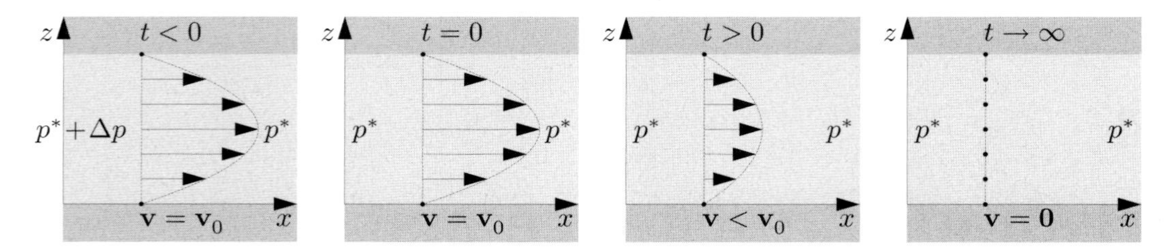

Fig. 4.1 A sketch of a liquid (light gray) performing a Poiseuille flow inside a channel (dark gray). For times $t < 0$ the flow is in steady state, and due to the overpressure Δp applied to the left, the flow profile is a characteristic parabola. Δp is suddenly turned off at $t = 0$, but inertia keeps up the flow. For $t > 0$ the fluid velocity diminishes due to viscous friction, and in the limit $t \to \infty$ the fluid comes to rest relative to the channel walls.

limit $t \to \infty$ the fluid will come to rest relative to the channel walls. As time passes the kinetic energy of the fluid at $t = 0$ is gradually transformed into heat by the viscous friction.

In the following, we calculate the rate of change of the kinetic energy at any instant $t > 0$, where the overpressure has been removed. We neglect any influence of gravitational and electrical forces, and we use the fact that for an incompressible fluid the continuity equation reads $\partial_j v_j = 0$. The kinetic energy of the fluid can be expressed as an integral over the space Ω occupied by the channel,

$$E_{\text{kin}} = \int_\Omega d\mathbf{r}\, \tfrac{1}{2}\,\rho\,\mathbf{v}^2 = \int_\Omega d\mathbf{r}\, \tfrac{1}{2}\,\rho\, v_i v_i, \tag{4.3}$$

where we use the index notation. The rate of change of E_{kin} is

$$\partial_t E_{\text{kin}} = \int_\Omega d\mathbf{r}\, \rho\, v_i \partial_t v_i. \tag{4.4}$$

We can express the time derivative $\partial_t v_i$ using the Navier–Stokes equation (2.30b),

$$\rho\,\partial_t v_i = -\rho\, v_j \partial_j v_i + \eta\,\partial_j\partial_j v_i. \tag{4.5}$$

The rate of change of the kinetic energy can thus be written as

$$\begin{aligned}
\partial_t E_{\text{kin}} &= \int_\Omega d\mathbf{r}\,\Big\{ -\rho\, v_i v_j \partial_j v_i + \eta\, v_i \partial_j \partial_j v_i \Big\} \\
&= \int_\Omega d\mathbf{r}\,\Big\{ -\partial_j \Big[v_j \big(\tfrac{1}{2}\rho\, v_i^2\big) - \eta\, v_i \partial_j v_i \Big] - \eta\,(\partial_j v_i)(\partial_j v_i) \Big\} \\
&= -\int_{\partial\Omega} da\, n_j \Big[v_j \big(\tfrac{1}{2}\rho\, v_i^2\big) - \eta\, v_i \partial_j v_i \Big] - \eta \int_\Omega d\mathbf{r}\, (\partial_j v_i)(\partial_j v_i).
\end{aligned} \tag{4.6}$$

As indicated in Fig. 4.2 the surface $\partial\Omega$ consists of three parts: the solid side wall $\partial\Omega_{\text{wall}}$, the open inlet $\partial\Omega_1$, and the open outlet $\partial\Omega_2$. The contribution to the surface integral in Eq. (4.6) from $\partial\Omega_{\text{wall}}$ is zero due to the no-slip boundary condition that ensures $v_i \equiv 0$ on solid walls. The two contributions from $\partial\Omega_1$ and $\partial\Omega_2$ exactly cancel each other. The reason is that the translation invariance of the Poiseuille-flow problem makes the expression

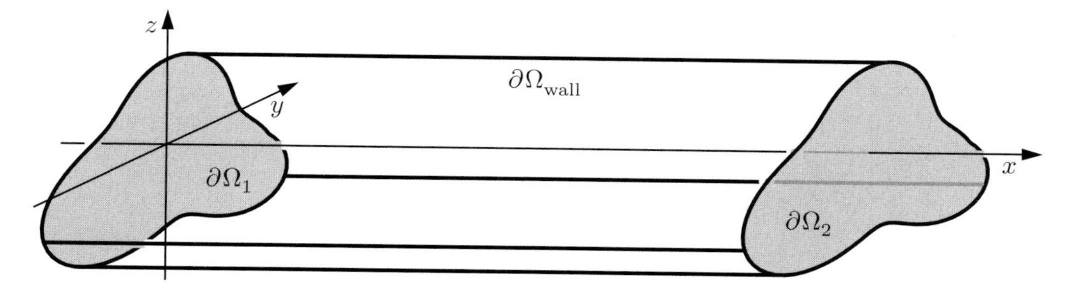

Fig. 4.2 A sketch of the geometry for calculating the viscous energy dissipation in a Poiseuille flow. The surface $\partial\Omega$ consists of three parts: the solid side wall $\partial\Omega_{\rm wall}$ (transparent), the open inlet $\partial\Omega_1$ to the left (light gray), and the open outlet $\partial\Omega_2$ to the right (light gray). The velocity field is parallel to the x direction, $\mathbf{v} = v_x\,\mathbf{e}_x$.

in the square bracket independent of x and hence it is the same on the two end surfaces, while the two normal vectors are opposite to each other $\mathbf{n}(\partial\Omega_1) = -\mathbf{e}_x = -\mathbf{n}(\partial\Omega_2)$. The viscous energy dissipation in a Poiseuille flow relaxing towards thermodynamical equilibrium is therefore given by the volume integral

$$\partial_t E_{\rm kin} = -\eta \int_\Omega d\mathbf{r}\,(\partial_j v_i)(\partial_j v_i) = -\eta \int_\Omega d\mathbf{r}\,\Big[(\partial_y v_x)^2 + (\partial_z v_x)^2\Big]. \tag{4.7}$$

In the last equality we have used the special form of the Poiseuille flow, $\mathbf{v} = v_x(y,z)\,\mathbf{e}_x$. We note that since the kinetic energy is diminishing in time, so that $\partial_t E_{\rm kin} < 0$, and since the integrand is always positive, the viscosity coefficient η must be positive.

Finally, we let $W_{\rm visc}$ denote the heat generated by the viscous friction. Thus, $\partial_t E_{\rm kin} = -\partial_t W_{\rm visc}$ and we can write

$$\partial_t W_{\rm visc} = -\partial_t E_{\rm kin} = \eta \int_\Omega d\mathbf{r}\,(\partial_j v_i)(\partial_j v_i) = \eta \int_\Omega d\mathbf{r}\,\Big[(\partial_y v_x)^2 + (\partial_z v_x)^2\Big]. \tag{4.8}$$

4.1.2 Viscous dissipation of energy in steady state

After having used the relaxing Poiseuille flow to obtain an expression for the rate of viscous dissipation of energy, $\partial_t W_{\rm visc}$, we now turn to the steady-state Poiseuille flow. Consider the usual case where the pressure $p(\partial\Omega_1)$ to the left on $\partial\Omega_1$ is higher than the pressure $p(\partial\Omega_2)$ to the right on $\partial\Omega_2$,

$$p(\partial\Omega_1) = p(\partial\Omega_2) + \Delta p. \tag{4.9}$$

For such a flow the velocity field is constant and consequently the kinetic energy of the fluid is constant. The rate $\partial_t W_{\rm visc}$ of heat generation by viscous friction is balanced by the mechanical power $\partial_t W_{\rm mech}$ put into the fluid by the pressure force,

$$\partial_t E_{\rm kin} = \partial_t W_{\rm mech} - \partial_t W_{\rm visc} = 0. \tag{4.10}$$

To calculate $W_{\rm mech}$ we note that in comparison with Eq. (4.5) the steady-state Navier–Stokes equation now contains a non-zero pressure gradient and no time derivative,

$$0 = -\rho\, v_j \partial_j v_i + \eta\, \partial_j \partial_j v_i - \partial_i p. \tag{4.11}$$

In analogy with Eq. (4.6) we can determine $W_{\rm mech}$ by multiplying the pressure term in Eq. (4.11) by v_i and integrating over volume,

$$\partial_t W_{\text{mech}} = \int_\Omega \mathrm{d}\mathbf{r}\, v_i\big(-\partial_i p\big) = -\int_\Omega \mathrm{d}\mathbf{r}\, \partial_i\big(v_i p\big) = -\int_{\partial\Omega} \mathrm{d}a\, n_i\big(v_i p\big). \tag{4.12}$$

As before the contribution from the solid walls at $\partial\Omega_{\text{wall}}$ is zero due to the no-slip boundary condition and only the inlet $\partial\Omega_1$ and outlet $\partial\Omega_2$ surfaces yield non-zero contributions. The surface normals are opposite, $\mathbf{n}(\partial\Omega_1) = -\mathbf{e}_x = -\mathbf{n}(\partial\Omega_2)$, and the pressure is constant at each end-face, so we get

$$\partial_t W_{\text{mech}} = p(\partial\Omega_1)\int_{\partial\Omega_1} \mathrm{d}a\, v_x - p(\partial\Omega_2)\int_{\partial\Omega_2} \mathrm{d}a\, v_x = \Delta p \int_{\partial\Omega_1} \mathrm{d}a\, v_x(y,z) = Q\,\Delta p. \tag{4.13}$$

The second equality is obtained by using the translation invariance $v_x(\partial\Omega_2) = v_x(\partial\Omega_1)$. The result for the viscous dissipation of energy in steady-state Poiseuille flow is thus, as expected, analogous to the expression for the electric power consumed by Joule heating in a resistor, $\partial_t W_{\text{elec}} = I\,\Delta V$,

$$\partial_t W_{\text{visc}} = \eta \int_\Omega \mathrm{d}\mathbf{r}\, \Big[\big(\partial_y v_x\big)^2 + \big(\partial_z v_x\big)^2\Big] = Q\,\Delta p. \tag{4.14}$$

See Exercise 4.2 for further examples of viscous power consumption in hydraulic resistors.

4.2 Hydraulic resistance of some straight channels

In this section we will list a selection of the hydraulic resistance of specific channels, such as the one shown in Fig. 4.3 and studied in Exercise 4.7. Using the results derived in Section 3.4 for the Poiseuille flow in straight channels, it is easy to list the hydraulic resistance R_{hyd} for a number of different cross-sections, as is done in Table 4.1. Next to the analytical expressions for R_{hyd} is given numerical values for R_{hyd}. These values are calculated using the viscosity of water and fixing the length L along the channel axis to be 1 mm. The length scales perpendicular to the axis are also of the order 100 µm.

The quoted results are all valid for the special case of a translation invariant (straight) channel. This symmetry led to the vanishing of the non-linear term $(\mathbf{v}\cdot\boldsymbol{\nabla})\mathbf{v}$ in the Navier–Stokes equation. However, to handle more general cases it would be very useful to find out when the results for R_{hyd} can be used. This analysis is carried out in the next section, where we shall learn that the dimensionless Reynolds number plays a central role.

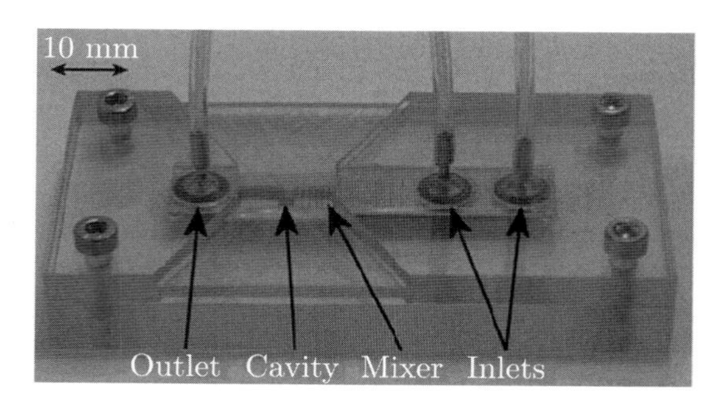

Fig. 4.3 A polymer-based microfluidic dye laser, in which the active part is a rectangular microchannel carrying a dye solution. For proper functioning the dye solution must have a certain flow rate, and hence it is crucial to know the hydraulic resistance of the channel, see Exercise 4.7 and also Fig. 16.4. Courtesy of Anders Kristensen, DTU Nanotech.

Table 4.1 A list of the hydraulic resistance for straight channels with different cross-sectional shapes. The numerical values are calculated using the following parameters: $\eta = 1$ mPa s (water), $L = 1$ mm, $a = 100$ µm, $b = 33$ µm, $h = 100$ µm, and $w = 300$ µm.

shape		R_{hyd} expression	R_{hyd} $[10^{11}\ \frac{\mathrm{Pa\,s}}{\mathrm{m}^3}]$	reference
circle	a	$\frac{8}{\pi}\,\eta L\,\frac{1}{a^4}$	0.25	Eq. (3.39b)
ellipse	b a	$\frac{4}{\pi}\,\eta L\,\frac{1+(b/a)^2}{(b/a)^3}\,\frac{1}{a^4}$	3.93	Eq. (3.38)
triangle	a a a	$\frac{320}{\sqrt{3}}\,\eta L\,\frac{1}{a^4}$	18.5	Eq. (3.46)
two plates	h w	$12\,\eta L\,\frac{1}{h^3 w}$	0.40	Eq. (3.30)
rectangle	h w	$\frac{12\,\eta L}{1-0.63(h/w)}\,\frac{1}{h^3 w}$	0.51	Eq. (3.58)
square	h h h h	$28.4\,\eta L\,\frac{1}{h^4}$	2.84	Exercise 4.4
parabola	h w	$\frac{105}{4}\,\eta L\,\frac{1}{h^3 w}$	0.88	Eq. (3.80)
arbitrary	$\mathcal{P}$ $\mathcal{A}$	$\approx 2\,\eta L\,\frac{\mathcal{P}^2}{\mathcal{A}^3}$	–	Eq. (3.27a)

4.3 Shape dependence of hydraulic resistance

Given the results in Table 4.1 of the hydraulic resistance R_{hyd} in some straight channels parallel to the x axis, it is natural to ask how R_{hyd} depends on the area $\mathcal{A}$,

$$\mathcal{A} \equiv \int_{\Omega} \mathrm{d}x\mathrm{d}y, \tag{4.15a}$$

and the perimeter $\mathcal{P}$,

$$\mathcal{P} \equiv \int_{\partial\Omega} \mathrm{d}\ell, \tag{4.15b}$$

of the cross-section Ω in the yz-plane with boundary $\partial\Omega$. A natural unit for the hydraulic resistance is R^*_{hyd}, which is given by dimensional analysis as

$$R^*_{\text{hyd}} \equiv \frac{\eta L}{\mathcal{A}^2}, \tag{4.16}$$

where L is the channel length and η the dynamic viscosity of the liquid. Typically, the fluid flow is subject to a no-slip boundary condition at the walls $\partial\Omega$ and thus the actual hydraulic resistance will depend on the perimeter as well as the cross-sectional area. This dependence can therefore be characterized by the dimensionless geometrical correction factor β given by

$$\beta \equiv \frac{R_{\text{hyd}}}{R^*_{\text{hyd}}} = \frac{\mathcal{A}^2}{\eta L}\,R_{\text{hyd}}. \tag{4.17}$$

For Poiseuille flow the relation between the pressure drop Δp, the velocity $v_x(y,z)$, and the geometrical correction factor β becomes

$$\Delta p = R_{\text{hyd}}Q = \beta R^*_{\text{hyd}}Q = \beta R^*_{\text{hyd}} \int_\Omega \mathrm{d}x\mathrm{d}y\, v_x(y,z), \tag{4.18}$$

where Q is the volume flow rate.

In lab-on-a-chip applications, where large surface-to-volume ratios are encountered, the problem of the bulk Poiseuille flow is typically accompanied by other surface-related physical or biochemical phenomena in the fluid. The list of examples includes surface chemistry, DNA hybridization on fixed targets, catalysis, interfacial electrokinetic phenomena such as electro-osmosis, electrophoresis and electroviscous effects as well as continuous edge-source diffusion. Though the phenomena are of very different nature, they have at least one thing in common; they are all to some degree surface phenomena and their strength and effectiveness depends strongly on the surface-to-volume ratio. It is common to quantify this by the dimensionless compactness $\mathcal{C}$ given by

$$\mathcal{C} \equiv \frac{\mathcal{P}^2}{\mathcal{A}}. \tag{4.19}$$

Below, we demonstrate a simple dependence of the geometrical correction factor β on the compactness $\mathcal{C}$ and our results thus point out a unified dimensionless measure of flow properties as well as the strength and effectiveness of surface-related phenomena central to lab-on-a-chip applications. Furthermore, our results allow for an easy evaluation of the hydraulic resistance for elliptical, rectangular, and triangular cross-sections with the geometrical measure $\mathcal{C}$ being the only input parameter. Above, we have emphasized microfluidic flows because here a variety of shapes are frequently encountered. However, our results are generally valid for all laminar flows.

Our main objective is to find the relation between the geometrical correction factor β and the compactness $\mathcal{C}$ for various families of geometries.

The family of elliptical cross-sections is special in the sense that R_{hyd} is known analytically for given semi-axis lengths a and b, see Table 4.1. An explicit expression for the geometrical correction factor β is obtained as follows

$$\beta(a,b) = \frac{R_{\text{hyd}}}{R^*_{\text{hyd}}} = \frac{\frac{4}{\pi}\,\eta L\,\frac{1+(b/a)^2}{(b/a)^3}\,\frac{1}{a^4}}{\eta L\,\frac{1}{(\pi ab)^2}} = 4\pi\Big(\frac{a}{b}+\frac{b}{a}\Big), \tag{4.20}$$

which for a circle yields $\beta(a,a) = 8\pi$.

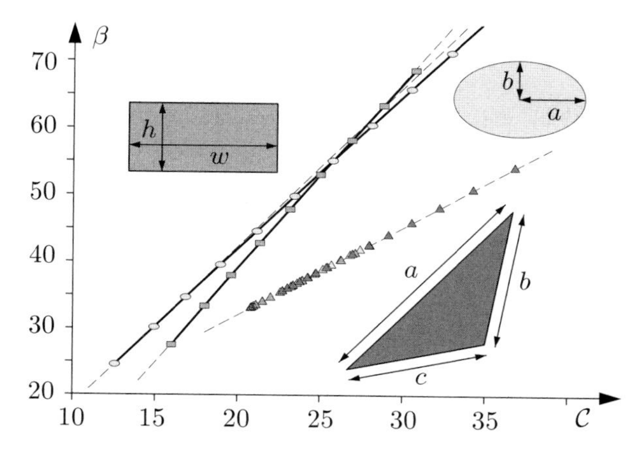

Fig. 4.4 The geometrical correction factor β of the hydraulic resistance versus compactness $\mathcal{C}$ for the elliptical, rectangular, and triangular classes. The solid lines are the exact results, and the dashed lines indicate Eqs. (4.22), (4.26), and (4.27). Numerical results from a finite-element simulation are also included (∘, □, and △). Note that in the case of triangles all classes (right, isosceles, and acute/obtuse scalene triangles marked by different grayscale triangles) fall on the same straight line. Adapted from Mortensen, Okkels and Bruus (2005).

By straightforward algebra we can express the line integral for the perimeter $\mathcal{P}$ as an integral over an angle θ, and the compactness $\mathcal{C}$ follows as

$$\mathcal{C}(\beta) = \frac{1}{2\pi^2}\left(\int_0^{\pi} \mathrm{d}\theta\sqrt{\beta + \sqrt{\beta^2 - (8\pi)^2}\cos\theta}\right)^2. \tag{4.21}$$

Expanding $\mathcal{C}(\beta)$ in β around $\beta = 8\pi$ and inverting we get

$$\beta(\mathcal{C}) = \frac{8}{3}\,\mathcal{C} - \frac{8\pi}{3} + \mathcal{O}([\mathcal{C} - 4\pi]^2). \tag{4.22}$$

In Fig. 4.4 we compare this approximate result (dashed line) with the exact solution (solid line), obtained from a numerical evaluation of Eq. (4.21). Results of a numerical finite-element solution of Poiseuille flow are also included (∘ points). As seen, there is a close-to-linear dependence of β on $\mathcal{C}$ as described by Eq. (4.22).

For the rectangular channel with a width-to-height ratio $\gamma = w/h$, we can by combining Eqs. (3.57c) and (4.17) obtain

$$\beta(\gamma) = \frac{\pi^3\gamma^2}{8}\left(\sum_{n,\mathrm{odd}}^{\infty} \frac{n\gamma}{\pi n^5} - \frac{2}{\pi^2 n^5}\tanh(n\pi\gamma/2)\right)^{-1}. \tag{4.23}$$

The compactness is easily found as

$$\mathcal{C}(\gamma) = \frac{\mathcal{P}^2}{\mathcal{A}} = \frac{(2w + 2h)^2}{wh} = 8 + 4\gamma + 4/\gamma. \tag{4.24}$$

Using the fact that $\tanh(x) \simeq 1$ for $x \gg 1$ we get

$$\beta(\gamma) \simeq \frac{12\pi^5\gamma^2}{\pi^5\gamma - 186\zeta(5)}, \quad \gamma \gg 1, \tag{4.25}$$

and by substituting $\gamma(\mathcal{C})$ into this expression and expanding $\mathcal{C}(\gamma)$ around $\gamma = 2$ with $\mathcal{C}(2) = 18$, we obtain again a linear relation between β and $\mathcal{C}$:

$$\beta(\mathcal{C}) \approx \frac{22}{7}\,\mathcal{C} - \frac{65}{3} + \mathcal{O}\big([\mathcal{C} - 18]^2\big). \tag{4.26}$$

In Fig. 4.4 we compare the exact solution, obtained by a parametric plot of Eqs. (4.23) and (4.24), to the approximate result, Eq. (4.26). Results of a numerical finite-element solution of Eq. (14.43a) are also included (□ points). As in the elliptical case, there is a close-to-linear dependence of β on $\mathcal{C}$ as described by Eq. (4.26).

For the equilateral triangle it follows from Table 4.1 that $\beta = 20\sqrt{3}$ and $\mathcal{C} = 12\sqrt{3}$. However, in the general case of a triangle with side lengths a, b, and c we are referred to numerical solutions of the Poiseuille flow. In Fig. 4.4 we show numerical results (△ points), from finite-element simulations, for scaling of right triangles, isosceles triangles, and acute/obtuse scalene triangles. The dashed line shows

$$\beta(\mathcal{C}) = \frac{25}{17}\,\mathcal{C} + \frac{40\sqrt{3}}{17}, \tag{4.27}$$

where the slope is obtained from a numerical fit. As seen, the results for different classes of triangles fall onto the same straight line. Since we have

$$\mathcal{C}(a,b,c) = \frac{8(a+b+c)^2}{\sqrt{\frac{1}{2}\left(a^2+b^2+c^2\right)^2 - \left(a^4+b^4+c^4\right)}} \tag{4.28}$$

the result in Eq. (4.27) allows for an easy evaluation of R_{hyd} for triangular channels.

Finally, by using the results given in Section 3.5 we can calculate the kth multipolar deformation of the circular cross-section and thereby extend the analytical results for Poiseuille flow beyond the few cases of regular geometries that we have treated above. By continuing the perturbation calculation to fourth order in the perturbation parameter α, we can obtain analytical expressions for both the velocity field and the boundary shape. This leads to analytical expressions for $\mathcal{A}$ and $\mathcal{P}$, which in turns results in the following expressions for β and $\mathcal{C}$:

$$\beta = 8\pi\left[1 + 2(k-1)\,\beta^2 + \frac{47 - 78k + 36k^2 - 4k^3}{8}\,\alpha^4\right] + \mathcal{O}(\alpha^6), \tag{4.29}$$

$$\mathcal{C} = 4\pi + 2\pi(k^2-1)\,\alpha^2. \tag{4.30}$$

The result only involves even powers of α since $\alpha \to -\alpha$ is equivalent to a shape rotation, which should leave β and $\mathcal{C}$ invariant, and as a consequence β depends linearly on $\mathcal{C}$ to fourth order in α,

$$\beta(\mathcal{C}) = \frac{8}{1+k}\,\mathcal{C} - 8\frac{3-k}{1+k}\,\pi + \mathcal{O}(\alpha^4). \tag{4.31}$$

Note that although derived for $k > 2$ this expression coincides with that of the ellipse, Eq. (4.22), for $k = 2$. Comparing, to second order in α, Eq. (4.29) with exact numerics we find that for α up to 0.4 the relative error is less than 0.2% and 0.5% for $k = 2$ and $k = 3$, respectively.

In summary, we have considered pressure-driven, steady-state Poiseuille flow in straight channels with various shapes, and found a close-to-linear relation between β and $\mathcal{C}$. Since the hydraulic resistance is $R_{\text{hyd}} \equiv \beta R^*_{\text{hyd}}$, we conclude that R_{hyd} depends linearly on $\mathcal{C}R^*_{\text{hyd}}$. Different classes of shape all display this linear relation, but the coefficients are non-universal. However, for each class only two points need to be calculated to fully specify the relation for the entire class. The difference is due to the smoothness of the boundaries. The elliptical and harmonic-perturbed classes have boundaries without any cusps, whereas the rectangular and triangular classes have sharp corners. The overall velocity profile tends to be convex and maximal near the center-of-mass of the channel. If the boundary is smooth the velocity in general goes to zero in a convex parabolic manner, whereas a concave parabolic dependence is generally found if the boundary has a sharp corner, as can be proved explicitly for the equilateral triangle Eq. (3.43). Since the concave drop is associated with a region of low velocity compared to the convex drop, geometries with sharp changes in the boundary tend to have a higher hydraulic resistance compared to smooth geometries with equivalent cross-sectional area. The results obtained here improves the rough estimate in Eq. (3.27a).

4.4 Reynolds number for systems with two length scales

The proper way to see if the non-linear term $(\mathbf{v}\cdot\boldsymbol{\nabla})\mathbf{v}$ in the Navier–Stokes equation can be neglected was discussed in Section 2.2.7. From this analysis we conclude that the solutions obtained for the ideal Poiseuille flows, where the non-linear term $(\mathbf{v}\cdot\boldsymbol{\nabla})\mathbf{v}$ is identically zero due to the symmetry, remains approximately valid if the Reynolds number is small, $Re \ll 1$.

In Section 2.2.7 we considered a system with only one characteristic length scale. However, most systems are characterized by more than one length, which leads to a more involved Reynolds number analysis. As an example, consider a section of length L and width w of the infinite, parallel-plate channel with height h shown in Fig. 4.5. The system is translation invariant in the y direction so that only the x and z co-ordinates enter in the following analysis. Although the system as shown is also translation invariant in the x direction, we perform the analysis as if this invariance is weakly broken rendering a non-zero vertical velocity v_z.

The two length scales entering the problem are the length L and the height h,

$$x = L\,\tilde{x}, \qquad z = h\,\tilde{z}, \tag{4.32}$$

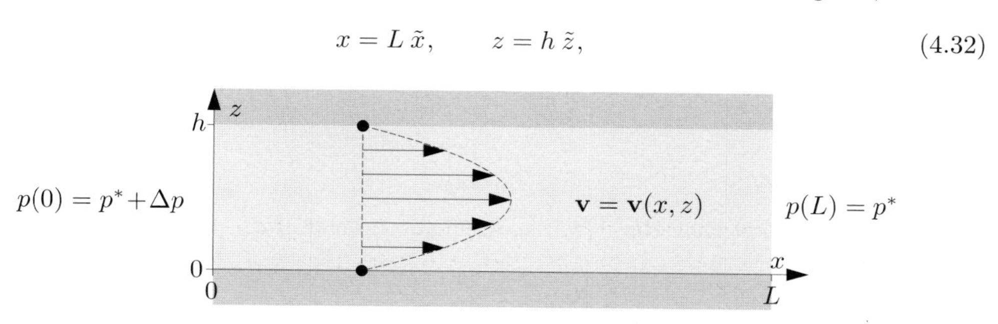

Fig. 4.5 A sketch in the xz-plane of an infinite, nearly-parallel-plate channel of height h. The system is translationally invariant in the y direction and fluid is flowing in the x direction due to a pressure drop Δp over a section of length L. The vertical velocity component is small, but not zero.

which results in the following spatial derivatives

$$\partial_x = \frac{1}{L}\,\tilde{\partial}_x \equiv \varepsilon\,\frac{1}{h}\,\tilde{\partial}_x, \qquad \partial_z = \frac{1}{h}\,\tilde{\partial}_z, \tag{4.33}$$

where we have introduced the aspect ratio ε defined by

$$\varepsilon \equiv \frac{h}{L} \ll 1. \tag{4.34}$$

The characteristic velocity in the x direction is given by the mean velocity $V_0 = Q/(wh)$, where Q is the flow rate through a section of width w and height h. The characteristic time T_0 is therefore given by

$$t = \frac{L}{V_0}\,\tilde{t} = T_0\,\tilde{t}. \tag{4.35}$$

From this follows the expressions for the two velocity components,

$$v_x = V_0\,\tilde{v}_x, \qquad v_z = \frac{h}{T_0}\,\tilde{v}_z = \varepsilon\,V_0\,\tilde{v}_z. \tag{4.36}$$

Finally, the characteristic pressure is given by the pressure drop P_0, see Table 4.1,

$$P_0 = R_\mathrm{hyd} Q \simeq \frac{\eta L}{h^3 w}\,Q = \frac{\eta V_0 L}{h^2}, \tag{4.37}$$

where for convenience we have dropped the numerical factor of 12.

If we follow the convention that the Reynolds number *Re* should contain the smallest length scale of the problem, here h, we define

$$Re \equiv \frac{\rho V_0 h}{\eta}. \tag{4.38}$$

Using the above-mentioned expressions we can rewrite the two-component Navier–Stokes equation and the continuity equation in terms of dimensionless variables. The result is

$$\varepsilon Re\Big(\tilde{\partial}_t + \tilde{v}_x\tilde{\partial}_x + \tilde{v}_z\tilde{\partial}_z\Big)\tilde{v}_x = -\tilde{\partial}_x\tilde{p} + \Big(\tilde{\partial}_z^2 + \varepsilon^2\tilde{\partial}_x^2\Big)\tilde{v}_x, \tag{4.39a}$$

$$\varepsilon^3 Re\Big(\tilde{\partial}_t + \tilde{v}_x\tilde{\partial}_x + \tilde{v}_z\tilde{\partial}_z\Big)\tilde{v}_z = -\tilde{\partial}_z\tilde{p} + \Big(\varepsilon^2\tilde{\partial}_z^2 + \varepsilon^4\tilde{\partial}_x^2\Big)\tilde{v}_z, \tag{4.39b}$$

$$\tilde{\partial}_x\tilde{v}_x + \tilde{\partial}_z\tilde{v}_z = 0. \tag{4.39c}$$

To first order in ε in the limit of high aspect ratios, $\varepsilon \to 0$, these equations become

$$\varepsilon Re\Big(\tilde{\partial}_t + \tilde{v}_x\tilde{\partial}_x + \tilde{v}_z\tilde{\partial}_z\Big)\tilde{v}_x = -\tilde{\partial}_x\tilde{p} + \tilde{\partial}_z^2\tilde{v}_x, \tag{4.40a}$$

$$0 = -\tilde{\partial}_z\tilde{p}, \tag{4.40b}$$

$$\tilde{\partial}_x\tilde{v}_x + \tilde{\partial}_z\tilde{v}_z = 0, \tag{4.40c}$$

and we can conclude that the effective Reynolds number Re_eff for this two-length-scale problem is

$$Re_\mathrm{eff} = \varepsilon\,Re = \frac{\rho V_0 h}{\eta}\,\frac{h}{L}. \tag{4.41}$$

This effective Reynolds number can therefore be arbitrarily small compared to the conventional Reynolds number given a sufficiently long channel. For a more detailed treatment of the effective Reynolds number see Section 14.1 and in particular Exercise 14.2.

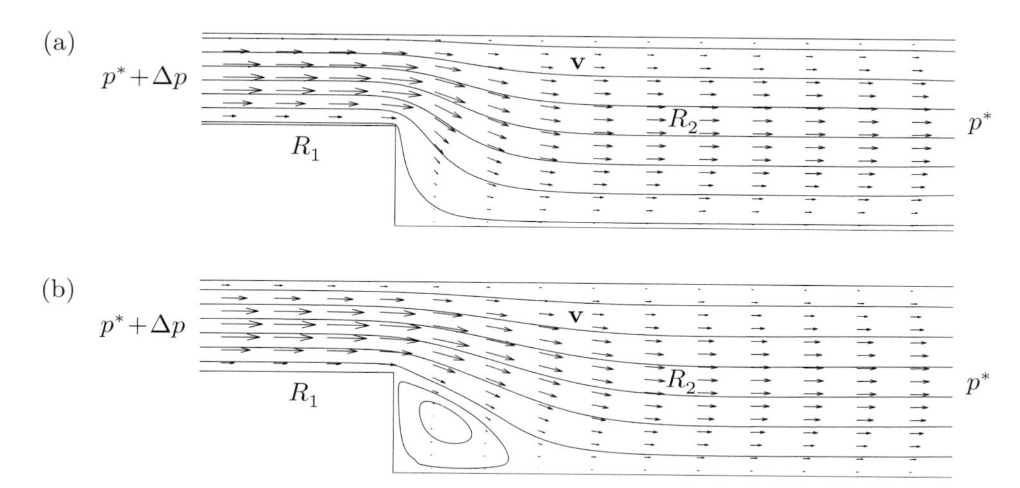

Fig. 4.6 (a) Two hydraulic resistors, R_1 and R_2, connected in series forming a backstep of height $h_2 - h_1$. The shown streamlines, see Section 14.2, and velocity vectors are calculated in COMSOL with $Re = 0.01 \ll 1$, so the Hagen–Poiseuille law is valid: $\Delta p \approx (R_1 + R_2)\, Q$. (b) Here, $Re = 100 \gg 1$ and a convection roll appears after the backstep. The large inertial forces make the Hagen–Poiseuille law invalid: $\Delta p \neq (R_1 + R_2)\, Q$. Numerical simulation courtesy of Martin Heller, DTU Nanotech.

4.5 Hydraulic resistance, two connected straight channels

When two straight channels of different dimensions are connected to form one long channel the translation invariance will in general be broken, and the expressions for the ideal Poiseuille flow no longer apply. However, we expect the ideal description to be approximately correct if the Reynolds number Re of the flow is sufficiently small. This is because a very small value of Re corresponds to a vanishingly small contribution from the non-linear term in the Navier–Stokes equation, a term that is strictly zero in ideal Poiseuille flows due to translational invariance.

The influence of the Reynolds number on the velocity field is illustrated in Fig. 4.6, where results of numerical simulations using COMSOL software are shown. Two infinite parallel-plate channels with heights h_1 and h_2 and hydraulic resistances R_1 and R_2 are joined in a series coupling forming a backstep of height $h_2 - h_1$. At low Reynolds number $Re = \rho V_0 h_1/\eta = 0.01$, panel (a), the transition from a perfect Poiseuille flow in R_1 is smooth and happens on a length scale shorter than h_1. At high Reynolds number $Re = \rho V_0 h_1/\eta = 100$, panel (b), the transition happens on a length scale larger than h_1, and a convection roll forms in the entrance region of R_2. This is a simple example of how it is a fair approximation to assume ideal Poiseuille flows in individual parts of a microfluidic network at low Reynolds numbers, whereas the approximation is dubious at high Reynolds numbers. In microfluidics the Reynolds number tends to be low due to the small length scales and low velocities, and the simple Hagen–Poiseuille law can be applied beyond the ideal situation of Fig. 3.5.

4.5.1 Two straight channels in series

Consider the series coupling of two hydraulic resistors as shown in Fig. 4.7. If we assume the validity of the Hagen–Poiseuille law for each of the resistors after they are connected,

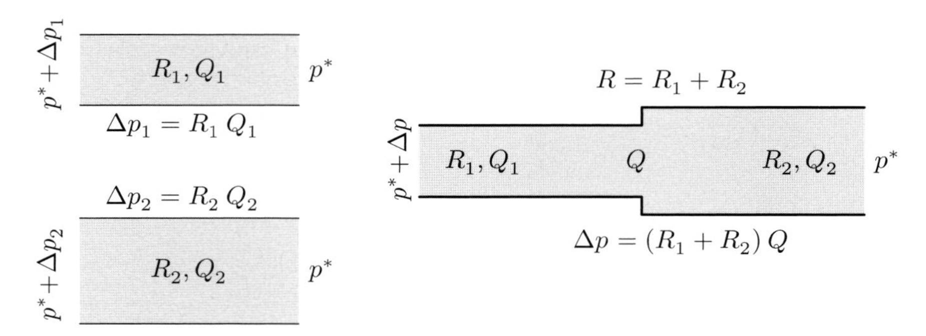

Fig. 4.7 The series coupling of two channels with hydraulic resistance R_1 and R_2. The simple additive law $R = R_1 + R_2$ is only valid in the limit of low Reynolds number, $Re \to 0$, and for long narrow channels.

then using the additivity of the pressure drop along the series coupling it is straightforward to show the law of additivity of hydraulic resistors in a series coupling, see Exercise 4.3,

$$R = R_1 + R_2. \tag{4.42}$$

Bearing in mind the discussion in the previous subsection, the additive law is only valid for low Reynolds numbers and for long and narrow channels.

4.5.2 Two straight channels in parallel

Consider the parallel coupling of two hydraulic resistors as shown in Fig. 4.8. If we assume the validity of the Hagen–Poiseuille law for each of the resistors after they are connected, then using the conservation of flow rate, i.e. $Q = Q_1 + Q_2$ in the parallel coupling it is straightforward to show the law of additivity of inverse hydraulic resistances in a parallel coupling, see Exercise 4.3,

$$R = \left(\frac{1}{R_1} + \frac{1}{R_2}\right)^{-1} = \frac{R_1\, R_2}{R_1 + R_2}. \tag{4.43}$$

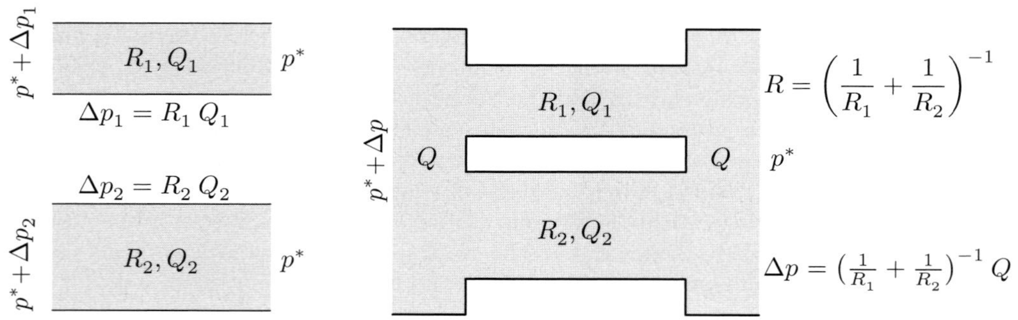

Fig. 4.8 The parallel coupling of two channels with hydraulic resistance R_1 and R_2. The additive law for the inverse resistances $R^{-1} = {R_1}^{-1} + {R_2}^{-1}$ is only valid in the limit of low Reynolds number, $Re \to 0$, and for long narrow channels far apart.

Exercise 4.5
Reynolds number of a man and of a bacterium
A living species of linear size L moving in water can typically move its own distance per second, implying a characteristic velocity $U = L/(1\ \text{s})$. Estimate the Reynolds number of a man and of a bacterium swimming in water. Comment on the result.

Exercise 4.6
Reynolds number in a two-length-scale system
Consider the two-length-scale system of Section 4.4. Prove that Eqs. (4.39a)–(4.39c) are correct forms of the dimensionless Navier–Stokes equation.

Exercise 4.7
The pressure needed to run a lab-on-chip laser
Consider the fluidic lab-on-a-chip dye laser shown in Fig. 4.3. The liquid used in the laser is ethanol with a viscosity at room temperature of $\eta = 1.197$ mPa s. The dimensions of the rectangular channel are length $L = 122$ mm, width $w = 300$ µm, and height $h = 10$ µm. For proper functioning the flow rate in the channel must be $Q = 10$ µL/h. Calculate the pressure needed to run this device properly.

Exercise 4.8
The hydraulic resistance of the cascade electro-osmotic micropump
The low-voltage cascade electro-osmotic micropump is introduced in Fig. 4.11 and treated further in Section 9.6. Consider such a pump with M identical stages, in which each multichannel segment contains N parallel-coupled narrow channels.

(a) Derive an expression for the total hydraulic resistance R_{hyd} of the M-stage cascade pump using the notation given in Fig. 4.11(b) for the hydraulic resistances in a single stage.

(b) Calculate the hydraulic resistance of a 15-stage, water-filled cascade pump with the geometry defined in Fig. 4.11(a). Neglect the relatively small resistance R_2 of the short and broad channel connecting the multichannel and the single-channel segments in a given stage.

Exercise 4.9
The hydraulic resistance of a slightly deformed cylindrical channel
Consider a Poiseuille flow in a slightly deformed cylindrical channel, where the radius $a(x)$ depends weakly on x. The result Eq. (3.67) to lowest order in the deformation α of another shape-deformed Poiseuille-flow problem indicates that the velocity field can be approximated by the unperturbed field fitted into the deformed channel. Therefore, for the deformed cylindrical channel we assume the x-dependent velocity field

$$v_x(x,y,z) \approx \frac{\partial_x p}{4\eta}\Big(a(x)^2 - y^2 - z^2\Big). \tag{4.53}$$

Use this to derive an approximate expression for the hydraulic resistance R_{hyd} for a deformed cylinder of length L with a weakly varying radius $a(x)$.

Exercise 4.10
A model of a microchannel with compliance
Consider the model Fig. 4.9(b) of a microchannel with compliance due to soft walls. Assume an oscillating inlet pressure $\Delta p(t) = \Delta p\ \mathrm{e}^{\mathrm{i}\omega t}$, and use the equivalent circuit diagram of Fig. 4.10(b) to analyze the flow rate $Q_1(t) = Q_1\ \mathrm{e}^{\mathrm{i}\omega t}$ in steady state.

(a) Find the hydraulic impedance Z_{hyd} of the microchannel by use of Fig. 4.10(b).

(b) Assume $R_1 = R_2 = R$ and find Q_1 in terms of a frequency-dependent pre-factor and the ratio $\Delta p/R$.

(c) Calculate Q_1 in the limits $\omega \to 0$ and $\omega \to \infty$ and discuss the result.

4.9 Solutions

Solution 4.1
The rate of dissipation of kinetic energy
Distinguish clearly between physical and mathematical arguments. Begin by a clear formulation of the physical assumptions and arguments that leads to the starting point of the analysis. Remember that a repeated index in the index notation implies a summation over that index: $\partial_j v_j \equiv \sum_{j=1}^{3} \partial_j v_j = \partial_x v_x + \partial_y v_y + \partial_z v_z$.

Solution 4.2
Viscous power consumption in hydraulic resistors
The power consumption in a circular and an infinite parallel-plate channel are denoted $P_{\bigcirc}$ and $P_{\parallel}$, respectively.

(a) $P = Q\Delta p = R_{\text{hyd}}Q^2$, so $P_{\bigcirc} = 8\eta L Q^2/(\pi a^4)$ and $P_{\parallel} = 12\eta L Q^2/(wh^3)$.

(b) In both cases $P \propto \eta L Q^2$. The cross-sectional geometry influences P slightly differently: $P_{\bigcirc} \propto a^{-4}$, while $P_{\parallel} \propto w^{-1}h^{-3}$, but in both cases the power increases significantly upon downscaling of the smallest transverse length, channel radius a or channel height h.

Solution 4.3
Series and parallel coupling of two hydraulic resistors
For a series coupling the flow rates in each resistor are identical, $Q_1 = Q_2 = Q$, while the partial pressure drops add up to the total pressure drop, $\Delta p = \Delta p_1 + \Delta p_2$. Using Hagen–Poiseuille's law on the latter relation yields

$$R_{\text{hyd}}Q = R_1 Q_1 + R_2 Q_2 = (R_1 + R_2)Q, \tag{4.54}$$

from which the desired result follows after division by Q.

For a parallel coupling the pressure drop over each resistor are identical, $\Delta p_1 = \Delta p_2 = \Delta p$, while the flow rates add up to the total flow rate, $Q = Q_1 + Q_2$. Using Hagen–Poiseuille's law on the latter relation yields

$$\frac{\Delta p}{R} = \frac{\Delta p_1}{R_1} + \frac{\Delta p_2}{R_2} = \Big(\frac{1}{R_1} + \frac{1}{R_2}\Big)\Delta p, \tag{4.55}$$

from which the desired result follows after division by Δp.

Solution 4.4
The hydraulic resistance of a square channel
The exact result for the flow rate in a square channel, where $h = w$, is obtained numerically from Eq. (3.57d) by summing a large but finite number of terms in the rapidly converging sum,

$$Q_{\text{exact}} = Q_0 \left[1 - \sum_{n,\text{odd}}^{\infty} \frac{1}{n^5}\,\frac{192}{\pi^5}\,\tanh\Big(\tfrac{1}{2}n\pi\Big)\right] = 0.4217\,Q_0. \tag{4.56}$$

(a) For a square channel, where $h = w$, the approximation Eq. (3.58) for the flow rate can be written as

$$Q_{\text{approx},1} \approx Q_0 \left[1 - 0.630\right] = 0.370\, Q_0. \tag{4.57}$$

The relative error is

$$\frac{Q_{\text{approx},1} - Q_{\text{exact}}}{Q_{\text{exact}}} = -0.1227. \tag{4.58}$$

(b) An improved approximation for Q in the square channel is obtained as follows. We note that $\tanh(\pi/2) = 0.9172$, while $\tanh(3\pi/2) = 0.9998$. It is thus a fair approximation to state that $\tanh(n\pi/2) = 1$ for $n \geq 3$, and we arrive at

$$\frac{Q_{\text{approx},2}}{Q_0} = 1 - \sum_{n,\text{odd}}^{\infty} \frac{1}{n^5}\frac{192}{\pi^5} \tanh\left(\tfrac{1}{2}n\pi\right) \tag{4.59}$$

$$\approx 1 - \frac{192}{\pi^5}\tanh\left(\tfrac{1}{2}\pi\right) - \frac{192}{\pi^5}\sum_{n=3,5,7}^{\infty}\frac{1}{n^5} \tag{4.60}$$

$$= 1 - \frac{192}{\pi^5}\sum_{n=1,3,5}^{\infty}\frac{1}{n^5} + \frac{192}{\pi^5} - \frac{192}{\pi^5}\tanh\left(\tfrac{1}{2}\pi\right) \tag{4.61}$$

$$= 1 - \left\{0.630 - \frac{192}{\pi^5}\left[1 - \tanh\left(\tfrac{1}{2}\pi\right)\right]\right\}. \tag{4.62}$$

Calculating the numerical value leads to

$$Q_{\text{approx},2} = 0.4220\, Q_0. \tag{4.63}$$

The relative error of this approximation is

$$\frac{Q_{\text{approx},2} - Q_{\text{exact}}}{Q_{\text{exact}}} = 0.0006. \tag{4.64}$$

Solution 4.5

Reynolds number of a man and of a bacterium

Since $U = L/(1\ \text{s})$ we find $Re = \rho U L/\eta = L^2 \times 10^6\ \text{m}^{-2}$. Now $L_{\text{man}} \approx 1$ m and $L_{\text{bact}} \approx 1$ µm imply $Re_{\text{man}} = 10^6$ and $Re_{\text{bact}} = 10^{-6}$. The motion in water of a man and of a bacterium is clearly dominated by inertia and viscous damping, respectively.

Solution 4.6

Reynolds number in a two-length-scale system

Once the dimensionless derivatives in Eq. (4.33) have been introduced, the result Eqs. (4.39a)–(4.39c) follow by direct substitution into the Navier–Stokes and continuity equations.

Solution 4.7

The pressure needed to run a lab-on-chip laser

The required flow rate is $Q = 10\ \mu\text{L/h} = 10 \times 10^{-9}\ \text{m}^3/(3600\ \text{s}) = 2.77 \times 10^{-12}\ \text{m}^3/\text{s}$ while the hydraulic resistance is $R_{\text{hyd}} = 12\eta L/[wh^3(1 - h/w)] = 5.97 \times 10^{15}\ \text{Pa s/m}^3$. This results in an operating pressure of $\Delta p = R_{\text{hyd}} Q = 16.6$ kPa.

Solution 4.8
The hydraulic resistance of the cascade electro-osmotic micropump

(a) The hydraulic resistance of the multichannel segment is given by the parallel-coupling rule Eq. (4.44) as $\frac{1}{N}\,R_1$. The total hydraulic resistance R_{hyd} is then given by the series-coupling rule Eq. (4.42) as

$$R_{\mathrm{hyd}} = M\left[\frac{1}{N}\,R_1 + R_2 + R_3\right]. \tag{4.65}$$

(b) All channels have rectangular cross-sections, and the corresponding expression for the hydraulic resistance is found in Table 4.1 of Section 4.2. For a single narrow channel in a multichannel segment we have $\eta = 1$ mPa s, $L = 800$ µm, $w = 20$ µm, $h = 5$ µm, and consequently $R_1 = 4.5\times 10^{15}$ Pa s/m^3. We neglect R_2, and $R_3 = 3.2\times 10^{13}$ Pa s/m^3 is found as was R_1 except now $w = 50$ µm and $h = 20$ µm. Finally, with $M = 15$ and $N = 10$ we find from Eq. (4.65) that $R_{\mathrm{hyd}} = 15\big[\frac{1}{10}45.0 + 0.32\big] \times 10^{14}$ Pa s/m$^3 = 7.2\times 10^{15}$ Pa s/m^3.

Solution 4.9
The hydraulic resistance of a slightly deformed cylindrical channel

As a and $\partial_x p$ now depend of x we write Eq. (3.39b) as $Q = (\pi/8\eta)a^4(x)\,\partial_x p(x)$. Hence

$$\Delta p = \int_0^L \mathrm{d}x\,\partial_x p = \left[\frac{8\eta}{\pi}\int_0^L \mathrm{d}x\,\frac{1}{a^4(x)}\right] Q, \quad \text{or} \quad R_{\mathrm{hyd}} = \frac{8\eta}{\pi}\int_0^L \mathrm{d}x\,\frac{1}{a^4(x)}. \tag{4.66}$$

Solution 4.10
A model of a microchannel with compliance

R_1 is in series with the parallel coupling of R_2 and C_{hyd} having the RC time $\tau = R_2 C_{\mathrm{hyd}}$.

(a)

$$Z_{\mathrm{hyd}} = R_1 + \big(R_2^{-1} + \mathrm{i}\omega C_{\mathrm{hyd}}\big)^{-1} = R_1 + R_2\big(1 + \mathrm{i}\omega\tau\big)^{-1}. \tag{4.67}$$

(b)

$$Q_1 = \frac{\Delta p}{Z_{\mathrm{hyd}}} = \frac{1}{1 + (1+\mathrm{i}\omega\tau)^{-1}}\,\frac{\Delta p}{R} = \frac{[2 + (\omega\tau)^2] + \mathrm{i}\omega\tau}{4 + (\omega\tau)^2}\,\frac{\Delta p}{R}. \tag{4.68}$$

(c) For $\omega \to 0$ we obtain $Q_1 = \frac{\Delta p}{2R}$, i.e. the channel wall is always fully expanded thus leading all flow through both resistors. For $\omega \to \infty$ we obtain $Q_1 = \frac{\Delta p}{R}$, i.e. no liquid flows through R_2 as it, after passing R_1, stays inside the ever-expanding/relaxing channel.

5.3 The diffusion equation

In the following we consider the diffusion of a single solute and therefore suppress the index α. If the velocity field $\mathbf{v}$ of the solvent is zero, convection is absent and Eq. (5.25) becomes the diffusion equation,

$$\partial_t c = D\,\nabla^2 c. \tag{5.26}$$

Simple dimensional analysis of this equation can already reveal some important physics. It is clear that if T_0 and L_0 denotes the characteristic time and length scale over which the concentration $c(\mathbf{r}, t)$ varies, then

$$L_0 = \sqrt{DT_0} \qquad \text{or} \qquad T_0 = \frac{L_0^2}{D}, \tag{5.27}$$

which resembles Eq. (5.8). The diffusion constant D thus determines how fast a concentration diffuses a certain distance. Values of D are listed in Table 8.1, and typically they are

$$D \approx 2 \times 10^{-9}\ \text{m}^2/\text{s}, \quad \text{small ions in water}, \tag{5.28a}$$
$$D \approx 5 \times 10^{-10}\ \text{m}^2/\text{s}, \quad \text{sugar molecules in water}, \tag{5.28b}$$
$$D \approx 4 \times 10^{-11}\ \text{m}^2/\text{s}, \quad \text{30-base-pair DNA molecules in water}, \tag{5.28c}$$
$$D \approx 1 \times 10^{-12}\ \text{m}^2/\text{s}, \quad \text{5000-base-pair DNA molecules in water}, \tag{5.28d}$$

which for diffusion across the typical microfluidic distance $L_0 = 100\ \mu\text{m}$ give the times

$$T_0(100\ \mu\text{m}) \approx 5\ \text{s}, \quad \text{small ions in water}, \tag{5.29a}$$
$$T_0(100\ \mu\text{m}) \approx 20\ \text{s}, \quad \text{sugar molecules in water}, \tag{5.29b}$$
$$T_0(100\ \mu\text{m}) \approx 250\ \text{s} \approx 4\ \text{min}, \quad \text{30-base-pair DNA molecules in water}, \tag{5.29c}$$
$$T_0(100\ \mu\text{m}) \approx 10^4\ \text{s} \approx 3\ \text{h}, \quad \text{5000-base-pair DNA molecules in water}. \tag{5.29d}$$

Let us now turn to some analytical solutions of the diffusion equation.

5.3.1 Limited point-source diffusion in 1D, 2D, and 3D

Let us first remind ourselves about the basic properties of the normal distribution $P(s)$ of a normalized dimensionless variable s, and its mean value $\langle s \rangle$ as well as its variance, or width of distribution, $\langle s^2 \rangle$:

$$P(s) \equiv \frac{1}{\sqrt{2\pi}}\,\mathrm{e}^{-\frac{1}{2}s^2}, \qquad \langle s \rangle = \int_{-\infty}^{\infty} \mathrm{d}s\; s\,P(s) = 0, \qquad \langle s^2 \rangle = \int_{-\infty}^{\infty} \mathrm{d}s\; s^2\,P(s) = 1. \tag{5.30}$$

These results will be useful in the following analysis of diffusion.

Consider 1D limited point-source diffusion, where a fixed number N_0 of ink molecules is injected at position $x = 0$ at time $t = 0$ in the middle of a infinitely thin and infinitely long water-filled tube aligned along the x axis. The initial point-like concentration acts as the source of the diffusion, and it can be written as a Dirac delta function[1]

[1]The Dirac delta function $\delta(x)$ is defined by: $\delta(x) = 0$ for $x \neq 0$ and $\int_{-\infty}^{\infty} \mathrm{d}x\, \delta(x) = 1$.

$$c(x, t = 0) = N_0\, \delta(x). \tag{5.31}$$

The ink molecules immediately begins to diffuse out into the water, and it is easy to show by inspection, see Exercise 5.4, that the solution to the diffusion equation (5.26), which in 1D reduces to $\partial_t c = D\partial_x^2 c$, given the initial condition Eq. (5.31) is

$$c(x, t) = N_0\, (4\pi D t)^{-\frac{1}{2}} \exp\Big[-\frac{x^2}{4Dt}\Big] = N_0\, P(s_x), \tag{5.32}$$

where we have introduced the normal distribution $P(s_x)$ of the dimensionless variable

$$s_x \equiv \frac{x^2}{2Dt}. \tag{5.33}$$

It is natural to define the square $\ell^2_{\mathrm{diff,1D}}$ of the 1D diffusion length $\ell_{\mathrm{diff,1D}}$ as the width of the distribution. So from Eqs. (5.30), (5.32) and (5.33) we get

$$\ell^2_{\mathrm{diff,1D}} \equiv \langle x^2 \rangle = 2Dt\, \langle s_x^2 \rangle = 2Dt. \tag{5.34}$$

Generalization of this result to 2D and 3D, see Exercise 5.4, with the initial conditions $c(x, y, t{=}0) = N_0\, \delta(x)\, \delta(y)$ and $c(x, y, z, t{=}0) = N_0\, \delta(x)\, \delta(y)\, \delta(z)$, respectively, gives

$$\begin{aligned} c(x, y, t) &= N_0 \left[(4\pi D t)^{-\frac{1}{2}} \exp\Big[-\frac{x^2}{4Dt}\Big]\right] \times \left[(4\pi D t)^{-\frac{1}{2}} \exp\Big[-\frac{y^2}{4Dt}\Big]\right] \\ &= N_0\, (4\pi D t)^{-1} \exp\Big[-\frac{x^2 + y^2}{4Dt}\Big] = N_0\, P(s_x)\, P(s_y), \end{aligned} \tag{5.35a}$$

and

$$c(x, y, z, t) = N_0\, (4\pi D t)^{-\frac{3}{2}} \exp\Big[-\frac{x^2 + y^2 + z^2}{4Dt}\Big] = N_0\, P(s_x)\, P(s_y)\, P(s_z), \tag{5.35b}$$

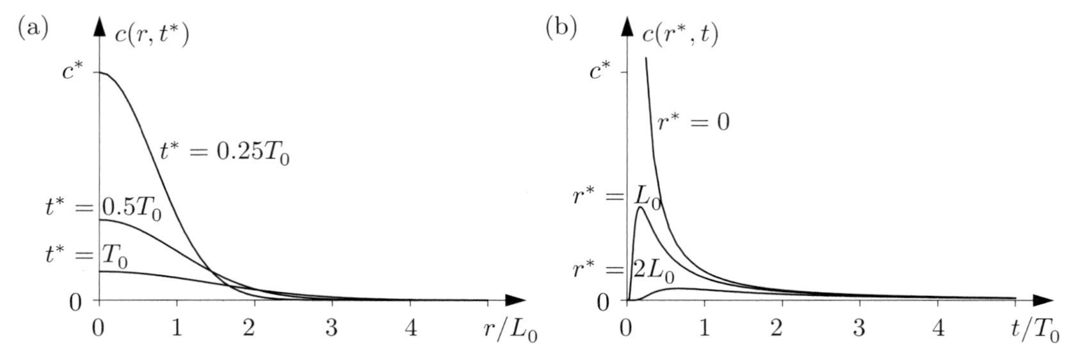

Fig. 5.2 The concentration $c(\mathbf{r}, t > 0)$ from Eq. (5.36) in the case of limited point-source diffusion. The length scale in the radial direction r has been chosen to be L_0, which fixes the time scale to be $T_0 = L_0^2/D$. (a) The r dependence of $c(r, t^*)$ for three given times $t^* = 0.25T_0$, $0.5T_0$, and T_0. (b) The time dependence of $c(r^*, t)$ for three given radial positions $r^* = 0$, L_0, and $2L_0$.

where we have introduced the two dimensionless variables

$$s_y \equiv \frac{y^2}{2Dt}, \qquad \text{and} \qquad s_z \equiv \frac{z^2}{2Dt}. \tag{5.36}$$

The 3D result Eq. (5.36) for $c(x,y,z,t)$ is presented in Fig. 5.2. In 2D and 3D ℓ^2_{diff} becomes

$$\ell^2_{\mathrm{diff,2D}} \equiv \langle r^2\rangle_{\mathrm{2D}} = \langle x^2+y^2\rangle \qquad = 2Dt\,\langle s_x^2+s_y^2\rangle \qquad = 4\,Dt, \tag{5.37a}$$

$$\ell^2_{\mathrm{diff,3D}} \equiv \langle r^2\rangle_{\mathrm{3D}} = \langle x^2+y^2+z^2\rangle = 2Dt\,\langle s_x^2+s_y^2+s_z^2\rangle = 6\,Dt, \tag{5.37b}$$

and we see that the diffusion lengths (not their squares) are $\sqrt{2}$ and $\sqrt{3}$ times larger in 2D and 3D, respectively, than that in 1D.

5.3.2 Limited planar-source diffusion

Another limited diffusion process is limited planar-source diffusion. Let the semi-infinite half-space $x > 0$ be filled with some liquid. Consider then an infinitely thin slab covering the yz-plane at $x = 0$ containing n_0 molecules per area that at time $t = 0$ begin to diffuse out into the liquid. With a factor 2 inserted to normalize the half-space integration, the initial condition is

$$c(\mathbf{r}, t=0) = n_0\, 2\delta(x), \tag{5.38}$$

which results in the solution

$$c(\mathbf{r}, t>0) = \frac{n_0}{(\pi Dt)^{\frac{1}{2}}}\, \exp\Big(-\frac{x^2}{4Dt}\Big). \tag{5.39}$$

5.3.3 Constant planar-source diffusion

We end by an example of diffusion with a constant source, i.e. an influx of solute is maintained at one of the boundary surfaces. Consider the same geometry as in the previous example, but change the boundary condition as follows. At time $t = 0$ a source filling the half-space

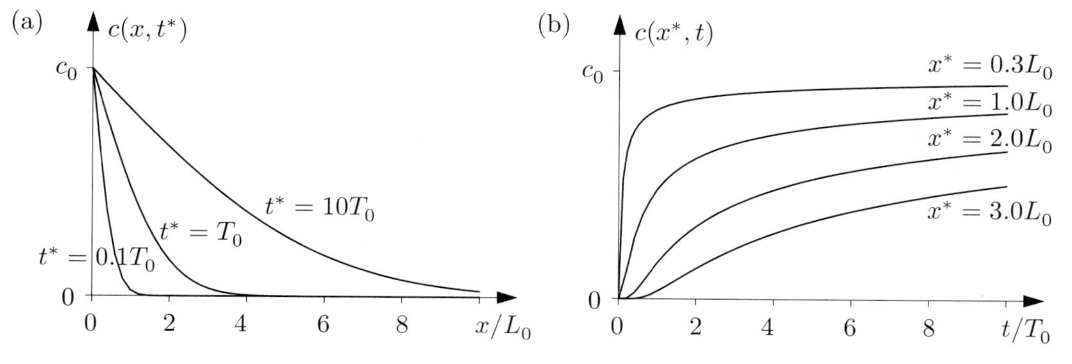

Fig. 5.3 The concentration $c(\mathbf{r}, t > 0)$ from Eq. (5.41) in the case of constant planar-source diffusion. The length scale in the x direction has been chosen to be L_0, which fixes the time scale to be $T_0 = L_0^2/D$. (a) The x dependence of $c(x, t^*)$ for three given times $t^* = 0.1T_0$, T_0, and $10T_0$. (b) The time dependence of $c(x^*, t)$ for four given positions $x^* = 0.3L_0$, L_0, $2L_0$, and $10L_0$.

$x < 0$ suddenly begins to provide an influx of molecules to the boundary plane $x = 0$ such that the density there remains constant at all later times,

$$c(x = 0, y, z, t > 0) = c_0. \tag{5.40}$$

By inspection it is straightforward to show that the solution can be written in terms of the complementary error function $\mathrm{erfc}(s) \equiv \frac{2}{\sqrt{\pi}} \int_s^\infty \mathrm{e}^{-u^2}\, du$,

$$c(\mathbf{r}, t > 0) = c_0 \,\mathrm{erfc}\left(\frac{x}{\sqrt{4Dt}}\right). \tag{5.41}$$

The constant planar-source diffusion is illustrated in Fig. 5.3.

5.3.4 Diffusion of momentum and the Navier–Stokes equation

It is not only mass that can diffuse as described above. Another important example is heat, but also for momentum there exists a diffusion equation. In fact, we have already through the Navier–Stokes equation worked with diffusion of mechanical momentum without noticing it. That the Navier–Stokes equation contains momentum diffusion becomes most clear if we consider the case of the decelerating Poiseuille flow discussed in Fig. 4.1 in Section 4.1, where the Navier–Stokes equation becomes very simple,

$$\rho \partial_t v_x = \eta\, \nabla^2 v_x. \tag{5.42}$$

In terms of the momentum density ρv_x we indeed obtain a diffusion equation like Eq. (5.26),

$$\partial_t(\rho v_x) = \nu\, \nabla^2(\rho v_x), \tag{5.43}$$

where the kinematic viscosity ν appears as the diffusion constant for momentum,

$$\nu \equiv \frac{\eta}{\rho} \quad (\approx 10^{-6}\ \mathrm{m^2/s}\ \text{for water}). \tag{5.44}$$

In analogy with Eq. (5.27) there exists a momentum diffusion time T_0 for diffusion by a characteristic length a, e.g. the radius of the microchannel,

$$T_0 = \frac{a^2}{\nu} \quad (\approx 10\ \mathrm{ms}\ \text{for water in a microchannel of radius } 100\ \mathrm{\mu m}). \tag{5.45}$$

The dimensionless ratio of the diffusivity ν of momentum relative to the diffusivity D of mass is denoted the Schmidt number Sc,

$$Sc = \text{Schmidt number} \equiv \frac{\nu}{D} = \frac{\eta}{\rho D}. \tag{5.46}$$

Note that the Schmidt number is an intrinsic property of the solution. This is in contrast to the Reynolds number that due to its dependence on the velocity is a property of the flow.

5.4 The H-filter: separating solutes by diffusion

Diffusion is an old, well-known and much used method to separate solutes with different diffusion constants. The method has also been employed in microfluidics, where the advantages of laminar flow and fast diffusion over small distances can be combined and utilized.

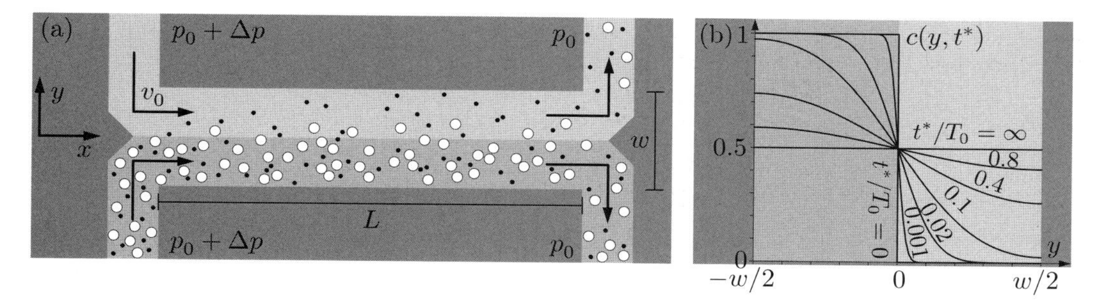

Fig. 5.4 (a) A top view in the xy-plane of a flat (height h) H-filter consisting of a central channel (length L and width w) with two inlet channels to the left and two outlet channels to the right. A pure buffer liquid (light gray) and a buffer liquid (gray) containing big (white) and small (black) solutes are introduced via one inlet each, and perform a pressure-driven, steady state, laminar flow with average velocity v_0. (b) Concentration profiles $c(y, t^*)$ in the central channel as a function of the transverse direction y at different positions $x^* = v_0 t^*$ along the channel.

Here, we shall briefly study one such example, the so-called H-filter, which was among some of the first commercial microfluidic products.

The name of the H-filter is derived from its geometrical appearance, see the xy-plane top view in Fig. 5.4(a). The legs of the H are the two inlet channels to the left, kept at pressure $p_0 + \Delta p$, and the two outlet channels to the right, kept at pressure p_0. The cross-bar of the H is the central channel where diffusion takes place. A pure buffer liquid (light gray) is introduced at one inlet, while another buffer liquid (gray) containing big (white) and small (black) solutes is introduced in another inlet. All channels are flat having the same width w and height $h \ll w$, the central channel has the length L, and the two buffer liquids are both taken to be water. Working with length scales in the micrometer range, say $h = 10$ µm, $w = 100$ µm and $L = 1$ mm, and a flow velocity below 1 mm/s, the flow is laminar, and the two buffer liquids do not mix, as indicated by the gray and light gray shading in Fig. 5.4(a). As seen in Fig. 3.9 the average velocity profile in a flat channel is constant across the width w except within a distance of $h/2$ from either side wall. We denote this velocity v_0.

Regarding the behavior in the H-filter of a given solute with diffusion constant D in the buffer, two time scales become relevant, namely the time τ_{conv} it takes to be convected downstream from the inlet to the outlet, and the time τ_{diff} it takes to diffuse across the half-width of the channel. They are given by

$$\tau_{\mathrm{conv}} = \frac{L}{v_0}, \tag{5.47a}$$

$$\tau_{\mathrm{diff}} = \frac{\left(\frac{1}{2}w\right)^2}{2D} = \frac{w^2}{8D}. \tag{5.47b}$$

For a solute with $\tau_{\mathrm{conv}} \ll \tau_{\mathrm{diff}}$ diffusion does not have time enough to act, and it will (largely) remain in its original buffer stream leaving the other buffer stream (relatively) pure, see the white particles in Fig. 5.4(a). For the case $\tau_{\mathrm{conv}} \gtrsim \tau_{\mathrm{diff}}$ the solute has time enough to diffuse across the central channel and the concentration of the solute will be the same in the two buffer streams, see the black particles in Fig. 5.4(a). Consequently, operating the H-filter with two solutes in one buffer stream and making sure that they fulfil $\tau_{\text{conv},1} \ll \tau_{\text{diff},1}$ and $\tau_{\text{conv},2} \gtrsim \tau_{\text{diff},2}$, it is possible to separate out solute 2 from solute 1, although arriving only at half the initial concentration. For a given choice of L, w and v_0 the critical value D^* of the diffusion constant, where complete mixing by diffusion happens, can be found by requiring $\tau_{\text{diff}}(D^*) = \tau_{\text{conv}}$. This gives

$$D^* = \frac{v_0 w^2}{8L}. \tag{5.48}$$

For $L = 1$ mm, $w = 100$ µm and $v_0 = 1$ mm/s we find $D^* = 1.3 \times 10^{-9}$ m^2/s, which according to Eq. (5.28) is close to the diffusion constant of small ions in water. Thus, it is possible to separate these from larger molecules using the H-filter.

It is important to realize that the inherent randomness of diffusion processes makes the separation obtained by the H-filter statistical in nature. One cannot expect to achieve 100% separation, since in the separated outlet stream a fraction α, the impurity fraction, of the solute concentration will be the unwanted slowly diffusing solute. However, by making a multistage setup with several H-filters in series, one can in principle come arbitrarily close to 100% separation of the fast-diffusing solute from the slow one, with the prize of halving the concentration at each stage.

Quantitative estimates of the impurity fraction α can be obtained by solving the diffusion equation (5.26). Here, we utilize the laminarity of microfluidics, which ensures that time is converted into position: the position x downstream in the central channel is given by the time t as $x = v_0 t$. In a slice across the channel of thickness Δx near $x = 0$ there are $N_0 = c_0 w h \Delta x$ solute molecules, where c_0 is the concentration in the inlet buffer. Since on average these molecules are convected downstream by the speed v_0 there will be the same number of molecules in each slice of thickness Δx, and the consecutive slices x correspond to consecutive time instants t. Hence, we study the evolution of the concentration profile $c(y,t)$, which is an example of limited-source diffusion, where the initial condition is the half-box profile $c(y,0) = c_0$, for $-w/2 < y < 0$, and $c(y,0) = 0$, for $0 < y < w/2$, which emerges right at the point $x = 0$, where the two buffer streams meet. The boundary condition is zero current at the side walls, which according to Fick's law Eq. (5.21) becomes $\partial_y c(\pm w/2, t) = 0$. The solution to this limited source diffusion problem is shown in Fig. 5.4(b). Note that due to symmetry around the centerline $y = 0$ the concentration remains $c(0,t) = \frac{1}{2}c_0$, and the solution for small times, where the side walls have not yet been reached by the diffusing molecules, the solution is identical to the constant planar-source diffusion shown in Fig. 5.3.

5.5 Taylor dispersion; a convection-diffusion example

In the following section we will study an example of combined convection and diffusion, which occurs if a concentration $c(\mathbf{r},t)$ of some solute is placed in a solution flowing with the non-zero velocity field $\mathbf{v}(\mathbf{r},t)$. The simplest case, which nevertheless turns out to be complicated, is obtained for the steady-state Poiseuille flow in a cylindrical microchannel, where $\mathbf{v} = v_x(r)\,\mathbf{e}_x$. The corresponding convection-diffusion equation (5.25) becomes

$$\partial_t c + v_x \partial_x c = D\left(\partial_r^2 c + \tfrac{1}{r}\,\partial_r c + \partial_x^2 c\right), \tag{5.49}$$

where

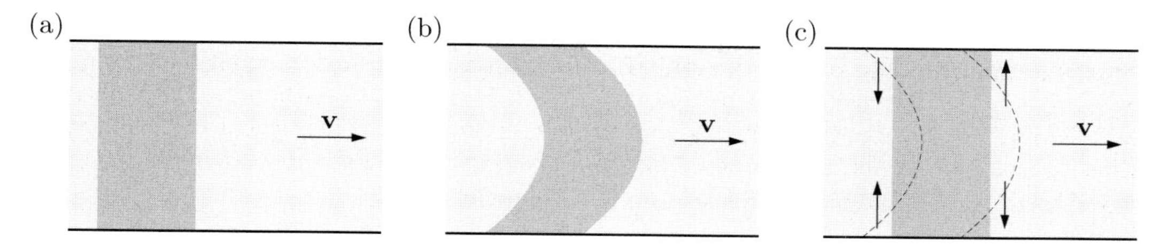

Fig. 5.5 A sketch of the Taylor dispersion problem in a cylindrical microchannel of radius a with a steady Poiseuille flow (horizontal arrow $\mathbf{v}$). (a) The initial flat concentration (dark gray) of the solute. (b) Neglecting diffusion the solute gets stretched out into a paraboloid-shaped plug. (c) With diffusion, indicated by the vertical arrows, the deformed concentration profile gets evened out.

$$v_x(r) = 2\Big(1 - \frac{r^2}{a^2}\Big)\, V_0, \tag{5.50}$$

so that $V_0 \equiv \frac{1}{\pi a^2}\int_0^a \mathrm{d}r\, 2\pi r\, v_x(r)$ is the average velocity of the Poiseuille flow.

In the Taylor dispersion problem, sketched in Fig. 5.5 we consider a homogeneous band of solute placed in the microchannel at $t = 0$ and study how this concentration profile disperses due to convection from the Poiseuille flow and due to diffusion from the concentration gradients.

If diffusion did not play any role the band of solute would become stretched into an increasingly longer paraboloid-shaped band due to the Poiseuille flow. However, diffusion is present and it counteracts the stretching: in the front end of the concentration profile diffusion brings solute particles from the high concentration near the center out towards the low concentration sides, whereas in the back end it brings solute particles from the high-concentration sides towards the low concentration near the center. As we shall see, the result is a quite evenly shaped plug moving downstream with a speed equal to the average Poiseuille flow velocity V_0.

5.5.1 Dimensional analysis and the Péclet number

To get a first insight into the problem we make a dimensional analysis of the convection-diffusion equation (5.49). The characteristic lengths over which the concentration changes in the radial and axial direction are denoted a and L_0, respectively. In microfluidics the radial length scale a is the radius or width of the channel, and it is often much smaller than L_0. The characteristic flow velocity is denoted V_0, but combining this with either of the two length scales and the diffusivity D we have four possible choices for the time scale T_0:

$$\tau_{\mathrm{diff}}^{\mathrm{rad}} = \frac{a^2}{D}, \quad \text{time to move the distance } a \text{ by radial diffusion} \tag{5.51a}$$

$$\tau_{\mathrm{diff}}^{\mathrm{ax}} = \frac{L_0^2}{D}, \quad \text{time to move the distance } L_0 \text{ by axial diffusion} \tag{5.51b}$$

$$\tau_{\mathrm{conv}}^{L} = \frac{L_0}{V_0}, \quad \text{time to move the distance } L_0 \text{ by axial convection} \tag{5.51c}$$

$$\tau_{\mathrm{conv}}^{a} = \frac{a}{V_0}, \quad \text{time to move the distance } a \text{ by axial convection.} \tag{5.51d}$$

It is customary to use $T_0 = \tau_{\text{conv}}^{a} = a/V_0$ as the characteristic time scale. With $x = L_0\tilde{x}$, $r = a\tilde{r}$, $t = T_0\tilde{t}$, and $v_x = V_0\tilde{v}_x$ the dimensionless convection-diffusion equation becomes

$$\frac{V_0}{a}\,\partial_{\tilde{t}}c + \frac{V_0}{L_0}\,\tilde{v}_x\partial_{\tilde{x}}c = \frac{D}{a^2}\left(\partial_{\tilde{r}}^2 c + \tfrac{1}{\tilde{r}}\,\partial_{\tilde{r}}c\right) + \frac{D}{L_0^2}\,\partial_{\tilde{x}}^2 c. \tag{5.52}$$

Introducing the mass diffusion Péclet number *Pé*, defined as

$$Pé \equiv \frac{\text{diffusion time}}{\text{convection time}} = \frac{\tau_{\text{diff}}^{\text{rad}}}{\tau_{\text{conv}}^{a}} = \frac{\frac{a^2}{D}}{\frac{a}{V_0}} = \frac{V_0\,a}{D}, \tag{5.53}$$

the diffusion-convection equation can be written as

$$Pé\,\partial_{\tilde{t}}c + Pé\,\frac{a}{L_0}\,\tilde{v}_x\partial_{\tilde{x}}c = \left(\partial_{\tilde{r}}^2 c + \frac{1}{\tilde{r}}\,\partial_{\tilde{r}}c\right) + \frac{a^2}{L_0^2}\,\partial_{\tilde{x}}^2 c. \tag{5.54}$$

For high Péclet numbers, where $\tau_{\text{conv}}^{a} \ll \tau_{\text{diff}}^{\text{rad}}$ and convection thus happens much faster than diffusion, the terms on the left-hand side of the convection-diffusion equation dominate, and we are in the convection-dominated regime. Conversely, for low Péclet numbers, where $\tau_{\text{diff}}^{\text{rad}} \ll \tau_{\text{conv}}^{a}$ and diffusion happens much faster than convection, the terms on the right-hand side dominate, and we are in the diffusion dominated regime. Note that due to the factor a^2/L_0^2 on the right-hand side the radial diffusion sets in at different time scales from those of the axial diffusion. This fact will be exploited in the Taylor dispersion model.

5.5.2 A heuristic treatment of Taylor dispersion

Before launching a more detailed calculation, let us consider a heuristic treatment of the Taylor dispersion. We are interested in estimating the effective diffusion constant D_{eff} for diffusion along the x axis of a long, narrow cylinder of radius a. For a given time interval t there is the ever present Brownian diffusion, which yields a contribution Dt to the square of the diffusion length along the x axis. However, due to the convection flow there is one more contribution.

For simplicity imagine the liquid of the cylinder parted into three concentric cylinder shells each of thickness $a/3$, the middle of which moves with the average flow velocity V_0. Let us now fix the time interval t so that it correspond to the time it takes to diffusive radially the distance $a/3$, i.e. $t \equiv a^2/(9D)$. We note that this radial diffusion is transformed into an axial motion by the flow, because a random jump form one liquid shell to its neighbor will result in an axial displacement of $\pm(V_0/2)\,t = \pm V_0 a^2/(18D)$ as the liquid shells move relative to each other approximately with the speed $V_0/2$.

The square of the axial diffusion length can therefore be written as the sum of the two above-mentioned contributions,

$$\ell_{\text{diff}}^2 \approx \left[\sqrt{Dt}\right]^2 + \left[\frac{V_0 t}{2}\right]^2 = \left[D + \frac{V_0^2 t}{4}\right]t = \left[D + \frac{V_0^2 a^2}{72D}\right]t. \tag{5.55}$$

Using the standard diffusion relation $\ell^2 = Dt$, Eq. (5.55) leads to the prediction of an effective axial diffusion constant D_{eff} given by

$$D_{\text{eff}} \approx D + \frac{V_0^2 a^2}{72D} = \left[1 + \frac{Pé^2}{72}\right]D. \tag{5.56}$$

With a minor correction of the numerical coefficient, this result is confirmed by Eqs. (5.65) and (5.70) below.

5.5.3 Taylor's model for dispersion in microfluidic channels

In Taylor's model for the dispersion problem sketched in Fig. 5.5, the convection-diffusion equation is studied in the reference frame moving with average speed V_0 of the imposed Poiseuille flow. Furthermore, we consider the limit of large times $t \gg \tau_{\text{diff}}^{\text{rad}}$, where the diffusion process has had time to act over the short radial distance a yielding a radially averaged concentration profile. The concentration profile in the axial direction changes over the long length scale $L_0 \gg a$, and the axial diffusion of this profile can be studied separately. In terms of the characteristic time scales Eq. (5.51) we can formulate the domain of validity of the model as two inequalities. The central time scale is the radial diffusion time $\tau_{\text{diff}}^{\text{rad}}$, which must be smaller than the long axial convection time τ_{conv}^{L}, to ensure the radial smearing of the concentration profile, i.e. $\tau_{\text{diff}}^{\text{rad}} \ll \tau_{\text{conv}}^{L}$. On the other hand, the short convection time τ_{conv}^{L} must be smaller than $\tau_{\text{diff}}^{\text{rad}}$ to ensure that convection does play a role, i.e. $\tau_{\text{conv}}^{a} \ll \tau_{\text{diff}}^{\text{rad}}$. From this rough argument we have established the domain of validity of Taylor's model,

$$\tau_{\text{conv}}^{a} \ll \tau_{\text{diff}}^{\text{rad}} \ll \tau_{\text{conv}}^{L} \qquad \Rightarrow \qquad 1 \ll Pé \ll \frac{L_0}{a}. \tag{5.57}$$

As we proceed with the actual solution of the convection-diffusion equation for Taylor's model, we shall show that these two inequalities are fulfilled, except for some numerical pre-factors that appear on the way.

As $L_0 \gg a$ in the large-time limit, we see that the axial diffusion term $(a^2/L_0^2)\partial_{\tilde{x}}^2 c$ in Eq. (5.52) can be neglected compared to the radial diffusion term, $\partial_{\tilde{r}}^2 c + (1/r)\partial_{\tilde{r}} c$. As a consequence, the only axial change in concentration follows from convection. It is therefore natural to make a co-ordinate transformation to a co-ordinate system (r, x') that moves along with the mean velocity V_0 of the paraboloid Poiseuille flow Eq. (5.50),

$$x' \equiv x - V_0\, t, \qquad \text{and} \qquad v_x'(r) = v_x(r) - V_0 = \Big(1 - 2\frac{r^2}{a^2}\Big) V_0. \tag{5.58}$$

In the moving co-ordinate system the convection-diffusion equation without the discarded axial diffusion becomes

$$\partial_t c + V_0\Big(1 - 2\frac{r^2}{a^2}\Big)\, \partial_{x'} c = D\,\Big(\partial_r^2 c + \tfrac{1}{r}\,\partial_r c\Big). \tag{5.59}$$

Moreover, in the time limit we are working, where radial diffusion dominates, it follows consistently that in the moving co-ordinate system the concentration profile is stationary (at average speed, convection does not contribute) and the axial gradient of the concentration is independent of r, i.e. $\partial_t c = 0$ and $\partial_{\tilde{x}} c = \partial_{\tilde{x}} c(x')$. The validity of this assumption can always be checked once the solution is obtained. The problem is thus reduced to

$$\Big(1 - 2\frac{r^2}{a^2}\Big) V_0\, \partial_{x'} c = D\,\Big(\partial_r^2 c + \tfrac{1}{r}\,\partial_r c\Big), \tag{5.60}$$

which, as $\partial_{x'} c$ is independent of r, is an ordinary differential equation for $c(r)$ with the solution

$$c(r,x') = \bar{c}(x') + \frac{a^2 V_0}{4D}\,\partial_{x'} c(x')\Big(-\frac{1}{3} + \frac{r^2}{a^2} - \frac{1}{2}\frac{r^4}{a^4}\Big), \tag{5.61}$$

where $\bar{c}(x')$ is the concentration averaged over the cross-section at x'. We can easily derive the condition for having an r-independent axial gradient of c by differentiating Eq. (5.61) with respect to x' and demand the second term to be negligibly small,

$$\partial_{x'} c = \partial_{x'} \bar{c}(x'), \quad \text{if} \quad \frac{a^2 V_0}{4D}\,\frac{1}{L_0} \ll 1. \tag{5.62}$$

The last equality can be expressed in terms of $P\acute{e}$ as

$$P\acute{e} \ll 4\,\frac{L_0}{a}. \tag{5.63}$$

5.5.4 The solution to the Taylor dispersion problem

By calculating the average current density $\bar{J}(x')$ through the cross-section at x' using Eqs. (5.58) and (5.61), we can derive Fick's law for the average concentration $\bar{c}(x')$ and read off the effective diffusion coefficient D_{eff} for the resulting 1D diffusion problem,

$$\bar{J}(x') = \frac{1}{\pi a^2}\int_0^a \mathrm{d}r\; 2\pi r\;\rho c(r,x') v'_x(r) = -\frac{a^2 V_0^2}{48D}\;\rho\partial_{x'}\bar{c} \equiv -D_{\text{eff}}\;\rho\partial_{x'}\bar{c}, \tag{5.64}$$

where D_{eff}, also known as the Taylor dispersion coefficient, is defined as

$$D_{\text{eff}} \equiv \frac{a^2 V_0^2}{48D} = \frac{P\acute{e}^2}{48}\,D. \tag{5.65}$$

Conservation of mass applied in the moving co-ordinate system yields

$$\rho\partial_t \bar{c} = -\partial_{x'}\bar{J} \quad \Rightarrow \quad \partial_t \bar{c} = D_{\text{eff}}\;\partial_{x'}^2 \bar{c}. \tag{5.66}$$

Using the result Eq. (5.39) for limited planar-source diffusion we can immediately write down the solution to the Taylor dispersion problem (transformed back to the unmoved co-ordinate system)

$$\bar{c}(x,t) = \frac{n_0}{(\pi D_{\text{eff}} t)^{\frac{1}{2}}}\;\exp\Big[-\frac{(x - V_0 t)^2}{4D_{\text{eff}} t}\Big]. \tag{5.67}$$

The result is only valid if molecular diffusion is negligible compared to dispersion,

$$D \ll D_{\text{eff}} \quad \Rightarrow \quad D \ll \frac{a^2 V_0^2}{48D} \quad \Rightarrow \quad \sqrt{48} \ll P\acute{e}. \tag{5.68}$$

Combining this with the earlier condition Eq. (5.63) we arrive at the domain of validity of the solution Eq. (5.67)

$$\sqrt{48} \ll P\acute{e} \ll 4\frac{L_0}{a}. \tag{5.69}$$

This inequality replaces the rougher estimate given in Eq. (5.57). In the improved theory by Aris (1956), the effective diffusion constant is found to be

$$D_{\text{eff}} = \left[1 + \frac{a^2 V_0^2}{48 D^2}\right] D = \left[1 + \frac{1}{48}\, Pé^2\right] D, \quad 1 \ll Pé. \tag{5.70}$$

Based on the effective diffusion coefficient D_{eff} we can introduce a time scale τ_{Taylor} for Taylor dispersion to broaden a sample to a width of w in a channel of radius a,

$$\tau_{\text{Taylor}} = \frac{w^2}{D_{\text{eff}}} = \frac{D}{48 V_0^2} \frac{w^2}{a^2}. \tag{5.71}$$

It is seen that for a given velocity the Taylor dispersion acts slower as the channel radius is decreased.

5.6 Exercises

Exercise 5.1
Constant-step random walk in 1D
Consider the random walk in 1D defined in Section 5.1 with constant step size $\Delta x_i = \pm\ell$.

(a) List all possible end-positions x_N for random walks with $N = 4$ steps, and note in how many ways each can be reached.

(b) Argue why the constant-step random walk with N steps can be described as a binomial distribution, and use this fact as an alternative way to calculate $\langle x_N \rangle$ and $\langle x_N^2 \rangle$.

Exercise 5.2
Constant-step, continuous-direction random walk in 2D and 3D
Consider a random walk in 2D like $\mathbf{R}_N$ of Eq. (5.11), but now allow for a step $\Delta\mathbf{r}_i$ of length ℓ in any direction, $\Delta\mathbf{r}_i = \ell\cos\theta_i\,\mathbf{e}_x + \ell\sin\theta_i\,\mathbf{e}_y$, given by the angle θ_i.

(a) Calculate the diffusion length ℓ_{diff}^{2D} in this model.

(b) Extend the model to 3D and calculate the corresponding diffusion length ℓ_{diff}^{3D}.

Exercise 5.3
The convection-diffusion equation
Verify that Eqs. (5.15) and (5.19) indeed lead to Eq. (5.20).

Exercise 5.4
Solutions to the diffusion equation
Study the various analytic solutions to the diffusion equation presented in Section 5.3 and prove that the solutions Eqs. (5.32), (5.39), and (5.41) indeed are solutions to the diffusion equation given the respective initial conditions Eqs. (5.31), (5.38), and (5.40).

Exercise 5.5
The Einstein relation linking diffusion to viscosity
Consider the Einstein relation Eq. (6.49) for the diffusion constant of a sphere.

(a) Estimate the diffusion constant D in water at room temperature for a rigid sphere with the same radius as a typical small ion. Comment on the result.

(b) Fluorescent latex spheres used for biodetection in lab-on-chip systems have typically a radius $a = 0.5$ µm. Estimate how long a time τ_{diff} it takes such a sphere to diffuse across a 100 µm wide water-filled microchannel at room temperature.

Exercise 5.6
Thermally induced jump rates for molecules in water

Locally, within a few atomic distances around a given H_2O molecule in water there is spatial order. The molecule is thus lying in the potential minimum created by the surrounding molecules. The molecule of mass M executes small harmonic oscillations of angular frequency $\omega = 2\pi f$. To jump to a neighboring site, a water molecule needs to overcome the potential barrier of height ΔE. It attempts to jump with the harmonic oscillator frequency f, but each attempt is only successful with the thermal probability factor $\exp(-\Delta E/k_\mathrm{B}T)$.

The frequency of the harmonic oscillations can be estimated as follows. Write the potential as $V(x) = \frac{1}{2}Kx^2$. The maximum of the barrier occurs at $x = \frac{1}{2}\,d$, so it is reasonable to put $V(\frac{1}{4}\,d) = \frac{1}{2}\,\Delta E$. Using this, show that the rate Γ for successful jumps is

$$\Gamma = \frac{2}{\pi d}\sqrt{\frac{\Delta E}{M}}\,\mathrm{e}^{-\Delta E/k_\mathrm{B}T}. \tag{5.72}$$

Exercise 5.7
The current density from thermally induced molecular jumping
Let the thickness of one molecular layer be denoted d.

(a) Argue that the particle current density in the x direction from a given layer situated at the plane $x = 0$ is $J_x(0) = \rho(0)d\,\Gamma$.

(b) Write a similar expression for the particle current density coming from the layer at $x = d$ going back to the first layer. Argue that the total current density is given by $J_x^\mathrm{tot} = J_x(0) - J_x(d) \approx -\partial_x\rho\, d^2\Gamma$.

(c) Show that this result gives the diffusion constant $D = d^2\Gamma$.

Exercise 5.8
A theoretical expression for the viscosity
Combine the results of the previous exercises and obtain the following expression for η:

$$\eta = \frac{k_\mathrm{B}T}{12ad}\sqrt{\frac{M}{\Delta E}}\,\mathrm{e}^{\Delta E/k_\mathrm{B}T}. \tag{5.73}$$

Compare this result with the experimental values for water given in Section A.1. Use the following parameter values: $a = 0.1$ nm, $d = 0.4$ nm, and M you determine yourself. The value for ΔE we estimate from the specific vaporization energy of water, $E_\mathrm{vap}/M = 2.26 \times 10^6$ J/kg (i.e. the latent heat for producing steam from boiling water at 100 °C): $\Delta E \approx \frac{1}{2}\,\frac{1}{2}\,E_\mathrm{vap}$. One factor $\frac{1}{2}$ is because the barrier height is roughly half the binding energy, and the other because on average there are two hydrogen bonds per molecule.

5.7 Solutions

Solution 5.1
Constant-step random walk in 1D
For each step in the constant step-size random walk in 1D there are two possibilities, so N steps result in 2^N paths. Any path has P positive steps $+\ell$, where $0 \le P \le N$, and $N - P$ negative steps $-\ell$, so the paths can end at $x_N/\ell = P - (N - P) = 2P - N$.

(a) For $N = 4$ steps there are $2^4 = 16$ paths ending at $x_4/\ell = -4, -2, 0, 2, 4$. The extreme points can each be reached in only one way: either all steps are negative or all are positive. The point $x_4/\ell = 2$ is reached after 3 positive and 1 negative step. There are four ways to place the negative step in the sequence, so the end point can be reached by four different paths. Similarly, there are four paths leading to $x_4/\ell = -2$. Now we have accounted for 10 of the 16 paths, which leaves six paths to end at $x_4/\ell = 0$.

(b) A N-step random walk consists of N consecutive and uncorrelated binary choices, $+\ell$ or $-\ell$. Let us consider $+\ell$ as the successful outcome occurring with probability $p = \frac{1}{2}$. This is the very definition of a binomial process with the probability distribution $f(P) = C_{N,P}\big(\frac{1}{2}\big)^N\big(1-\frac{1}{2}\big)^{(N-P)} = \frac{1}{2^N}C_{N,P}$ for successful outcome. here $C_{N,P} = N!/[P!(N-P)!]$ is the binomial coefficient. Well-known results are $\langle P\rangle = Np = \frac{1}{2}N$ and $\langle(P-\langle P\rangle)^2\rangle = Np(1-p) = \frac{1}{4}N$. As $x_N/\ell = 2P - N$ we obtain directly $\langle x_N\rangle = 0$ and $\langle x_N^2\rangle = N\ell^2$.

Solution 5.2
Constant-step, continuous-direction random walk in 2D and 3D
The ith step as $\Delta\mathbf{r}_i = \ell\mathbf{e}_i$, where $\mathbf{e}_i$ is a unit vector pointing in an arbitrary direction.

(a) $\langle\mathbf{R}_N\rangle = \langle\sum_i^N \mathbf{e}_i\rangle = \ell\sum_i^N\langle\mathbf{e}_i\rangle = \mathbf{0}$, as the unit vectors have random directions. Hence, the diffusion length in this model is given by $\big(\ell_{\text{diff}}^{2D}\big)^2 = \langle\mathbf{R}_N^2\rangle = \ell^2\langle\sum_i^N\mathbf{e}_i\cdot\sum_j^N\mathbf{e}_j\rangle = \ell^2\sum_{i,j}^N\langle\mathbf{e}_i\cdot\mathbf{e}_j\rangle = \ell^2\sum_i^N\langle\mathbf{e}_i\cdot\mathbf{e}_i\rangle = N\ell^2$. All the offdiagonal scalar-products average to zero due to the random direction of the unit vectors.

(b) We did not use the dimension in the previous argument, so the diffusion length in 3D is the same for the given model, $\ell_{\text{diff}}^{3D} = \sqrt{N}\ell$.

Solution 5.3
The convection-diffusion equation
We keep the total mass current density $\rho\mathbf{v}$ together as a unit, thus separating it from the solute concentration c_α. Carrying out the differentiation in Eq. (5.19) we get

$$\rho\partial_t c_\alpha + c_\alpha\partial_t\rho = -(\rho\mathbf{v})\cdot\boldsymbol{\nabla}c_\alpha - c_\alpha\boldsymbol{\nabla}\cdot(\rho\mathbf{v}) - \boldsymbol{\nabla}\cdot\mathbf{J}_\alpha^{\text{diff}}. \tag{5.74}$$

By using Eq. (5.15) $c_\alpha\partial_t\rho$ cancels $-c_\alpha\boldsymbol{\nabla}\cdot(\rho\mathbf{v})$, and we arrive at Eq. (5.20).

Solution 5.4
Solutions to the diffusion equation
To simplify the calculations below we note that for a function of the form $f(s) = As^p$ we can write its derivative as $\partial_s f(s) = \frac{p}{s}f(s)$ and $\partial_s\exp\big[f(s)\big] = (\partial_s f(s))\exp\big[f(s)\big] = \frac{p}{s}f(s)\exp\big[f(s)\big]$.

(a) Limited point-source diffusion. In 1D we have $\partial_t c = \big(-\frac{1}{2t} + \frac{x^2}{4Dt^2}\big)c$, $\partial_x c = -\frac{x}{2Dt}c$, and $\partial_x^2 c = \big(-\frac{1}{2Dt} + \frac{x^2}{4D^2t^2}\big)c$. From this we get $\partial_t c = D\partial_x^2 c$ as wanted. In 3D we can either use separation of variables in Cartesian co-ordinates, or use spherical polar co-ordinates, see Section C.3. For the latter, without angular dependence, we have $\nabla^2 c = \partial_r^2 c + \frac{2}{r}\partial_r c$. From Eq. (5.36) we find $\partial_r c = -\frac{r}{2Dt}c$, and thus $\partial_r^2 c = \big(-\frac{1}{2Dt} + \frac{r^2}{4D^2t^2}\big)c$, as well as $\partial_t c = \big(-\frac{3}{2}\frac{1}{t} + \frac{r^2}{4Dt^2}\big)c$. Hence, $\partial_t c = D\big(\partial_r^2 + \frac{2}{r}\partial_r\big)c$. The initial condition, Eq. (5.31), is fulfilled if $c(\mathbf{r}\neq\mathbf{0},0) = 0$ and $\int d\mathbf{r}\,c(\mathbf{r},0) = N_0$. Now, for $\mathbf{r}\neq\mathbf{0}$ Eq. (5.36) gives $c(\mathbf{r},t)\to 0$ exponentially fast for $t\to 0$, while for any $t>0$ we have $\int d\mathbf{r}\,c(\mathbf{r},t) = N_0(4\pi Dt)^{-\frac{3}{2}}\int_0^\infty dr\,4\pi r^2\exp\big(-\frac{r^2}{4Dt}\big) = N_0\frac{4}{\sqrt{\pi}}\int_0^\infty du\,u^2e^{-u^2} = N_0\frac{4}{\sqrt{\pi}}\Gamma\big(\frac{3}{2}\big) = N_0$.

(b) Limited planar-source diffusion. From Eq. (5.39) we find $\partial_x c = -\frac{x}{2Dt}c$, and thus $\partial_x^2 c = \big(-\frac{1}{2Dt} + \frac{x^2}{4D^2t^2}\big)c$, as well as $\partial_t c = \big(-\frac{1}{2}\frac{1}{t} + \frac{x^2}{4Dt^2}\big)c$. Hence, $\partial_t c = D\partial_x^2 c$. The initial condition Eq. (5.38) is fulfilled if $c(x>0,0) = 0$ and $\int_0^\infty dx\,c(x,0) = n_0$. Now, for

$x > 0$ Eq. (5.39) gives $c(x,t) \to 0$ exponentially fast for $t \to 0$, while for any t we have $\int_0^\infty \mathrm{d}x\, c(x,0) = n_0(\pi Dt)^{-\frac{1}{2}} \int_0^\infty \mathrm{d}x\, \exp\big(-\frac{x^2}{4Dt}\big) = n_0 \frac{2}{\sqrt{\pi}} \int_0^\infty \mathrm{d}u\, \mathrm{e}^{-u^2} = n_0 \frac{2}{\sqrt{\pi}} \Gamma\big(\frac{1}{2}\big) = n_0$.

(c) Constant planar-source diffusion. From Eq. (5.41) and the associated footnote for $\mathrm{erfc}(s)$ we find $\partial_x c = -c_0 \frac{1}{\sqrt{\pi D}} t^{-\frac{1}{2}} \exp\big(-\frac{x^2}{4Dt}\big)$, and thus $\partial_x^2 c = c_0 \frac{1}{\sqrt{\pi D}} \frac{x}{2D} t^{-\frac{3}{2}} \exp\big(-\frac{x^2}{4Dt}\big)$, as well as $\partial_t c = c_0 \frac{x}{2\sqrt{\pi D}} t^{-\frac{3}{2}} \exp\big(-\frac{x^2}{4Dt}\big)$. Hence, $\partial_t c = D\partial_x^2 c$. The boundary condition Eq. (5.40) is fulfilled by Eq. (5.41) since $c(0,t) = c_0 \frac{2}{\sqrt{\pi}} \int_0^\infty \mathrm{d}u\, \mathrm{e}^{-u^2} = c_0 \frac{2}{\sqrt{\pi}} \Gamma\big(\frac{1}{2}\big) = c_0$.

Solution 5.5
The Einstein relation linking diffusion to viscosity
We use $k_\mathrm{B}T = 4.14 \times 10^{-21}$ J and $\eta = 10^{-3}$ Pa s in the Einstein relation (6.49).

(a) For a hydrated ion we take the radius $a = 0.1$ nm and arrive at $D = 2.2 \times 10^{-9}$ m^2/s. According to Eq. (5.28a) this is very close to the experimental value.

(b) Taking $a = 0.5$ µm we find $D = 4.4 \times 10^{-13}$ m^2/s. With this diffusion constant the time it takes to diffuse $L = 100$ µm is $\tau_\mathrm{diff} = L^2/D = 2.3 \times 10^4$ s $= 6.3$ h.

Solution 5.6
Thermally induced jump rates for molecules in water
The rate for successful jumps is the product of the attempt rate f with the probability of success $\exp(-\Delta E/k_\mathrm{B}T)$, i.e. $\Gamma = f\exp(-\Delta E/k_\mathrm{B}T)$. So we just need to determine f, which is given by the oscillation frequency $f = \frac{1}{2\pi}\omega = \frac{1}{2\pi}\sqrt{K/M}$, where K is the force constant of the harmonic potential in which the molecule of mass M is moving. From the energy estimate we have $\frac{1}{2}\Delta E = V\big(\frac{1}{2}d\big) = \frac{1}{2}K\big(\frac{1}{4}d\big)^2$, which leads to $K = 16\frac{\Delta E}{d^2}$ and thus $f = \frac{2}{\pi d}\sqrt{\frac{\Delta E}{M}}$. Consequently $\Gamma = \frac{2}{\pi d}\sqrt{\frac{\Delta E}{M}} \exp(-\Delta E/k_\mathrm{B}T)$.

Solution 5.7
The current density from thermally induced molecular jumping
The surface density of molecules in one molecular layer of thickness d is ρd.

(a) The jumping rate Γ gives an estimate of how often a molecule jumps to one side, and consequently the current density is $J_x = \rho d\,\Gamma$.

(b) The total molecular current density in a plane between two neighboring molecular layers is given by the difference of the current densities coming from the two sides, i.e. $J_x^\mathrm{tot} = J_x(0) - J_x(d) = [\rho(0) - \rho(d)]\, d\,\Gamma = -\frac{1}{d}[\rho(d) - \rho(0)]\, d^2\Gamma \approx -\partial_x\rho\, d^2\Gamma$.

(c) Applying Fick's law, $J_x = -D\partial_x\rho$, to the previous result gives $D = d^2\Gamma$.

Solution 5.8
A theoretical expression for the viscosity
The Einstein relation and the previous analysis of the diffusion constant yields

$$\eta = \frac{k_\mathrm{B}T}{6\pi Da} = \frac{k_\mathrm{B}T}{6\pi d^2\Gamma a} = \frac{k_\mathrm{B}T}{6\pi d^2\Gamma a} = \frac{k_\mathrm{B}T}{12ad}\sqrt{\frac{M}{\Delta E}}\, \mathrm{e}^{\Delta E/k_\mathrm{B}T}. \tag{5.75}$$

Inserting into this expression the parameters listed in the exercise as well as the mass of a water molecule, $M = 18 \times 1.67 \times 10^{-27}$ kg yields $\eta = 0.7$ mPa s at room temperature.

6

Time-dependent flow

We continue the study of time-dependent phenomena and turn our attention to unsteady flow problems. Four selected topics will be treated: the onset of a Couette flow, the transient decay of a Poiseuille flow in a cylindrical microfluidic channel when a constant driving pressure is removed abruptly, the steady state of an incompressible fluid with harmonically oscillating boundary conditions, and the accelerated motion of a spherical body in a fluid. Not surprisingly, the mathematical treatment becomes more complex than previously.

6.1 Starting a Couette flow

In Section 3.3 we found the steady-state solution of the Couette flow in the gap of height h between a stationary bottom plate and a parallel top plate moving with the constant velocity $v_0\,\mathbf{e}_x$, see Fig. 3.3. Now we analyze how this steady-state velocity field is established given an instant change of speed of the top plate from zero to v_0 at time $t = 0$,

$$v_x(h,t) = \begin{cases} 0, & \text{for } t < 0, \\ v_0, & \text{for } t > 0. \end{cases} \tag{6.1}$$

The governing equations in steady state, Eq. (3.13), changes in the unsteady case to

$$\mathbf{v}(\mathbf{r},t) = v_x(z,t)\,\mathbf{e}_x, \tag{6.2a}$$

$$\eta\,\partial_z^2 v_x(z,t) = \rho\,\partial_t v_x(z,t). \tag{6.2b}$$

For $t > 0$ the boundary and initial conditions for $v_x(z,t)$ are

$$v_x(0,t) = 0, \qquad v_x(h,t) = v_0, \qquad v_x(z,0) = 0, \qquad v_x(z,\infty) = v_0\,\frac{z}{h}. \tag{6.3}$$

We solve this linear partial differential equation by the method of eigenfunction expansion introduced in Section 1.6. As eigenfunctions we use products of z- and t-dependent functions $v_n(z)T_n(t)$ to separate the variables,

$$v_x(z,t) = v_x(z,\infty) - \sum_n c_n\,v_n(z)\,T_n(t) = v_0\,\frac{z}{h} - \sum_n c_n\,v_n(z)\,T_n(t), \tag{6.4}$$

where the known steady-state solution $v_x(z,\infty)$ is taken explicitly into account, so that the eigenfunction expansion must vanish for $t \to \infty$ as well as for $z = 0$ and $z = h$. Inserting a single product term in the Navier–Stokes equation (6.2b) leads to

$$\frac{\partial_z^2 v_n(z)}{v_n(z)} = \frac{1}{\nu}\,\frac{\partial_t T_n(t)}{T_n(t)}, \tag{6.5}$$

where $\nu = \eta/\rho$ is the kinematic viscosity introduced in Eq. (5.44). Since the z variable appears solely on the left-hand side and the t variable on the right-hand side, this equation has non-trivial solutions only if the two sides equal the same constant, denoted $-\lambda_n$,

$$\frac{\partial_z^2 v_n(z)}{v_n(z)} = -\lambda_n = \frac{1}{\nu}\,\frac{\partial_t T_n(t)}{T_n(t)}. \tag{6.6}$$

The solutions for $v_n(z)$ and $T_n(t)$ are seen to be of the form

$$v_n(z) \propto \sin\Big(\sqrt{\lambda_n}\, z\Big), \tag{6.7a}$$

$$T_n(t) \propto \mathrm{e}^{-\lambda_n t}. \tag{6.7b}$$

The separation constant λ_n is determined by the conditions $T_n(\infty) = 0$, so $\lambda_n > 0$, and $v_n(0) = v_n(h) = 0$, so $\sqrt{\lambda_n} = n\pi/h$, where $n = 1, 2, 3, \ldots$. Finally, the expansion constants c_n are determined by the condition $v_x(z, 0) = 0$ or

$$\sum_{n=1}^{\infty} c_n \, \sin\Big(n\pi\,\frac{z}{h}\Big) = v_x(z,\infty) = v_0\,\frac{z}{h}. \tag{6.8}$$

This is a classical Fourier expansion problem, which is solved in detail in Section 12.1 for the mathematical equivalent problem of reaching thermal equilibrium in the presence of a temperature difference. Here, we just present the solution for the velocity field $v_x(z,t)$,

$$v_x(z,t) = v_0\left[\frac{z}{h} - \frac{2}{\pi}\sum_{n=1}^{\infty}\frac{1}{n}\,(-1)^{n+1}\,\sin\Big(n\pi\frac{z}{h}\Big)\,\exp\Big(-n^2\pi^2\frac{\nu}{h^2}\,t\Big)\right]. \tag{6.9}$$

From the full solution we see that due to the exponential decay the terms with large values of n quickly approaches zero. The last term to vanish is the one with $n = 1$, so we can establish the following approximation:

$$v_x(z,t) \approx v_0\left[\frac{z}{h} - \frac{2}{\pi}\sin\Big(\pi\frac{z}{h}\Big)\,\exp\Big(-\pi^2\frac{\nu}{h^2}\,t\Big)\right], \quad \text{for } t > \tau_1 \equiv \frac{h^2}{\pi^2\nu}. \tag{6.10}$$

It is seen that the steady-state Couette flow is established on a time scale given by τ_1. For times t larger than a few times τ_1 the deviations from the steady-state flow is exponential small. For water in a parallel-plate channel of height $h = 100$ µm we find

$$\tau_1 = \frac{h^2}{\pi^2\nu} \approx 1\ \text{ms}, \tag{6.11}$$

which is in agreement with the simple estimate leading to the momentum diffusion time T_0 in Eq. (5.45).

6.2 Stopping a Poiseuille flow by viscous forces

We continue with the analogous problem of starting and stopping a Poiseuille flow. We treat the latter case in the text, and leave the former case as an exercise for the reader. In Section 3.4.4 we analyzed the steady-state Poiseuille flow in a channel with a circular cross-section. In the following we study in detail what was already alluded to in Fig. 4.1, namely how such a flow decays and stops, when the pressure drop suddenly vanishes. We consider a finite section of length L of the infinite channel, and we assume that for time $t < 0$ a fully developed steady-state Poiseuille flow was present, driven by the pressure drop $p(0) = p^* + \Delta p$ and $p(L) = p^*$. Then at $t = 0$ the overpressure Δp is suddenly removed such that for $t > 0$ the pressure is given by $p(0) = p(L) = p^*$. In reality, the new pressure is not established instantly, but it is set up with the speed of sound in the liquid, typically of the order 10^3 m/s. However, this is much faster than the velocities obtained by the liquid, so it is a good approximation to assume that the new pressure is set up instantly in accordance with the usual assumption of incompressibility of the liquid. In the following we calculate how the velocity field of the liquid evolves in time for $t > 0$.

Due to the cylindrical symmetry the non-linear term $\rho(\mathbf{v}\cdot\boldsymbol{\nabla})\mathbf{v}$ in the Navier–Stokes equation remains zero, but in contrast to the steady-state version Eq. (3.41) we must now keep the explicit time derivative. Since $\Delta p = 0$ we arrive at

$$\rho\,\partial_t v_x(r,t) - \eta\left[\partial_r^2 + \frac{1}{r}\,\partial_r\right] v_x(r,t) = 0. \tag{6.12}$$

The boundary and initial conditions for $v_x(r,t)$ are

$$v_x(a,t) = 0, \qquad \partial_r v_x(0,t) = 0, \qquad v_x(r,0) = \frac{\Delta p}{4\eta L}\left(a^2 - r^2\right), \qquad v_x(r,\infty) = 0, \tag{6.13}$$

where we have utilized the fact that starting from the steady-state solution, Eq. (3.42a), the liquid ends at rest. Note that the azimuthal angle ϕ does not enter the problem.

The disappearance of the non-linear term and the appearance of a zero right-hand side makes Eq. (6.12) a homogeneous linear differential equation. As for the Couette flow, we do not set out to find the solution $v_x(r,t)$ directly, but instead we seek some simpler solutions $u_n(r,t)$, which can be used in an eigenfunction expansion

$$v_x(r,t) = \sum_n \tilde{c}_n\, u_n(r,t), \tag{6.14}$$

where $\tilde{c}_n$ are expansion coefficients to be determined. One particular class of solutions $u_n(r,t)$ to Eq. (6.12) can be found by separation of the variables using the following trial solution,

$$u_n(r,t) \equiv T_n(t)\,\tilde{u}_n(r). \tag{6.15}$$

Inserting this into Eq. (6.12) and dividing by $T_n(t)\,\tilde{u}_n(r)$ yields

$$\frac{1}{T_n(t)}\,\partial_t T_n(t) = \frac{\nu}{\tilde{u}_n(r)}\left[\partial_r^2 + \frac{1}{r}\,\partial_r\right]\tilde{u}_n(r). \tag{6.16}$$

As above, the t-dependent left-hand side can only equal the r-dependent right-hand side if the two sides equal the same constant $-\lambda_n$. Thus, we arrive at

$$\partial_t T_n(t) = -\lambda_n\, T_n(t), \tag{6.17a}$$

$$\Big[\partial_r^2 + \frac{1}{r}\,\partial_r\Big]\tilde{u}_n(r) = -\frac{\lambda_n}{\nu}\,\tilde{u}_n(r). \tag{6.17b}$$

The solutions to these standard differential equations are

$$T_n(t) = \exp\big(-\lambda_n t\big), \tag{6.18a}$$

$$\tilde{u}_n(r) = \tilde{c}_n^{(0)} J_0\Big(\sqrt{\tfrac{\lambda_n}{\nu}}\, r\Big) + \tilde{c}_n^{(1)} Y_0\Big(\sqrt{\tfrac{\lambda_n}{\nu}}\, r\Big), \tag{6.18b}$$

where $\tilde{c}_n^{(0)}$ and $\tilde{c}_n^{(1)}$ are constants, and where J_0 and Y_0 are Bessel functions of the first and second kind, respectively, both of order zero.[1]

To narrow down the possible solutions we use three of the four boundary conditions in Eq. (6.13). From $v_x(r,\infty) = 0$ it follows that $T_n(\infty) = 0$ and thus $\lambda_n > 0$. From $\partial_r v_x(0,t) = 0$ it follows that $\partial_r \tilde{u}_n(0) = 0$, so the Bessel function $Y_0(r)$, which diverges for $r \to 0$, must be excluded and thus $\tilde{c}_n^{(1)} = 0$. From $v_x(a,t) = 0$ it follows that $\tilde{u}_n(a) = 0$ and thus

$$\sqrt{\tfrac{\lambda_n}{\nu}}\, a = \gamma_n, \quad \text{where } J_0(\gamma_n) \equiv 0, \quad n = 1,2,3,\ldots \tag{6.19}$$

Here, we have introduced the countable number of roots γ_n of the Bessel function J_0.[2] This provides us with a complete set of basis functions that can be used to express any solution of Eq. (6.12) in the form of a Fourier–Bessel eigenfunction expansion

$$v_x(r,t) = \sum_{n=1}^{\infty} \tilde{c}_n^{(0)} J_0\Big(\gamma_n\, \tfrac{r}{a}\Big)\, \exp\Big(-\gamma_n^2 \tfrac{\nu}{a^2}\, t\Big). \tag{6.20}$$

The unknown coefficients $\tilde{c}_n^{(0)}$ are determined by the third boundary condition in Eq. (6.13) for $u(r,0)$,

$$u(r,0) = \sum_{n=1}^{\infty} \tilde{c}_n^{(0)} J_0\Big(\gamma_n\, \tfrac{r}{a}\Big) \equiv \frac{\Delta p}{4\eta L}\,\big(a^2 - r^2\big). \tag{6.21}$$

Introducing the dimensionless co-ordinate $\rho = r/a$, multiplying Eq. (6.21) by $\rho J_0\big(\gamma_m\, \rho\big)$, integrating over ρ, and using the orthogonality relation

$$\int_0^1 \mathrm{d}\rho\, \rho\, J_0\big(\gamma_m\, \rho\big) J_0\big(\gamma_n\, \rho\big) = \tfrac{1}{2}\, \big[J_1(\gamma_m)\big]^2\, \delta_{mn} \tag{6.22}$$

for the Bessel functions $J_0\big(\gamma_n\, \rho\big)$, we can calculate the coefficient $\tilde{c}_m^{(0)}$,

$$\tilde{c}_m^{(0)} = \frac{a^2 \Delta p}{2\eta L}\, \frac{1}{\big[J_1(\gamma_m)\big]^2} \int_0^1 \mathrm{d}\rho\, \big(\rho - \rho^3\big)\, J_0\big(\gamma_m\, \rho\big) = \frac{2a^2 \Delta p}{\eta L}\, \frac{1}{\gamma_m^3\, J_1(\gamma_m)}. \tag{6.23}$$

Note that the Bessel function J_1 of order 1 now appears.[3]

[1] The solution, Eq. (6.18b), is the cylindrical co-ordinate analog of the Cartesian case:

$$\partial_x^2 \tilde{u}_n(x) = -\tfrac{\lambda_n}{\nu}\, \tilde{u}_n(x) \quad \Rightarrow \quad \tilde{u}_n(x) = \tilde{c}_n^{(0)} \sin\Big(\sqrt{\tfrac{\lambda_n}{\nu}}\, x\Big) + \tilde{c}_n^{(1)} \cos\Big(\sqrt{\tfrac{\lambda_n}{\nu}}\, x\Big).$$

[2] The first four roots of $J_0(\gamma_i) = 0$ are $\gamma_1 = 2.405$, $\gamma_2 = 5.520$, $\gamma_3 = 8.654$, and $\gamma_4 = 11.792$.

[3] The integral in Eq. (6.23) is calculated by using $\int \mathrm{d}x\, x J_0(x) = x J_1(x)$ and the recursive formula $\int \mathrm{d}x\, x^k J_0(x) = x^k J_1(x) + (k-1)x^{k-1} J_0(x) - (k-1)^2 \int \mathrm{d}x\, x^{k-2} J_0(x)$.

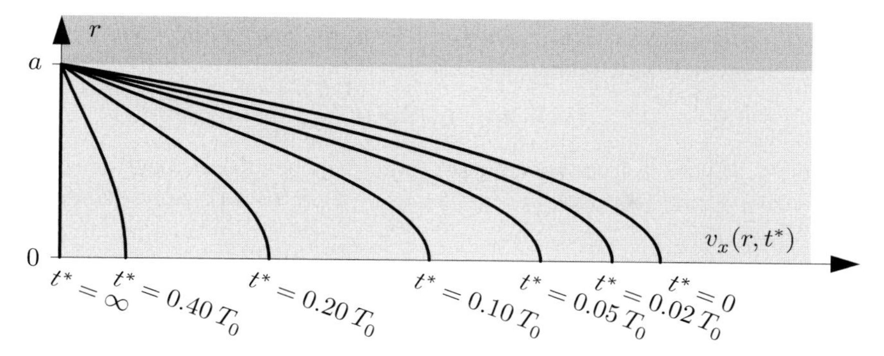

Fig. 6.1 The evolution in time of the velocity profile $v_x(r,t)$ in a cylindrical channel with radius a for a decelerating Poiseuille flow due to the abrupt disappearance of the driving pressure Δp at $t = 0$. The time is expressed in units of the momentum diffusion time $T_0 = a^2/\nu$. The velocity profile is shown at seven different times t^* spanning from the paraboloid velocity profile at $t^* = 0$ to the zero-velocity field at $t^* = \infty$. The infinite sum in Eq. (6.24) has been truncated at $n = 19$.

The final result for the velocity field $v_x(r,t)$ of a stopping Poiseuille flow can now be obtained by combining Eqs. (6.20) and (6.23),

$$v_x(r,t) = \frac{a^2\Delta p}{4\eta L}\sum_{n=1}^{\infty}\frac{8}{\gamma_n^3\,J_1(\gamma_n)}\,J_0\Big(\gamma_n\,\tfrac{r}{a}\Big)\,\exp\Big(-\gamma_n^2\tfrac{\nu}{a^2}\,t\Big). \tag{6.24}$$

The velocity profile at different times during the evolution of the full Poiseuille flow paraboloid is shown in Fig. 6.1. Note that the time scale t_{acc} characteristic for the deceleration basically is the momentum diffusion time $T_0 = a^2/\nu$, but more accurately it is determined by the exponential factor containing the smallest Bessel function root $\gamma_1 = 2.405$. In the case of water in a typical microfluidic channel with radius $a = 100$ µm we get

$$t_{\mathrm{acc}} = \frac{1}{\gamma_1^2}\,\frac{a^2}{\nu} = \frac{1}{\gamma_1^2}\,T_0 \approx 2\times 10^{-3}\ \mathrm{s}. \tag{6.25}$$

6.3 Flow induced by slowly oscillating boundaries

In microfluidic systems the liquids can be manipulated by different harmonically oscillating forces at the boundaries of the system. In Section 14.5 non-zero slip velocities are established by surface electrodes with an AC voltage bias, and in Section 15.4 velocities are induced by piezo-acoustic elements mounted on the channel walls. So, it is of interest to study general aspects of flow induced by slowly oscillating boundaries.

We postpone the study of effects related to the non-zero compressibility to Chapter 15 on acoustofluidics, and confine our analysis here to fluids moving with velocities v much slower than the speed of sound c_{a} of the medium, i.e. at very low Mach numbers $M = v/c_{\mathrm{a}} \ll 1$. So in the following all fluids are incompressible. Moreover, for simplicity, we consider fluids occupying the half-infinite space $z > 0$ bounded by the xy-plane at $z = 0$. On this boundary the fluid is given a slip velocity $\mathbf{v}_s$ in the form of a standing wave in the x direction varying harmonically in time,

$$\mathbf{v}(x,y,0,t) = \mathbf{v}_s(x,y,t) \equiv v_0\cos(kx)\,\mathrm{e}^{-\mathrm{i}\omega t}\,\mathbf{e}_x, \tag{6.26}$$

where $k = 2\pi/\lambda$ is a constant wave number corresponding to the wavelength λ of the standing wave. Far away from the boundary at $z = 0$ the fluid is at rest,

$$\mathbf{v}(x, y, \infty, t) \equiv \mathbf{0}. \tag{6.27}$$

The parameters v_0, k, and ω from the velocity boundary condition together with the fluid parameters, density ρ and viscosity η, can be used to introduce dimensionless variables marked by a tilde,

$$\mathbf{v} = v_0\, \tilde{\mathbf{v}}, \tag{6.28a}$$

$$\mathbf{r} = \frac{1}{k}\, \tilde{\mathbf{r}}, \tag{6.28b}$$

$$t = \frac{1}{\omega}\, \tilde{t}, \tag{6.28c}$$

$$p = k v_0 \eta\, \tilde{p}. \tag{6.28d}$$

The parameters also define two dimensionless numbers characteristic of the system, the Reynolds number *Re* measuring the importance of inertia relative to viscous forces, and a variant of the Péclet number, measuring the momentum diffusion time relative to the convection time by oscillation,

$$Re = \frac{v_0}{k\nu}, \tag{6.29a}$$

$$P\acute{e} = \frac{\omega}{k^2\nu}. \tag{6.29b}$$

In the rest of this section we use dimensionless variables, so it is convenient simply to drop the tilde. Given the harmonic time dependence the dimensionless Navier–Stokes equation becomes

$$-\mathrm{i}P\acute{e}\,\mathbf{v} + Re(\mathbf{v}\cdot\boldsymbol{\nabla})\mathbf{v} = -\boldsymbol{\nabla}p + \nabla^2\mathbf{v}, \tag{6.30}$$

which for the low Reynolds numbers in microfluidics reduces to the Stokes equation (2.42),

$$\nabla^2\mathbf{v} + \mathrm{i}P\acute{e}\,\mathbf{v} = \boldsymbol{\nabla}p, \quad \text{for } Re \ll 1. \tag{6.31}$$

All fields have the same trivial time dependence $\exp(-\mathrm{i}t)$, which therefore is suppressed in the following. There is complete translation invariance along the y axis, so neither the velocity field nor the pressure field depend on y, and the y component of the velocity is zero. This leaves us with a two-dimensional incompressible flow, which most conveniently can be solved by introducing the so-called stream function $\psi(x, z)$, by which the non-zero velocity components are given by

$$\begin{pmatrix} v_x(x,z) \\ v_z(x,z) \end{pmatrix} = \begin{pmatrix} \partial_z\psi(x,z) \\ -\partial_x\psi(x,z) \end{pmatrix}. \tag{6.32}$$

The stream function guarantees the fulfilment of the continuity equation,

$$\boldsymbol{\nabla}\cdot\mathbf{v} = \partial_x v_x + \partial_z v_z = \partial_x\partial_z\psi + \partial_z(-\partial_x\psi) = 0. \tag{6.33}$$

The first step on the way to determine the stream function $\psi(x, z)$ passes via the pressure field. Given the continuity equation (6.33) and the commutivity of the divergence operator and the Laplace operator, the result of taking the divergence of the Stokes equation (6.31) is, as expected from Eq. (2.45), a Laplace equation for the pressure,

$$\nabla^2 p = 0. \tag{6.34}$$

Since the system is driven by the standing wave given by Eq. (6.26), the dimensionless x dependence of p must contain a linear combination of $\cos(x)$ and $\sin(x)$, and consequently to fulfill the Laplace equation, the z dependence must contain a factor $\exp(-z)$. We are therefore left to guess a stream function of the form, which keeps a simple standing wave along x and allows a more complicated dependence of z,

$$\psi(x,z) = \cos(x)\, f(z)\, \mathrm{e}^{-z}. \tag{6.35}$$

This stream function leads to velocity components of the form

$$v_x(x,z) = \cos(x)\,\big[f'(z) - f(z)\big]\,\mathrm{e}^{-z}, \tag{6.36a}$$

$$v_z(x,z) = \sin(x)\, f(z)\,\mathrm{e}^{-z}, \tag{6.36b}$$

and the velocity boundary conditions therefore imply

$$f(0) = 0, \qquad f'(0) = 1, \qquad f(\infty) = 0. \tag{6.37}$$

The left-hand side of the Navier–Stokes equation (6.31) results in

$$\nabla^2 v_x + \mathrm{i}P\acute{e}\, v_x = \cos(x)\,\big[f'''(x) - 3f''(x) + (2+\mathrm{i}P\acute{e})f'(z) - \mathrm{i}P\acute{e}\, f(z)\big]\,\mathrm{e}^{-z}, \tag{6.38a}$$

$$\nabla^2 v_z + \mathrm{i}P\acute{e}\, v_z = \sin(x)\,\big[f''(x) - 2f'(x) + \mathrm{i}P\acute{e}\, f(z)\big]\,\mathrm{e}^{-z}. \tag{6.38b}$$

For this to equal a pressure gradient, it is reasonable to guess at

$$p(x,z) = A\,\sin(x)\,\mathrm{e}^{-z}, \tag{6.39}$$

which transforms the Navier–Stokes equation into

$$f'''(x) - 3f''(x) + (2+\mathrm{i}P\acute{e})f'(z) - \mathrm{i}P\acute{e}\, f(z) - A = 0, \tag{6.40a}$$

$$f''(x) - 2f'(x) + \mathrm{i}P\acute{e}\, f(z) + A = 0. \tag{6.40b}$$

A simple differential equation for $f(z)$ is obtained by addition of these two equations,

$$f'''(x) - 2f''(z) + \mathrm{i}P\acute{e}\, f'(z) = 0. \tag{6.41}$$

This equation is easily solved by the *ansatz* $f(z) = B\big[1 - \mathrm{e}^{\alpha z}\big]$, and it is found that $\alpha = 1 - \sqrt{1 - \mathrm{i}P\acute{e}}$.

Collecting all partial results we end with the dimensionless expressions for the stream function, the velocity components and the pressure,

$$\delta \equiv \sqrt{1 - \mathrm{i}P\acute{e}}, \tag{6.42a}$$

$$\psi(x,z) = \frac{1}{1-\delta}\,\cos(x)\,\Big[\mathrm{e}^{-\delta z} - \mathrm{e}^{-z}\Big], \tag{6.42b}$$

$$v_x(x,z) = \frac{-1}{1-\delta}\,\cos(x)\,\Big[\delta\mathrm{e}^{-\delta z} - \mathrm{e}^{-z}\Big], \tag{6.42c}$$

$$v_z(x,z) = \frac{1}{1-\delta}\,\sin(x)\,\Big[\mathrm{e}^{-\delta z} - \mathrm{e}^{-z}\Big], \tag{6.42d}$$

$$p(x,z) = \frac{\mathrm{i}P\acute{e}}{1-\delta}\,\sin(x)\,\mathrm{e}^{-z}. \tag{6.42e}$$

This result can be a little difficult to interpret, so to obtain a simpler result, we find its limiting value as we let $Pé \to 0$, e.g. by letting the frequency approach zero,

$$\nabla^2 \mathbf{v} = \boldsymbol{\nabla} p, \quad \text{for } Re, Pé \ll 1, \tag{6.43a}$$

$$\psi(x,z) = \cos(x)\, z\, \mathrm{e}^{-z}, \tag{6.43b}$$

$$v_x(x,z) = \cos(x)\,(1-z)\, \mathrm{e}^{-z}, \tag{6.43c}$$

$$v_z(x,z) = \sin(x)\, z\, \mathrm{e}^{-z}, \tag{6.43d}$$

$$p(x,z) = \sin(x)\, 2\, \mathrm{e}^{-z}. \tag{6.43e}$$

It is not difficult to realize that the resulting flow field contains oscillating vortex structures. The alternating flow direction along the x axis at a given time was explicitly stated by the boundary condition. The vortices rotate around vortex centers $\mathbf{r}_m$ defined by the criterion $\mathbf{v}(\mathbf{r}_m) = \mathbf{0}$, and this is seen to be fulfilled for $\mathbf{r}_m = (m\pi, 1)$, where $m = 0, \pm1, \pm2, \pm3, \ldots$. Fluid leaves the boundary along lines at $x = (m + \frac{1}{2})\pi$ and enters it along lines at $x = (m - \frac{1}{2})\pi$. In Fig. 14.6 is shown a vector plot of the resulting velocity field.

By using Fourier series and the linearity of the equations, the result obtained can be generalized from the cosine-wave boundary condition $v_x(x,y,0,t) = \cos(x)\,\mathrm{e}^{-\mathrm{i}t}$ to any periodic function $v(x)\,\mathrm{e}^{-\mathrm{i}t} = \sum_{n=1}^{\infty} v_n \cos(nx)\,\mathrm{e}^{-\mathrm{i}t}$ with periodicity of unity, see Exercise 6.1.

6.4 Accelerated motion of a spherical body in a liquid

As mentioned in Section 3.7 many applications of lab-on-a-chip systems involve the motion of small objects, such as magnetic beads, fluorescent markers, or biological cells, inside the microfluidic channels. In the following, we study two aspects beyond steady-state motion of spherical bodies, namely simple acceleration and Brownian motion. As before we restrict our analysis to rigid bodies.

6.4.1 A spherical body approaching steady state in a liquid

We begin by studying the acceleration of a sphere with radius a and mass $(4\pi/3)a^3\rho_{\mathrm{sph}}$ as it approaches steady state in a fluid. Initially, both the sphere and the fluid are at complete rest. Suddenly, at time $t = 0$ a constant external force $F_{\mathrm{ext}}\,\mathbf{e}_x$ begins to act on the sphere. As the force is constant all motion in the following takes place along the direction given by $\mathbf{e}_x$, and the resulting velocity of the sphere is denoted $\mathbf{u}(t) = u(t)\,\mathbf{e}_x$. The equation of motion for the sphere becomes

$$\tfrac{4}{3}\pi a^3 \rho_{\mathrm{sph}} \partial_t u = -6\pi\eta a\, u + F_{\mathrm{ext}}. \tag{6.44}$$

The solution to this standard differential equation is

$$u(t) = \frac{F_{\mathrm{ext}}}{6\pi\eta a} - u_0 \exp\Big(-\frac{9\eta}{2\rho_{\mathrm{sph}}a^2}\,t\Big), \tag{6.45}$$

where u_0 is an integration constant that needs to be specified by the boundary conditions. If the sphere is at rest for $t = 0$ then

$$u(t) = \frac{F_{\mathrm{ext}}}{6\pi\eta a}\Big[1 - \exp\Big(-\frac{9\eta}{2\rho_{\mathrm{sph}}a^2}\,t\Big)\Big]. \tag{6.46}$$

The characteristic time scale τ_{acc} in the exponential is seen to be very small for a microsystem. For a cell the density almost equals that of water, while $a \approx 5\ \mu\mathrm{m}$. This yields

$$\tau_{\mathrm{acc}} = \frac{2\rho_{\mathrm{sph}} a^2}{9\eta} \approx 5\ \mu\mathrm{s}. \tag{6.47}$$

Thus, in a viscous environment inertial forces are indeed negligible, and for the case of the microsphere it is reasonable to assume that it is always moving in local steady state.

6.4.2 A diffusing spherical body and the Einstein relation

Another accelerated motion of a sphere is the random diffusion or Brownian motion. Here, we shall derive the very useful Einstein relation, which gives the diffusion constant D of the Brownian motion in terms of the parameters of the liquid and the sphere.

A sphere of radius a moving with velocity $\mathbf{v}$ through a liquid of viscosity η experiences the Stokes drag force $\mathbf{F}_{\mathrm{drag}}$. Consider a position-dependent solution of density $\rho(\mathbf{r})$ of spherical molecules in the same liquid. Due to gradients in the density these molecules will diffuse according to Fick's law, $\mathbf{J} = -D\,\boldsymbol{\nabla}\rho$. Since the chemical potential μ by definition is the free energy of the last added molecule, the force $\mathbf{F}_{\mathrm{diff}}$ driving the diffusion is given by minus the gradient of μ,

$$\mathbf{F}_{\mathrm{diff}} = -\boldsymbol{\nabla}\mu. \tag{6.48}$$

In steady state the forces from diffusion and drag exactly balance each other. Combining this force balance with Fick's law and the, in Appendix D derived, thermodynamic relation $\mu(T,\rho) = \mu_0 + k_{\mathrm{B}}T\ln(\rho/\rho_0)$, where the subscript 0 refers to some constant standard concentration, we arrive at the Einstein relation,

$$D = \frac{k_{\mathrm{B}}T}{6\pi a\eta}. \tag{6.49}$$

Here, k_{B} is Boltzmann's constant, and it is useful to note that at room temperature

$$k_{\mathrm{B}}T = 1.3805\times10^{-23}\ \mathrm{J/K}\times 300\ \mathrm{K} = 4.14\times10^{-21}\ \mathrm{J}. \tag{6.50}$$

For a microbead with $a = 0.5\ \mu\mathrm{m}$ frequently used in microfluidics and for an ion-sized bead with $a = 0.1\ \mathrm{nm}$ diffusing in water at 300 K, we find from Eq. (6.49)

$$D_{\mathrm{bead}}(0.5\ \mu\mathrm{m}, 300\ \mathrm{K}) = 4.4\times10^{-13}\ \mathrm{m^2\,s^{-1}}, \tag{6.51a}$$

$$D_{\mathrm{bead}}(0.1\ \mathrm{nm}, 300\ \mathrm{K}) = 2.2\times10^{-9}\ \mathrm{m^2\,s^{-1}}. \tag{6.51b}$$

6.5 Other time-dependent flows

There will be more examples of time-dependent phenomena in the coming chapters. In Sections 7.3 and 7.4 we study transient flow in capillary rise and capillary pumps driven by the surface tension. This theme is taken up again under nanofluidics in Section 17.2, and in the same chapter, Section 17.3, the transient squeeze flow is studied in the context of nanoimprint lithography. A number of time-dependent phenomena driven by electric forces are given in Sections 8.3.3 and 14.5 concerning electrolytes as well as in Section 10.6 involving dielectric forces on suspended particles. Thermal transients are treated briefly in Section 12.1, and finally the acousto- and optofluidic wave equation is the topic of Chapters 15 and 16, respectively.

6.6 Exercises

Exercise 6.1
Slowly oscillating boundary with an arbitrarily shaped standing wave
The solution to the flow problem involving an oscillating boundary with a cosine-shaped standing wave Eq. (6.26) is presented in Eq. (6.43) using dimensionless variables in the limit of low Reynolds and Péclet number.

(a) Change the wave number from k to an integer n times k and write down the solution using the full physical dimensions.

(b) Change the boundary condition from a single cosine-shaped standing wave to a general standing wave shape given by the Fourier series

$$\mathbf{v}(x,y,0,t) = \left[\sum_{n=1}^{\infty} v_n \cos(nkx)\right] \mathrm{e}^{-\mathrm{i}\omega t}\,\mathbf{e}_x, \tag{6.52}$$

and write down the corresponding solution to the flow problem Eq. (6.43a).

Exercise 6.2
Starting a Poiseuille flow in a circular channel
Consider the setup given in Section 6.2 for the decelerating Poiseuille flow in a circular channel, but invert the problem, so that the starting point is a liquid at complete rest for $t < 0$ and no pressure drop, $p(0) = p(L) = p^*$. Then suddenly at $t = 0$ a constant pressure drop is applied such that $p(0) = p^* + \Delta p$ and $p(L) = p^*$ for $t > 0$. Determine the velocity field $u_x(r,t)$ for the accelerating Poiseuille flow.

Exercise 6.3
Starting a Poiseuille flow in a parallel-plate channel
Redo the previous exercise but change the cross-section of the channel from circular to that of an infinite parallel-plate channel of height h.

6.7 Solutions

Solution 6.1
Slowly oscillating boundary with an arbitrarily shaped standing wave
The transformation between dimensionless and physical variables is given by Eq. (6.28), except now k is changed into nk.

(a) Using physical variables and changing k to nk, Eq. (6.43) becomes

$$\mathbf{v}_n(x,z,t) = \begin{pmatrix} v_n \cos(nkx)\,(1-nkz)\,\mathrm{e}^{-nkz} \\ v_n \sin(nkx)\,nkz\,\mathrm{e}^{-nkz} \end{pmatrix} \mathrm{e}^{-\mathrm{i}\omega t}, \tag{6.53a}$$

$$p_n(x,z,t) = 2nkv_n\eta \sin(nkx)\,\mathrm{e}^{-nkz}\,\mathrm{e}^{-\mathrm{i}\omega t}. \tag{6.53b}$$

(b) In the limit of low Reynolds and Péclet numbers the Navier–Stokes equation is linear, and we can obtain the general solutions $\mathbf{v}$ and p for the velocity and pressure field by adding the partial solutions with wave number nk given in (a) above,

$$\mathbf{v}(x,y,z,t) = \sum_{n=1}^{\infty} \mathbf{v}_n(x,z,t), \quad \text{and} \quad p(x,y,z,t) = \sum_{n=1}^{\infty} p_n(x,z,t), \tag{6.54}$$

which by inspection is seen to fulfil the boundary condition Eq. (6.52).

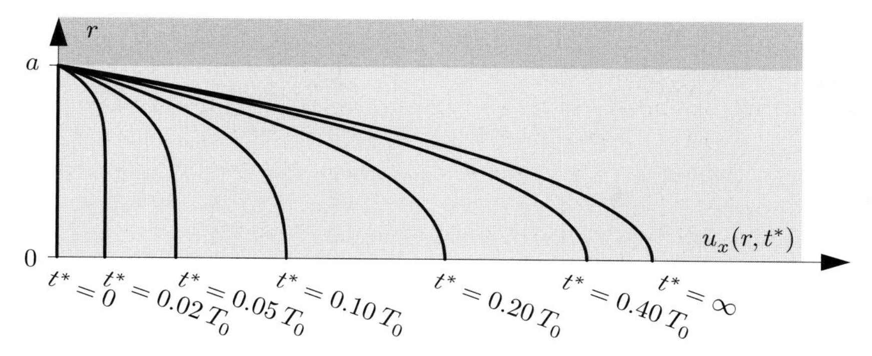

Fig. 6.2 The evolution in time of the velocity profile $u_x(z,t)$ in a circular channel with radius a for a Poiseuille flow under acceleration due to the abrupt appearance of the driving pressure Δp at $t = 0$. The time is expressed in units of the momentum diffusion time $T_0 = a^2/\nu$. The velocity profile is shown at seven different times t^* spanning from the zero-velocity profile at $t^* = 0$ to the fully developed paraboloid shape at $t^* = \infty$.

Solution 6.2
Starting a Poiseuille flow in a circular channel
The Navier–Stokes equation becomes an inhomogeneous, linear partial differential equation

$$\rho\,\partial_t u_x(r,t) - \eta\left[\partial_r^2 + \frac{1}{r}\,\partial_r\right]u_x(r,t) = \frac{\Delta p}{L}. \tag{6.55}$$

A particular solution to this equation fulfilling the no-slip boundary conditions is of course the well-known steady-state solution, $u_\infty(r) = \frac{\Delta p}{4\eta L}\big(a^2 - r^2\big)$. Moreover, $v_x(r,t)$ of Eq. (6.24) is a solution of the corresponding homogeneous equation, Eq. (6.12), with the initial condition $v_x(r,0) = u_\infty(r)$. Therefore, $u_x(r,t) = u_\infty - v_x(r,t)$ is a solution to our problem, Eq. (6.55), with the appropriate boundary/initial conditions $u_x(a,t) = 0$, $\partial_r u_x(0,t) = 0$, $u_x(r,0) = 0$, and $u_x(r,\infty) = u_\infty$. The explicit expression for $u_x(r,t)$ is

$$u_x(r,t) = \frac{a^2\Delta p}{4\eta L}\left[1 - \frac{r^2}{a^2} - \sum_{n=1}^{\infty}\frac{8}{\gamma_n^3\,J_1(\gamma_n)}\,J_0\Big(\gamma_n\,\tfrac{r}{a}\Big)\,\exp\Big(-\gamma_n^2\tfrac{\nu}{a^2}\,t\Big)\right]. \tag{6.56}$$

In Fig. 6.2 is shown the time evolution of the velocity profile from rest to the fully developed paraboloid flow.

Solution 6.3
Starting Poiseuille flow in a parallel-plate channel
The analysis of the time evolution of a Poiseuille flow for an infinite parallel-plate channel begins with the Navier–Stokes equation analogous to Eq. (6.55),

$$\rho\,\partial_t u_x(z,t) - \eta\,\partial_z^2 v_x(z,t) = \frac{\Delta p}{L}. \tag{6.57}$$

The boundary conditions for $u_x(z,t)$ are

$$u_x(h,t) = 0, \qquad u_x(0,t) = 0, \qquad u_x(z,0) = 0, \qquad u_x(z,\infty) = \frac{\Delta p}{2\eta L}\,(h-z)z, \tag{6.58}$$

where we have utilized that the steady-state solution, Eq. (3.29), will be reached in the limit $t \to \infty$. In analogy with the previous exercise the full solution $u_x(z,t)$ can therefore be written as a sum

$$u_x(z,t) \equiv u_x(z,\infty) - v_x(z,t) \tag{6.59}$$

of the particular solution $u_x(z,\infty)$ and a general solution $v_x(z,t)$ to the corresponding homogeneous equation. When inserting Eq. (6.59) into Eq. (6.57) we obtain the homogeneous differential equation that $v_x(z,t)$ has to satisfy, compare with Eq. (6.12),

$$\partial_t v_x(z,t) - \nu\, \partial_z^2 v_x(z,t) = 0, \tag{6.60}$$

with $\nu = \eta/\rho$. The boundary conditions for $v_x(z,t)$ follow from Eqs. (6.58) and (6.59),

$$v_x(h,t) = 0, \qquad v_x(0,t) = 0, \qquad v_x(z,0) = \frac{\Delta p}{2\eta L}\,(h-z)z, \qquad v_x(z,\infty) = 0. \tag{6.61}$$

To proceed, we do not set out to find the solution $v_x(z,t)$ directly, but instead we seek some simpler solutions $u_n(z,t)$, which can be used in an eigenfunction expansion

$$v_x(z,t) = \sum_n \tilde{c}_n\, u_n(z,t), \tag{6.62}$$

where $\tilde{c}_n$ are some expansion coefficients. One particular class of solutions $u_n(z,t)$ to Eq. (6.60) can be found by separation of the variables using the following trial solution,

$$u_n(z,t) \equiv T_n(t)\, \tilde{u}_n(z). \tag{6.63}$$

Inserting this into Eq. (6.60) and dividing by $T_n(t)\,\tilde{u}_n(z)$ yields

$$\frac{1}{T_n(t)}\, \partial_t T_n(t) = \frac{\nu}{\tilde{u}_n(z)}\, \partial_z^2 \tilde{u}_n(z). \tag{6.64}$$

The t-dependent left-hand side can only equal the z-dependent right-hand side if the two sides equal the same constant $-\lambda_n$. Thus, we arrive at

$$\partial_t T_n(t) = -\lambda_n\, T_n(t), \tag{6.65a}$$

$$\partial_z^2 \tilde{u}_n(z) = -\frac{\lambda_n}{\nu}\, \tilde{u}_n(z). \tag{6.65b}$$

The solutions to these standard differential equations are

$$T_n(t) = \exp\big(-\lambda_n t\big), \tag{6.66a}$$

$$\tilde{u}_n(z) = \tilde{c}_n^{(0)} \sin\Big(\sqrt{\tfrac{\lambda_n}{\nu}}\, z\Big) + \tilde{c}_n^{(1)} \cos\Big(\sqrt{\tfrac{\lambda_n}{\nu}}\, z\Big), \tag{6.66b}$$

where $\tilde{c}_n^{(0)}$ and $\tilde{c}_n^{(1)}$ are constants.

To narrow down the possible solutions we use three of the four boundary conditions in Eq. (6.61). From $v_x(z,\infty) = 0$ follows $T_n(\infty) = 0$ and thus $\lambda_n > 0$. The sine-term is identical zero for $z = 0$, so no-slip $v_x(0,t) = 0$ at $z = 0$ can only be maintained if the cosine-term is

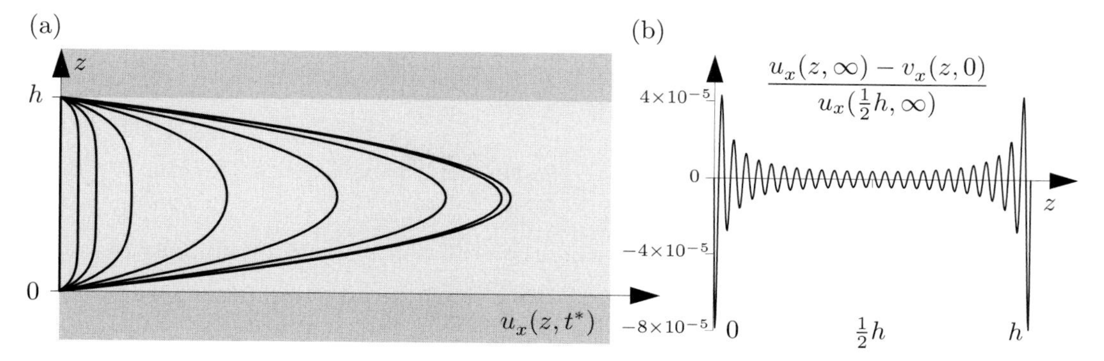

Fig. 6.3 (a) The evolution in time of the velocity profile $u_x(z,t)$ in an infinite parallel-plate channel with height h for a Poiseuille flow under acceleration due to the abrupt appearance of the driving pressure Δp at $t=0$. The time is expressed in units of the momentum diffusion time $T_0 = h^2/\nu$. The velocity profile is shown at seven different times t^* spanning from the zero-velocity profile at $t^* = 0$, through $t^*/T_0 = 0.005, 0.01, 0.02, 0.05, 0.1, 0.2$, and 0.4 to the fully developed parabolic shape at $t^* = \infty$. (b) The relative difference between the full parabolic velocity field $u_x(z,\infty)$, taking into account only the terms in Eq. (6.68) with $n < 50$, and $v_x(z,0) = \frac{\Delta p}{2\eta L}\,(h-z)z$.

excluded by putting $\tilde{c}_n^{(1)} = 0$. Further, no-slip at $z = h$ imposes the following constraint on the argument of the sine-term,

$$\sqrt{\tfrac{\lambda_n}{\nu}}\,h = n\pi, \quad n = 1,2,3,\ldots. \tag{6.67}$$

Here, $n\pi$ is the countable number of roots of the sine function. This provides us with a complete set of basis functions that can be used to express any solution of Eq. (6.60) in the form of a Fourier sine series

$$v_x(z,t) = \sum_{n=1}^{\infty} \tilde{c}_n^{(0)} \sin\Big(n\pi\,\tfrac{z}{h}\Big)\,\exp\Big(-n^2\pi^2\tfrac{\nu}{h^2}\,t\Big). \tag{6.68}$$

The unknown coefficients $\tilde{c}_n^{(0)}$ are determined by the third boundary condition in Eq. (6.61) for $v_x(z,0)$,

$$v_x(z,0) = \sum_{n=1}^{\infty} \tilde{c}_n^{(0)} \sin\Big(n\pi\tfrac{z}{h}\Big) \equiv \frac{\Delta p}{2\eta L}\,(h-z)z. \tag{6.69}$$

Introducing the dimensionless co-ordinate $\zeta = z/h$, multiplying Eq. (6.69) by $\sin\big(m\pi\,\zeta\big)$, integrating over ζ, and using the orthogonality relation

$$\int_0^1 d\zeta\,\sin\big(m\pi\,\zeta\big)\sin\big(n\pi\,\zeta\big) = \tfrac{1}{2}\,\delta_{mn} \tag{6.70}$$

for the sine functions $\sin(n\pi\zeta)$, we can calculate the coefficient $\tilde{c}_m^{(0)}$,

$$\tilde{c}_m^{(0)} = \frac{h^2\Delta p}{\eta L}\int_0^1 d\zeta\,\big(\zeta-\zeta^2\big)\sin\big(m\pi\,\zeta\big) = \begin{cases} 0, & m \text{ even}, \\ \dfrac{4h^2\Delta p}{\eta L}\,\dfrac{1}{(m\pi)^3}, & m \text{ odd}. \end{cases} \tag{6.71}$$

Note that only odd values of m yield non-zero contributions.[4]

The final result for the velocity field $u_x(z,t)$ of a starting Poiseuille flow can now be obtained by combining Eqs. (6.59), (6.68), and (6.71),

$$u_x(z,t) = \frac{h^2 \Delta p}{2\eta L}\left[\left(1-\frac{z}{h}\right)\frac{z}{h} - \sum_{n,\text{odd}}^{\infty} \frac{8}{(n\pi)^3} \sin\left(n\pi\,\tfrac{z}{h}\right) \exp\left(-n^2\pi^2\tfrac{\nu}{h^2}\,t\right)\right]. \tag{6.72}$$

The velocity profile at different times during the evolution of the full parabolic Poiseuille flow is shown in Fig. 6.3.

[4]The integral in Eq. (6.71) is calculated by using the two relations $\int \mathrm{d}x\; x\sin x = \sin x - x\cos x$ and $\int \mathrm{d}x\; x^2 \sin x = 2x\sin x + (2-x^2)\cos x$.

7 Capillary effects

One of the characteristic features of microfluidics is the dominance of surface effects due to the large surface to bulk ratio on the micrometer scale. A prominent class of surface effects are known as capillary effects, named after the latin word *capillus* for hair, since, as we shall see, they are particularly strong in microchannels having bore diameters equal to or less than the width of a human hair, which is about 50 µm.

The capillary effects can be understood by studying Gibbs free energy G, the energy of systems where the thermodynamic control parameters are pressure p, temperature T, and particle number N. In particular, we shall be interested in equilibrium or quasi-equilibrium situations, where the Gibbs free energy per definition is at a minimum. As an example, let the system under consideration consist of two subsystems divided by a free surface at equilibrium. The total Gibbs energy G of the system is then given as a sum of several energy contributions G_i such as the free energy of each of the two subsystems and the free energy of the surface. Let the free surface of the system be given in terms of some variable ξ such as position, volume, or geometrical shape, and let the equilibrium value be given by $\xi = \xi_0$. Variations $\xi = \xi_0 + \delta\xi$ away from the equilibrium value ξ_0 must result in a vanishing variation δG of the free energy, because if G could vary, the system would spontaneously change ξ_0 to obtain a lower free energy contradicting the assumption that ξ_0 is the equilibrium value. This can be formulated mathematically as

$$\delta G = \partial_\xi G \, \delta\xi = \left(\sum_i \partial_\xi G_i \right) \delta\xi = 0, \quad \text{(at equilibrium)}. \tag{7.1}$$

This expression will be used in the following to establish the governing equations for capillary effects.

7.1 Surface tension

A central concept in the theory of surfaces is the surface tension. The surface tension depends on the two materials on each side of the surface.

7.1.1 Definition of surface tension

The surface tension γ of an interface[1] is defined as the Gibbs free energy per area for fixed pressure and temperature,

[1] In the literature, surface tension is normally denoted γ or σ. To avoid confusion with the stress tensor, γ will be used throughout this book.

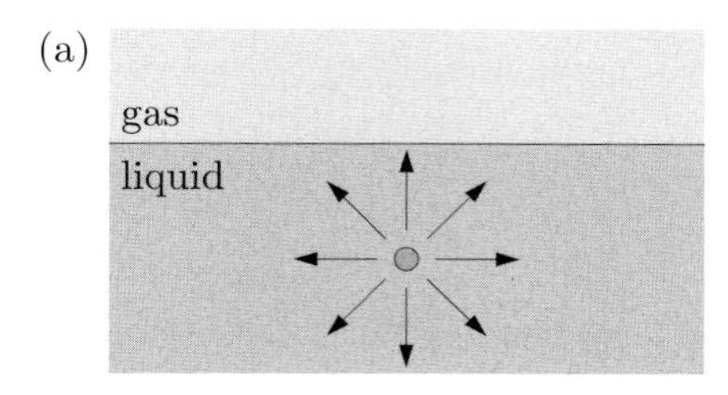

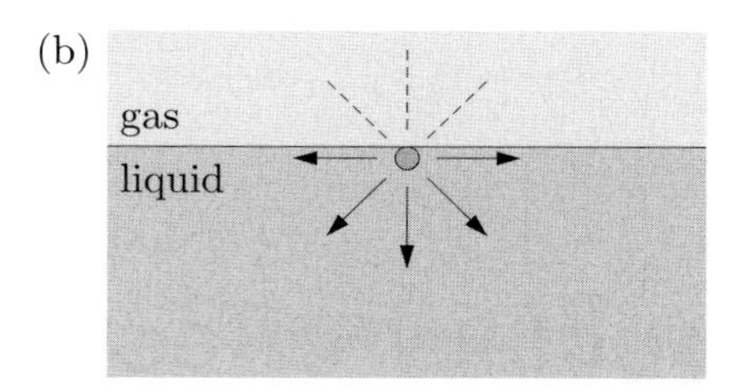

Fig. 7.1 The origin of surface tension for a liquid/gas interface. (a) A molecule in the bulk of the liquid forms (nonpermanent) chemical bonds (arrows) with the neighboring molecules surrounding it. (b) A molecule at the surface of the liquid misses the chemical bonds in the direction of the surface (dashed lines). Consequently, the energy of surface molecules is higher than that of bulk molecules, and the formation of such an interface costs energy.

$$\gamma \equiv \left(\frac{\partial G}{\partial \mathcal{A}}\right)_{p,T}. \tag{7.2}$$

The SI unit of γ is therefore

$$[\gamma] = \mathrm{J\,m^{-2}} = \mathrm{N\,m^{-1}} = \mathrm{Pa\,m}. \tag{7.3}$$

A microscopic model for surface tension between a liquid and a gas is sketched in Fig. 7.1. A molecule in the bulk forms (nonpermanent or fluctuating) chemical bonds with the neighboring molecules thus gaining a certain amount of binding energy. A molecule at the surface cannot form as many bonds since there are almost no molecules in the gas. This lack of chemical bonds results in a higher energy for the surface molecules. This is exactly the surface tension: it costs energy to form a surface. Using this model it is easy to estimate the order of magnitude of surface tension for a liquid/gas interface. A molecule in the bulk has roughly six nearest neighbors (think of a cubic geometry). A surface molecule has only five, missing the one above it in the gas. The area covered by a single molecule is roughly $\mathcal{A} \approx (0.3\ \mathrm{nm})^2$, see Fig. 1.2, while a typical intermolecular bond ΔE in a liquid is of the order of a couple of thermal energies, $\Delta E \approx 2k_\mathrm{B}T \approx 0.8 \times 10^{-20}$ J, see Eq. (6.50). This yields

$$\gamma \approx \frac{2k_\mathrm{B}T}{\mathcal{A}} = \frac{0.8 \times 10^{-20}\ \mathrm{J}}{(3 \times 10^{-10}\ \mathrm{m})^2} \approx 90\ \mathrm{mJ\,m^{-2}}. \tag{7.4}$$

The measured value for the water/air interface at 20 ℃ is 72.9 mJ/m^2, see Table 7.1.

Surface tension can also be interpreted as a force per length having the unit N/m $=$ J/m^2. This can be seen by considering a flat rectangular surface of length L and width w. If we keep the width constant while stretching the surface the amount ΔL from L to $L + \Delta L$, an external force F must act to supply the work $\Delta G = F\Delta L$ necessary for creating the new surface area $w\Delta L$ containing the energy $\Delta G = \gamma w \Delta L$,

$$\frac{F}{w} = \frac{1}{w}\frac{\Delta G}{\Delta L} = \frac{1}{w}\frac{\gamma\, w\Delta L}{\Delta L} = \gamma. \tag{7.5}$$

7.1.2 The Young–Laplace pressure across curved interfaces

An important consequence of a non-zero surface tension is the presence of the so-called Young–Laplace pressure drop Δp_surf across a curved interface in thermodynamical equilibrium. The expression for Δp_surf is derived using the energy minimum condition Eq. (7.1).

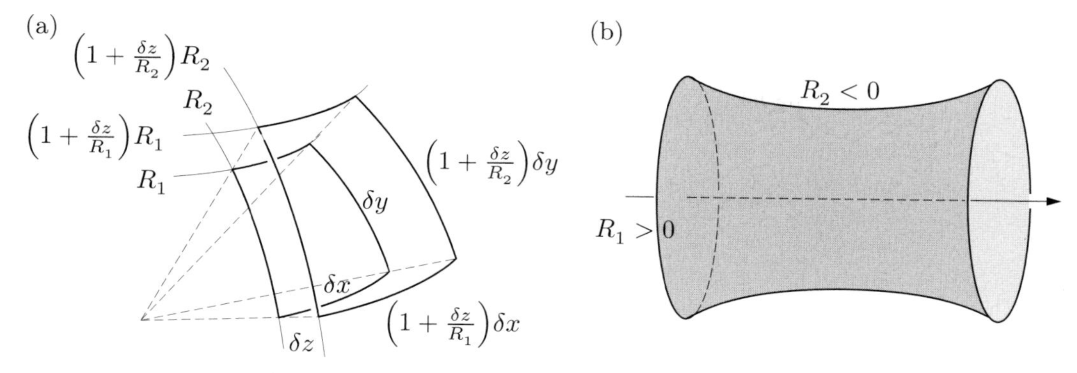

Fig. 7.2 (a) The displacement by the amount δz of a small section of a curved surface with area $\mathcal{A} = \delta x\, \delta y$. The local radii of curvature changes from R_i to $R_i + \delta z = (1 + \delta z/R_i)R_i$, $i = 1, 2$ thus changing the area from $\mathcal{A}$ to $(1+\delta z/R_1)(1+\delta z/R_2)\mathcal{A}$. (b) A sketch of a soap film suspended by two circular frames with open ends. The pressures inside and outside are equal, so $\Delta p_\mathrm{surf} = 0$ implying $1/R_1 + 1/R_2 = 0$. Here, the radius of curvature is positive, $R_1 > 0$, in the azimuthal direction and negative, $R_2 < 0$, in the axial direction.

Consider a small piece of the curved surface with the area $\mathcal{A} = \delta x\, \delta y$ in equilibrium as sketched in Fig. 7.2(a). We now study the consequences of expanding the area through a small displacement δz in the direction parallel to the local normal vector of the surface. The two local radii of curvature in the x and the y directions thus change from R_i to $R_i + \delta z = (1 + \delta z/R_i)R_i$, $i = 1, 2$. The side lengths δx and δy are changed similarly, leading to a change in area from $\mathcal{A}$ to $(1+\delta z/R_1)(1+\delta z/R_2)\mathcal{A}$. Neglecting terms of order $(\delta z)^2$ the area has therefore been enlarged by the amount $\delta\mathcal{A}$ given by

$$\delta\mathcal{A} \approx \left(\frac{\delta z}{R_1} + \frac{\delta z}{R_2}\right)\mathcal{A}. \tag{7.6}$$

If we disregard any influence of gravity there will only be two contributions to the change δG of the free energy of the system: an increase in surface energy G_surf due to an increased area, and a decrease in pressure-volume energy G_pV due to the increase in volume. In this case Eq. (7.1) becomes

$$\delta G = \delta G_\mathrm{surf} + \delta G_\mathrm{pV} = \gamma\, \delta\mathcal{A} - \big[\mathcal{A}\, \delta z\big]\, \Delta p_\mathrm{surf} = 0. \tag{7.7}$$

Inserting Eq. (7.6) in this expression and isolating the pressure drop yields the Young–Laplace equation

$$\Delta p_\mathrm{surf} = \left(\frac{1}{R_1} + \frac{1}{R_2}\right)\gamma. \tag{7.8}$$

It is important to note the sign convention used here: the pressure is highest in the convex medium, i.e. the medium where the centers of the curvature circles are placed. An example illustrating more complex signs of the curvatures is shown in Fig. 7.2(b), where a thin soap film supported by two coaxial, circular frames with open ends is analyzed. The open ends result in equal pressures inside and outside, whence the Young–Laplace pressure drop is zero, $\Delta p_\mathrm{surf} = 0$, which by Eq. (7.8) implies a vanishing mean curvature, $1/R_1 + 1/R_2$.

Table 7.1 Measured values of the surface tension γ at liquid/vapor interfaces and of the static contact angle θ at liquid–solid–air contact lines for clean surfaces. All values are at 20 ℃.

liquid	γ [mJ/m^2]	liquid	solid	θ
water	72.9	water	glass	0°
mercury	486.5	water	gold	0°
benzene	28.9	water	Si wafer	22°
methanol	22.5	water	PMMA	72°
ethanol	23.0	water	teflon	115°
glycerol	63.0	ethanol	glass	0°
blood	~60.0	mercury	glass	140°

The solution is a film with a positive curvature in the azimuthal direction and a negative curvature in the axial direction.

When using the Navier–Stokes equation to analyze the flow of two immiscible fluids, 1 and 2, the Young–Laplace pressure appears as a boundary condition at the interface between the two fluids. In the direction of the surface normal $\mathbf{n}$ the difference between the stresses $\sigma^{(1)}$ and $\sigma^{(2)}$ of the two fluids, see Eq. (2.26), must equal Δp_{surf} to avoid the existence of unphysical forces of infinite magnitude,

$$-\left(p^{(1)} - p^{(2)}\right) n_i + \left(\sigma_{ik}^{(1)} - \sigma_{ik}^{(2)}\right) n_k = \left(\frac{1}{R_1} + \frac{1}{R_2}\right) \gamma^{(12)}. \tag{7.9}$$

The mean curvature $\kappa(\mathbf{r})$ of a surface in a point $\mathbf{r}$ is seen to play a central role in the theory of surface tension. In more precise terms it is defined as

$$\kappa(\mathbf{r}) \equiv \kappa_1(\mathbf{r}) + \kappa_2(\mathbf{r}) = \frac{1}{R_1(\mathbf{r})} + \frac{1}{R_2(\mathbf{r})}, \tag{7.10}$$

where κ_1 and κ_2 are the curvatures of two curves on the surface that intersect at the point $\mathbf{r}$ perpendicular to each other, and where $R_1(\mathbf{r})$ and $R_2(\mathbf{r})$ are their respective radii of curvature. For a curve $\mathbf{r}(s)$ described by the parameter s, the curvature $\kappa(s)$ is defined in differential geometry by

$$\kappa(s) = \frac{|\mathbf{r}'(s) \times \mathbf{r}''(s)|}{|\mathbf{r}'(s)|^3}, \tag{7.11}$$

where the prime denotes differentiation with respect to s. If one interprets s as time, the expression for the curvature can by understood from a physical point of view, since $\mathbf{r}'$ and $\mathbf{r}''$ can be identified with the instantaneous velocity $\mathbf{v}$ and acceleration $\mathbf{a}$, respectively. For a particle in a circular orbit of radius $\mathbf{r}$ we know from classical mechanics that $|\mathbf{a}| = |\mathbf{v}|^2/r$ or $1/r = |\mathbf{a}|/|\mathbf{v}|^2$, but $1/r$ is exactly the curvature, so Eq. (7.11) really states that $\kappa = |\mathbf{e}_v \times \mathbf{a}|/|\mathbf{v}|^2$, where $\mathbf{e}_v \equiv \mathbf{v}/|\mathbf{v}|$ is a unit vector in the tangential direction. The sign of the curvature is given by the direction of the vector $\mathbf{r}'(s) \times \mathbf{r}''(s)$.

Alternatively, the mean radius of curvature can be defined as the surface divergence of the surface normal $\mathbf{n}$,

$$\kappa = \boldsymbol{\nabla}_s \cdot \mathbf{n}, \tag{7.12}$$

where $\boldsymbol{\nabla}_s = \boldsymbol{\nabla} - \mathbf{n}(\mathbf{n}\cdot\boldsymbol{\nabla})$ is the gradient operator along the surface.

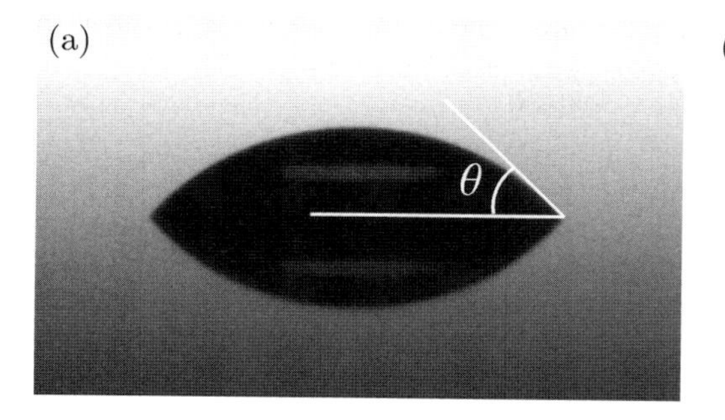

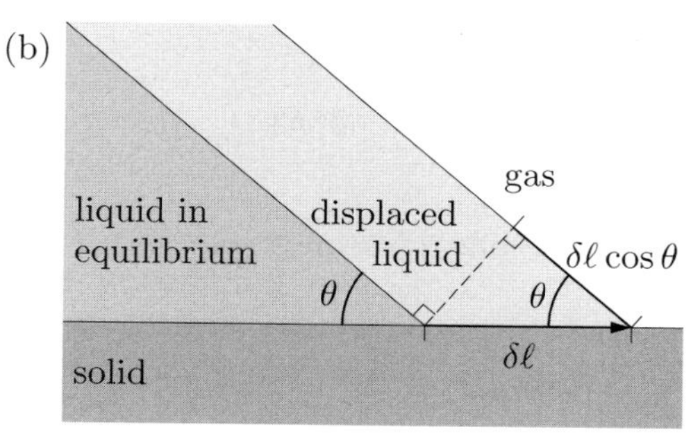

Fig. 7.3 (a) The contact angle θ is defined as the angle between the solid/liquid and the liquid/gas interfaces at the contact line. The picture is taken from a measurement of the contact angle of a water drop on a pure (and reflecting) silicon dioxide substrate in air showing $\theta = 52.3°$. (b) A sketch of the small displacement $\delta\ell$ of the contact line away from the equilibrium position. The changes of the interface areas are proportional to $+\delta\ell$, $+\delta\ell\cos\theta$, and $-\delta\ell$ for the solid/liquid, liquid/gas, and solid/gas interfaces, respectively.

7.2 Contact angle

Another fundamental concept in the theory of surface effects in microfluidics is the contact angle that appears at the contact line between three different phases, typically the solid wall of a channel and two immiscible fluids inside that channel. The two concepts, contact angle and surface tension, allow for understanding the capillary forces that act on two-fluid flows inside microchannels in lab-on-a-chip systems.

7.2.1 Definition of the contact angle

The contact angle θ is defined as the angle between the solid/liquid and the liquid/gas interfaces at the contact line where three immiscible phases meet, as illustrated in Fig. 7.3(a). In equilibrium θ is determined by the three surface tensions γ_{sl}, γ_{lg}, and γ_{sg} for the solid/liquid, liquid/gas and solid/gas interfaces by Young's equation to be discussed in the following subsection. Some typical values for contact angles θ are listed in Table 7.1.

Whereas the contact angle is well defined in equilibrium it turns out to depend in a complicated way on the dynamical state of a moving contact line. One can for example observe that the contact angle at the advancing edge of a moving liquid drop on a substrate is different from that at the receding edge.

7.2.2 Young's equation; surface tensions and contact angle

To derive an expression for the contact angle in equilibrium we again use the free-energy minimum condition Eq. (7.1). We consider the system sketched in Fig. 7.3(b), where in equilibrium a flat interface between a liquid and a gas forms the angle θ with the surface of a solid substrate. Imagine now that the liquid/gas interface is tilted an infinitesimal angle around an axis parallel to the contact line and placed far away from the substrate interface. As a result the contact line is moved the distance $\delta\ell$ while keeping the contact angle θ. To order $\delta\ell$ the only change in free energy comes from the changes in interface areas near the contact line. It is easy to see from Fig. 7.3(b) that the changes of the interface areas are proportional to $+\delta\ell$, $+\delta\ell\cos\theta$, and $-\delta\ell$ for the solid/liquid, liquid/gas, and solid/gas interfaces, respectively. The energy balance at equilibrium, Eq. (7.1), for the Gibbs energy per unit length $\frac{1}{w}\,\delta G$ along the contact line becomes,

$$\frac{1}{w}\,\delta G = \gamma_{\mathrm{sl}}\delta\ell + \gamma_{\mathrm{lg}}\delta\ell\cos\theta - \gamma_{\mathrm{sg}}\delta\ell = 0, \tag{7.13}$$

which after simple rearrangements gives Young's equation for the contact angle θ,

$$\cos\theta = \frac{\gamma_{\mathrm{sg}} - \gamma_{\mathrm{sl}}}{\gamma_{\mathrm{lg}}}. \tag{7.14}$$

Systems with contact angles $\theta < 90°$ are called hydrophilic (water loving), while those with $\theta > 90°$ are called hydrophobic (water fearing).

7.3 Capillary length and capillary rise

In the previous discussion we have neglected gravity, an approximation that turns out to be very good in many cases for various microfluidic systems. Consider for example an incompressible liquid of volume Ω with a free liquid/air interface $\partial\Omega$. The equilibrium shape of the liquid will be determined by minimizing the free energy G consisting of the surface energy and the gravitational potential energy of the bulk,

$$G_{\mathrm{min}} = \min_{\Omega}\left\{\gamma\int_{\partial\Omega} da \;+\; \rho\,g\int_{\Omega} d\mathbf{r}\; z\right\}, \tag{7.15}$$

under the constant-volume constraint $\int_\Omega d\mathbf{r} = \text{const}$. Here, the gravitational acceleration is taken in the negative z direction, $\mathbf{g} = -g\,\mathbf{e}_z$. The equilibrium shape for a free liquid drop in zero gravity is a sphere, since the sphere has the minimal area for a given volume.

We see from Eq. (7.15) that the shape problem is governed by a characteristic length, the so-called capillary length ℓ_{cap},

$$\ell_{\mathrm{cap}} \equiv \sqrt{\frac{\gamma}{\rho\,g}}, \tag{7.16}$$

which for the water/air interface at 20 ℃ takes the value

$$\ell_{\mathrm{cap}}^{\mathrm{water/air}} = \sqrt{\frac{0.073\ \mathrm{J/m^2}}{1000\ \mathrm{kg/m^3}\ 9.81\ \mathrm{m/s^2}}} = 2.7\ \mathrm{mm}. \tag{7.17}$$

Since $a \ll \ell_{\mathrm{cap}} \Rightarrow \rho\,g \ll \gamma/a^2$, gravity does not influence the shape of free water/air interfaces in microfluidic systems of sizes a well below 1 mm. In Fig. 7.4(a) is shown that insects on the sub-3-mm scale benefit from the large surface tension to gravity ratio.

The understanding of the concept of capillary length can also be used to analyze the so-called capillary rise that happens in narrow, vertically standing microchannels. Capillary rise can be observed as sketched in Fig. 7.4(b) by dipping one end of a narrow open-ended tube into some liquid. The liquid will rise inside the tube until it reaches equilibrium at some height H above the zero level $z = 0$ defined as the flat liquid level far away from the tube. The task is to determine H.

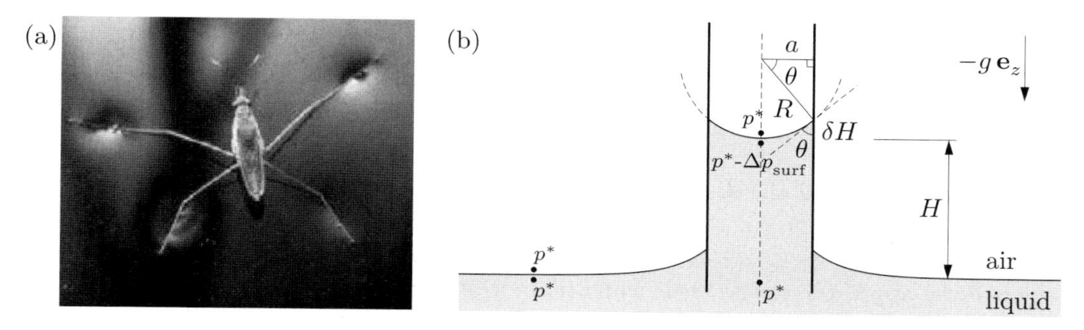

Fig. 7.4 (a) The importance of surface tension for microsystems illustrated by an insect able to walk on water. The gravitational force is balanced by the surface tension of the water/air interface. (b) Capillary rise in a vertically standing cylindrical microchannel.

7.3.1 Capillary rise height

For simplicity we consider a vertically placed microtube with a circular cross-section of radius $a \ll \ell_{\mathrm{cap}}$. The vertical direction is denoted $\mathbf{e}_z$ and gravity is $\mathbf{g} = -g\,\mathbf{e}_z$. The contact angle of the tube–liquid–air system is denoted θ and the surface tension of the liquid/air interface is called γ. Because $a \ll \ell_{\mathrm{cap}}$ and because the tube is circular the liquid/air surface of minimal energy inside the tube will be spherical. Thus, the two radii of curvature are identical, and from the geometry of Fig. 7.4(b) we find

$$R_1 = R_2 \equiv R = \frac{a}{\cos\theta}. \tag{7.18}$$

Because the liquid/air interface is curved, there will be a non-zero Young–Laplace pressure drop Δp_{surf} across it. Following the sign convention of Eq. (7.8) the pressure is higher in the convex air volume just above the interface as compared to the pressure $p_{\mathrm{liq}}(H)$ in the concave liquid volume just below the interface. Since the pressure of the air is standard atmospheric pressure p_0, we find

$$p_{\mathrm{liq}}(H) = p_0 - \Delta p_{\mathrm{surf}} = p_0 - \frac{2\gamma}{R} = p_0 - \frac{2\gamma}{a}\,\cos\theta. \tag{7.19}$$

The pressure $p_{\mathrm{liq}}(0)$ at $z = 0$ inside the liquid far away from the tube is p_0 because the Young–Laplace pressure across a flat surface is zero, but according to Eq. (3.3) it is also given in terms of the hydrostatic pressure generated by the liquid above $z = 0$,

$$p_0 = p_{\mathrm{liq}}(0) = p_{\mathrm{liq}}(H) + \rho g H. \tag{7.20}$$

Combining Eqs. (7.19) and (7.20) yields the equilibrium height H of the capillary rise,

$$H = \frac{2\gamma}{\rho g a}\,\cos\theta = 2\,\frac{\ell_{\mathrm{cap}}^2}{a}\,\cos\theta = \frac{2}{\rho g a}\,(\gamma_{\mathrm{sg}} - \gamma_{\mathrm{sl}}). \tag{7.21}$$

Quite significant rise heights can be obtained in microchannels. From Table 7.1 we find $H = 4.2$ cm for water in a 100 µm radius PMMA polymer channel, and $H = 42$ cm for $a = 10$ µm.

Because it is relatively easy to measure accurately the geometrical quantities a, H, and $\cos\theta$, Eq. (7.21) is one of the most accurate ways to measure surface tension,

$$\gamma = \frac{\rho g}{2}\,\frac{aH}{\cos\theta}. \tag{7.22}$$

7.3.2 Capillary rise time

After having established the equilibrium height H that the meniscus of the liquid reaches by capillary rise inside a vertically placed tube with circular cross-section, we shall now calculate the approximate rise time. Let $L(t)$ be the height of the liquid column inside the tube at time t. Equilibrium is reached as $t \to \infty$ so $L(\infty) = H$. By mass conservation the speed $\mathrm{d}L/\mathrm{d}t$ by which the liquid rises must be given by the average velocity $V_0 = Q/(\pi a^2)$ of the vertical liquid flow inside the tube of radius a. If, for simplicity, we assume that the liquid flow is a fully developed Poiseuille flow, we can express the flow rate Q by Eq. (3.39b) and obtain

$$\frac{\mathrm{d}L(t)}{\mathrm{d}t} = V_0 = \frac{Q}{\pi a^2} \approx \frac{a^2\Delta p(t)}{8\eta}\,\frac{1}{L(t)}. \tag{7.23}$$

The pressure drop $\Delta p(t)$ between $z = 0$ and $z = L(t)$ induced by viscous friction in the rising liquid column must equal the Young–Laplace pressure drop across the meniscus minus the decreasing hydrostatic pressure of the liquid column,

$$\Delta p(t) = \Delta p_{\mathrm{surf}} - \rho\, g\, L(t). \tag{7.24}$$

When inserting this into Eq. (7.23) with the explicit expression Eq. (7.19) for Δp_{surf} we obtain a first-order ordinary differential equation for the rise height $L(t)$,

$$\frac{\mathrm{d}L(t)}{\mathrm{d}t} = \frac{\gamma}{8\eta}\left[2a\cos\theta\,\frac{1}{L(t)} - \frac{\rho\, g a^2}{\gamma}\right] = \frac{\rho g a^2}{8\eta}\left[\frac{H}{L(t)} - 1\right]. \tag{7.25}$$

To facilitate the analysis the differential equation is made dimensionless,

$$t = \tau_{\mathrm{cap}}\,\tilde{t}, \qquad \text{where} \quad \tau_{\mathrm{cap}} \equiv \frac{8\eta H}{\rho g a^2}, \tag{7.26a}$$

$$L = H\,\tilde{L}, \tag{7.26b}$$

$$\frac{\mathrm{d}\tilde{L}(\tilde{t})}{\mathrm{d}\tilde{t}} = \frac{1}{\tilde{L}(\tilde{t})} - 1, \quad \tilde{L}(0) = 0, \quad \tilde{L}(\infty) = 1. \tag{7.26c}$$

At small times $\tilde{t} \ll 1$ we have $\tilde{L} \ll 1$, so $d\tilde{L}/d\tilde{t} \approx 1/\tilde{L}$. This is easily integrated to give

$$\tilde{L}(\tilde{t}) = \sqrt{2\tilde{t}}, \qquad \tilde{t} \ll 1. \tag{7.27}$$

For large times $\tilde{t} \to \infty$ we have $\tilde{L} \to 1$ from below. Thus, we can write $\tilde{L} = 1 - \delta\tilde{L}$, where $0 \le \delta\tilde{L} \ll 1$. Inserting this in Eq. (7.26c) we get $-d(\delta\tilde{L})/d\tilde{t} = 1/(1-\delta\tilde{L}) - 1 \approx \delta\tilde{L}$, which implies $\delta\tilde{L} \propto \exp(-\tilde{t})$. We therefore arrive at

$$\tilde{L}(\tilde{t}) = 1 - A\,\exp\big(-\tilde{t}\,\big), \qquad \tilde{t} \gg 1, \tag{7.28}$$

where A is some undetermined constant. Going back to physical dimensions we can conclude that the meniscus in capillary rise initially advances as the square root of time, but on the time scale $\tau_{\rm cap}$ it crosses over to approach the equilibrium height H asymptotically as an exponential saturation with the same time scale $\tau_{\rm cap}$ as the characteristic time in the exponent. Recalling from the discussion of Eq. (7.21) that for water in a PMMA tube of radius $a = 100$ µm we have $H = 4.2$ cm, the value for $\tau_{\rm cap}$ in this case becomes

$$\tau_{\rm cap}^{\rm water/air} = 3.4 \text{ s}. \tag{7.29}$$

This value is good news, because the whole calculation of the capillary rise time was made under the assumption that the Poiseuille flow profile was fully developed, and as we know from Eq. (6.25) this profile is established on the much smaller time scale of 2 ms. The result for the capillary rise time is thus consistent with the assumption of the calculation.

7.3.3 Capillary rise and dimensionless numbers

We end the section on capillary rise by mentioning three dimensionless numbers that often are used to characterize the phenomenon.

When some characteristic length scale a is established for a system, the Bond number Bo of the system can be introduced,

$$Bo = \frac{\text{gravitational force}}{\text{surface tension force}} = \frac{\rho g a^2}{\gamma} = \frac{a^2}{\ell_{\rm cap}^2}. \tag{7.30}$$

Note that $Bo = 1$ if the characteristic length scale equals the capillary length, $a = \ell_{\rm cap}$. Surface tension dominates over gravitation when $Bo \ll 1$ or equivalently, when the characteristic size a of the system is much smaller than the capillary length, $a \ll \ell_{\rm cap}$.

When some characteristic velocity V_0 is imposed on the system, the capillary number Ca can be introduced,

$$Ca = \frac{\text{viscous force}}{\text{surface tension force}} = \frac{\eta V_0}{\gamma}. \tag{7.31}$$

Note that $Ca = 1$ if the imposed velocity equals the intrinsic viscosity-surface velocity, $V_0 = \gamma/\eta$.

Finally, given two dimensionless numbers their ratio will also be a dimensionless number. The ratio of Ca and Bo is denoted the Stokes number $N_{\rm St}$, and it can be introduced when both a length scale and a velocity scale is given,

$$N_{\rm St} = \frac{\text{viscous force}}{\text{gravitational force}} = \frac{Ca}{Bo} = \frac{\eta V_0}{\rho g a^2}. \tag{7.32}$$

7.4 Capillary pumps

If a microchannel is placed horizontally along the x axis as shown in Fig. 7.5, the gravitational force cannot balance the capillary forces, so the capillary "rise" or capillary flow will continue as long as there is a channel for the liquid to propagate in. The theory for the position $L(t)$ of the meniscus in this case is analogous to the theory of capillary rise treated in the previous section, except that gravity now drops out of the equations. The position $L(t) = 0$ is defined as the entrance of the microchannel at the input reservoir, which is so wide that no Young–Laplace pressure drop is present there, i.e. $p(x{=}0) = p_0$.

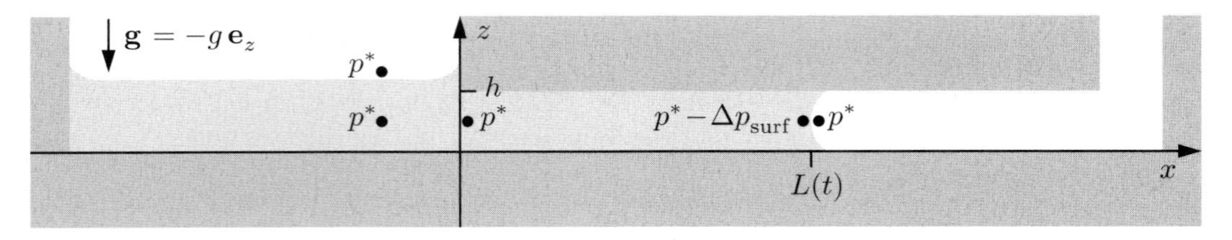

Fig. 7.5 A sketch of the principle of a capillary pump (dark gray). The curved meniscus at position $L(t)$ results in an uncompensated Young–Laplace underpressure $-\Delta p_{\mathrm{surf}}$ that drives the liquid (light gray) to the right in the microchannel. Notice all the points where the pressure is (approximately) equal to the atmospheric pressure p_0 of the air (white). Note that the reservoir to the left is so wide that gravity dominates and results in a flat surface with a zero Young–Laplce pressure drop.

7.4.1 Capillary-pump advancement times

We are going to apply the capillary-pump analysis for microfluidic channels with flat rectangular cross-sections of width w and height $h \ll w$. Hence, we shall use the Hagen–Poiseuille result $Q = h^3 w \Delta p/(12\eta L)$ of Eq. (3.30). The pressure drop Δp between the entrance at $x = 0$ and the advancing meniscus at $x = L(t)$ is constant and simply given by the Young–Laplace pressure drop,

$$\Delta p = \Delta p_{\mathrm{surf}} = \frac{2\gamma}{h}\cos\theta = \frac{2}{h}\left(\gamma_{\mathrm{sg}} - \gamma_{\mathrm{sl}}\right). \tag{7.33}$$

In analogy with Eq. (7.23) the speed $\mathrm{d}L(t)/\mathrm{d}t$ by which the front of the liquid is advancing through the microchannel is determined by mass conservation of the flow in the tube. Assuming a fully developed Poiseuille flow profile at $x = 0$, we find at $x = L(t)$ that

$$\frac{\mathrm{d}L(t)}{\mathrm{d}t} = V_0 = \frac{Q}{wh} \approx \frac{h^2 \Delta p_{\mathrm{surf}}}{12\eta}\,\frac{1}{L(t)}. \tag{7.34}$$

This differential equation is easily integrated by separation of $L\,\mathrm{d}L$ and $\mathrm{d}t$. Introducing the characteristic time τ_{adv} for the capillary advancement of the meniscus,

$$\tau_{\mathrm{adv}} \equiv \frac{6\eta}{\Delta p_{\mathrm{surf}}} = \frac{3\eta h}{\gamma\cos\theta} = \frac{3\eta h}{\gamma_{\mathrm{sg}} - \gamma_{\mathrm{sl}}} \quad \text{(parallel-plate channel)}, \tag{7.35}$$

the solution can be written as

$$L(t) = \sqrt{\frac{h\gamma\cos\theta}{3\eta}\,t} = h\sqrt{\frac{t}{\tau_{\mathrm{adv}}}}, \quad \text{(parallel-plate channel)}. \tag{7.36}$$

This result is analogous to the small-time behavior Eq. (7.27) of capillary rise.

The above analysis is easily redone for a circular channel of radius a, and as result Eq. (7.36) and τ_{adv} are slightly changed,

$$L(t) = \sqrt{\frac{a\gamma\cos\theta}{2\eta}\,t} = a\sqrt{\frac{t}{\tau_{\mathrm{adv}}}}, \quad \tau_{\mathrm{adv}} \equiv \frac{4\eta}{\Delta p_{\mathrm{surf}}} = \frac{2\eta a}{\gamma\cos\theta} \quad \text{(circular channel)}. \tag{7.37}$$

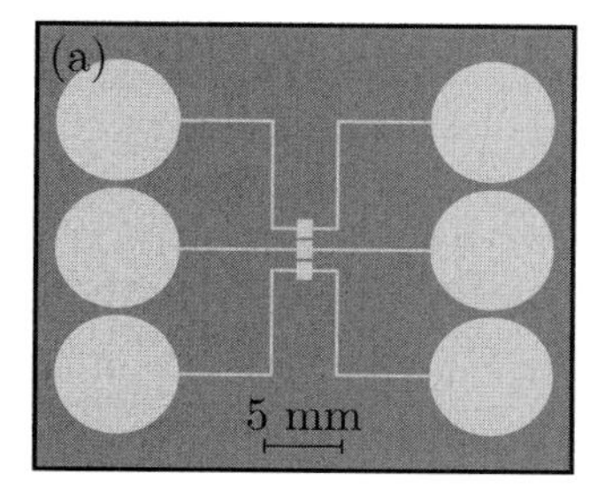

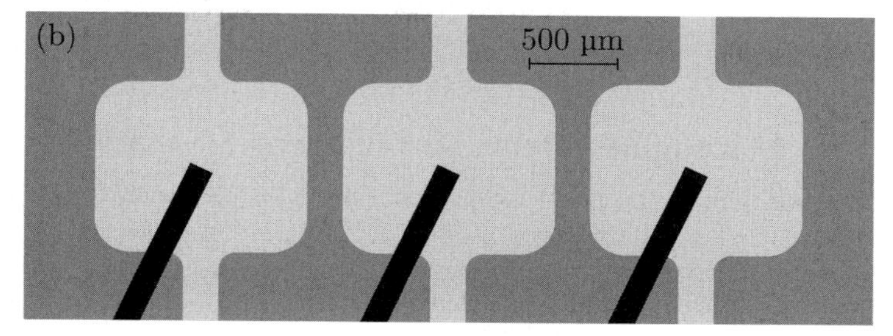

Fig. 7.6 Sketch of a biosensor with a capillary-force pump fabricated in 20 µm thick PMMA (dark gray) by the group of Anja Boisen at MIC, DTU. (a) The design of the 35 mm × 30 mm chip, which contains six circular reservoirs (light gray) of radius $r = 4$ mm and three channels (light gray) of width $w = 200$ µm connecting them pair-wise. (b) The center of the chip (rotated 90°) where each of the three channels widens into a 1.2 mm × 1.0 mm rectangular measuring site (light gray). The biosensing at each site is done by resistive readout of the strain in a flexible cantilever (black) dipped into the liquid.

7.4.2 A biosensor chip with a capillary-force pump

As an example of the use of capillary pumps in lab-on-a-chip systems we shall study the biosensor chip developed by the group of Boisen at MIC, see Fig. 7.6.

The core of the system is the use of micrometer-scale cantilevers that have been coated with specific biomolecules. Such cantilevers can be used as biosensors when they are immersed into a liquid biochemical solution. The principle of operation is simple: When biochemical reactions take place at the surface of the cantilever mechanical surface stresses are induced. The cantilever bends due to these stresses, and the bending can be detected by a piezo-resistive readout built into the cantilever. By careful selection of the biocoating, the cantilever can be designed to respond selectively to certain biomolecules.

The chip is constructed by spinning a polymer layer, here PMMA of height $h = 20$ µm, onto a glass plate. By photolithography, six circular reservoirs of radius $r = 4$ mm are etched into the PMMA-layer and three channels with rectangular cross-sections of width $w = 200$ µm and height $h = 20$ µm connect them pairwise. The mask design of the chip is shown in Fig. 7.6(a). At the center of the chip each of the three channels widens into a 1.2 mm × 1.0 mm rectangular measuring site, see Fig. 7.6(b), where the cantilever probes are going to be dipped into the liquid. The whole chip is covered by a second glass plate to seal off the microfluidic channels, but holes are provided for liquid handling at the six reservoirs and at the three measuring sites. Using a simple pipette the biochemical liquid is injected into one of the large reservoirs. By capillary forces the liquid is sucked into the microchannel leading from the reservoir to the measuring site.

To apply Eqs. (7.35) and (7.36) we define the beginning of the capillary channel, $x = 0$, at the reservoir inlet reservoir. The distance from a center reservoir or a corner reservoir to the corresponding measuring site is $L_1 = 8$ mm and $L_2 = 15$ mm, respectively. Using the physical parameter values in Table 7.1 for a water–PMMA–air system we find the time t_{arriv} it takes the liquid to arrive at the measuring sites to be

$$t_{\mathrm{arriv}} \approx 1 \text{ s.} \tag{7.38}$$

The "powerless" capillary pump systems is thus both adequate and useful for the task of delivering liquids at specific points on the chip.

7.5 Marangoni effect; surface-tension gradients

In establishing Eq. (7.9) for the matching condition for the Navier–Stokes equation at the interface between two immiscible fluids, we have assumed that the surface tension is a constant. However, there are many cases where the surface tension in fact is varying in space. In particular, gradients in the concentration of surfactants (such as soap) at the interface and temperature gradients implies gradients in the surface tension γ.

Just as gradients in the pressure field imply a gradient force per volume, $-\boldsymbol{\nabla} p$, so does a gradient in the surface tension imply a gradient force per area, $+\boldsymbol{\nabla}\gamma$. The difference in sign between the two gradient forces is due to the fact that pressure forces tend to maximize volume, whereas surface-tension forces tend to minimize area. The surface-tension gradient force is known as the Marangoni force,

$$\mathbf{f}_{\mathrm{Maran}} \equiv \boldsymbol{\nabla}\gamma. \tag{7.39}$$

Adding the Marangoni force to Eq. (7.9) yields a more general matching condition,

$$-\left(p^{(1)} - p^{(2)}\right) n_i + \left(\sigma_{ik}^{(1)} - \sigma_{ik}^{(2)}\right) n_k = \left(\frac{1}{R_1} + \frac{1}{R_2}\right) \gamma^{(12)} + \partial_i \gamma^{(12)}. \tag{7.40}$$

One can get an idea of the size of temperature-induced Marangoni forces by noting that heating a water/air interface by 5 ℃ from 20 ℃ to 25 ℃ will lower the surface tension by 0.8 $\mathrm{mJ/m^2}$ from 72.9 $\mathrm{mJ/m^2}$ to 72.1 $\mathrm{mJ/m^2}$. The shorter a distance over which one can maintain this temperature gradient the stronger the Marangoni force; but in all cases only in microsystems one can hope for a sufficiently large effect compared to other forces.

The Marangoni force can be used as a microscale propulsion system, as some bacteria actually do in Nature. The principle is simple: If one emits some surfactant that lowers the surface tension behind a little body then the body will be pushed forward, as the interface tries to minimize the region of high surface tension (without surfactant) while maximizing the region of low surface tension (with surfactant). One can build a little boat illustrating this principle by attaching a piece of soap at the end of a stick. As the soap dissolves the stick moves forward.

7.6 Exercises

Exercise 7.1
Surface tension interpreted as force per length
Make a sketch clarifying the geometry of the stretched-surface argument, which in Section 7.1.1 led to Eq. (7.5). Derive this equation carefully using your sketch.

Exercise 7.2
The Young–Laplace pressure in a flat channel with equal contact angles
Argue why the Young–Laplace pressure drop indeed is given by Eq. (7.33) for a liquid/gas interface inside a flat and very wide rectangular channel with the same contact angle for the top and bottom plate.

Exercise 7.3
The Young–Laplace pressure in a flat channel with different contact angles
Find the expression for the Young–Laplace pressure drop across a liquid/gas interface with surface tension γ inside a flat and very wide rectangular channel of height h, where the contact angle for the bottom and top plate are given by θ_1 and θ_2, respectively.

Exercise 7.4
Droplets on substrates with various contact angles
Consider liquid droplets on a solid substrate in air for the following three cases: water on gold, water on platinum, and mercury on glass. Use the values of the physical parameters given in Table 7.1 and make a sketch of the resulting shapes assuming small droplets $a \ll \ell_{\mathrm{cap}}$ in all three cases.

Exercise 7.5
Capillary rise for mercury
Consider mercury (Hg) as the liquid in a capillary rise experiment using a glass tube of radius $a = 100$ µm. The relevant physical parameters for Hg and glass are $\rho = 1.36 \times 10^4$ kg/m^3, $\theta = 140°$, and $\gamma = 0.487$ J/m^2.

(a) Go through the arguments leading to the expression Eq. (7.21) for the capillary rise height H. Hint: make a sketch like Fig. 7.4(b) and be careful with the sign of $\cos\theta$.

(b) Determine the value of H for the mercury–glass system.

Exercise 7.6
Alternative formula for the capillary rise height
Consider the expression Eq. (7.21) for the capillary rise height H.

(a) Use Young's equation to show the second expression for H,

$$H = \frac{2}{\rho g a}\,(\gamma_{\mathrm{sg}} - \gamma_{\mathrm{sl}}). \tag{7.41}$$

(b) Use the constant-energy argument, $\delta G = 0$, at equilibrium to prove Eq. (7.41) considering a displacement δH away from the equilibrium position H and the corresponding change δG_{grav} in gravitational energy and δG_{surf} in surface energy. Hint: Note that the shape of the meniscus remains unchanged during the displacement and hence does not contribute to any change in the Gibb's free energy.

Exercise 7.7
The expressions for the capillary rise time
Consider Section 7.3.2, where the dynamics of the capillary rise is treated.

(a) Derive the dimensionless differential equation Eq. (7.26c) for the position $\tilde{L}(\tilde{t})$ of the rising meniscus during capillary rise.

(b) Verify that the solutions Eqs. (7.27) and (7.28) are correct.

Exercise 7.8
Dimensionless numbers for capillary rise
Consider Section 7.3.3, where the three dimensionless numbers Bo, Ca, and N_{St} are defined.

(a) Calculate the values of these three dimensionless numbers for the same physical parameter values that led to the estimate for $\tau_{\mathrm{cap}}^{\mathrm{water/air}}$ in Eq. (7.29), and use $V_0 = H/\tau_{\mathrm{cap}}$.

(b) Discuss the significance of the obtained values.

Exercise 7.9
Liquids advancing by capillary forces in horizontal microchannels
Consider Section 7.4.1, where the the capillary pump advancement times are discussed.

(a) State the assumptions leading to Eq. (7.36) for the position of the advancing meniscus in a capillary pump.

(b) Check that the solution Eq. (7.36) is consistent with the assumptions.

Exercise 7.10
Capillary pump with circular cross-section
In Section 7.4.1 the capillary pump with rectangular cross-section is analyzed. Redo the analysis for a capillary pump with a circular cross-section and prove Eq. (7.37) for the advancement time τ_{adv}.

Exercise 7.11
Advancement times in the capillary pump
Apply the results of Section 7.4 to the following problems.

(a) Calculate more precisely the arrival times $t^{(1)}_{\mathrm{arriv}}$ and $t^{(2)}_{\mathrm{arriv}}$ for the liquids advancing from a corner reservoir and a center reservoir, respectively, in the biosensor chip described in Section 7.4.2.

(b) Use Table 7.1 to predict how the advancement times would alter if the walls of the capillary pump were changed from PMMA to gold and to platinum.

(c) Discuss the consequences for the functionality of a capillary pump if $\theta = 90°$ and if $\theta > 90°$.

Exercise 7.12
The sign of the Marangoni force
Make a sketch of a surface with a varying surface tension γ. Argue why the sign in the Marangoni force, $\mathbf{f}_{\mathrm{Maran}} \equiv +\boldsymbol{\nabla}\gamma$, is positive.

7.7 Solutions

Solution 7.1
Surface tension interpreted as force per length
During the stretch the force F acts over the distance ΔL thus performing the work $\Delta W = F\Delta L$ on the surface. Assuming that energy is not dissipated all external work is transformed into surface energy $\Delta\mathcal{G}_{\mathrm{surf}} = 2\gamma w\Delta L$, where the factor of 2 is because the stretched film has a surface both on the front and on the back. Thus, $W = \Delta\mathcal{G}_{\mathrm{surf}}$, which implies $\gamma = F/(2w)$.

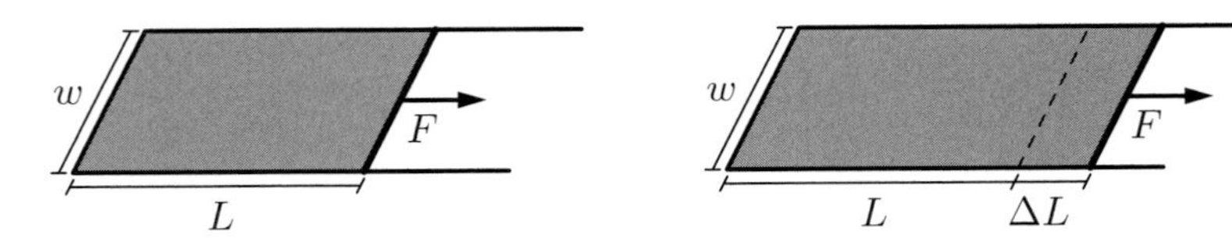

Fig. 7.7 Sketch of the geometry for a film of initial length L being stretched the amount ΔL.

Solution 7.2
The Young–Laplace pressure in a flat channel with equal contact angles

The curvature in the wide transverse direction of the flat channel is of the order $2/w$, so inclusion of the radii of curvature in both directions leads to a the Young–Laplace pressure of the form $\Delta p_{\text{surf}} = \gamma(2/h + 2/w)\cos\theta = (2\gamma/h)(1 + h/w)\cos\theta \approx (2\gamma/h)\cos\theta$, which is valid for $h \gg w$.

Solution 7.3
The Young–Laplace pressure in a flat channel with different contact angles

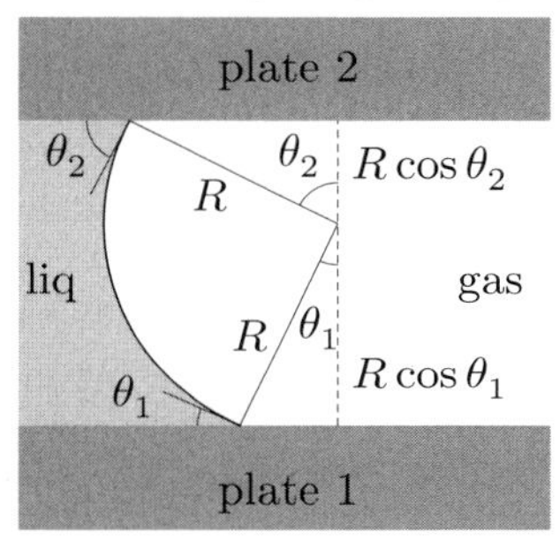

Fig. 7.8 The geometry of the liquid meniscus.

The distance between the bottom and top plates of the flat channel is denoted h. Let θ_1 and θ_2 be the contact angle at the bottom and top plates, respectively, then, as shown to the left, $h = R\cos\theta_1 + R\cos\theta_2$. Assuming that the only non-zero radius of curvature R is perpendicular to the plates, the Young–Laplace pressure is given by

$$\Delta p_{\text{surf}} = \frac{\gamma}{R} = \frac{\gamma}{h}(\cos\theta_1 + \cos\theta_2) = \frac{2\gamma}{h}\frac{\cos\theta_1 + \cos\theta_2}{2}. \quad (7.42)$$

Solution 7.4
Droplets on substrates with various contact angles

For $a \ll \ell_{\text{cap}}$ the surface tension dominates over the gravitational force. As a consequence all droplet shapes are sections of spheres.

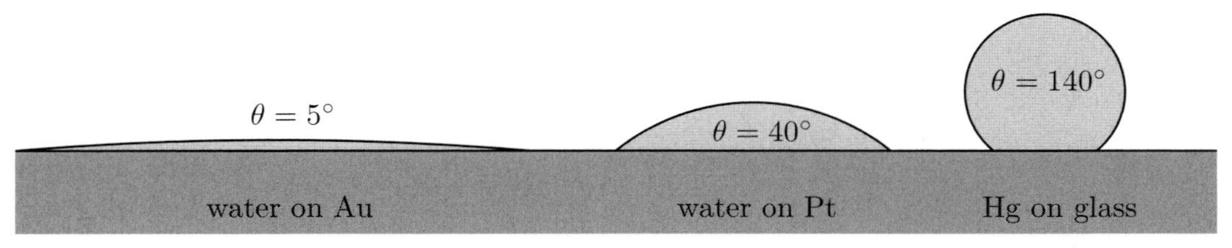

Fig. 7.9 Sketch of drop shapes in the cases of almost complete wetting, wetting and non-wetting. To be visible at all, the contact angle for water on gold has been set to 5° instead of 0°.

Solution 7.5
Capillary rise for mercury

For mercury, $\theta = 140^\circ > 90^\circ$. As a consequence the liquid surface in a sketch corresponding to Fig. 7.4(b) will curve the other way, and Δp_{surf} changes sign. So the underpressure becomes an overpressure leading to capillary fall instead of rise. The arguments leading to Eq. (7.21) still hold, so $H = 2\gamma\cos\theta/(\rho g a) = -55.7$ mm.

Solution 7.6
Alternative formula for the capillary rise height
The second expression for H in Eq. (7.21) clearly shows that capillary rise is due to a gain in energy obtained when liquid covers an area of a solid previously covered by gas.

(a) Young's equation states that $\gamma\cos\theta = \gamma_{\mathrm{sg}} - \gamma_{\mathrm{sl}}$, hence the desired result follows by simple substitution.

(b) A vertical displacement δH changes the gravitational energy by the amount $\delta G_{\mathrm{grav}} = mg\delta H = (\rho\pi a^2 H)g\delta H$, while the corresponding change in surface energy is $\delta G_{\mathrm{surf}} = (\gamma_{\mathrm{sl}} - \gamma_{\mathrm{sg}})\mathcal{A} = (\gamma_{\mathrm{sl}} - \gamma_{\mathrm{sg}})2\pi a\delta H$. From $\delta G_{\mathrm{grav}} + \delta G_{\mathrm{surf}} = 0$ follows the desired result, $2(\gamma_{\mathrm{sl}} - \gamma_{\mathrm{sg}}) = \rho g a H$.

Solution 7.7
The expressions for the capillary rise time
Equation (7.25) follows directly from the assumption of a fully developed quasi-steady Poiseuille flow, Eq. (7.23), driven by the constant Young–Laplace pressure minus the time-dependent hydrostatic pressure, Eq. (7.24).

(a) The dimensionless form of Eq. (7.25) is obtained by multiplying the equation by $8\eta/(\rho g a^2)$ and substituting L by $H\tilde{L}$. This yields $\tau_{\mathrm{cap}}\mathrm{d}\tilde{L}/\mathrm{d}t = \tilde{L}^{-1} - 1$. Finally, substituting t by $\tau_{\mathrm{cap}}\tilde{t}$ leads to the result $\mathrm{d}\tilde{L}/\mathrm{d}\tilde{t} = \tilde{L}^{-1} - 1$, Eq. (7.26c).

(b) For $\tilde{t} \ll 1$ we have $\tilde{L} \approx \sqrt{2\tilde{t}} \ll 1$, Eq. (7.27). Hence, $\mathrm{d}\tilde{L}/\mathrm{d}\tilde{t} = \sqrt{2}\frac{1}{2}/\sqrt{\tilde{t}} = 1/\sqrt{2\tilde{t}} = 1/\sqrt{\tilde{L}} \approx 1/\sqrt{\tilde{L}} - 1$.

For $\tilde{t} \gg 1$ we have $\tilde{L} \approx 1 - \exp(-\tilde{t})$ and $\exp(-\tilde{t}) \ll 1$, Eq. (7.28). Consequently, $\mathrm{d}\tilde{L}/\mathrm{d}\tilde{t} = \exp(-\tilde{t}) = [1 + \exp(-\tilde{t})] - 1 \approx 1/[1 - \exp(-\tilde{t})] - 1 = 1/\tilde{L} - 1$.

Solution 7.8
Dimensionless numbers for capillary rise
The parameters used in Eq. (7.29) are $a = 100$ µm, $\rho = 10^3$ kg/m^3, $g = 9,81$ m/s, $\eta = 10^{-3}$ Pa s, and $\gamma = 0.073$ J/m^2. Moreover, $V_0 = H/\tau_{\mathrm{cap}} \approx 1$ cm/s.

(a) From the parameter values follow $Bo = 1.3\times10^{-3}$, $Ca = 1.4\times10^{-5}$, and $N_{\mathrm{St}} = 0.1$.

(b) Since $Bo \ll 1$ and $Ca \ll 1$ surface tension dominates over both gravity and viscosity. And since $N_{\mathrm{St}} \ll 1$ gravity dominates over viscosity.

Solution 7.9
Liquids advancing by capillary forces in horizontal microchannels
We discuss the consistency of Eq. (7.36) with the underlying assumptions.

(a) The main assumptions are (i) translation invariance along the x axis, (ii) quasi-steady Poiseuille flow, i.e. the acceleration term $\rho\partial_t v_x$ is negligible in the Navier–Stokes eqaution, and (iii) the circulation rolls in the front of the liquid string have a negligible influence on the flow. These rolls must exist to ensure a smooth transition from the Poiseuille flow parabola in the bulk of the liquid and the curved meniscus at the front moving at constant speed.

(b) Assuming that Eq. (7.36), stating $L \propto t^{1/2}$, is correct, we find $v = \partial_t L = \frac{1}{2}L/t$ and $\partial_t v = -\frac{1}{4}L/t^2 = -\frac{1}{2}v/t$. Moreover, if Poiseuille flow is present in the majority of the liquid string, $\eta\partial_z^2 v = \Delta p/L = [12\eta L/(wh^3)]Q/L = (12\eta/h^2)v$. So the acceleration term can be neglected if $|\rho\partial_t v| \ll |\eta\partial_z^2 v|$, i.e. if $\rho v/t \ll (12\eta/h^2)v$ or $h^2/(12\nu) \ll t$, where ν is the kinematic viscosity. For $h = 100$ µm and $\nu = 10^{-6}$ m^2/s we find it necessary to demand that $t \gg 0.8$ ms to ensure that the solution is consistent with assumption (ii).

Since hydrodynamics do not contain an intrinsic length scale, the circulation rolls in the front of the liquid string must have the size h. Hence, they can be neglected if they are much smaller than the entire liquid string, $h \ll L$, which means $1 \ll \sqrt{t/\tau_{\mathrm{adv}}}$ or $t \gg \tau_{\mathrm{adv}}$. Using in Eq. (7.35) the usual parameters for water as well as $h = 100\ \mu\mathrm{m}$, we find $\tau_{\mathrm{adv}} \approx 10\ \mu\mathrm{s}$. Thus, the solution is consistent with assumption (*iii*) if $t \gg 10\ \mu\mathrm{s}$, but this is already ensured if assumption (*ii*) holds.

Solution 7.10
Capillary pump with circular cross-section

For a circular cross-section with radius a, Eq. (7.33) becomes $\Delta p_{\mathrm{surf}} = 2\gamma\cos\theta/a$ while Eq. (7.34) changes to $\mathrm{d}L/\mathrm{d}t = a^2\Delta p_{\mathrm{surf}}/(8\eta L)$. This in turn modifies Eq. (7.35) to $\tau_{\mathrm{adv}} = 4\eta/\Delta p_{\mathrm{surf}} = 2\eta a/\gamma\cos\theta$ as stated in Eq. (7.37).

Solution 7.11
Advancement times in the capillary pump

The channels under consideration have $h = 20\ \mu\mathrm{m}$ and $w = 200\ \mu\mathrm{m}$ and thus an aspect ratio $h/w = 0.1$. The correction factor for the hydraulic resistance becomes

$$1 - 0.630\frac{h}{w} = 0.937. \tag{7.43}$$

Since this correction factor, according to Eq. (3.58), appears opposite the viscosity η it is natural to define an effective viscosity η_{eff} by

$$\eta_{\mathrm{eff}} \equiv \frac{\eta}{1 - 0.630\frac{h}{w}} = 1.067\eta. \tag{7.44}$$

The finite width of the channel implies an apparent increase in the viscosity (or in reality in the hydraulic resistance).

(a) The capillary advancement time τ_{adv} for the PMMA–water–air system can now be estimated from Eq. (7.35) as

$$\tau_{\mathrm{adv}} = \frac{3\eta_{\mathrm{eff}}h}{\gamma\cos\theta} \tag{7.45}$$

$$= \frac{3\,(1.067\times 10^{-3})\,(2\times 10^{-5})}{0.0729\cos(73.7^\circ)}\ \mathrm{s} \tag{7.46}$$

$$= 3.129\times 10^{-6}\ \mathrm{s}. \tag{7.47}$$

From Eq. (7.36) it follows that

$$t = \left(\frac{L}{h}\right)^2 \tau_{\mathrm{adv}}. \tag{7.48}$$

With $L_1 = 8$ mm and $L_2 = 15$ mm the advancement times t_1 and t_2 become

$$t_1 = 0.500\ \mathrm{s}, \qquad t_2 = 1.760\ \mathrm{s}. \tag{7.49}$$

(b) Changing from PMMA to platinum or gold leads to a lowering of the contact angle θ. This increases $\cos\theta$ and thus decreases $\tau_{\mathrm{adv}} \propto 1/\cos\theta$. The pump becomes more efficient by this change in material with the gold being the best choice if fast pumping is wanted.

(c) In the limit $\theta \rightarrow 90°$ we obtain $\cos\theta \rightarrow 0$, and consequently $\tau_{\mathrm{adv}} \rightarrow \infty$. The pump ceases to work. For $\theta > 90°$ it costs surface energy for the liquid to enter the pump, so in fact the pump will force out liquid initially present in the channels. It thus acts as a pump in the reverse direction.

Solution 7.12
The sign of the Marangoni force
Consider the sketch to the right shown in Solution 7.1, and assume that the surface tension γ varies as a function of x (the L-direction). The external force F acting at $x = L$ is shown, $F(L) = \gamma(L)w$. Similarly, an antiparallel external force in must act at $x = 0$ given by $F(0) = -\gamma(0)w$, hence the total surface tension force acting on the area is $F_{\mathrm{tot}} = F(L) - F(0) = [\gamma(L) - \gamma(0)]w$. Thus, the surface tension force per area, the Marangoni force, becomes $f_{\mathrm{Maran}} = F_{\mathrm{tot}}/wL = [\gamma(L) - \gamma(0)]/L \rightarrow +\partial_x\gamma$, for $L \rightarrow 0$.

8 Electrohydrodynamics

In many lab-on-a-chip applications the motions of the liquids or the solutes are controlled electrically. Therefore it is highly relevant to study electrohydrodynamics, i.e. the coupling of electromagnetism and hydrodynamics. Using this wide definition, electrohydrodynamics comprises a wide range of phenomena such as the electrical properties of liquids *per se*, electrochemistry, and electrokinetics.

One obvious way to couple electromagnetism to hydrodynamics is through the electric body force $\rho_{\mathrm{el}}\mathbf{E}$ in the Navier–Stokes equation, as we have seen in Eq. (2.30b),

$$\rho\Big(\partial_t \mathbf{v} + (\mathbf{v}\cdot\boldsymbol{\nabla})\mathbf{v}\Big) = -\boldsymbol{\nabla}p + \eta\nabla^2\mathbf{v} + \rho\,\mathbf{g} + \rho_{\mathrm{el}}\mathbf{E}, \tag{8.1}$$

for a liquid with a non-zero charge density ρ_{el} in an external electric field $\mathbf{E}$.

In this chapter, we shall only deal with electromagnetic phenomena in the electrostatic regime, i.e. we are disregarding any magnetic and radiative effects. In accordance with the continuum hypothesis of Section 1.3.2 the governing equations are the Maxwell equations for continuous media, where the electric field $\mathbf{E}$, the displacement field $\mathbf{D}$, the polarization field $\mathbf{P}$, the electrical current density $\mathbf{J}_{\mathrm{el}}$, and the electrical potential ϕ have all been averaged locally over their microscopic counterparts. The fundamental equations are:

$$\boldsymbol{\nabla}\times\mathbf{E} = \mathbf{0}, \tag{8.2a}$$

$$\boldsymbol{\nabla}\cdot\mathbf{D} = \boldsymbol{\nabla}\cdot(\epsilon\mathbf{E}) = \rho_{\mathrm{el}}, \tag{8.2b}$$

$$\mathbf{D} = \epsilon_0\mathbf{E} + \mathbf{P} = \epsilon\mathbf{E}, \tag{8.2c}$$

$$\mathbf{J}_{\mathrm{el}} = \sigma_{\mathrm{el}}\mathbf{E}. \tag{8.2d}$$

Due to Eq. (8.2a) the $\mathbf{E}$-field can be written as (minus) the gradient of a potential ϕ. If ϵ is constant this gradient leads to the Poisson equation when inserted into Eq. (8.2b),

$$\mathbf{E} = -\boldsymbol{\nabla}\phi, \tag{8.3a}$$

$$\nabla^2\phi(\mathbf{r}) = -\frac{1}{\epsilon}\,\rho_{\mathrm{el}}(\mathbf{r}). \tag{8.3b}$$

These equations will be used in the following analysis of electrohydrodynamic phenomena in microfluidics.

8.1 Polarization and dipole moments

Polarization effects play an important role in microfluidics, so it seems appropriate to review the basic theory of polarization. Consider a small particle, i.e. a biological cell or a small part

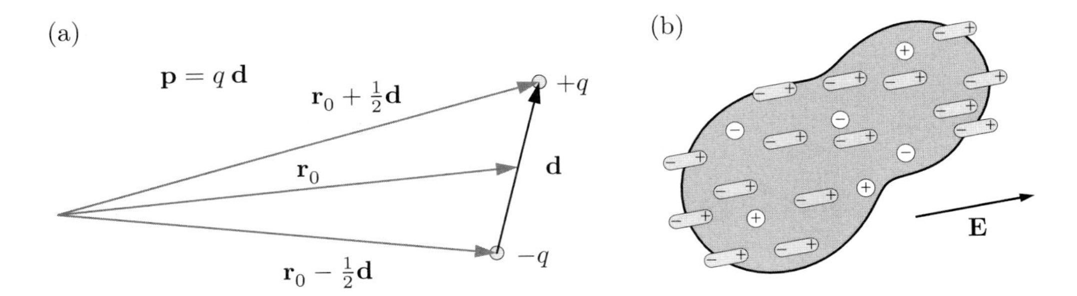

Fig. 8.1 (a) The simple point dipole consisting of a charge $+q$ separated from a charge $-q$ by the distance $\mathbf{d}$. (b) The dipole moments (light gray ovals) and external charges (white circles) inside a body Ω (dark gray). Polarization charge is left behind in the body when the dipole moments sticks out at the surface of the body.

of some liquid, having the electric charge density ρ_{el}, which occupies the region Ω in space centered around the point $\mathbf{r}_0$. General positions inside the particle are denoted $\mathbf{r}_0+\mathbf{r}$. If an external electrical field $\mathbf{E}$ is imposed on the system the ith component F_i^{el} of the electrical force $\mathbf{F}_{\mathrm{el}} = \int_\omega d\mathbf{r}\, \rho_{\mathrm{el}}\mathbf{E}$ acting on the particle is given by Taylor expanding the fields in Ω,

$$F_i^{\mathrm{el}} = \int_\Omega d\mathbf{r}\, \rho_{\mathrm{el}}(\mathbf{r}_0+\mathbf{r})E_i(\mathbf{r}_0+\mathbf{r}) \approx \int_\Omega d\mathbf{r}\, \rho_{\mathrm{el}}(\mathbf{r}_0+\mathbf{r})\Big[E_i(\mathbf{r}_0)+r_j\partial_jE_i(\mathbf{r}_0)\Big] = QE_i(\mathbf{r}_0)+p_j\partial_jE_i(\mathbf{r}_0), \tag{8.4}$$

where we have introduced the charge Q and electric dipole moment $\mathbf{p}$ of the particle as

$$Q \equiv \int_\Omega d\mathbf{r}\, \rho_{\mathrm{el}}(\mathbf{r}_0+\mathbf{r}), \tag{8.5a}$$

$$\mathbf{p} \equiv \int_\Omega d\mathbf{r}\, \rho_{\mathrm{el}}(\mathbf{r}_0+\mathbf{r})\,\mathbf{r}. \tag{8.5b}$$

As expected, there is an electrical force when the charge Q of the region Ω is non-zero, but note that a force is also present even when $Q=0$ if both the dipole moment $\mathbf{p}$ and the electric-field gradient tensor $\boldsymbol{\nabla}\mathbf{E}$ are non-zero. The forces in the latter case are denoted dielectric forces, and they play a central role in the discussion of dielectrophoresis in Chapter 10.

A particularly simple example of a dipole moment is the two-point-charge dipole, or in short the point dipole, sketched in Fig. 8.1(a). It is defined by the charge distribution

$$\rho_{\mathrm{el}}(\mathbf{r}_0+\mathbf{r}) = +q\,\delta\big(+\tfrac{1}{2}\mathbf{d}-\mathbf{r}\big) - q\,\delta\big(-\tfrac{1}{2}\mathbf{d}-\mathbf{r}\big), \tag{8.6}$$

which, when inserted into Eq. (8.5b), results in the dipole moment

$$\mathbf{p} = q\mathbf{d}. \tag{8.7}$$

The polarization vector $\mathbf{P}(\mathbf{r}_0)$ appearing in Eq. (8.2c) is defined as the dipole moment density in a small region Ω^* surrounding $\mathbf{r}_0$ as the volume $\mathrm{Vol}(\Omega^*)$ is taken to zero,

$$\mathbf{P}(\mathbf{r}_0) \equiv \lim_{\mathrm{Vol}(\Omega^*)\to 0}\left[\frac{1}{\mathrm{Vol}(\Omega^*)}\int_{\Omega^*} d\mathbf{r}\, \rho_{\mathrm{el}}(\mathbf{r}_0+\mathbf{r})\,\mathbf{r}\right]. \tag{8.8}$$

The divergence $\boldsymbol{\nabla}\cdot\mathbf{P}$ of the polarization can be interpreted as the polarization charge density. This is shown by considering the arbitrarily shaped body Ω sketched in Fig. 8.1(b), which contains a number of dipoles in the polarizable medium as well as some external charges not part of the medium. In the bulk of the body the charges from the dipole moments cancel each other, but at the surface part of the dipole charges go outside the body. Since $\mathbf{d}\cdot\mathbf{n}da$ describes the volume of a dipole $q\mathbf{d}$ sticking out of the surface element da with unit normal vector $\mathbf{n}$ the amount of polarization charge Q_{pol} left behind in the body is given by

$$Q_{\mathrm{pol}} = -\int_{\partial\Omega} da\, \mathbf{n}\cdot\Big(\frac{q\mathbf{d}}{\mathrm{Vol}(\Omega^*)}\Big) = -\int_{\partial\Omega} da\, \mathbf{n}\cdot\mathbf{P} = -\int_{\Omega} d\mathbf{r}\, \boldsymbol{\nabla}\cdot\mathbf{P}. \tag{8.9}$$

This result holds for any region Ω so the polarization charge density ρ_{pol} can be defined as

$$\rho_{\mathrm{pol}} \equiv -\boldsymbol{\nabla}\cdot\mathbf{P}, \tag{8.10}$$

but this allows for a simple expression for the density ρ_{ext} of the external charges:

$$\rho_{\mathrm{ext}} = \rho_{\mathrm{tot}} - \rho_{\mathrm{pol}} = \epsilon_0\boldsymbol{\nabla}\cdot\mathbf{E} + \boldsymbol{\nabla}\cdot\mathbf{P} = \boldsymbol{\nabla}\cdot\big(\epsilon_0\mathbf{E} + \mathbf{P}\big). \tag{8.11}$$

Thus, by defining the displacement field as $\mathbf{D} \equiv \epsilon_0\mathbf{E} + \mathbf{P}$, this leads to Eq. (8.2b), and we have learned that ρ_{el} should not comprise the polarization charge density ρ_{pol}. For liquids and isotropic solids the polarization is proportional to the electrical field, and the following expressions introducing the susceptibility χ and the relative dielectric constant ϵ_{r} can be used:

$$\mathbf{D} = \epsilon_0\mathbf{E} + \mathbf{P} = \epsilon_0\mathbf{E} + \epsilon_0\chi\mathbf{E} = \epsilon_0(1+\chi)\mathbf{E} = \epsilon_0\epsilon_{\mathrm{r}}\mathbf{E} = \epsilon\mathbf{E}. \tag{8.12}$$

We leave the topic of electric dipoles, polarization, and dielectric effects for now, but shall return to it in Chapter 10, when we study dielectrophoretic handling of charge-neutral particles. In the following, we shall instead focus on the electric effects related to electric monopoles.

8.2 Electrokinetic effects

Having established the fundamental equations for electrohydrodynamics we move on to the first example, electrophoresis. This is one of four electrokinetic phenomena that are important in microfluidics. They all involve the motion of liquids relative to charged surfaces. The terminology in use is

1. **Electrophoresis** – the movement of a charged surface (of say dissolved or suspended material) relative to a stationary liquid induced by an applied electric field.
2. **Electro-osmosis** – the movement of liquid relative to a stationary charged surface (of say a capillary tube) induced by an applied electric field.
3. **Sedimentation potential** – the electric potential created when charged particles are made to move relative to a stationary liquid.
4. **Streaming potential** – the electric potential created when a liquid is made to move relative to a charged surface.

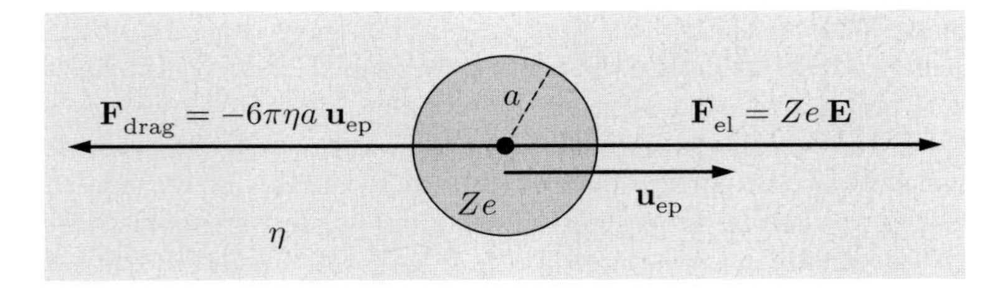

Fig. 8.2 The principle of electrophoresis. A spherical particle of charge Ze and radius a moves in a low-conductivity liquid with viscosity η under the influence of an applied electrical field $\mathbf{E}$. The motion becomes steady at the velocity $\mathbf{u}_{\mathrm{ep}}$, when the Stokes drag force $\mathbf{F}_{\mathrm{drag}}$ balances the electrical driving force $\mathbf{F}_{\mathrm{el}}$.

8.2.1 Electrophoresis

In the following, we study how an applied electrical field $\mathbf{E}$ influences a spherical particle of radius a and charge Ze in a stationary liquid of low electrical conductivity, say de-ionized water. This is a particularly simple case of electrophoresis. The low conductivity of the liquid implies the lack of ions that otherwise would have accumulated around the charged particle and partly neutralized its charge, an effect known as electrical screening. Therefore, as sketched in Fig. 8.2, the electric force is simply

$$\mathbf{F}_{\mathrm{el}} = Ze\mathbf{E}, \tag{8.13}$$

where e is the elementary charge and Z the integer valence number. From Section 6.4.1 we know that on the short time scale of a few µs, the charged particle reaches steady-state motion, here the electrophoretic velocity $\mathbf{u}_{\mathrm{ep}}$, due to viscous drag. In this situation the Stokes drag force Eq. (3.127), $\mathbf{F}_{\mathrm{drag}} = -6\pi\eta a\, \mathbf{u}_{\mathrm{ep}}$, balances $\mathbf{F}_{\mathrm{el}}$,

$$\mathbf{F}_{\mathrm{tot}} = \mathbf{F}_{\mathrm{el}} + \mathbf{F}_{\mathrm{drag}} = 0 \quad \Rightarrow \quad \mathbf{u}_{\mathrm{ep}} = \frac{Ze}{6\pi\eta a}\,\mathbf{E} \equiv \mu_{\mathrm{ion}}\,\mathbf{E}. \tag{8.14}$$

The dependence of the resulting drift velocity $\mathbf{u}_{\mathrm{ep}}$ on particle charge and size makes electrophoresis usable in biochemistry for sorting of proteins and DNA fragments. The sample under consideration is dissolved in water and inserted in one end of a tube with electrodes at each end. A voltage difference is applied to the electrodes, and the part of the sample that arrives first at the other end of the tube contains the smallest and most charged particles.

8.2.2 Ionic mobility and conductivity

From Eq. (8.14) we see that the terminal velocity $\mathbf{u}_{\mathrm{ep}}$ is proportional to the applied electrical field $\mathbf{E}$. The proportionality constant is called the ionic mobility μ_{ion},

$$\mu_{\mathrm{ion}} \equiv \frac{Ze}{6\pi\eta a}. \tag{8.15}$$

This simple theoretical estimate based on a macroscopic continuum model is in remarkable agreement with measured values of the ionic mobility of ions having a radius in the sub-nm range and moving in water. The radius a, however, is not the bare ionic radius $a \approx 0.05$ nm

Table 8.1 Experimental values for ionic mobility and diffusivity for small ions in aqueous solutions at small concentrations. Note how H^+ and OH^- have significantly different values due to their special modes of propagation by exchange of electrons with the neutral water molecules. Data are reproduced from Atkins (1994).

ions at $T = 25°C$	H^+	Ag^+	K^+	Li^+	Na^+	Br^-	Cl^-	F^-	I^-	OH^-
mobility μ_{ion} $[10^{-8}\ m^2\ (V\,s)^{-1}]$	36.2	6.42	7.62	4.01	5.19	8.09	7.91	5.70	7.96	20.6
diffusivity D_{ion} $[10^{-9}\ m^2\ s^{-1}]$	9.31	1.65	1.96	1.03	1.33	2.08	2.03	1.46	2.05	5.30

but instead the somewhat larger so-called hydrated radius $a \approx 0.2$ nm. This is due to the fact that ions in aqueous solutions accumulate approximately one atomic layer of water molecules. For $Z = 1$, $\eta = 1$ mPa s, and $a = 0.2$ nm we get

$$\mu_{ion} \approx 4 \times 10^{-8}\ \mathrm{m^2\ (V\,s)^{-1}}. \tag{8.16}$$

The experimental values for μ_{ion} are shown in Table 8.1.

The ionic mobility μ_{ion} is directly related to the ionic conductivity σ_{ion} as seen by combining Eqs. (8.2d) and (8.15),

$$\mu_{ion}\mathbf{E} = \mathbf{u}_{ep} = \frac{1}{Zec_{ion}}\,\mathbf{J}_{el} = \frac{\sigma_{ion}}{Zec_{ion}}\,\mathbf{E}, \tag{8.17}$$

so

$$\sigma_{ion} = Zec_{ion}\mu_{ion} \approx \frac{Z\,c_{ion}}{1\ \mathrm{mM}} \times 10^{-2}\ \mathrm{S\,m^{-1}}, \tag{8.18}$$

where the ionic concentration in the numerical example has been normalized to $c_{ion} = 1$ mM. The numerical result is in good agreement with experimental values for σ_{ion}.

8.3 The Debye layer near charged surfaces

The next electrohydrodynamic topic is a study of the electric potential and charge distribution in an electrolyte, i.e. an aqueous solution of ions, in equilibrium near a charged surface. The results obtained in this section will form the basis for our analysis in the following chapter of the electrokinetic effect called electro-osmosis and its applications to micropumps.

Consider an electrolyte in contact with a solid surface, either in the form of the walls of the microfluidic channel in which the liquid flows or in the form of a particle suspended in the liquid. Depending on the chemical composition of the solid and of the electrolyte chemical processes at the surface will result in a charge transfer between the electrolyte and the wall. As a result the wall and the electrolyte gets oppositely charged, while maintaining global charge neutrality. In Fig. 8.3(a) is sketched how the ions are distributed in the electrolyte after the charge transfer has taken place.

8.3.1 The continuum model of the Debye layer

The ions having the opposite charge of the solid, the counterions, are attracted to the solid, while the other ions, the co-ions, are repelled. For weak solutions the co- and counterions can

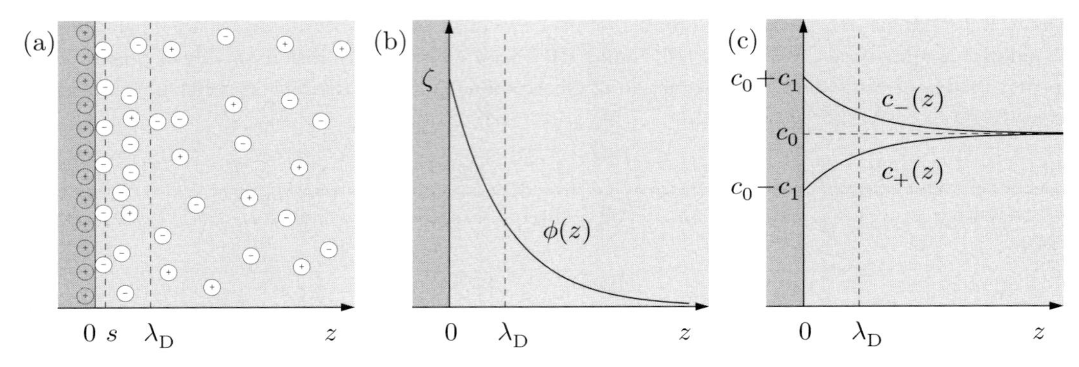

Fig. 8.3 (a) The ionic structure in thermal equilibrium of the Debye layer in an electrolyte (light gray, $z > 0$) near a solid surface in the xy-plane after charge transfer between the solid (dark gray, $z < 0$) and the electrolyte has taken place. For $0 < z < s$ lies the single layer of immobile counterions, the Stern layer. For $s < z < \lambda_\mathrm{D}$ follows the diffuse mobile layer of predominantly counterions. For $z > \lambda_\mathrm{D}$ the electrolyte is charge neutral. (b) The simple continuous field model for the electric potential $\phi(z)$ in the Debye layer. The potential at the Stern layer next to the surface takes the value $\phi(0) = \zeta$, while it decays to zero in the bulk on the length scale given by the Debye length λ_D. (c) The corresponding ionic densities $c_+(z)$ and $c_-(z)$ in the Debye layer.

be modeled as an ideal gas. In the case of zero temperature a complete charge cancellation, i.e. perfect electric shielding, would occur at the surface, however, at finite temperature thermal motion counteracts this behavior. The governing equation for the continuum ideal-gas description of the ionic concentrations $c_\pm(\mathbf{r})$ comes from the thermodynamic expression for the chemical potential $\mu_\pm(\mathbf{r})$, the free energy of the last added ion, see Appendix D,

$$\mu_\pm(\mathbf{r}) = \mu_0 + k_\mathrm{B}T \ln\left(\frac{c_\pm(\mathbf{r})}{c_0}\right) \pm Ze\phi(\mathbf{r}), \tag{8.19}$$

where μ_0 and c_0 is the chemical potential and ionic density, respectively, in the absence of the electric potential. Note that, for simplicity, the ionic valences are opposite, $\pm Z$.

Thermodynamic equilibrium implies that the chemical potential is constant throughout the system, because if it were varying the system could gain energy by reorganizing its constituents. We therefore have that $\mu_\pm(\mathbf{r}) = \text{const}$, and therefore $\boldsymbol{\nabla}\mu_\pm(\mathbf{r}) \equiv 0$, which yields

$$k_\mathrm{B}T\boldsymbol{\nabla} \ln\left(\frac{c_\pm(\mathbf{r})}{c_0}\right) = \mp Ze\boldsymbol{\nabla}\phi(\mathbf{r}). \tag{8.20}$$

In the following, we assume, as indicated in Fig. 8.3, that infinitely far away from the surface the two ionic concentrations approach the same unperturbed value c_0 and the electrical potential goes to zero, while at the surface the potential takes the value ζ, known in the literature as the zeta-potential:

$$c_\pm(\infty) = c_0, \qquad \phi(\infty) = 0, \qquad \phi(\text{surf}) = \zeta. \tag{8.21}$$

With these boundary conditions Eq. (8.20) is easily integrated,

$$c_\pm(\mathbf{r}) = c_0 \exp\left[\mp \frac{Ze}{k_\mathrm{B}T}\,\phi(\mathbf{r})\right], \tag{8.22}$$

which results in the charge density ρ_{el},

$$\rho_{\mathrm{el}}(\mathbf{r}) = Ze\big[c_+(\mathbf{r}) - c_-(\mathbf{r})\big] = -2Zec_0 \sinh\left[\frac{Ze}{k_{\mathrm{B}}T}\,\phi(\mathbf{r})\right]. \tag{8.23}$$

Expressing the charge density in terms of the potential using the Poisson equation (8.3b) leads to a differential equation, the so-called Poisson–Boltzmann equation, for the electrical potential,

$$\nabla^2\phi(\mathbf{r}) = 2\frac{Zec_0}{\epsilon} \sinh\left[\frac{Ze}{k_{\mathrm{B}}T}\,\phi(\mathbf{r})\right], \tag{8.24}$$

which can be solved numerically or in some special cases analytically. One analytical solution can be obtained in the case of a planar surface in the xy-plane at $z = 0$ and the electrolyte occupying the $z > 0$ half-space. Due to translation symmetry in the xy-plane the problem becomes one-dimensional and ϕ depends only on z, the direction perpendicular to the surface plane. The resulting, so-called Gouy–Chapman solution, is

$$\phi(z) = \frac{4k_{\mathrm{B}}T}{Ze}\,\mathrm{arctanh}\left[\tanh\left(\frac{Ze\zeta}{4k_{\mathrm{B}}T}\right)\exp\left(-\frac{z}{\lambda_{\mathrm{D}}}\right)\right], \tag{8.25}$$

where

$$\lambda_{\mathrm{D}} \equiv \sqrt{\frac{\epsilon k_{\mathrm{B}}T}{2(Ze)^2c_0}} \tag{8.26}$$

is the so-called Debye length to be derived in the next section. In Exercise 8.2 we prove that Eq. (8.25) indeed is a solution to the Debye-layer problem Eq. (8.24). For weak, binary electrolytes with an ionic concentration around 1 mM = 1 mol/m^3 and a dielectric constant equal that of water, $\epsilon = 78\epsilon_0$, we find the following value of λ_{D} at room temperature:

$$\lambda_{\mathrm{D}} \approx \sqrt{\frac{1\ \mathrm{mM}}{Z^2\,c_0}} \times 9.6\ \mathrm{nm}. \tag{8.27}$$

8.3.2 The Debye–Hückel approximation for the Debye layer

To gain insight into the physics of the Debye layer and to build up our intuition, we shall now study the so-called Debye–Hückel approximation. The approximation is valid when the electrical energy is small compared to the thermal energy, i.e. in the

$$\text{Debye–Hückel limit} \quad Ze\zeta \ll k_{\mathrm{B}}T. \tag{8.28}$$

In this limit, i.e. for zeta-potentials much less than 26 mV at room temperature, we can employ the Taylor expansion $\sinh(u) \approx u$ in Eq. (8.24) and obtain the simple equation.

$$\nabla^2\phi(\mathbf{r}) = 2\frac{(Ze)^2c_0}{\epsilon k_{\mathrm{B}}T}\,\phi(\mathbf{r}) \equiv \frac{1}{\lambda_{\mathrm{D}}^2}\,\phi(\mathbf{r}), \tag{8.29}$$

which explains why λ_{D}, given by Eq. (8.26), is introduced.

A planar surface in the xy-plane at $z = 0$ is the first special case that we solve analytically. Eq. (8.29) becomes the simple second-order ordinary differential equation,

$$\partial_z^2 \phi(z) = \frac{1}{\lambda_\mathrm{D}^2}\, \phi(z), \tag{8.30}$$

which, given the boundary conditions Eq. (8.21), has the exponential solution,

$$\phi(z) = \zeta \, \exp\left[-\frac{z}{\lambda_\mathrm{D}}\right] \quad (z > 0, \text{ single-plate wall}). \tag{8.31}$$

The charge density ρ_el in the Debye layer corresponding to the potential Eq. (8.31) is found by using the Poisson equation (8.3b),

$$\rho_\mathrm{el}(z) = -\epsilon \partial_z^2 \phi(z) = -\frac{\epsilon\zeta}{\lambda_\mathrm{D}^2}\, \exp\left[-\frac{z}{\lambda_\mathrm{D}}\right] \quad (z > 0, \text{ single-plate wall}). \tag{8.32}$$

The ionic densities $c_-(z)$ and $c_+(z)$ are found directly from Eq. (8.22) in the Debye–Hückel approximation by Taylor-expanding the exponential function,

$$c_\pm(z) = c_0 \left[1 \mp \frac{Ze\zeta}{k_\mathrm{B}T}\, \exp\left[-\frac{z}{\lambda_\mathrm{D}}\right]\right] \quad (z > 0, \text{ single-plate wall}). \tag{8.33}$$

It is seen how the density of co-ions is suppressed near the surface, while that of the counterions are enhanced. This result is sketched in Fig. 8.3(c). Note that Eq. (8.33) has been derived under the assumption limit $Ze\zeta \ll k_\mathrm{B}T$, so the ionic densities are always positive.

The infinite parallel-plate channel with surfaces at $z = \pm h/2$ is the second special case that we solve analytically. As before the potential ϕ only depends on z and the Poisson–Boltzmann equation is given by Eq. (8.30), but now the boundary conditions are $\phi(\pm h/2) = \zeta$. As this problem is symmetric about $z = 0$ the solution involves $\cosh(x/\lambda_\mathrm{D})$ rather than $\exp(-x/\lambda_\mathrm{D})$, and the resulting potential is seen to be

$$\phi(z) = \zeta\, \frac{\cosh\left(\frac{z}{\lambda_\mathrm{D}}\right)}{\cosh\left(\frac{h}{2\lambda_\mathrm{D}}\right)} \quad \left(-\frac{h}{2} < z < \frac{h}{2}, \text{ parallel-plate channel}\right). \tag{8.34}$$

As above, the charge density ρ_el, see Fig. 9.1, follows from the Poisson equation,

$$\rho_\mathrm{el}(z) = -\frac{\epsilon\zeta}{\lambda_\mathrm{D}^2}\, \frac{\cosh\left(\frac{z}{\lambda_\mathrm{D}}\right)}{\cosh\left(\frac{h}{2\lambda_\mathrm{D}}\right)} \quad \left(-\frac{h}{2} < z < \frac{h}{2}, \text{ parallel-plate channel}\right). \tag{8.35}$$

The circular-shaped channel with surfaces at radius $r = a$ is the third and last analytical solution presented here. Employing the boundary condition $\phi(a) = \zeta$, the symmetry of the problem, as in Eq. (3.41), dictates the use of cylindrical co-ordinates without angular dependences. The Poisson–Boltzmann equation becomes

$$\left[\partial_r^2 + \frac{1}{r}\,\partial_r\right]\phi(r) = \frac{1}{\lambda_\mathrm{D}^2}\, \phi(r) \quad (0 < r < a, \text{ circular channel}). \tag{8.36}$$

This is recognized as the modified Bessel differential equation of order zero, so the solution involves the modified Bessel function of order 0,

$$\phi(r) = \zeta\,\frac{I_0\!\left(\frac{r}{\lambda_\mathrm{D}}\right)}{I_0\!\left(\frac{a}{\lambda_\mathrm{D}}\right)} \qquad (0 < r < a,\ \text{circular channel}). \tag{8.37}$$

The charge density $\rho_\mathrm{el}(r)$ follows from the Poisson and Poisson–Boltzmann equations as

$$\rho_\mathrm{el}(r) = -\epsilon\nabla^2\phi(r) = -\frac{\epsilon}{\lambda_\mathrm{D}^2}\,\phi(r) = -\frac{\epsilon\zeta}{\lambda_\mathrm{D}^2}\frac{I_0\!\left(\frac{r}{\lambda_\mathrm{D}}\right)}{I_0\!\left(\frac{a}{\lambda_\mathrm{D}}\right)} \qquad (0 < r < a,\ \text{circular channel}). \tag{8.38}$$

8.3.3 Surface charge and the Debye-layer capacitance

The Debye layer acts as an electrical capacitor since it accumulates electrical charge as a response to the electrical potential difference ζ between the surface and the bulk, see Fig. 8.4. In the following, we study this property.

One way to obtain the capacitance of the Debye layer is by integrating the charge density $\rho_\mathrm{el}(x)$ of Eq. (8.32) along the z direction from the surface at $z = 0$ to infinity, $z = \infty$. The result q_liq is the charge per area $\mathcal{A}$ that is contained in the liquid in the direction perpendicular to any small area $\mathcal{A}$ on the surface,

$$q_\mathrm{liq} = \int_0^\infty \mathrm{d}z\,\rho_\mathrm{el}(z) = \int_0^\infty \mathrm{d}z\left[-\frac{\epsilon\zeta}{\lambda_\mathrm{D}^2}\exp\left[-\frac{z}{\lambda_\mathrm{D}}\right]\right] = -\frac{\epsilon}{\lambda_\mathrm{D}}\,\zeta. \tag{8.39}$$

From this linear relation between charge per area and applied potential difference we can immediately read off the capacitance per area C_D of the Debye layer in thermal equilibrium,

$$C_\mathrm{D} \equiv \frac{\epsilon}{\lambda_\mathrm{D}}. \tag{8.40}$$

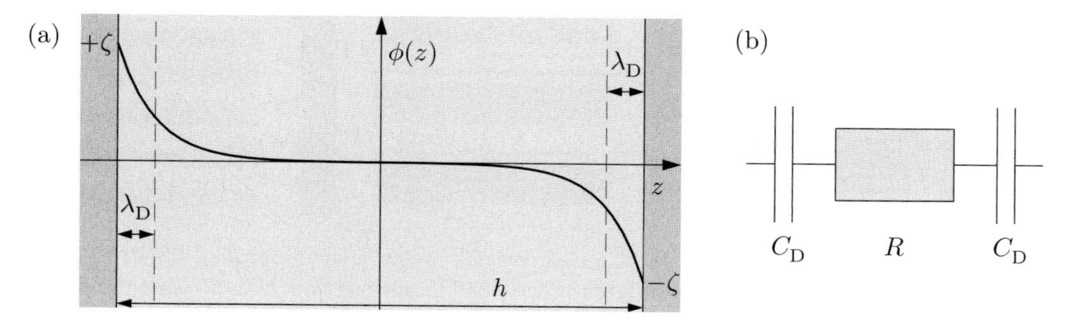

Fig. 8.4 (a) An electrolyte (light gray) occupying the space of height h between a set of parallel-plate metallic electrodes (dark gray). When applying a voltage difference 2ζ between the two electrodes, a Debye layer of width λ_D builds up on each of them. (b) The electrical equivalent diagram of the system shown in panel (a) consisting of one capacitor C_D for each Debye layer and one resistor R for the bulk electrolyte.

Using $\lambda_\mathrm{D} = 9.6$ nm and $\epsilon = 78\epsilon_0$ we find the value

$$C_\mathrm{D} = 0.072\ \mathrm{F\,m^{-2}}. \tag{8.41}$$

We can check the above result by calculating the surface charge per area q_surf that is accumulated on the surface. This is done by using a standard Gauss-box argument. Imagine a flat box of surface area $\mathcal{A}$ and a vanishingly small thickness placed parallel to the surface such that the surface lies inside the box. The symmetry of the problem dictates that there is only a non-zero electric field through the area $\mathcal{A}$ on the liquid side of the surface and, moreover, the electric field is perpendicular to this surface. Thus, the total charge inside the box is $\mathcal{A}q_\mathrm{surf} = \epsilon E\mathcal{A}$, and we have

$$q_\mathrm{surf} = \epsilon E = -\epsilon\partial_z\phi(0) = \frac{\epsilon}{\lambda_\mathrm{D}}\,\zeta. \tag{8.42}$$

As expected, this result is exactly the opposite of charge per area in the liquid, so indeed the Debye layer acts as a charge-neutral capacitor, where the solid surface and the electrolyte are the two "plates" of the capacitor.

Since the Debye layer acts as a capacitor and since the electrolyte has a finite conductivity σ_el or resistivity $1/\sigma_\mathrm{el}$ we should be able to ascribe a characteristic RC time τ_{RC} to the system. We consider an electrolyte sandwiched between a set of parallel-plate metallic electrodes, see Fig. 8.4. The distance between the electrodes is denoted h. When one electrode is biased by the voltage $+\zeta$ and the other by $-\zeta$ a Debye layer builds up on each of them. The equivalent diagram of the system consist of a series coupling of one capacitor for each Debye layer and one resistor for the bulk electrolyte. The RC time for this system is now found as

$$\tau_{RC} = RC = \left(\frac{h}{\sigma_\mathrm{el}\mathcal{A}}\right)\left(\frac{1}{2}\,\frac{\epsilon}{\lambda_\mathrm{D}}\,\mathcal{A}\right) = \frac{\epsilon}{2\sigma_\mathrm{el}}\,\frac{h}{\lambda_\mathrm{D}}. \tag{8.43}$$

The value of the RC time is found by using our standard values for the parameters: $\lambda_\mathrm{D} =$ 9.6 nm, $h = 100$ µm, $\epsilon = 78\epsilon_0$, and $\sigma_\mathrm{el} = 10^{-3}$ S/m,

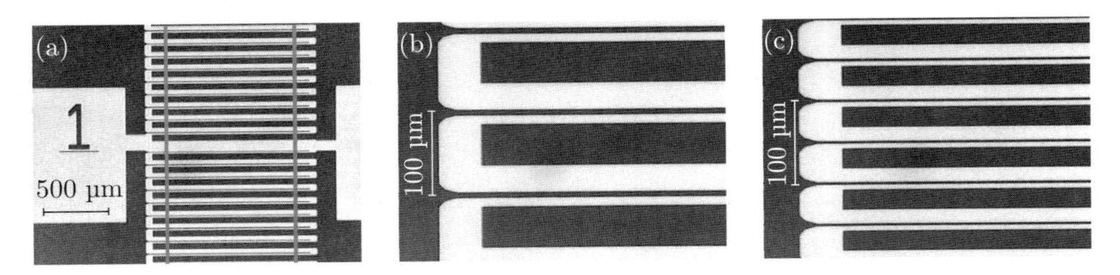

Fig. 8.5 (a) An optical micrograph of an asymmetric array of interdigitated Ti/Pt-electrodes (black) placed on top of a glass-based substrate (white). The vertical gray lines indicate the position of the side walls of a 33 µm high and 1 mm wide microfluidic channel situated directly on top of the electrode array to form the active part of the AC electro-osmotic micropump shown in Fig. 14.4. (b) A zoom-in on an electrode array with a repetition period of 100 µm. The width of the narrow and wide electrodes is 8 µm and 51 µm, respectively. (c) A zoom-in on an another electrode array with a repetition period of 50 µm. Here, the width of the narrow and wide electrodes is 4 µm and 26 µm, respectively. Micrographs courtesy of Misha Marie Gregersen, DTU Nanotech.

$$\tau_{RC} = 3.6 \text{ ms}. \tag{8.44}$$

For processes slower than 3.6 ms enough time is available for establishing the Debye layer. However, the Debye layer cannot follow faster processes. In AC experiments this translates into a characteristic frequency ω_c of the electrolyte including the Debye layer,

$$\omega_\mathrm{c} = \frac{2\pi}{\tau_{RC}} = 1.7 \times 10^3 \text{ rad/s}. \tag{8.45}$$

For frequencies higher than a few kHz the Debye layer is not established. In Fig. 8.5 is shown a microfluidic device with surface electrodes, fabricated in the author's group at MIC, to study and to utilize the RC time effects of the Debye layer in electrolytes. When biasing the electrodes with an AC voltage, the device pumps an electrolyte through the microchannel.

8.3.4 Electrophoresis and Debye-layer screening

In Section 8.2.1 we studied electrophoresis in the case of a non-conducting liquid. This case was simple since the charge of the particle suspended in the liquid did not suffer any electrical screening. The opposite limit, where the liquid is a highly conducting electrolyte, is also a simple case. Here, the particle charge is completely screened by the ions of the electrolyte within a distance of the Debye screening length λ_D. Since the effective charge of the particle in this case is zero it cannot move by electrophoresis, but only by dielectric forces, a process called dielectrophoresis.

Let us solve the Debye-layer problem for a charged spherical particle of radius a. The problem is spherically symmetric, so the potential due to the particle can only depend on the radial co-ordinate r, $\phi(\mathbf{r}) = \phi(r)$. The boundary conditions for $\phi(r)$ are

$$\phi(a) = \zeta, \qquad \phi(\infty) = 0. \tag{8.46}$$

After employing the Debye–Hückel approximation, the Poisson–Boltzmann equation (8.24) becomes

$$\frac{1}{r^2}\,\partial_r\Big(r^2\,\partial_r\phi\Big) = \frac{1}{\lambda_\mathrm{D}^2}\,\phi \quad \text{(spherical polar co-ordinates)}, \tag{8.47}$$

where the Laplace operator in spherical co-ordinates is simplified due to the lack of angular dependence in the problem. This differential equation is solved by the standard substitution $\psi(r) \equiv r\,\phi(r)$, since Eq. (8.47) is then transformed into the simpler equation

$$\partial_r^2\psi = \frac{1}{\lambda_\mathrm{D}^2}\,\psi \quad \text{(spherical symmetry)}, \tag{8.48}$$

with the straightforward exponential solutions $\psi(r) \propto \exp(\pm r/\lambda_\mathrm{D})$. Going back from $\psi(r)$ to $\phi(r)$ and employing the boundary conditions Eq. (8.46) yields the solution

$$\phi(r) = \zeta\,\frac{a}{r}\,\exp\left[\frac{a-r}{\lambda_\mathrm{D}}\right] \quad (r > a, \text{ spherical symmetry}). \tag{8.49}$$

This solution, which has the form of a modified or screened Coulomb potential, corresponds to a collection of counterions around the charged particle. This charge collection is known as a screening cloud. Just a few times the Debye length λ_D away for the particle, its charge cannot be observed, it is completely screened. For strong electrolytes, where λ_D is very small, the originally charged particle becomes charge-neutral for all practical purposes.

In the intermediate case of moderate Debye length, the electrophoresis problem becomes complicated. The motion of the particle distorts the screening cloud, which becomes asymmetric, resulting in very complex interactions between the electrolyte, the screening cloud and the particle. This topic is beyond the scope of this chapter.

In the case of very long Debye lengths, i.e. for non-conducting liquids, we recover the simple unscreened charged particle studied in Section 8.2.1.

8.4 Further reading

Two classic textbooks on the fundamentals of electrostatics and electrodynamics of continuous media are Jackson (1975) and Landau, Lifshitz, and Pitaevskiĭ, L. P. (1984). Among the books more focused on electrokinetics and electric handling of suspensions Probstein (1994) and Jones (1995) are very useful. For a recent review with a comprehensive list of references regarding the electrokinetics of Debye layers and diffuse-charge dynamics the reader should consult Bazant, Thornton and Ajdari (2004).

8.5 Exercises

Exercise 8.1
The dielectric force and the point dipole
Consider the electric force $\mathbf{F}_\mathrm{el}$ on a particle.

(a) Write the dielectrophoretic force in vector notation instead of the index notation employed in Eq. (8.4).

(b) Verify the expression $\mathbf{p} = q\mathbf{d}$, Eq. (8.7), for the point dipole, and write in vector notation the dielectric force acting on it.

Exercise 8.2
The analytic solution for the Debye-layer potential in 1D
In this exercise we prove that $\phi(x)$ given in Eq. (8.25) indeed is a solution to the Debye-layer problem for a planar charged surface.

(a) Use the following substitutions $x \equiv \lambda_\mathrm{D}\tilde{x}$, $\phi \equiv \zeta\tilde{\phi}$ and $\alpha \equiv Ze\zeta/(k_\mathrm{B}T)$ to show that the Poisson–Boltzmann equation (8.24) can be written in the dimensionless form

$$\partial_{\tilde{x}}^2 \tilde{\phi}(\tilde{x}) = \frac{1}{\alpha}\,\sinh\big[\alpha\tilde{\phi}(\tilde{x})\big]. \tag{8.50}$$

(b) Show that Eq. (8.50) can be rewritten by use of the substitution $u(\tilde{x}) \equiv \alpha\tilde{\phi}(\tilde{x})$ to the form

$$u'' = \sinh(u) \quad \Rightarrow \quad \Big(\tfrac{1}{2}\big[u'\big]^2\Big)' = \Big(\cosh(u)\Big)', \quad u = \alpha\tilde{\phi}, \tag{8.51}$$

where prime means derivative with respect to $\tilde{x}$.

(c) Use the boundary condition $u(\infty) = u'(\infty) = 0$ and the physical insight $u'(\tilde{x}) < 0$ to argue that Eq. (8.51) can be integrated once to yield the result

$$u' = -\sqrt{2\cosh(u) - 2} = -2\sinh\big(\tfrac{1}{2}u\big), \quad u = \alpha\tilde{\phi}, \tag{8.52}$$

where the hyperbolic relation $\cosh(u) = 2\sinh^2(u/2) + 1$ has been used.

(d) Now introduce the new function $v(\tilde{x}) = u(\tilde{x})/2 = \alpha\tilde{\phi}(\tilde{x})/2$ with the boundary conditions $v(0) = \alpha/2$ and $v(\infty) = 0$. Show that Eq. (8.52) can be rewritten and integrated by separation of the variables v and $\tilde{x}$ as follows:

$$\int_{\frac{\alpha}{2}}^{v(\tilde{x})} \frac{\mathrm{d}v}{\sinh(v)} = -\int_0^{\tilde{x}} \mathrm{d}\tilde{x} \quad \Rightarrow \quad \left[\log\left[\tanh\left(\frac{v}{2}\right)\right]\right]_{\frac{\alpha}{2}}^{v(\tilde{x})} = -\tilde{x}, \quad v = \frac{\alpha}{2}\,\tilde{\phi}. \tag{8.53}$$

(e) Show that Eq. (8.53) leads to the following form of the final expression for $\tilde{\phi}(\tilde{x})$,

$$\tilde{\phi}(\tilde{x}) = \frac{4}{\alpha}\,\tanh^{-1}\left[\tanh\left(\frac{\alpha}{4}\right)\mathrm{e}^{-\tilde{x}}\right]. \tag{8.54}$$

Exercise 8.3
The low-voltage limit of the Debye-layer potential in 1D
Consider the Debye–Hückel approximation in the 1D case of Section 8.3.2.

(a) Show by Taylor expansion of Eq. (8.25) that the exact solution and the Debye–Hückel approximation, Eq. (8.31), agree in the low-voltage limit $Ze\zeta \ll k_{\mathrm{B}}T$.

(b) At which value of ζ is the relative error made in the Debye–Hückel approximation 10% of the exact result?

Exercise 8.4
The analytic solution for the Debye-layer potential inside a cylindrical channel
Prove that $\phi(r)$ given in Eq. (8.37) is the solution to the Poisson–Boltzmann equation (8.36) for the straight channel with circular cross-section of radius a given the boundary conditions $\phi(a) = \zeta$ and $\partial_r\phi(0) = 0$ (no cusps in the potential along the center axis).

Exercise 8.5
Surface charge in the Debye layer of the parallel-plate channel
Determine within the Debye–Hückel approximation the surface charge density q_{surf} of the parallel-plate channel of height h, and verify in analogy with the single-wall results Eqs. (8.39) and (8.42) that q_{surf} is opposite to the charge density per area q_{liq} in the electrolyte.

Exercise 8.6
The simple model for the RC time of the Debye layer
Consider the equivalent circuit diagram in Fig. 8.4(b).

(a) Express the resulting capacitance in terms of C_{D}.

Let $V_{\mathrm{ext}}(t)$ be some externally applied voltage driving current $I(t)$ from one wall to the other. We assume that the intrinsic zeta-potential is zero.

(b) Find the differential equation for $I(t)$ and discuss the role of the RC time $\tau_{RC} = RC$ in the solution.

(c) Derive Eq. (8.43) and discuss the physical assumptions made and how the various parameters influence τ_{RC}.

Exercise 8.7
The Debye-layer potential of a charged sphere
Consider the charged sphere of radius a immersed in an electrolyte as described in Section 8.3.4.

(a) Check that the Laplace operator in spherical co-ordinates, when no angular dependence is present, is the one employed in Eq. (8.47).

(b) Verify that the substitution $\psi(r) = r\phi(r)$ indeed transforms Eq. (8.47) into Eq. (8.48).

(c) Prove that the solution Eq. (8.49) is correct, and calculate the charge density $\rho_{\mathrm{el}}(r)$ of the screening cloud surrounding the sphere.

(d) Compare the physical implications of the form of $\phi(r)$ in Eq. (8.49) with the case of an unscreened charged particle.

8.6 Solutions

Solution 8.1
The dielectric force and the point dipole

Consider the electric force $\mathbf{F}_{\mathrm{el}}$ on a particle.

(a) In vector notation Eq. (8.4) becomes $\mathbf{F}_{\mathrm{el}} = Q\mathbf{E} + (\mathbf{p}\cdot\boldsymbol{\nabla})\mathbf{E}$.

(b) We first note that $\int \mathrm{d}\mathbf{r}\, f(\mathbf{r})\delta(\mathbf{r}^* - \mathbf{r}) = f(\mathbf{r}^*)$. Hence, inserting Eq. (8.6) into Eq. (8.5b) yields $\mathbf{p} = q\frac{1}{2}\mathbf{d} - q\big(-\frac{1}{2}\big)\mathbf{d} = q\mathbf{d}$. From (a) we then obtain $\mathbf{F}_{\mathrm{el}} = q(\mathbf{d}\cdot\boldsymbol{\nabla})\mathbf{E}$.

Solution 8.2
The analytic solution for the Debye-layer potential in 1D

Only a few details need to be filled in solving this exercise concerning Eq. (8.25).

(a) Using the substitutions $x \equiv \lambda_{\mathrm{D}}\tilde{x}$, $\phi \equiv \zeta\tilde{\phi}$ and $\alpha \equiv Ze\zeta/(k_{\mathrm{B}}T)$ is straightforward. It remains to note the $\partial_x = (\partial_x\tilde{x})\partial_{\tilde{x}} = \lambda_D^{-1}\partial_{\tilde{x}}$, and thus $\partial_x^2 = \lambda_D^{-2}\partial_{\tilde{x}}^2$.

(b) Using $u(\tilde{x}) \equiv \alpha\tilde{\phi}(\tilde{x})$ gives $u'' = \sinh(u)$. Moreover, $\big(\frac{1}{2}[u']^2\big)' = u'u''$, while $[\cosh(u)]' = \sinh(u)\, u'$, which upon division by u' proves the implication in Eq. (8.51).

(c) Integration of Eq. (8.51) gives $[u']^2 = 2\cosh(u) + \text{const}$. At $\tilde{x} = \infty$ this reads $[u'(\infty)]^2 = 2\cosh[\alpha\tilde{\phi}(\infty)] + \text{const} = 2 + \text{const}$, so the boundary condition $u'(\infty) = 0$ requires $\text{const} = -2$. When isolating u' the negative sign must be used to ensure $u' < 0$.

(d) Using $v = u/2$, Eq. (8.52) becomes $v' = -\sinh(v)$ or $\mathrm{d}v/\mathrm{d}\tilde{x} = \sinh(v)$. Separation of the variables leads to $\mathrm{d}v/\sinh(v) = \mathrm{d}\tilde{x}$, and employing the boundary condition $v(0) = \alpha/2$ leads to the integral Eq. (8.53). We note that $\big[\log[\tanh(v/2)]\big]' = [1/\tanh(v/2)][1/\cosh^2(v/2)]v'/2 = 1/[2\sinh(v/2)\cosh(v/2)]v' = v'/\sinh(v)$, so Eq. (8.53) is integrated correctly.

(e) Since $\Big[\log(z)\Big]_a^b = \log(a) - \log(b) = \log(a/b)$, Eq. (8.53) becomes $\log\big[\tanh(\alpha\tilde{\phi}/4)/\tanh(\alpha/4)\big] = -\tilde{x}$, and Eq. (8.54) results once $\tilde{\phi}$ is isolated.

Solution 8.3
The low-voltage limit of the Debye-layer potential in 1D

For $s = Ze\zeta/(4k_{\mathrm{B}}T) \ll 1$ we have the following Taylor expansions: $\tanh(s) \approx s - s^3/3$ and $\operatorname{arctanh}(s) \approx s + s^3/3$.

(a) Given the above, and noting that $0 < \exp(-z/\lambda_{\mathrm{D}}) < 1$ for $z > 0$, a Taylor expansion of Eq. (8.25) to lowest order in s becomes $\phi(z) \approx \frac{4k_{\mathrm{B}}T}{Ze}\,\frac{Ze\zeta}{4k_{\mathrm{B}}T}\,\mathrm{e}^{-z/\lambda_{\mathrm{D}}} = \zeta\,\mathrm{e}^{-z/\lambda_{\mathrm{D}}}$.

(b) There is not a unique answer to the question, as the error of an approximating function is not well defined. One possibility is to note that the Gouy–Chapman and the Debye–Hückel solutions agree at $z = 0$ and $z = \infty$, so it is natural to study the function values at $z = \lambda_{\mathrm{D}}$, which is the only distinctive length scale of the problem, and here $\mathrm{e}^{-z/\lambda_{\mathrm{D}}}$ is just $\frac{1}{\mathrm{e}}$. The Taylor expansion to third order in s becomes $\operatorname{arctanh}[\tanh(s)/\mathrm{e}] \approx [\tanh(s)/\mathrm{e}] + [\tanh(s)/\mathrm{e}]^3/3 \approx [s - s^3/3]/\mathrm{e} + [s^3/\mathrm{e}^3]/3 = [1 - s^2(1 - 1/\mathrm{e}^2)/3]\, s/\mathrm{e}$. The relative error is approximately below 10% when the first correction to the leading term is below 10%, i.e. when $s^2(1-1/e^2)/3 < 0.1$ or $s < 0.6$, implying $\zeta < 2.4\frac{k_\mathrm{B}T}{Ze} = 61\ \mathrm{mV}/Z$.

Solution 8.4
The analytic solution for the Debye-layer potential inside a cylindrical channel
If we introduce the dimensionless radial co-ordinate $s = r/\lambda_\mathrm{D}$ and thus $\partial_s = \lambda_\mathrm{D}\partial_r$, the Poisson–Boltzmann equation (8.36) straightforwardly rewritten to Bessel's modified differential equation of order zero: $s^2\partial_s^2\psi + s\partial_s\psi - s^2\psi = 0$, where $\psi(s) \equiv \phi(\lambda_\mathrm{D}s)$. The general solution is the linear combination of the modified Bessel functions of order zero, I_0 and K_0, of the first and second kind, respectively: $\psi(s) = c_1 I_0(s) + c_2 K_0(s)$. Since $K_0(s) \to \infty$ for $s \to 0$, its derivative is also diverging, and this function must be discarded due to the second boundary condition $\partial_s\psi(0) = 0$, i.e. $c_2 = 0$. The first boundary condition requires $\zeta = \phi(a) = \psi(a/\lambda_\mathrm{D}) = c_1 I_0(a/\lambda_\mathrm{D})$, and consequently $c_1 = \zeta/I_0(a/\lambda_\mathrm{D})$. In conclusion $\psi(s) = \zeta\, I_0(s)/I_0(a/\lambda_\mathrm{D})$.

Solution 8.5
Surface charge in the Debye layer of the parallel-plate channel
Due to symmetry the two plates at $z = \pm h/2$ have the same surface charge density, but when using the Gauss box argument involving $q_\mathrm{surf} = \epsilon\mathbf{E}\cdot\mathbf{n}$, the electric field $\mathbf{E} = -\partial_z\phi\mathbf{e}_z$ is multiplied with the surface normal $\mathbf{n}(-h/2) = +\mathbf{e}_z$ or $\mathbf{n}(h/2) = -\mathbf{e}_z$. Using this and $\partial_s\cosh(s) = \sinh(s)$ we find

$$q_\mathrm{surf}\Big(\pm\frac{h}{2}\Big) = \pm\epsilon\partial_z\phi\Big(\pm\frac{h}{2}\Big) = \pm\frac{\sinh\Big(\pm\frac{h}{2\lambda_\mathrm{D}}\Big)}{\cosh\Big(\frac{h}{2\lambda_\mathrm{D}}\Big)}\frac{\epsilon\zeta}{\lambda_\mathrm{D}} = \tanh\Big(\frac{h}{2\lambda_\mathrm{D}}\Big)\frac{\epsilon\zeta}{\lambda_\mathrm{D}}. \tag{8.55}$$

The charge per area q_liq in the liquid is found to balance the surface charges by using the Poisson equation,

$$q_\mathrm{liq} = -\epsilon\int_{-\frac{h}{2}}^{\frac{h}{2}} \mathrm{d}z\,\partial_z^2\phi = -\epsilon\partial_z\phi\big(\tfrac{h}{2}\big) - (-\epsilon)\partial_z\phi\big(-\tfrac{h}{2}\big) = -\Big[q_\mathrm{surf}\Big(\frac{h}{2}\Big) + q_\mathrm{surf}\Big(-\frac{h}{2}\Big)\Big]. \tag{8.56}$$

Solution 8.6
The simple model for the *RC* time of the Debye layer
Consider the equivalent circuit diagram in Fig. 8.4(b).

(a) The resulting capacitance per area $\mathcal{A}$ is given by the two Debye-layer capacitors in series, i.e. $C/\mathcal{A} = (1/C_\mathrm{D} + 1/C_\mathrm{D})^{-1} = C_\mathrm{D}/2$. So $C = \mathcal{A}C_\mathrm{D}/2$.

(b) Let $V_\mathrm{ext}(t)$ be the total voltage drop across the series coupling from left to right, which also defines the positive direction of the current $I(t)$. Using the capacitor equation $V_\mathrm{D} = q_\mathrm{surf}/C_\mathrm{D}$ for the voltage drop across one Debye layer, and Ohm's law $V_\mathrm{R} = RI$ for the voltage drop across the resistor, we find $V_\mathrm{ext} = 2q_\mathrm{surf}/C_\mathrm{D} + RI$. Dividing by R and taking the time derivative we arrive at $\partial_t I + \frac{1}{\tau_{RC}}I = \partial_t V_\mathrm{ext}/R$, where we have used $I = \mathcal{A}\partial_t q_\mathrm{surf}$ and introduced $\tau_{RC} \equiv RC = R\mathcal{A}C_\mathrm{D}/2$. It is seen that τ_{RC} is the characteristic time scale for charging the Debye layers.

(c) We just found $\tau_{RC} = R\mathcal{A}C_\mathrm{D}/2$. Since $R = h/(\mathcal{A}\sigma_\mathrm{el})$ for a resistor of length h, area $\mathcal{A}$, and resistivity $1/\sigma_\mathrm{el}$, and since $C_\mathrm{D} = \epsilon/\lambda_\mathrm{D}$ we obtain $\tau_{RC} = [h/(\mathcal{A}\sigma_\mathrm{el})][\mathcal{A}\epsilon/(2\lambda_\mathrm{D})] = \epsilon h/(2\sigma_\mathrm{el}\lambda_\mathrm{D})$. We see that fast charging occurs for small channel heights h, large Debye lengths

λ_D, and large conductivity σ_{el}. As was explicitly stated when discussing the basic equation for the chemical potential Eq. (8.20), the main assumption leading to the RC model is that at each moment during the charging the Debye layer is close to thermodynamic equilibrium.

Solution 8.7
The Debye-layer potential of a charged sphere
Spherical polar co-ordinates are treated in Section C.3.2.

(a) For spherical symmetric scalar fields, where no angular dependence is present, only the first term in Eq. (C.33) is non-zero, i.e. $\nabla^2\phi = \frac{1}{r^2}\partial_r(r^2\partial_r\phi)$ as used in Eq. (8.47).

(b) First, we write Eq. (8.47) as $\frac{1}{r}\partial_r(r^2\partial_r\phi) = r\phi/\lambda_D^2 = \psi/\lambda_D^2$, where $\phi(r) = \psi(r)/r$ is used on the right-hand side. For the left-hand side we get $r^2\partial_r\phi = r^2(\partial_r\psi/r - \psi/r^2) = r\partial_r\psi - \psi$. So $\frac{1}{r}\partial_r(r\partial_r\psi - \psi) = \frac{1}{r}(r\partial_r^2\psi + \partial_r\psi - \partial_r\psi) = \partial_r^2\psi$. In conclusion $\partial_r^2\psi = \psi/\lambda_D^2$.

(c) Equation (8.49) yields $\phi(a) = \zeta$ and $\phi(\infty) = 0$, thus fulfilling the required boundary conditions. Moreover, multiplying the equation by r leads to $\psi(r) = c\exp(-r/\lambda_D)$, where the constant is $c = a\zeta\exp(a/\lambda_D)$, which clearly is a solution to the Poisson–Boltzmann equation (8.48). The charge density is given by

$$\rho_{el}(r) = -\epsilon\nabla^2\phi = -\frac{\epsilon}{\lambda_D^2}\phi = -\frac{\epsilon\zeta}{\lambda_D^2}\frac{a}{r}\exp\left[\frac{a-r}{\lambda_D}\right]. \tag{8.57}$$

(d) The potential around an unscreened charged particle is the well-known long-range Coulomb potential $\zeta a/r$. For the screened charged particle the range of the Coulomb potential has been shortened by the appearance of the exponential factor $\exp[(a-r)/\lambda_D]$ having the characteristic screening length λ_D.

9 Electroosmosis

Electro-osmosis is a non-equilibrium effect, where a liquid is brought to move relative to a charged surface by an applied external potential gradient $\boldsymbol{\nabla}\phi_{\mathrm{ext}}$. Therefore, to obtain a complete electrohydrodynamical transport theory, we need to supplement the diffusion-convection equation (5.17) with an electrically induced current density $\mathbf{J}^{\mathrm{el}}$.

9.1 Electrohydrodynamic transport theory

The symbol $\mathbf{J}^{\mathrm{el}}_{\alpha}$ usually refers to the electrical current density of ion α, so we introduce the symbol $\tilde{\mathbf{J}}^{\mathrm{el}}_{\alpha} = \mathbf{J}^{\mathrm{el}}_{\alpha}/(Z_{\alpha}e)$ for the particle current density of ion α having the valence Z_{α}. Likewise, the particle current densities due to convection and diffusion are the usual mass current densities divided by the ionic mass m_{α}, $\tilde{\mathbf{J}}^{\mathrm{conv}}_{\alpha} = \mathbf{J}^{\mathrm{conv}}_{\alpha}/m_{\alpha}$ and $\tilde{\mathbf{J}}^{\mathrm{diff}}_{\alpha} = \mathbf{J}^{\mathrm{diff}}_{\alpha}/m_{\alpha}$. Combining Eqs. (5.17) and (8.2d) we obtain one of the governing equations for transport in electrohydrodynamics, the so-called Nernst–Planck equation for the current density $\tilde{\mathbf{J}}_{\alpha}$,

$$\tilde{\mathbf{J}}_{\alpha} \equiv \tilde{\mathbf{J}}^{\mathrm{conv}}_{\alpha} + \tilde{\mathbf{J}}^{\mathrm{diff}}_{\alpha} + \tilde{\mathbf{J}}^{\mathrm{el}}_{\alpha} = c_{\alpha}\mathbf{v} - D_{\alpha}\boldsymbol{\nabla}c_{\alpha} - \frac{\sigma^{\mathrm{el}}_{\alpha}}{Z_{\alpha}e}\boldsymbol{\nabla}\phi, \tag{9.1}$$

where c_{α} is the particle density, or concentration, of ion α. Naturally, in the spirit of linear response theory, the gradients appearing in this equation (note that $\mathbf{v}$ can be thought of arising from pressure gradients) relate only to pressure p_{ext}, concentration c_{ext}, and potential ϕ_{ext} applied externally on top of the equilibrium fields p_{eq}, c_{eq}, and ϕ_{eq},

$$p = p_{\mathrm{eq}} + p_{\mathrm{ext}}, \tag{9.2a}$$

$$c = c_{\mathrm{eq}} + c_{\mathrm{ext}}, \tag{9.2b}$$

$$\phi = \phi_{\mathrm{eq}} + \phi_{\mathrm{ext}}. \tag{9.2c}$$

The reason is that in equilibrium (in the center-of-mass system) there are no current densities, even though the fields themselves may be non-zero. We have already used this fact implicitly in Section 3.4.1, where for a horizontally placed channel the gravitational force was balanced by minus the gradient of the hydrostatic pressure $p_{\mathrm{eq}}(z) = -\rho g z$, and in Section 8.3.1, where the electrical force in the Debye layer was balanced by minus the gradient of a concentration-dependent pressure, see Eq. (8.20).

These considerations will be used in the following when analyzing the electro-osmotic effect, which is based on moving the ions in the Debye layer by an external potential.

9.2 Ideal electro-osmotic flow

The principle of electro-osmotic (EO) flow is shown in Fig. 9.1. Two metallic electrodes are situated at each end of a channel, in which charge separation at the walls has led to the

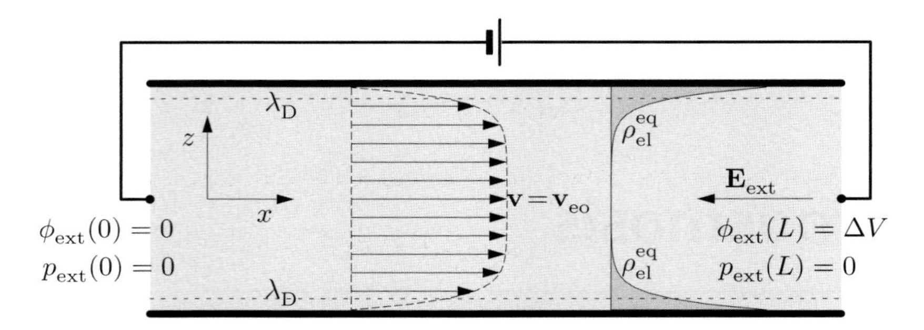

Fig. 9.1 The velocity profile $\mathbf{v}$ (dashed line and arrows) and the negative Debye-layer charge density profile ρ_{el}^{eq} (dark gray and full line) in an ideal electro-osmotic (EO) flow inside a cylindrical channel of radius a and positively charged walls (thick horizontal lines). The EO flow is induced by the external potential difference $\Delta\phi_{ext} = \Delta V$ resulting in the homogeneous electric field $\mathbf{E}_{ext}$. Note how the velocity profile reaches the constant value v_{eo} at a distance of a few times the Debye length λ_D from the walls. No pressure drop is present along the channel in this ideal case.

formation of an equilibrium Debye layer. When a DC potential difference $\Delta V = \Delta\phi_{ext}$ is applied over the electrodes the resulting electrical field $\mathbf{E}_{ext}$,

$$\mathbf{E}_{ext} \equiv -\boldsymbol{\nabla}\phi_{ext}, \tag{9.3}$$

exerts a body force $\rho_{el}^{eq}\mathbf{E}_{ext}$ on the Debye layer, which begins to move and then by viscous drag pulls the charge neutral bulk liquid along. If no electrochemical processes occur at the electrodes the motion stops after a very short time (of the order µs) when the electrodes are screened by the formation of a Debye layer around them. If, however, electrochemical processes, e.g. electrolysis, can take place at the electrodes such a charge build-up is prevented, and electrical currents can flow in the system and thus move the liquid by viscous drag. In the following we derive an expression for the resulting electro-osmotic velocity field in the liquid.

Given the equilibrium charge density $\rho_{el}^{eq}(\mathbf{r})$ of the Debye layer the Navier–Stokes equation to be used for analyzing EO flows is

$$\rho\Big(\partial_t\mathbf{v} + (\mathbf{v}\cdot\boldsymbol{\nabla})\mathbf{v}\Big) = -\boldsymbol{\nabla}p_{ext} + \eta\nabla^2\mathbf{v} - \rho_{el}^{eq}\boldsymbol{\nabla}\phi_{ext}. \tag{9.4}$$

Note that by assumption the external potential does not introduce any changes in the charge density.

We define an ideal EO flow by the following four conditions: (i) the ζ potential is constant along the wall, (ii) the electrical field is homogeneous, (iii) the flow is in steady state, and (iv) the Debye length is much smaller than the radius or half-width a of the channel, $\lambda_D \ll a$.

As the first explicit example we study the infinite parallel-plate channel with the positively charged walls placed parallel to the xy-plane at $z = -h/2$ and $z = h/2$. The external electrical field is applied in the negative x direction, $\mathbf{E} = -E\mathbf{e}_x$, and the external pressure gradient is put to zero. The symmetry of this ideal EO flow setup dictates the following structure of the fields,

$$\nabla\phi_{\mathrm{ext}}(\mathbf{r}) = -\mathbf{E} = E\,\mathbf{e}_x, \tag{9.5a}$$

$$\nabla p_{\mathrm{ext}}(\mathbf{r}) = \mathbf{0}, \tag{9.5b}$$

$$\mathbf{v}(\mathbf{r}) = v_x(z)\,\mathbf{e}_x, \tag{9.5c}$$

and only the x component of the steady-state Navier–Stokes equation is non-trivial,

$$0 = \eta\partial_z^2\, v_x(z) + \Big[\epsilon\partial_z^2\phi_{\mathrm{eq}}(z)\Big]\, E. \tag{9.6}$$

The solution almost presents itself by rewriting this equation as

$$\partial_z^2\left[v_x(z) + \frac{\epsilon E}{\eta}\,\phi_{\mathrm{eq}}(z)\right] = 0. \tag{9.7}$$

Employing the boundary conditions

$$v_x\Big(\pm\tfrac{h}{2}\Big) = 0, \tag{9.8}$$

we obtain the solution

$$v_x(z) = \Big[\zeta - \phi_{\mathrm{eq}}(z)\Big]\,\frac{\epsilon E}{\eta}. \tag{9.9}$$

The two-wall potential $\phi_{\mathrm{eq}}(z)$ is given by Eq. (8.34), which combined with Eq. (9.9) gives

$$v_x(z) = \left[1 - \frac{\cosh\left(\frac{z}{\lambda_{\mathrm{D}}}\right)}{\cosh\left(\frac{h}{2\lambda_{\mathrm{D}}}\right)}\right] v_{\mathrm{eo}}, \tag{9.10}$$

where we have introduced the EO velocity v_{eo} defined by the Helmholtz–Smoluchowski relation as

$$v_{\mathrm{eo}} \equiv \frac{\epsilon\zeta}{\eta}\, E. \tag{9.11}$$

This expression is analogous to electrophoresis and ionic mobility, Eq. (8.14), and we quite naturally define the EO mobility μ_{eo} as

$$\mu_{\mathrm{eo}} \equiv \frac{v_{\mathrm{eo}}}{E} = \frac{\epsilon\zeta}{\eta}. \tag{9.12}$$

Typical values for EO flow are

$$\zeta \approx 100\ \mathrm{mV}, \qquad \mu_{\mathrm{eo}} \approx 7\times 10^{-8}\ \mathrm{m^2\,(V\,s)^{-1}}, \qquad v_{\mathrm{eo}} \approx 1\ \mathrm{mm\,s^{-1}}. \tag{9.13}$$

Unfortunately, we note that $e\zeta/k_{\mathrm{B}}T \approx 4$, and consequently the simplifying Debye-Hückel approximation $e\zeta/k_{\mathrm{B}}T \approx 1$ cannot be expected to be valid. However, qualitatively the results obtained using this approximation are generally quite good.

For the ideal EO flow we obtain the simple velocity profile

$$\mathbf{v}(\mathbf{r}) \approx v_{\mathrm{eo}}\, \mathbf{e}_x = -\mu_{\mathrm{eo}}\, \mathbf{E}, \quad \text{for } \lambda_{\mathrm{D}} \ll \tfrac{1}{2}\, h. \tag{9.14}$$

The corresponding flow rate, the so-called free EO flow rate Q_{eo}, for a section of width w is given by

$$Q_{\mathrm{eo}} = \int_0^w \mathrm{d}y \int_{-h/2}^{h/2} \mathrm{d}z\, v_x(y,z) = v_{\mathrm{eo}}\, wh, \quad \text{for } \lambda_{\mathrm{D}} \ll \tfrac{1}{2}\, h, \tag{9.15}$$

while for general values of λ_{D}/h, but still within the Debye-Hückel approximation, we get the following expression for Q_{eo} as shown in Exercise 9.1,

$$Q_{\mathrm{eo}} = wh\, v_{\mathrm{eo}}\, f(s_{\mathrm{o}}), \tag{9.16}$$

where the dimensionless variable s_{o} and function $f(s_{\mathrm{o}})$ are given by

$$s_{\mathrm{o}} \equiv \frac{h}{2\lambda_{\mathrm{D}}}, \tag{9.17a}$$

$$f(s_{\mathrm{o}}) \equiv \left[1 - \frac{1}{s_{\mathrm{o}}}\, \tanh(s_{\mathrm{o}})\right]. \tag{9.17b}$$

It is also relevant to calculate the electric current I_{eo}. It consists of two parts, the conductive current $I_{\mathrm{eo}}^{\mathrm{cond}}$ due to the ionic conductivities $\sigma_{\mathrm{ion}}^{\pm}$ and the electric field E, and the convective current $I_{\mathrm{eo}}^{\mathrm{conv}}$ due to the charge density $\rho_{\mathrm{el}}(z)$ moved by the velocity field $v_x(z)$, and as shown in Exercise 9.1 we obtain

$$\begin{aligned} I_{\mathrm{eo}} = I_{\mathrm{eo}}^{\mathrm{cond}} + I_{\mathrm{eo}}^{\mathrm{conv}} &= wh(\sigma_{\mathrm{ion}}^{+} + \sigma_{\mathrm{ion}}^{-})(-E) + 2w \int_0^{\frac{h}{2}} \mathrm{d}z\, \rho_{\mathrm{el}}(z)\, v_x(z) \\ &= -wh(\sigma_{\mathrm{ion}}^{+} + \sigma_{\mathrm{ion}}^{-})E - \epsilon\zeta\, v_{\mathrm{eo}}\, \frac{wh}{\lambda_{\mathrm{D}}^2} \left[\frac{\lambda_{\mathrm{D}}}{h} \tanh\Big(\frac{h}{2\lambda_{\mathrm{D}}}\Big) - \frac{1}{2}\mathrm{sech}^2\Big(\frac{h}{2\lambda_{\mathrm{D}}}\Big)\right] \\ &= -wh(\sigma_{\mathrm{ion}}^{+} + \sigma_{\mathrm{ion}}^{-})E \left\{1 + \frac{\epsilon^2\zeta^2}{\lambda_{\mathrm{D}} h(\sigma_{\mathrm{ion}}^{+} + \sigma_{\mathrm{ion}}^{-})\eta} \left[\tanh\Big(\frac{h}{2\lambda_{\mathrm{D}}}\Big) - \frac{h}{2\lambda_{\mathrm{D}}}\mathrm{sech}^2\Big(\frac{h}{2\lambda_{\mathrm{D}}}\Big)\right]\right\}. \end{aligned} \tag{9.18}$$

Introducing the dimensionless function $g(s_{\mathrm{o}})$,

$$g(s_{\mathrm{o}}) \equiv \frac{1}{s_{\mathrm{o}}}\, \tanh(s_{\mathrm{o}}) - \mathrm{sech}^2(s_{\mathrm{o}}), \tag{9.19}$$

and the dimensionless pre-factor α, defined by

$$\alpha \equiv \frac{\epsilon^2\zeta^2}{2\lambda_{\mathrm{D}}^2(\sigma_{\mathrm{ion}}^{+} + \sigma_{\mathrm{ion}}^{-})\eta} \approx 2.6, \tag{9.20}$$

where the value is obtained for $\zeta \approx 100$ mV, $\epsilon \approx 78\epsilon_0$, $\lambda_{\mathrm{D}}^2\sigma_{\mathrm{ion}}^{\pm}/c \approx 5\ \mathrm{S\,m^{-1}\,M^{-1}}$, and $\eta = 1$ mPa s, we can write the convection current $I_{\mathrm{eo}}^{\mathrm{conv}}$ in terms of $I_{\mathrm{eo}}^{\mathrm{cond}}$ as

$$I_{\mathrm{eo}}^{\mathrm{conv}} = \alpha\, g(s_{\mathrm{o}})\, I_{\mathrm{eo}}^{\mathrm{cond}} \tag{9.21}$$

The total electric current in an ideal EO flow can thus be written as

$$I_{\mathrm{eo}} = I_{\mathrm{eo}}^{\mathrm{cond}} + I_{\mathrm{eo}}^{\mathrm{conv}} = \big[1 + \alpha\, g(s_{\mathrm{o}})\big]\, I_{\mathrm{eo}}^{\mathrm{cond}} = \big[1 + \alpha\, g(s_{\mathrm{o}})\big]\, wh\big(\sigma_{\mathrm{ion}}^{+} + \sigma_{\mathrm{ion}}^{-}\big)E. \tag{9.22}$$

We note that $0 < g(s_{\mathrm{o}}) < 1$ for all $s_{\mathrm{o}} > 0$, and that for a normal microfluidic channel of height $h = 100\ \mu\mathrm{m}$ containing an electrolyte of concentration $c = 1\ \mathrm{mM}$ we get that $I_{\mathrm{eo}}^{\mathrm{conv}} \lesssim 5\times 10^{-4}\, I_{\mathrm{eo}}^{\mathrm{cond}}$. In microfluidic electro-osmotic flows without superimposed pressure-driven flow, the electric charge transport is normally completely dominated by conduction.

Analytical results for an ideal EO flow within the Debye-Hückel approximation can also be given for a cylindrical channel of circular cross-section with radius a. The equilibrium potential $\phi_{\mathrm{eq}}(r)$ is given by Eq. (8.37), and the velocity field has the structure $\mathbf{v} = v_x(r)\,\mathbf{e}_x$ and obeys the boundary conditions

$$\partial_r v_x(0) = 0, \tag{9.23a}$$

$$v_x(a) = 0. \tag{9.23b}$$

Otherwise, the analysis carries through exactly as for the infinite parallel-plate geometry and we arrive at the result

$$v_x(r) = \left[1 - \frac{I_0\!\left(\frac{r}{\lambda_{\mathrm{D}}}\right)}{I_0\!\left(\frac{a}{\lambda_{\mathrm{D}}}\right)}\right] v_{\mathrm{eo}}. \tag{9.24}$$

Also here, $\mathbf{v} \approx v_{\mathrm{eo}}\,\mathbf{e}_x$ for $\lambda_{\mathrm{D}} \ll a$, so the free EO flow rate becomes

$$Q_{\mathrm{eo}} = \int_0^{2\pi} \mathrm{d}\theta \int_0^a \mathrm{d}r\; r v_x(r,\theta) = v_{\mathrm{eo}}\,\pi a^2, \quad \text{for } \lambda_{\mathrm{D}} \ll a. \tag{9.25}$$

9.3 Debye-layer overlap

Although in the following we mostly are dealing with ideal EO flow in the limit $\lambda_{\mathrm{D}} \ll a$, we shall in this section briefly discuss what happens if the Debye length λ_{D} becomes comparable with the transverse length scale a, and the Debye layers from various part of the wall overlap at the center of the channel. According to our standard value Eq. (8.27) for λ_{D} this will happen for a cylindrical channel with radius a of the order 10 nm. With modern nanotechnology it is in fact possible to make such channels intentionally, and for some fine-masked porous materials such dimensions actually occur in nature.

In Fig. 9.2(a) are shown three normalized EO flow profiles obtained by plotting $v_x(r)/v_{\mathrm{eo}}$ of Eq. (9.24) with the value $\lambda_{\mathrm{D}}/a = 0.01$, 0.1, and 1. It is seen how, for small values of λ_{D}/a, a flat nearly constant velocity profile is obtained as stated in the previous section. As λ_{D}/a increases to 0.1, a rounded profile results still being flat near the center of the channel. When λ_{D} becomes comparable to a the profile has changed into a paraboloid shape. The latter result is easily verified by Taylor expanding Eq. (9.24) in a/λ_{D}, which gives

$$v_x(r) \approx \frac{a^2}{4\lambda_{\mathrm{D}}^2}\left[1 - \frac{r^2}{a^2}\right] v_{\mathrm{eo}} + \mathcal{O}\Big(\big(a/\lambda_{\mathrm{D}}\big)^3\Big). \tag{9.26}$$

The expansion also shows that the EO flow profile gets heavily suppressed as λ_{D} is increased beyond a as is evident from the pre-factor $a^2/\lambda_{\mathrm{D}}^2$, see Fig. 9.2(b).

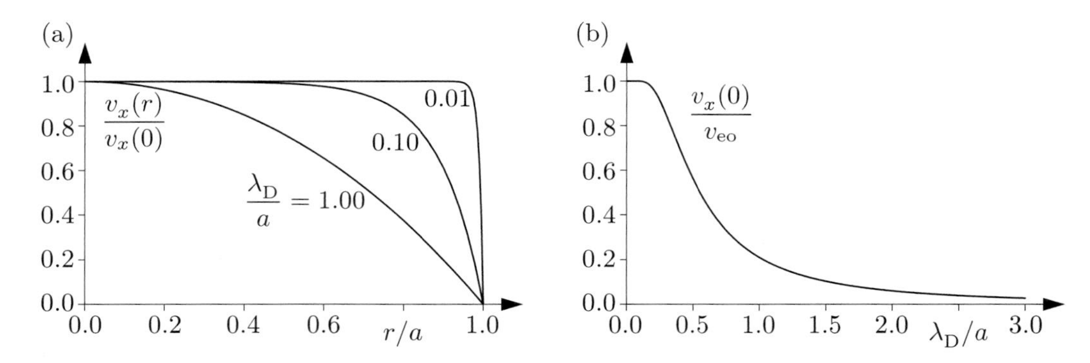

Fig. 9.2 (a) The normalized EO flow profile $v_x(r)/v_x(0)$ for a cylindrical channel of radius a with three different values of the Debye length: nearly constant ($\lambda_D/a = 0.01$), rounded ($\lambda_D/a = 0.1$), and parabolic ($\lambda_D/a = 1$). (b) The maximal velocity in the channel, $v_x(0)$ in units of v_{eo} as a function of λ_D/a. Note that $v_x(0)/v_{eo} \approx 1$ for $\lambda_D/a < 0.1$ while $v_x(0)/v_{eo} \approx a^2/(4\lambda_D^2)$ for $\lambda_D/a \gg 1$.

In conclusion, when the Debye screening length is large compared to the transverse dimension of the channel, the screening of the charges on the wall becomes incomplete, and the electrical potential does not vary much across the channel. In this case, the Debye layers at different positions at the wall overlap in the center and prevent the maintenance of a charge-neutral bulk liquid. Together with the Navier–Stokes equation this further implies that the velocity profile likewise does not vary much, and as a result the no-slip boundary condition is felt strongly even at the center of the channel. The electro-osmotic flow is therefore strongly suppressed in the limit of Debye-layer overlap, i.e. when $\lambda_D \geq a$.

9.4 Ideal EO flow with backpressure

In Section 9.2 we have seen how a non-equilibrium EO flow can be generated by applying an electrical potential difference $\Delta\phi_{ext}$ along a channel where an equilibrium Debye layer exists. For an ideal EO flow without any external pressure the flow rate Q_{eo} is given in Eq. (9.25). Here, we study how the flow rate depends on an externally applied pressure difference Δp_{ext}, i.e. we study the capability of an EO microchannel to work as a micropump.

The setup is sketched in Fig. 9.3. A cylindrical channel of radius a and length L is oriented along the x axis between $x = 0$ and $x = L$. As before, the walls and the Debye layer are positively and negatively charged, respectively. The external electrical potential ϕ_{ext} and pressure p_{ext}, both assumed to depend linearly on x, are applied as follows:

$$\phi_{ext}(x=0) = 0, \qquad \phi_{ext}(x=L) = \Delta V, \qquad -\boldsymbol{\nabla}\phi_{ext} = \mathbf{E} = -\frac{\Delta V}{L}\,\mathbf{e}_x, \tag{9.27a}$$

$$p_{ext}(x=0) = 0, \qquad p_{ext}(x=L) = \Delta p, \qquad -\boldsymbol{\nabla}p_{ext} = -\frac{\Delta p}{L}\,\mathbf{e}_x. \tag{9.27b}$$

The Navier–Stokes equation for this highly symmetric problem, where $\mathbf{v}(\mathbf{r}) = v_x(r)\mathbf{e}_x$, reduces to the following linear differential equation for the x component:

$$0 = \eta\nabla^2\, v_x(r) + \Big[\epsilon\nabla^2\phi_{eq}(r)\Big]\frac{\Delta V}{L} - \frac{\Delta p}{L}, \tag{9.28}$$

with the boundary conditions

9.7 Further reading

The field of electro-osmotic micropumps is presently very active. Besides the background literature on electrohydrodynamics and electrokinetics mentioned in Section 8.4 the reader can benefit from studying the following papers and references therein. For experimental work on porous EO micropumps see the review paper by Laser and Santiago (2004) and Brask, Bruus and Kutter (2005). For various, mainly experimental, aspects of AC EO effects see Brown *et al.* (2000), Ajdari (2000), Studer *et al.* (2004). Kirby and Hasselbrink (2004), Cahill *et al.* (2004), and Ramos *et al.* (2005). These can be supplemented by recent theory papers on the topic by Olesen, Bruus and Ajdari (2006) and Kilic, Bazant and Ajdari (2007).

9.8 Exercises

Exercise 9.1
EO flow in an infinite parallel-plate channel
Use the explicit expression Eq. (9.10) for the velocity field and the fact that $\int \mathrm{d}s \cosh(s) = \sinh(s)$ to calculate the flow rate Q_{eo} for an ideal electro-osmotic flow in an infinite parallel-plate channel.

Exercise 9.2
The flow rate and pressure capability of an infinite parallel-plate EO pump
Consider an infinite parallel-plate EO pump of length L, width w and height h with the zeta-potential ζ on the top and bottom walls. The pumping liquid has the viscosity η. The applied voltage drop along the pump driving the EO flow is ΔV, and it is assumed to give rise to a homogeneous electrical field.

(a) Derive an expression for the zero-pressure free EO flow rate Q_{eo}.

(b) Derive an expression for the zero-flow pressure capability p_{eo}.

Exercise 9.3
EO flow in a cylindrical channel with circular cross-section
Consider the cylindrical channel with circular cross-section of radius a studied in Section 9.2 with the zeta-potential ζ on the wall, $\phi_{\mathrm{eq}}(a) = \zeta$.

(a) Write down the relevant component of the Navier–Stokes equation and check that the flow field $v_x(r)$ Eq. (9.24) is a solution to this equation fulfilling the no-slip boundary condition.

(b) In Eq. (9.25) is given the flow rate Q_{eo} in the limit $\lambda_{\mathrm{D}} \ll a$. Derive the expression valid for any value of λ_{D}.

Exercise 9.4
The internal pressure in a single-stage zero-voltage EO pump
Consider the single-stage zero-voltage EO pump studied in the first half of Section 9.6.

(a) Derive Eqs. (9.41a) and (9.41b) and show that they imply the expression for the internal pressure p_c Eq. (9.42a).

(b) Discuss p_c in the free-flow case $\Delta p = 0$ and in the zero-EO-flow case $Q^*_{\mathrm{eo}} = 0$. Try to relate the discussion to other known cases.

Exercise 9.5
The geometry of the many-channel EO pump
Consider the ensemble of N identical, parallel cylindrical channels of inner radius a and wall thickness w forming a close-packed hexagonal lattice. Prove Eq. (9.38) relating the open area $\mathcal{A}_{\text{open}}$ available for fluid flow to the total cross-sectional area $\mathcal{A}_{\text{tot}}$ of the ensemble.

Exercise 9.6
A multistage EO cascade pump with a total voltage drop of zero
Consider as in Section 9.6 a multistage EO cascade pump consisting of M identical copies of a single-stage zero-voltage EO pump coupled in series. The multistage pump is placed between $x = 0$ and $x = M\,2L$. The jth stage lies between $x = (j-1)\,2L$ and $x = j\,2L$. Each stage has the voltages zero at its ends and V at the center $x = (j-\frac{1}{2})2L$. The corresponding pressures are denoted p_{j-1} and p_j at the ends and p_j^{c} in the center.

(a) Find the flow-rate–pressure characteristic of this multistage, porous EO pump, i.e. express flow rate Q as a function of the total pressure drop $\Delta p = p_M - p_0$ and the parameters R_{hyd}^*, Q_{eo}^*, N, and M.

(b) Determine the EO pressure p_{eo}, i.e. $p_M - p_0$ when $Q = 0$, and the flow rate Q_{eo} in the case of free EO flow, i.e. when $p_M - p_0 = 0$.

Exercise 9.7
Experimental realization of a multistage EO cascade pump
Multistage EO cascade pumps with 6 and 15 stages have been developed by Takamura *et al.* (2003), see the design geometry in Fig. 4.11(b). Each stage consists of a ten-channel EO pump followed by a single-channel EO pump as described in Section 9.6.

(a) Use the experimental results shown in Fig. 9.6(b) to estimate the values of the EO mobility μ_{eo} and the zeta-potential ζ assuming that the liquid is pure water.

(b) Determine the values of the parameters L, N, M, R_{hyd}^*.

(c) Calculate p_{eo} and Q_{eo} for a 6- and a 15-stage pump and compare with the experimental results.

9.9 Solutions

Solution 9.1
EO flow in an infinite parallel-plate channel
Combining Eqs. (9.10) and (9.15) and applying the two substitutions $s \equiv z/\lambda_{\text{D}}$ as well as $s_{\text{o}} \equiv h/(2\lambda_{\text{D}})$, we obtain

$$
\begin{aligned}
Q_{\text{eo}} &= \int_0^w \mathrm{d}y \int_{-h/2}^{h/2} \mathrm{d}z\, v_x(y,z) = w v_{\text{eo}}\, 2\lambda_{\text{D}} \int_0^{s_{\text{o}}} \mathrm{d}s \left[1 - \frac{\cosh(s)}{\cosh(s_{\text{o}})}\right] \\
&= 2w\lambda_{\text{D}}\, v_{\text{eo}} \left[s_{\text{o}} - \frac{\sinh(s_{\text{o}})}{\cosh(s_{\text{o}})}\right] = wh\, v_{\text{eo}} \left[1 - \frac{1}{s_{\text{o}}}\tanh(s_{\text{o}})\right].
\end{aligned}
\tag{9.49}
$$

Solution 9.2
The flow rate and pressure capability of an infinite parallel-plate EO pump
The magnitude of the electric field is given by $E = \Delta V/L$.

(a) From Eq. (9.15) it follows that $Q_{\text{eo}} = v_{\text{eo}} wh = \frac{\epsilon\zeta wh}{\eta L}\Delta V$.

(b) From Eq. (9.35b) it follows that $p_{\text{eo}} = R_{\text{hyd}} Q_{\text{eo}} = \frac{12\eta L}{wh^3}\frac{\epsilon\zeta wh}{\eta L}\Delta V = \frac{12\epsilon\zeta}{h^2}\Delta V$.

Solution 9.3
EO flow in a cylindrical channel with circular cross-section
In analogy with Eq. (9.7) the non-linear term vanishes and only the x component of the Navier–Stokes equation is non-trivial, $(\partial_r^2 + \frac{1}{r}\partial_r)[v_x(r) + (\epsilon E/\eta)\phi_{\mathrm{eq}}(r)] = 0$.

(a) Using $\phi(r)$ from Eq. (8.37) as the equilibrium potential we can rewrite the velocity field v_x given in Eq. (9.24) as $v_x(r) = [1 - \phi_{\mathrm{eq}}(r)/\zeta]\, v_{\mathrm{eo}} = v_{\mathrm{eo}} - (\epsilon E/\eta)\phi_{\mathrm{eq}}(r)$. When this is inserted into the Navier–Stokes equation above, the two terms containing $\phi_{\mathrm{eq}}(z)$ cancel and we arrive at $(\partial_r^2 + \frac{1}{r}\partial_r)v_{\mathrm{eo}} = 0$, which is true as v_{eo} is a constant. Moreover, the boundary condition is fulfilled since $v_x(a) = [1 - \phi_{\mathrm{eq}}(a)/\zeta]\, v_{\mathrm{eo}} = [1 - \zeta/\zeta]\, v_{\mathrm{eo}} = 0$.

(b) Due to the axisymmetry the EO flow rate is given by $Q_{\mathrm{eo}} = 2\pi \int_0^a \mathrm{d}r\; r\; v_x(r)$, where $v_x(r)$ is given by Eq. (9.24). With the variable substitution $s = r/\lambda_{\mathrm{D}}$ and using that $\int_0^{s_\mathrm{o}} \mathrm{d}s\; s\, I_0(s) = s_\mathrm{o}\, I_1(s_\mathrm{o})$ we find by integration that

$$Q_{\mathrm{eo}} = \pi a^2\, v_{\mathrm{eo}} \left[1 - 2\frac{\lambda_{\mathrm{D}}}{a}\frac{I_1\big(\frac{a}{\lambda_{\mathrm{D}}}\big)}{I_0\big(\frac{a}{\lambda_{\mathrm{D}}}\big)}\right]. \tag{9.50}$$

Solution 9.4
The internal pressure in a single-stage zero-voltage EO pump

(a) This is done by straightforward algebra, no tricks are involved.

(b) For $\Delta p = 0$ we discuss three cases. (i) The two sections are identical, $\alpha = 1$, which makes $Q = 0$ and $p_c = R^*_{\mathrm{hyd}} Q^*_{\mathrm{eo}} = R_{\mathrm{hyd},1} Q_{\mathrm{eo},1} = R_{\mathrm{hyd},2} Q_{\mathrm{eo},2}$; both sections enter on an equal footing. (ii) Section 2 has a vanishing hydraulic resistance, $\alpha \ll 1$, which makes $Q = \alpha^2 Q^*_{\mathrm{eo}} = Q_{\mathrm{eo},1}$ and $p_c = R^*_{\mathrm{hyd}} Q^*_{\mathrm{eo}} = R_{\mathrm{hyd},2} Q_{\mathrm{eo},2}$; section 1 delivers the flow rate, while section 2 creates the pressure. (iii) Section 1 has a vanishing hydraulic resistance, $\alpha \gg 1$, which makes $Q = -Q^*_{\mathrm{eo}} = Q_{\mathrm{eo},2}$ and $p_c = \alpha^{-2} R^*_{\mathrm{hyd}} Q^*_{\mathrm{eo}} = R_{\mathrm{hyd},1} Q_{\mathrm{eo},2}$; section 2 delivers the flow rate, while section 1 creates the pressure.

For $Q^*_{\mathrm{eo}} = 0$ the circuit is just a normal series coupling of $R_{\mathrm{hyd},1}$ and $R_{\mathrm{hyd},2}$. The total resistance is $R_{\mathrm{hyd}} = R_{\mathrm{hyd},1} + R_{\mathrm{hyd},2} = (1 + \alpha^{-4}) R^*_{\mathrm{hyd}}$ so that $Q = \Delta p / R_{\mathrm{hyd}} = \frac{1}{1+\alpha^{-4}} \Delta p / R^*_{\mathrm{hyd}} = \frac{\alpha^4}{1+\alpha^4} \Delta p / R^*_{\mathrm{hyd}}$ in accordance with Eq. (9.42b). The central pressure is then $p_c = R_{\mathrm{hyd},1} Q = \alpha^{-4} R^*_{\mathrm{hyd}} \times \frac{\alpha^4}{1+\alpha^4} \Delta p / R^*_{\mathrm{hyd}} = \frac{1}{1+\alpha^4} \Delta p$ in agreement with Eq. (9.42a).

Solution 9.5
The geometry of the many-channel EO pump
When close-packing N circular disks of radius $a + w$ we note that each disk has six neighbors. The centers of the central disk and any two neighbors that touch each other form an equilateral triangle with side length $2(a+w)$ and area $\sqrt{3}(a+w)^2$. One third of this area is designated to each of the three disks, and since the central disk participates in six such areas, it takes up an area of $6 \times \frac{1}{3} \times \sqrt{3}(a+w)^2 = 2\sqrt{3}(a+w)^2$. Thus, in total, the N disks take up an area $\mathcal{A}_{\mathrm{tot}} = 2\sqrt{3}(a+w)^2 N = (2\sqrt{3}/\pi)(1 + w/a)^2 \times N\pi a^2 = (2\sqrt{3}/\pi)(1 + w/a)^2 \mathcal{A}_{\mathrm{open}}$, since the open area of each disk is πa^2.

Solution 9.6
A multistage EO cascade pump with a total voltage drop of zero
Due to mass conservation the flow rate Q is the same in each of the M stages of the pump.

(a) Using the single-stage result Eq. (9.47b), we can write the flow through the jth stage of the multistage EO pump as

$$Q = \frac{N^2-1}{N^2+1}Q^*_{\rm eo} - \frac{1}{N^2+1}\,\frac{p_j - p_{j-1}}{R^*_{\rm hyd}}, \qquad j = 1, 2, \ldots, M. \tag{9.51}$$

Adding these M equations, utilizing $\sum_{j=1}^{M}(p_j - p_{j-1}) = p_M - p_0 = \Delta p$, $\sum_{j=1}^{M} Q = MQ$, and $\sum_{j=1}^{M} Q^*_{\rm eo} = MQ^*_{\rm eo}$, and finally dividing by M, we arrive at the Q–p characteristics,

$$Q = \frac{N^2-1}{N^2+1}\,Q^*_{\rm eo} - \frac{1}{M(N^2+1)}\,\frac{\Delta p}{R^*_{\rm hyd}}. \tag{9.52}$$

(b) From the result above we easily find the EO pressure cabability at $Q = 0$ to be $p_{\rm eo} = M(N^2-1)R^*_{\rm hyd}Q^*_{\rm eo} \to MR_{\rm hyd,1}Q_{\rm eo,1}$, for $N \to \infty$. Likewise, the free EO flow rate at $\Delta p = 0$ becomes $Q_{\rm eo} = \frac{N^2-1}{N^2+1}\,Q^*_{\rm eo} \to Q_{\rm eo,2}$, for $N \to \infty$.

Solution 9.7

Experimental realization of a multistage EO cascade pump

The experimental results of the pump are given in Fig. 9.6(b). The top panel is the zero-flow pressure capability $p_{\rm eo}$, while the bottom panel is the EO velocity $v_{\rm eo}$.

(a) In the figure, we read off $(\Delta V,\ v_{\rm eo}) = (25\ \text{V},\ 500\ \mu\text{m/s})$, while $L = 800\ \mu\text{m}$, and thus from Eq. (9.12)

$$\mu_{\rm eo} \equiv \frac{v_{\rm eo}}{E} = \frac{v_{\rm eo}}{\Delta V/L} = \frac{5\times10^{-4}\ \text{m/s}\times 8\times10^{-4}\ \text{m}}{25\ \text{V}} = 1.6\times10^{-8}\ \text{m}^2/(\text{V s}). \tag{9.53}$$

From Eq. (9.12) we also get

$$\zeta = \frac{\eta\mu_{\rm eo}}{\epsilon} = \frac{10^{-3}\ \text{Pa s}\times 1.6\times10^{-8}\ \text{m}^2/(\text{V s})}{78.0\times 8.85\times10^{-12}\ \text{F/m}} = 23.2\ \text{mV}. \tag{9.54}$$

(b) The parameters are: $L = 800\ \mu\text{m}$, $N = \sqrt{10}$, $M = 6$ or 15, and with $\eta = 10^{-3}$ Pa s, $w = 5\times10^{-5}$, and $h = 2\times10^{-5}$m we obtain

$$R^*_{\rm hyd} = \frac{12\eta L}{wh^3(1-0.63h/w)} = 3.21\times10^{13}\ \text{Pa s/m}^3 = 32.1\ \text{Pa s/nL}. \tag{9.55}$$

(c) At $\Delta V = 10$ V the EO flow $Q_{\rm eo}$ rate is found by multiplying the cross-sectional area with the observed EO velocity read from the graph:

$$Q_{\rm eo} = v_{\rm eo}wh = 2\times10^{-4}\ \text{m/s}\times 5\times10^{-5}\ \text{m}\times 2\times10^{-5}\ \text{m} = 0.2\ \text{nL/s}. \tag{9.56}$$

At the same voltage the pressure per stage is given by Eq. (9.48a) as

$$p_{\rm eo} = \left(N^2-1\right)R^*_{\rm hyd}Q_{\rm eo} = 9\times 3.21\times10^{13}\times 2\times10^{-13}\ \text{Pa} = 58\ \text{Pa}. \tag{9.57}$$

For the 6- and 15-stage pumps the predicted pressures are

$$p_{\rm eo}(6) = \ 6\times 58\ \text{Pa} = 348\ \text{Pa}, \tag{9.58a}$$

$$p_{\rm eo}(15) = 15\times 58\ \text{Pa} = 870\ \text{Pa}. \tag{9.58b}$$

The measured pressures are 150 Pa and 350 Pa, i.e. a factor of 2.5 lower than predicted. It may be due to a hydraulic leak or an uneven distribution of voltage drops that the pump cannot sustain as high a pressure as predicted.

10
Dielectrophoresis

Dielectrophoresis (DEP) is the movement of a charge-neutral particle in a fluid induced by an inhomogeneous electric field. This driving field can be either DC or AC, see Jones (1995).

We begin our analysis by considering a DC field $\mathbf{E}$. Moreover, we shall exclusively work with linear media such that the polarization $\mathbf{P}$ of the dielectric fluid is given by Eq. (8.12)

$$\mathbf{P} = \epsilon_0 \chi\, \mathbf{E}, \tag{10.1}$$

where χ is the susceptibility, and such that the induced dipole moment $\mathbf{p}$ of the dielectric particle is

$$\mathbf{p} = \alpha\, \mathbf{E}, \tag{10.2}$$

where α is the polarizability.

According to Eq. (8.4) a dielectric force $\mathbf{F}_{\mathrm{dip}}$ acts on a dipole moment $\mathbf{p}$ situated in an inhomogeneous electric field $\mathbf{E}$, i.e. a field with a non-zero gradient tensor $\boldsymbol{\nabla}\mathbf{E}$,

$$\mathbf{F}_{\mathrm{dip}} = (\mathbf{p}\cdot\boldsymbol{\nabla})\mathbf{E}. \tag{10.3}$$

Before launching a rigorous analysis we shall build our intuition regarding induced polarization by presenting some heuristic arguments.

10.1 Induced polarization and dielectric forces; heuristically

As sketched in Fig. 10.1 we consider a dielectric sphere with dielectric constant ϵ_2 placed in a dielectric fluid with dielectric constant ϵ_1. An inhomogeneous electric field $\mathbf{E}$ is imposed by charging a spherical electrode to the left and a planar electrode to the right.

From Eq. (8.12),

$$\mathbf{D} = \epsilon_0\mathbf{E} + \mathbf{P} = \epsilon_0(1+\chi)\mathbf{E} = \epsilon\mathbf{E}, \tag{10.4}$$

it follows that when an electric field $\mathbf{E}$ is applied to a medium with a large dielectric constant ϵ, the medium will acquire a large polarization $\mathbf{P}$ and consequently contain many dipoles $\mathbf{p}$. This is sketched in Fig. 10.1(a1), where the medium (light gray) with the smaller dielectric constant ϵ_1 contains a few polarization charges at its surfaces, while the sphere (dark gray) with the larger dielectric constant $\epsilon_2 > \epsilon_1$ contains more charges at its surface.

In Fig. 10.1(b1), the situation is reversed. Now, the medium (dark gray) has the larger dielectric constant ϵ_1 and many polarization charges at its surfaces, while the sphere (light gray) has the smaller dielectric constant $\epsilon_2 < \epsilon_1$ and fewer polarization charges.

In Figs. 10.1(a2) and (b2) only the unpaired surface charges of panel (a1) and (b1) are shown, which makes it easy to draw the direction of the dipole moment $\mathbf{p}$ of the dielectric

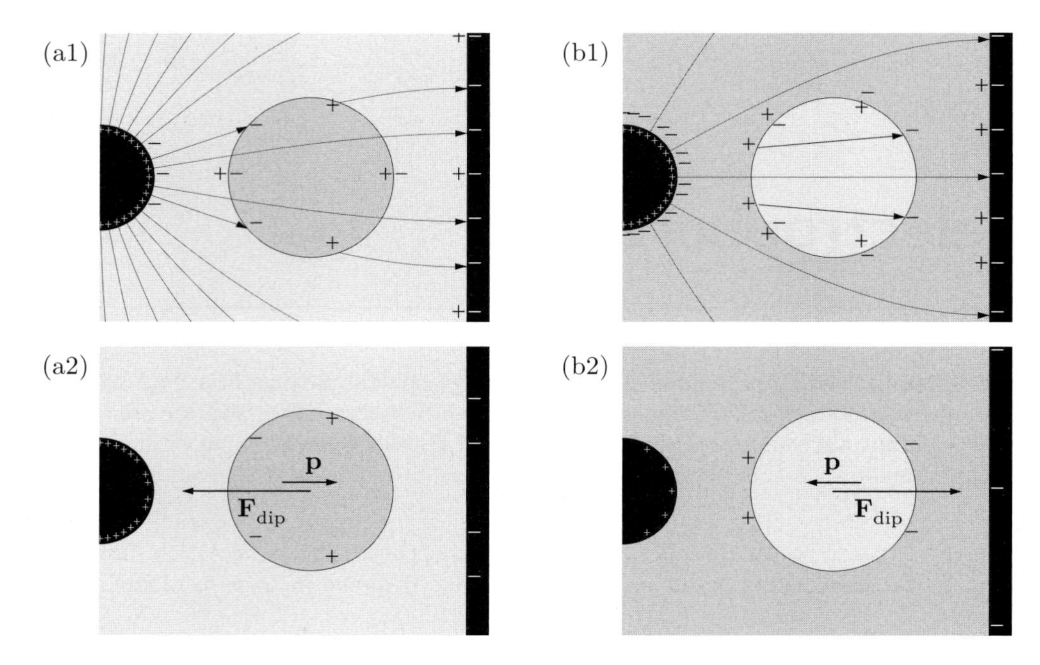

Fig. 10.1 Sketch supporting the heuristic argument for the direction of the electric dipole moment **p** induced in a dielectric sphere with dielectric constant ϵ_2 by the inhomogeneous electrical field **E**. The sphere is placed in a dielectric fluid with dielectric constant ϵ_1 and the dielectric force acting on the sphere is denoted $\mathbf{F}_{\mathrm{dip}}$. (a1) The particle is more polarizable than the fluid, i.e. $\epsilon_2 > \epsilon_1$. Here, the fluid could be vacuum. (b1) The particle is less polarizable than the fluid, i.e. $\epsilon_2 < \epsilon_1$. (a2) and (b2) The effective charges and directions of **p** and $\mathbf{F}_{\mathrm{dip}}$ corresponding to (a1) and (b1), respectively.

sphere. Since, by construction, the gradient of the electric field points to the region with highest density of electrical field lines, i.e. to the left, it is also easy by use of Eq. (10.3) to deduce the direction of the dielectric force $\mathbf{F}_{\mathrm{dip}}$.

For $\epsilon_1 < \epsilon_2$ the dielectric force pulls the dielectric particle towards the region of strong **E**-field (to the left), while for $\epsilon_1 > \epsilon_2$ the particle is pushed away from this region (towards the right).

10.2 A point dipole in a dielectric fluid

The first step in our more rigorous analysis is to determine the electrical potential $\phi_{\mathrm{dip}}(\mathbf{r})$ arising from a point dipole $\mathbf{p} = q\mathbf{d}$ placed at the center of the co-ordinate system in a dielectric fluid with dielectric constant ϵ.

$$\mathbf{p} = q\mathbf{d}, \qquad \begin{cases} +q \text{ at } +\frac{1}{2}\mathbf{d}, \\ -q \text{ at } -\frac{1}{2}\mathbf{d}. \end{cases} \tag{10.5}$$

From an observation point **r** the distance to the dipole charges $+q$ and $-q$ are $|\mathbf{r} - \mathbf{d}/2|$ and $|\mathbf{r} + \mathbf{d}/2|$, respectively. The potential $\phi_{\mathrm{dip}}(\mathbf{r})$ from a point dipole, where $d \ll r$, therefore becomes

$$\phi_{\rm dip}(\mathbf{r}) = \frac{+q}{4\pi\epsilon}\frac{1}{|\mathbf{r}-\mathbf{d}/2|} + \frac{-q}{4\pi\epsilon}\frac{1}{|\mathbf{r}+\mathbf{d}/2|} \approx \frac{1}{4\pi\epsilon}\frac{\mathbf{p}\cdot\mathbf{r}}{r^3} = \frac{p}{4\pi\epsilon}\frac{\cos\theta}{r^2}, \tag{10.6}$$

where θ is the angle between the dipole $\mathbf{p}$ and the observation point vector $\mathbf{r}$. If a given potential $\phi_{\rm tot}(\mathbf{r})$ contains a component of the form $B\cos\theta/r^2$,

$$\phi_{\rm tot}(\mathbf{r}) = B\,\frac{\cos\theta}{r^2} + \phi_{\rm rest}(\mathbf{r}), \tag{10.7}$$

it implies that a dipole of strength

$$p = 4\pi\epsilon\, B \tag{10.8}$$

is located at the center of the co-ordinate system. This result will be used in the following when calculating the induced dipole moment of a dielectric sphere placed in a dielectric fluid.

10.3 A dielectric sphere in a dielectric fluid; induced dipole

In Fig. 10.2(a) is shown a dielectric fluid with dielectric constant ϵ_1, which is penetrated by a homogeneous electric field $\mathbf{E}_0 = -\boldsymbol{\nabla}\phi_0$ generated by charging some capacitor plates at $x = \pm\infty$. In spherical polar co-ordinates (r,θ,φ), using the x axis and not the z axis as the polar axis, the unperturbed potential ϕ_0 is given by

$$\phi_0(r,\theta,\varphi) = -E_0\, x(r,\theta,\varphi) = -E_0\, r\cos\theta. \tag{10.9}$$

A dielectric sphere of radius a and dielectric constant ϵ_2 is then placed in the fluid as shown in Fig. 10.2(b). The electric field polarizes the sphere resulting in a distortion of the electrical field, which now becomes $\mathbf{E} = -\boldsymbol{\nabla}\phi$, where the potential ϕ is given by one function ϕ_1 outside the sphere and another function ϕ_2 inside,

$$\phi(r,\theta,\varphi) = \begin{cases} \phi_1(r,\theta), & \text{for } r > a, \\ \phi_2(r,\theta), & \text{for } r < a, \end{cases} \tag{10.10}$$

where we notice that the system is rotationally symmetric around the x axis so that $\phi(\mathbf{r}) = \phi(r,\theta)$ does not depend on the azimuthal angle φ.

The boundary conditions at the surface of the sphere, $r = a$, are the usual ones for electrostatics: the normal component $\mathbf{D}\cdot\mathbf{e}_r$ of $\mathbf{D}$ and the tangential component $\mathbf{E}\cdot\mathbf{e}_\theta$ of $\mathbf{E}$ must be continuous across the surface of the sphere at $r = a$. So at $r = 0$, $r = a$, and $r = \infty$ we have in total four boundary conditions:

$$\phi_2(0,\theta) \text{ is finite}, \tag{10.11a}$$

$$\phi_1(a,\theta) = \phi_2(a,\theta), \tag{10.11b}$$

$$\epsilon_1\partial_r\phi_1(a,\theta) = \epsilon_2\partial_r\phi_2(a,\theta), \tag{10.11c}$$

$$\phi_1(r,\theta) \underset{r\to\infty}{\longrightarrow} -E_0 r\cos\theta. \tag{10.11d}$$

Both the fluid and the sphere are dielectric media without external charges, hence $\rho_{\rm el} = 0$ in Eq. (8.3b) and the potential obeys the Laplace equation

$$\nabla^2\phi(\mathbf{r}) = 0. \tag{10.12}$$

The general solution to the Laplace equation in spherical co-ordinates with no dependence on the azimuthal angle φ can be expressed in terms of the Legendre polynomials P_l as

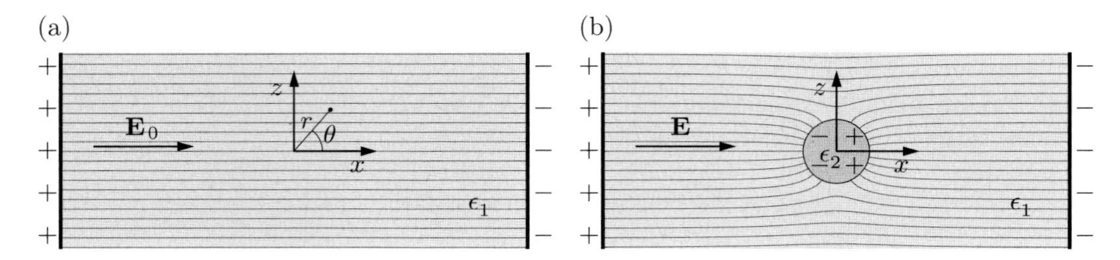

Fig. 10.2 (a) A dielectric fluid with a dielectric constant ϵ_1 penetrated by an unperturbed homogeneous electric field $\mathbf{E}_0 = -\boldsymbol{\nabla}\phi_0$, where $\phi_0(r,\theta,\varphi) = -E_0 x = -E_0 r\cos\theta$. (b) A dielectric sphere of radius a and dielectric constant $\epsilon_2 > \epsilon_1$ placed in the dielectric fluid. The electric field polarizes the sphere and a perturbed electric field $\mathbf{E} = -\boldsymbol{\nabla}\phi$ results.

$$\phi(r,\theta) = \sum_{l=0}^{\infty} \left[A_l r^l + B_l r^{-(l+1)} \right] P_l(\cos\theta). \tag{10.13}$$

Because of the boundary condition Eq. (10.11d), which forces $\phi(\mathbf{r})$ to be proportional to the first Legendre polynomial $P_1(\cos\theta) = \cos\theta$ it is reasonable to employ a trial solution containing only the $l = 1$ terms in Eq. (10.13). Thus, using boundary conditions Eqs. (10.11a) and (10.11d) we arrive at a trial solution of the form

$$\phi_1(r,\theta) = -E_0 r\cos\theta + B\,\frac{\cos\theta}{r^2}, \text{ for } r > a, \tag{10.14a}$$

$$\phi_2(r,\theta) = Ar\cos\theta, \text{ for } r < a. \tag{10.14b}$$

The remaining two boundary conditions, Eqs. (10.11b) and (10.11c), yield two equations for the two unknown coefficients A and B,

$$-E_0 a + \frac{1}{a^2}\,B = a\,A, \tag{10.15a}$$

$$-E_0 - \frac{2}{a^3}\,B = \frac{\epsilon_2}{\epsilon_1}\,A, \tag{10.15b}$$

which are easily solved to give

$$A = \frac{-3\epsilon_1}{\epsilon_2 + 2\epsilon_1}\,E_0, \tag{10.16a}$$

$$B = \frac{\epsilon_2 - \epsilon_1}{\epsilon_2 + 2\epsilon_1} a^3\,E_0. \tag{10.16b}$$

In conclusion, the trial solution works and results in the following solutions for the electrical potentials ϕ_1 and ϕ_2:

$$\phi_1(\mathbf{r}) = -E_0 r\cos\theta + \frac{\epsilon_2 - \epsilon_1}{\epsilon_2 + 2\epsilon_1}\,a^3 E_0\,\frac{\cos\theta}{r^2} = \phi_0(\mathbf{r}) + \phi_{\mathrm{dip}}(\mathbf{r}), \text{ for } r > a, \tag{10.17a}$$

$$\phi_2(\mathbf{r}) = \frac{-3\epsilon_1}{\epsilon_2 + 2\epsilon_1}\,E_0 r\cos\theta = \frac{3\epsilon_1}{\epsilon_2 + 2\epsilon_1}\,\phi_0(\mathbf{r}), \text{ for } r < a. \tag{10.17b}$$

Equation (10.17b) shows that the potential ϕ_2 inside the sphere is merely proportional to the unperturbed potential ϕ_0. However, Eq. (10.17a) reveals a richer structure in the potential outside the sphere in dielectric fluid: here the unperturbed potential has been supplemented with a dipole potential ϕ_{dip}. In an applied electric field the dielectric sphere acquires an induced dipole moment $\mathbf{p}$, which according to Eqs. (10.7) and (10.8) has the value

$$\mathbf{p} = 4\pi\epsilon_1 \frac{\epsilon_2 - \epsilon_1}{\epsilon_2 + 2\epsilon_1}\, a^3 \mathbf{E}_0. \tag{10.18}$$

The fraction in the pre-factor plays a significant role, and it has therefore been given a name, the Clausius–Mossotti factor $K(\epsilon_1, \epsilon_2)$,

$$K(\epsilon_1, \epsilon_2) \equiv \frac{\epsilon_2 - \epsilon_1}{\epsilon_2 + 2\epsilon_1}. \tag{10.19}$$

Note how the exact result in Eq. (10.18) confirms the heuristic picture: when the sphere is more dielectric than the liquid, $\epsilon_2 > \epsilon_1$, the induced dipole moment $\mathbf{p}$ and the unperturbed field $\mathbf{E}_0$ are parallel, while they become antiparallel when the sphere is less dielectric than the liquid, $\epsilon_2 < \epsilon_1$. We also see that the induced dipole moment vanishes if the sphere and the fluid have the same dielectric constant, $\epsilon_2 = \epsilon_1$.

This result Eq. (10.18) is very useful, since it provides us with a simple way to calculate the dielectric forces acting on a dielectric sphere immersed in a dielectric fluid.

10.4 The dielectrophoretic force on a dielectric sphere

The exact theory for the dielectric force $\mathbf{F}_{\text{dip}}$ on a dielectric sphere of finite radius a is complicated. The reason is that while it is straightforward to calculate the induced dipole moment $\mathbf{p}$ in a homogeneous external electrical field $\mathbf{E}_0$, as shown in the previous section, the calculation becomes more involved in an inhomogeneous field. In technical terms we need to take higher multipole moments into account besides the dipole moment. If, however, the radius a of the sphere is much smaller than the distance ℓ over which the external electrical field varies we can still use Eq. (10.18) for the induced dipole. As shown below, this follows from a Taylor expansion (here just taken to first order) of the external electrical field $\mathbf{E}_0(\mathbf{r})$ around the center co-ordinate $\mathbf{r}_0$ of the sphere,

$$\mathbf{E}_0(\mathbf{r}) \approx \mathbf{E}_0(\mathbf{r}_0) + \big[(\mathbf{r} - \mathbf{r}_0)\cdot\boldsymbol{\nabla}\big]\, \mathbf{E}_0(\mathbf{r}_0) = \mathbf{E}_0(\mathbf{r}_0) + \mathcal{O}(a/\ell). \tag{10.20}$$

In this expression the value of the gradient term is of the order a/ℓ since $|\mathbf{r} - \mathbf{r}_0| < a$ and $\boldsymbol{\nabla}\mathbf{E}_0(\mathbf{r}_0) \approx (1/\ell)\mathbf{E}_0(\mathbf{r}_0)$. Clearly, the induced dipole moment $\mathbf{p}$ could also depend on the gradient $\boldsymbol{\nabla}\mathbf{E}_0(\mathbf{r}_0)$, but this correction would also be suppressed by the same factor of a/ℓ, so Eq. (10.18) for the dipole moment is generalized to

$$\mathbf{p} \approx a^3 4\pi\epsilon_1 K(\epsilon_1, \epsilon_2)\mathbf{E}_0(\mathbf{r}_0) + a^4 \big[\mathbf{f}_1(\epsilon_1, \epsilon_2)\cdot\boldsymbol{\nabla}\big]\mathbf{E}_0(\mathbf{r}_0) = a^3 4\pi\epsilon_1 K(\epsilon_1, \epsilon_2)\mathbf{E}_0(\mathbf{r}_0) + \mathcal{O}(a/\ell). \tag{10.21}$$

The vector function $\mathbf{f}_1(\epsilon_1, \epsilon_2)$ appearing above is a generalized Clausius–Mossotti function. Combining Eqs. (10.3), (10.20), and (10.21) we arrive at

$$\mathbf{F}_{\rm dip}(\mathbf{r}_0) = \big[\mathbf{p}(\mathbf{r}_0)\cdot\boldsymbol{\nabla}\big]\mathbf{E}_0(\mathbf{r}_0) + \mathcal{O}(a/\ell)$$
$$= 4\pi\epsilon_1\,\frac{\epsilon_2-\epsilon_1}{\epsilon_2+2\epsilon_1}\,a^3\big[\mathbf{E}_0(\mathbf{r}_0)\cdot\boldsymbol{\nabla}\big]\mathbf{E}_0(\mathbf{r}_0) + \mathcal{O}(a/\ell)$$
$$= 2\pi\epsilon_1\,\frac{\epsilon_2-\epsilon_1}{\epsilon_2+2\epsilon_1}\,a^3\boldsymbol{\nabla}\Big[\mathbf{E}_0(\mathbf{r}_0)^2\Big] + \mathcal{O}(a/\ell). \tag{10.22}$$

In the last equality we have used $\boldsymbol{\nabla}\big[\mathbf{E}^2\big] = 2\mathbf{E}\cdot\boldsymbol{\nabla}\mathbf{E}$, which is valid in electrostatics where $\boldsymbol{\nabla}\times\mathbf{E} = \mathbf{0}$ (see Exercise 10.1), valid also in AC if $\ell \ll c/\omega$.

We shall use Eq. (10.22) in the following. This kind of dipole force is often called a dielectrophoretic force $\mathbf{F}_{\rm DEP}$, and using the Clausius–Mossotti factor it is written as

$$\mathbf{F}_{\rm DEP}(\mathbf{r}_0) = 2\pi\epsilon_1\,K(\epsilon_1,\epsilon_2)\,a^3\boldsymbol{\nabla}\Big[\mathbf{E}_0(\mathbf{r}_0)^2\Big]. \tag{10.23}$$

The direction of the DEP force is governed by the direction of the gradient of the square of the electrical field. Since $\mathbf{E}_0$ only appears as $\mathbf{E}_0^2$ the sign of the DEP force is independent of the sign of $\mathbf{E}_0$ but is given by the sign of the Clausius–Mossotti factor $K(\epsilon_1,\epsilon_2)$.

10.5 Dielectrophoretic particle trapping in microfluidics

The dielectrophoretic (DEP) force $\mathbf{F}_{\rm DEP}$ can be used to trap dielectric particles suspended in microfluidic channels. The principle is quite simple. An inhomogeneous electric field is created in a microchannel by charging carefully shaped metal electrodes at the walls of the channel. Dielectric particles suspended in the liquid flowing through the microchannel will be attracted to the electrodes, and if the DEP force $\mathbf{F}_{\rm DEP}$ is stronger than the viscous drag force $\mathbf{F}_{\rm drag}$,

$$|\mathbf{F}_{\rm DEP}| > |\mathbf{F}_{\rm drag}|, \tag{10.24}$$

the particles will get trapped by the electrodes.

To exemplify the technique and get some analytical expressions for the forces involved we shall study the particularly simple geometry shown in Fig. 10.3. The microfluidic channel is rectangular with length L, width w, and height h as in Section 3.4.6, which in the flat and wide channel limit $h \ll w$ can be approximated by the infinite parallel-plate channel of Section 3.4.2. The origin of the co-ordinate system is placed at the center of the floor wall such that $-L/2 < x < L/2$, $-w/2 < y < w/2$, and $0 < z < h$. A pressure drop of Δp along the channel results in the flow profile $\mathbf{v} = v_x(z)\mathbf{e}_x$, Eq. (3.29),

$$v_x(z) = \frac{\Delta p}{2\eta L}\,(h-z)z = 6\Big(1-\frac{z}{h}\Big)\,\frac{z}{h}\,v_0, \tag{10.25}$$

where v_0 is the average flow velocity, such that the flow rate is given by $Q = v_0\,wh$.

The dielectric particles suspended in the liquid have radius a. Neglecting finite-size effects from the channel walls we can therefore approximate the drag force acting on a sphere trapped at the position $\mathbf{r}$ by Eq. (3.127)

$$\mathbf{F}_{\rm drag} \approx 6\pi\eta\,a\,\mathbf{v}(\mathbf{r}). \tag{10.26}$$

The inhomogeneous electric field is created by applying a potential $\phi = \Delta V$ to a spherical metallic electrode of radius r_0 situated at the floor at $\mathbf{r} = \mathbf{0}$ and the potential $\phi = 0$ to a

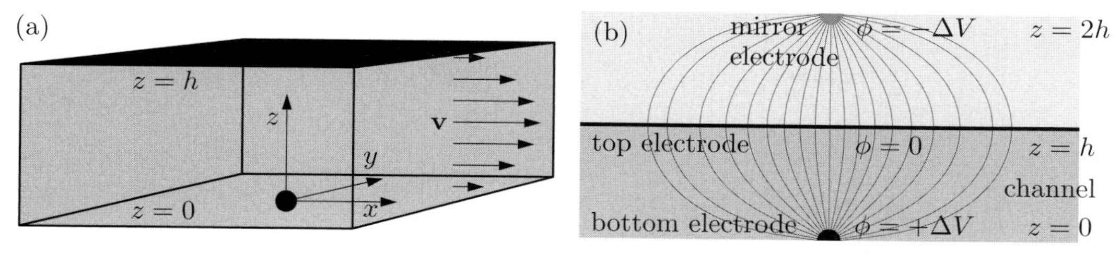

Fig. 10.3 (a) An example of a DEP trap in a rectangular microfluidic channel of dimensions $L \times w \times h$ to catch dielectric particles suspended in a liquid flow with velocity profile $\mathbf{v}$. An inhomogeneous electric field $\mathbf{E}$ is created by applying a voltage difference ΔV between the (semi-)spherical electrode at the bottom of the microchannel and the planar electrode covering the top. Through the DEP force the bottom electrode will attract the suspended dielectric particles. (b) The electrical field lines calculated by the method of image charges for the potential $\phi(\mathbf{r})$. The spherical bottom electrode at $\mathbf{r} = \mathbf{0}$ has the potential $\phi(\mathbf{0}) = \Delta V$. The planar top electrode of potential $\phi(h\mathbf{e}_z) = 0$ can be realized by placing a mirror electrode with the potential $\phi(2h\mathbf{e}_z) = -\Delta V$ at $\mathbf{r} = 2h\mathbf{e}_z$.

planar metallic electrode covering the ceiling at the plane $\mathbf{r} = h\,\mathbf{e}_z$, see Fig. 10.3(a). We assume the liquid has a vanishing conductivity so that we can disregard the formation of Debye screening layers near the electrodes. By the mirror-image charge method it is easy to construct the electrical potential $\phi(\mathbf{r})$ of this configuration,

$$\phi(\mathbf{r}) = \frac{r_0}{|\mathbf{r}|}\,\Delta V - \frac{r_0}{|\mathbf{r} - 2h\mathbf{e}_z|}\,\Delta V, \tag{10.27}$$

which clearly by symmetry has $\phi(\mathbf{r} = h\mathbf{e}_z) \equiv 0$. The trapping of particles takes place close to the spherical electrode, i.e. $|\mathbf{r}| \ll h$. In this region, the electrical field is given approximately as

$$\mathbf{E}(\mathbf{r}) = -\boldsymbol{\nabla}\phi(\mathbf{r}) \approx \frac{r_0 \Delta V}{r^2}\,\mathbf{e}_r, \ \text{for}\ r_0 < |\mathbf{r}| \ll h, \tag{10.28}$$

where $\mathbf{e}_r$ is the radial vector pointing away from the spherical electrode.

With the electrical field Eq. (10.28) at hand it is an easy task to derive an expression for the DEP force, Eq. (10.23),

$$\mathbf{F}_{\mathrm{DEP}}(\mathbf{r}) = 2\pi\epsilon_1\,\frac{\epsilon_2 - \epsilon_1}{\epsilon_2 + 2\epsilon_1}\,a^3 \boldsymbol{\nabla}\left[\frac{(\Delta V)^2 r_0^2}{r^4}\right] = -8\pi\,\frac{\epsilon_2 - \epsilon_1}{\epsilon_2 + 2\epsilon_1}\,\frac{a^3 r_0^2}{r^5}\,\epsilon_1 (\Delta V)^2\,\mathbf{e}_r. \tag{10.29}$$

The maximal DEP force $F_{\mathrm{DEP}}^{\max}$ is achieved when the particle is as close to the spherical electrode as possible, $r = r_{\min} = r_0 + a$. If we denote the electrode radius by $r_0 = \Gamma a$, we obtain the minimal distance

$$r_{\min} = (1 + \Gamma)a, \quad \Gamma \equiv \frac{r_0}{a}. \tag{10.30}$$

From this follows an estimate for the maximal DEP force:

$$F_{\mathrm{DEP}}^{\max} \equiv |\mathbf{F}_{\mathrm{DEP}}(r_{\min})| = 8\pi\,\frac{\epsilon_2 - \epsilon_1}{\epsilon_2 + 2\epsilon_1}\,\frac{\Gamma^2}{(1 + \Gamma)^5}\,\epsilon_1 (\Delta V)^2. \tag{10.31}$$

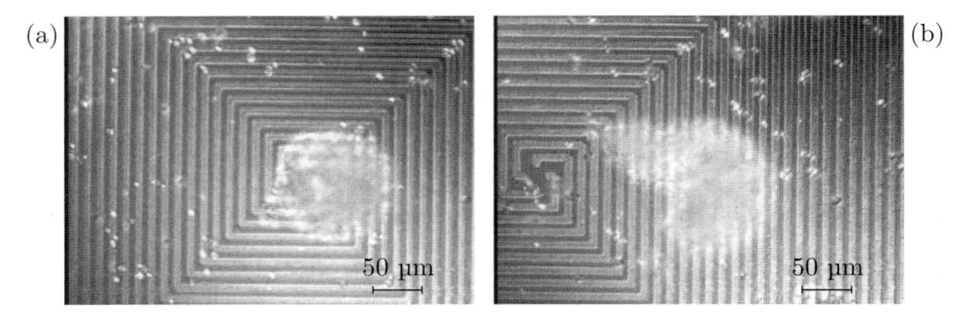

Fig. 10.4 An actual DEP trap for biological cells. (a) A cloud of yeast cells are caught at the center of the spiral electrode, where the gradients are largest. (b) The cells are carried away by the liquid flow after being released by removing the applied voltage on the electrodes. Micrographs courtesy of Anders Wolff, DTU Nanotech.

The average flow velocity at the position $z = r_0 + a = (1+\Gamma)a$ follows from Eq. (10.25)

$$v_x(r_0 + a) = 6\Big(1 - (1+\Gamma)\frac{a}{h}\Big)\,(1+\Gamma)\,\frac{a}{h}\,v_0 \approx 6(1+\Gamma)\,\frac{a}{h}\,v_0. \tag{10.32}$$

The drag force on a particle placed at $r = r_{\mathrm{min}} = r_0 + a$ follows from Eqs. (10.26) and (10.32),

$$|\mathbf{F}_{\mathrm{drag}}(r_0 + a)| \approx 6\pi\eta\, a\, v_x(r_0 + a) = 36\pi(1+\Gamma)\,\frac{\eta a^2}{h}\,v_0. \tag{10.33}$$

The largest average velocity v_0^{max} that still allows for trapping of particles at the spherical electrode is found from the condition $F_{\mathrm{drag}} = F_{\mathrm{DEP}}^{\mathrm{max}}$, which results in

$$v_0^{\mathrm{max}} = \frac{2}{9}\,\frac{\epsilon_2 - \epsilon_1}{\epsilon_2 + 2\epsilon_1}\,\frac{\Gamma^2}{(1+\Gamma)^6}\,\frac{h\epsilon_1(\Delta V)^2}{\eta a^2}. \tag{10.34}$$

To obtain trapping we need a liquid with a dielectric constant smaller than that of the particle. Let us therefore use the liquid benzene with $\epsilon_1 = 2.28\,\epsilon_0$ and $\eta = 0.65$ mPa s and pyrex glass particles with $\epsilon_2 = 6.0\,\epsilon_0$. The length scales are set to $a = r_0 = 5$ μm and $h = 100$ μm, while the applied voltage drop is $\Delta V = 10$ V. With these parameters we find

$$v_0^{\mathrm{max}} = 3.0\ \mathrm{cm/s}. \tag{10.35}$$

It is encouraging that this value is so high. From a theoretical point of view it ought to be possible to design DEP traps for dielectric particles, and indeed this turns out to be the case experimentally. In Fig. 10.4 is shown an actual DEP trap for biological cells fabricated in the group of Anders Wolff at DTU Nanotech.

10.6 The AC dielectrophoretic force on a dielectric sphere

So far we have only considered DC voltages driving the DEP trap. There are, however, many advantages in using AC voltage bias instead. One advantage is that any charge monopoles (ions) in the system will not change their mean position being influenced by AC electric fields. Another and related advantage is that the creation of permanent Debye screening layers at the electrodes is avoided. A third advantage is that in the AC mode the DEP trap will also work even if the liquid and the particle have non-zero conductivities, $\sigma_{\mathrm{el},1}$ and $\sigma_{\mathrm{el},2}$. Finally, as we shall see, under AC drive the Clausius–Mossotti factor depends on the driving frequency ω and it can even change its sign, which allow us to control *in situ* whether the DEP force should be attractive or repulsive.

In the following we shall study a simple harmonic time variation $\exp(-\mathrm{i}\omega t)$ (meaning that we must take the real part at the end). In that case the applied potential $\phi(\mathbf{r},t)$ and the associated electrical field $\mathbf{E}(\mathbf{r},t) = -\boldsymbol{\nabla}\phi(\mathbf{r},t)$ have the forms

$$\phi(\mathbf{r},t) \equiv \phi(\mathbf{r})\,\mathrm{e}^{-\mathrm{i}\omega t}, \tag{10.36a}$$

$$\mathbf{E}(\mathbf{r},t) \equiv \mathbf{E}(\mathbf{r})\,\mathrm{e}^{-\mathrm{i}\omega t}. \tag{10.36b}$$

When dealing with such harmonic time dependencies, say,

$$A(t) = \mathrm{Re}\big[A_0\mathrm{e}^{-\mathrm{i}\omega t}\big], \tag{10.37a}$$

$$B(t) = \mathrm{Re}\big[B_0\mathrm{e}^{-\mathrm{i}\omega t}\big], \tag{10.37b}$$

where A_0 and B_0 are constant complex amplitudes, it is useful to know how to calculate the time average $\langle A(t)B(t)\rangle$ over one full period τ:

$$\langle A(t)B(t)\rangle \equiv \frac{1}{\tau}\int_0^\tau \mathrm{d}t\; A(t)B(t) = \frac{1}{2}\,\mathrm{Re}\Big[A_0B_0^*\Big]. \tag{10.38}$$

The proof of this expression, where B_0^* denotes the complex conjugate of B_0, is left as an exercise for the reader.

We now move on to generalize the expression for the DEP force taking AC fields and conductivity into account. The starting point is the general boundary condition for the radial component $E_r(r,\theta) = -\partial_r\phi(a,\theta)$ at the surface of the dielectric sphere,

$$\epsilon_1 E_{r,1}(a,\theta,t) - \epsilon_2 E_{r,2}(a,\theta,t) = q_{\mathrm{surf}}. \tag{10.39}$$

For perfect dielectrics the surface charge density q_{surf} is zero, as stated in Eq. (10.11c), but now with non-zero conductivities and AC fields it becomes non-zero and in fact time dependent. The time derivative of q_{surf} is given by charge conservation and Ohm's law,

$$\partial_t q_{\mathrm{surf}}(t) = J_{r,1}(a,\theta,t) - J_{r,2}(a,\theta,t) = \sigma_{\mathrm{el},1}E_{r,1}(a,\theta,t) - \sigma_{\mathrm{el},2}E_{r,2}(a,\theta,t). \tag{10.40}$$

Taking the time derivative of Eq. (10.39) using $\mathbf{E}$-fields of the form Eq. (10.36b), substituting Eq. (10.40) into the result, and multiplying with i/ω, we arrive at

$$\Big(\epsilon_1 - \mathrm{i}\frac{\sigma_{\mathrm{el},1}}{\omega}\Big)E_{r,1}(a,\theta) = \Big(\epsilon_2 - \mathrm{i}\frac{\sigma_{\mathrm{el},2}}{\omega}\Big)E_{r,2}(a,\theta). \tag{10.41}$$

We see that if we define a complex dielectric function $\epsilon(\omega)$ as

$$\epsilon(\omega) \equiv \epsilon - \mathrm{i}\frac{\sigma}{\omega}, \tag{10.42}$$

then the boundary condition in the AC case, Eq. (10.41), is seen to have the same mathematical form as the boundary condition, Eq. (10.11c), in the DC case. We can therefore use the result Eq. (10.23) directly just using the complex dielectric functions in the Clausius–Mossotti factor, which is where the boundary condition has been used,

$$\mathbf{F}_{\text{DEP}}(\mathbf{r}_0, t) = 2\pi\epsilon_1 \frac{\epsilon_2(\omega) - \epsilon_1(\omega)}{\epsilon_2(\omega) + 2\epsilon_1(\omega)}\, a^3\nabla\Big[\mathbf{E}(\mathbf{r}_0, t)^2\Big]. \tag{10.43}$$

Note that the ϵ_1 in the pre-factor is the dielectric constant and not the dielectric function. To obtain the real time-averaged DEP force $\langle\mathbf{F}_{\text{DEP}}\rangle$ for the complex result Eq. (10.43) we use Eq. (10.38) with $A(t) = K\big[\epsilon_1(\omega), \epsilon_2(\omega)\big]\mathbf{E}(\mathbf{r}_0, t)$ and $B(t) = \mathbf{E}(\mathbf{r}_0, t)$. The result is

$$\langle\mathbf{F}_{\text{DEP}}(\mathbf{r}_0, \omega)\rangle = 2\pi\epsilon_1\, \text{Re}\left[\frac{\epsilon_2(\omega) - \epsilon_1(\omega)}{\epsilon_2(\omega) + 2\epsilon_1(\omega)}\right] a^3\nabla\Big[\mathbf{E}_{\text{rms}}(\mathbf{r}_0)^2\Big]. \tag{10.44}$$

Here, we have introduced the usual root-mean-square value $\mathbf{E}_{\text{rms}} = \mathbf{E}/\sqrt{2}$.

The expression Eq. (10.44) can be used to get a good first insight into the dielectrophoretic forces in the AC mode. One interesting result is to find the critical frequency ω_{c} at which the sign of $\langle\mathbf{F}_{\text{DEP}}(\mathbf{r}_0, \omega)\rangle$ changes. This is found from the Clausius–Mossotti function by demanding $\text{Re}\big\{[\epsilon_2(\omega_{\text{c}}) - \epsilon_1(\omega_{\text{c}})][\epsilon_2(\omega_{\text{c}}) + 2\epsilon_1(\omega_{\text{c}})]^*\big\} = 0$, which yields

$$\omega_{\text{c}} = \sqrt{\frac{(\sigma_{\text{el},1} - \sigma_{\text{el},2})(\sigma_{\text{el},2} + 2\sigma_{\text{el},1})}{(\epsilon_2 - \epsilon_1)(\epsilon_2 + 2\epsilon_1)}}. \tag{10.45}$$

Let us calculate a characteristic value for ω_{c} for a biological cell, consisting mainly of the cytoplasm, in water. We use the following parameters: $\sigma_{\text{el},2} = 0.1$ S/m and $\epsilon_2 = 60.0\epsilon_0$ for the cell, and $\sigma_{\text{el},1} = 0.01$ S/m and $\epsilon_1 = 78.0\epsilon_0$ for water. The value obtained is

$$\omega_{\text{c}} = 1.88 \times 10^8 \text{ rad/s}. \tag{10.46}$$

The frequency-dependent DEP force can be used to separate, e.g. living cells from dead cells and cancer cells from normal cells. The different cells have different electrical properties, and consequently they have different critical frequencies ω_{c} determining at which frequencies ω they are caught by the DEP electrode and at which they are expelled by it.

10.7 Exercises

Exercise 10.1
The gradient of $\mathbf{E}^2$ in electrostatics
In electrostatics $\boldsymbol{\nabla}\times\mathbf{E} = \mathbf{0}$. Show that this leads to

$$\boldsymbol{\nabla}\big[\mathbf{E}^2\big] = 2(\mathbf{E}\cdot\boldsymbol{\nabla})\mathbf{E}. \tag{10.47}$$

Hint: write the ith component of $\boldsymbol{\nabla}\big[\mathbf{E}^2\big]$ in index notation as $\partial_i E_j E_j$ and use that $\boldsymbol{\nabla}\times\mathbf{E} = \mathbf{0}$ implies $\partial_i E_j = \partial_j E_i$ for $i \neq j$.

Exercise 10.2
The potential arising from a point dipole
Consider the point dipole $\mathbf{p} = q\mathbf{d}$ located at the origin of the co-ordinate system as defined in Eq. (10.5). Prove Eq. (10.6) expressing the far-field potential $\phi_{\text{dip}}(\mathbf{r})$ due to this dipole. Hint: perform a Taylor expansion using the fact that $d \ll r$.

Exercise 10.3
Two particular solutions to the Laplace equation
In the case of a zero charge density the electrical potential $\phi(\mathbf{r})$ fulfills the Laplace equation $\nabla^2\phi = 0$. Prove that $\phi_1(\mathbf{r}) = Ar\cos\theta$ and $\phi_2(\mathbf{r}) = B\cos\theta/r^2$ are solutions to this equation in spherical co-ordinates (r,θ,φ), see Eq. (C.33).

Exercise 10.4
A dielectric sphere in a dielectric fluid
Consider the dielectric sphere in a dielectric fluid as defined in Section 10.3.

(a) Prove that for the trial solution, Eqs. (10.14a) and (10.14b), the boundary conditions Eqs. (10.11b) and (10.11c) lead to Eqs. (10.15a) and (10.15b) for the coefficients A and B.

(b) Prove that Eqs. (10.17a) and (10.17b) indeed provide a solution to the charge-free electrostatic problem of a perfect dielectric sphere placed in a perfect dielectric fluid.

(c) Use the solution for $\phi_2(\mathbf{r})$ to draw the electrical field lines *inside* the dielectric sphere of Fig. 10.2(b) (where $\epsilon_1 < \epsilon_2$).

Exercise 10.5
A sphere with a dielectric constant smaller than that of the surrounding fluid
In Fig. 10.2(b) is shown the electric field lines in the case of a sphere with a dielectric constant ϵ_2 larger than that of the surrounding fluid, $\epsilon_2 > \epsilon_1$. Sketch the electrical field lines in the opposite case $\epsilon_2 < \epsilon_1$.

Exercise 10.6
The simple dielectrophoretic (DEP) trap
Consider the simple DEP trap sketched in Fig. 10.3.

(a) Discuss to what extent Eq. (10.27) in fact gives a potential satisfying the boundary conditions $\phi(r = r_0) = \Delta V$ and $\phi(z = h) = 0$.

(b) Make a sketch of a sphere trapped near the spherical electrode. Include the forces acting on the sphere, and discuss the validity of Eq. (10.34).

(c) Check the units in Eq. (10.34) and the value $v_0^{\max} = 3.0$ cm/s quoted in Eq. (10.35).

Exercise 10.7
The time average of a product of time-dependent functions
Consider the real physical quantities $A(t)$ and $B(t)$ with harmonic time variation,

$$A(t) = \mathrm{Re}\big[A_0 \mathrm{e}^{-i\omega t}\big], \qquad B(t) = \mathrm{Re}\big[B_0 \mathrm{e}^{-i\omega t}\big], \tag{10.48}$$

where A_0 and B_0 are complex amplitudes. Prove that the time average $\langle A(t)B(t)\rangle$ over one full period τ is given by

$$\langle A(t)B(t)\rangle \equiv \frac{1}{\tau}\int_0^\tau \mathrm{d}t\, A(t)B(t) = \frac{1}{2}\,\mathrm{Re}\big[A_0 B_0^*\big]. \tag{10.49}$$

Hint: rewrite $A(t)B(t)$ using that $\mathrm{Re}\big[Z\big] = \frac{1}{2}\big[Z + Z^*\big]$ for any complex number Z.

Exercise 10.8
The AC DEP force on a dielectric sphere
Consider the dielectric sphere in an AC electric field presented in Section 10.6.

(a) Show how Eqs. (10.39) and (10.40) lead to the AC boundary condition Eq. (10.41) at the surface of the sphere for the normal component of the electric field.

(b) Prove that the frequency-dependent Clausius–Mossotti factor becomes zero at the critical frequency $\omega = \omega_\mathrm{c}$ given in Eq. (10.45). Hint: rewrite the complex fraction so that its denominator becomes real.

(c) Plot the frequency-dependent Clausius–Mossotti factor

$$K(\omega) = \mathrm{Re}\left[\frac{\epsilon_2(\omega) - \epsilon_1(\omega)}{\epsilon_2(\omega) + 2\epsilon_1(\omega)}\right] \tag{10.50}$$

in the interval $0 < \omega < 3\omega_\mathrm{c}$. Use the parameter values given after Eq. (10.45) for a biological cell, consisting mainly of the cytoplasm, in water.

10.8 Solutions

Solution 10.1
The gradient of $\mathbf{E}^2$ in electrostatics
Using index notation we get $\big(\boldsymbol{\nabla}\big[\mathbf{E}^2\big]\big)_i = \partial_i(E_jE_j) = 2E_j\partial_iE_j$. We would like to exchange the i and j indices on the last two terms. If $i = j$ this is trivially true, but it is also true for $i \neq j$ if $\boldsymbol{\nabla}\times\mathbf{E} = \mathbf{0}$, since, e.g. $0 = (\boldsymbol{\nabla}\times\mathbf{E})_x = \partial_yE_z - \partial_zE_y = 0$. Thus, we get

$$\big(\boldsymbol{\nabla}\big[\mathbf{E}^2\big]\big)_i = \partial_i(E_jE_j) = 2E_j\partial_iE_j = 2E_j\partial_jE_i = \big[2(\mathbf{E}\cdot\boldsymbol{\nabla})\mathbf{E}\big]_i. \tag{10.51}$$

Solution 10.2
The potential arising from a point dipole
We introduce the function $f(\mathbf{s}) = \frac{1}{|\mathbf{r}+\mathbf{s}|} = \big[(\mathbf{r}+\mathbf{s})\cdot(\mathbf{r}+\mathbf{s})\big]^{-\frac{1}{2}}$, where $\mathbf{r}$ is some constant vector and $s \ll r$. A first-order Taylor expansion of $f(\mathbf{s})$ around $\mathbf{s} = \mathbf{0}$ becomes $f(\mathbf{s}) \approx f(\mathbf{0}) + \mathbf{s}\cdot\boldsymbol{\nabla}_\mathbf{s} f(\mathbf{0})$. Since $\boldsymbol{\nabla}_\mathbf{s} f(\mathbf{s}) = -\frac{1}{2}\frac{1}{|\mathbf{r}+\mathbf{s}|^3}2(\mathbf{r}+\mathbf{s})$ we get $f(\mathbf{s}) \approx= \frac{1}{r} - \frac{\mathbf{s}\cdot\mathbf{r}}{r^3}$. In terms of f the dipole potential can be written as $\phi_\mathrm{dip} = \frac{+q}{4\pi\epsilon}f\big(-\frac{\mathbf{d}}{2}\big) + \frac{-q}{4\pi\epsilon}f\big(\frac{\mathbf{d}}{2}\big)$, which upon insertion of the Taylor expansion for f becomes $\phi_\mathrm{dip} \approx \frac{+q}{4\pi\epsilon}\big[\frac{1}{r} - \frac{-\mathbf{d}\cdot\mathbf{r}}{2r^3}\big] + \frac{-q}{4\pi\epsilon}\big[\frac{1}{r} - \frac{\mathbf{d}\cdot\mathbf{r}}{2r^3}\big] = \frac{1}{4\pi\epsilon}\frac{(q\mathbf{d})\cdot\mathbf{r}}{r^3}$.

Solution 10.3
Two particular solutions to the Laplace equation
For $S(r,\theta,\phi) = r\cos\theta$ Eq. (C.33) gives

$$\nabla^2(r\cos\theta) = \frac{1}{r^2}\partial_r(r^2\cos\theta) + \frac{1}{r^2\sin\theta}\partial_\theta(-r\sin^2\theta) = \frac{2}{r}\cos\theta - \frac{2\sin\theta\,\cos\theta}{r\sin\theta} = 0. \tag{10.52}$$

For $S(r,\theta,\phi) = \cos\theta/r^2$ Eq. (C.33) gives

$$\nabla^2\Big(\frac{\cos\theta}{r^2}\Big) = \frac{1}{r^2}\partial_r\Big(\frac{-2\cos\theta}{r}\Big) + \frac{1}{\sin\theta r^2}\partial_\theta\Big(\frac{-\sin^2\theta}{r^2}\Big) = \frac{2\cos\theta}{r^4} - \frac{2\cos\theta}{r^4} = 0. \tag{10.53}$$

Solution 10.4
A dielectric sphere in a dielectric fluid

(a) Using the trial solution Eqs. (10.14a) and (10.14b), in which all terms contain the factor $\cos\theta$, the boundary conditions Eqs. (10.11b) and (10.11c) do not affect the angular dependence. Thus, all $\cos\theta$ factors are cancelled out, and we obtain Eq. (10.15a) by inserting $r = a$. Equation (10.15b) is obtained after taking the partial derivatives ∂_r, which brings down the powers 1 and -2, and then inserting $r = a$.

(b) Equations (10.17a) and (10.17b) are superpositions of functions ϕ that in Exercise 10.3 were shown to be solutions of the Laplace equation $\nabla^2\phi = 0$. Hence they describe a charge-free electrostatic situation. The boundary conditions were checked in (a) above.

(c) Equation (10.17b shows that the potential ϕ_2 inside the sphere is proportional to the unperturbed potential ϕ_0, and thus from $\mathbf{E} = -\boldsymbol{\nabla}\phi$ it follows that $\mathbf{E}_2 = [3/(2+\epsilon_2/\epsilon_1)]\mathbf{E}_0$. Since $\epsilon_1 < \epsilon_2$ the density of the resulting homogeneous field lines is smaller than that of $\mathbf{E}_0$.

Solution 10.5
A sphere with a dielectric constant smaller than that of the surrounding fluid

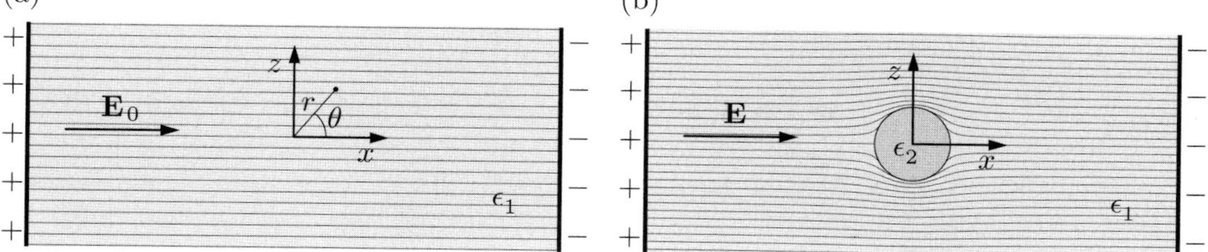

Fig. 10.5 For $\epsilon_2 < \epsilon_1$, e.g. a polymer sphere in water, the liquid is more polarizable than the sphere, and the field lines tend to avoid entering the sphere.

Solution 10.6
The simple dielectrophoretic (DEP) trap
In the following we use the parameter values given after Eq. (10.34).

(a) Putting $\mathbf{r} = r_0\mathbf{e}$ and Taylor expanding in r_0/h in the second term of Eq. (10.27) yields

$$\phi(r_0\mathbf{e}) = \frac{r_0\Delta V}{|r_0\mathbf{e}|} - \frac{r_0\Delta V}{|r_0\mathbf{e} - 2h\mathbf{e}_z|} = \Delta V - \frac{\Delta V}{|\mathbf{e} - 2(h/r_0)\mathbf{e}_z|} \approx \Big[1 + \mathcal{O}\Big(\frac{r_0}{h}\Big)\Big]\,\Delta V. \qquad (10.54)$$

With $r_0 = 5$ µm and $h = 100$ µm we estimate the relative error to be $r_0/h = 5\%$.

(b) Consider a configuration where the sphere is lying on the bottom of the channel to the right of the electrode and touching it. The DEP force is then $\mathbf{F}_{\mathrm{DEP}} = -F_{\mathrm{DEP}}^{\mathrm{max}}\mathbf{e}_r$, where $\mathbf{e}_r$ is the unit vector pointing from the center of the electrode to the center of the sphere. The drag force is $\mathbf{F}_{\mathrm{drag}} = F_{\mathrm{drag}}\mathbf{e}_x$, while the normal force from the bottom of the channel is $\mathbf{F}_N = -F_N\mathbf{e}_z$. In the x direction the force balance become $F_{\mathrm{drag}} = F_{\mathrm{DEP}}^{\mathrm{max}}\cos\theta$.

With $a = r_0 = 5$ µm we find from Eq. (10.28) that the electric field varies from $\Delta V/r_0$ to $\Delta V/(9r_0)$ across the sphere. Thus, the electric field varies significantly on a length scale corresponding to the diameter of the sphere, $\ell = 2r_0$, and the correction term $a/\ell = 1/2$ in Eq. (10.20) is in fact not small. If the Stokes drag law is valid the assumption for the drag force is not bad: With minor deviations of order $r_0/h = 5\%$ the velocity field varies linearly across the sphere, and the resulting drag force is well estimated by the average value evaluated at the center position of the sphere. However, as the sphere is close to the wall, the assumption of the walls being far away, see Section 3.7, is violated, and the Stokes drag law is not valid. In conclusion, the estimate for the v_0^{max} can only be trusted as an order-of-magnitude estimate.

(c) From $[\epsilon_1] = \mathrm{F/m}$ it follows that $[\epsilon_1(\Delta V)^2] = (\mathrm{F/m})\,\mathrm{V}^2 = (\mathrm{FV})\,\mathrm{V/m} = \mathrm{C\,V/m} = \mathrm{J/m} = \mathrm{N}$, and since $[\eta\, a^2] = \mathrm{Pa\,s\,m}^2 = \mathrm{N\,s}$, we arrive at

$$[v_0^{\mathrm{max}}] = \frac{\mathrm{m\,N}}{\mathrm{N\,s}} = \frac{\mathrm{m}}{\mathrm{s}}. \tag{10.55}$$

Solution 10.7
The time average of a product of time-dependent functions
Rewriting $A(t) = \frac{1}{2}[A_0\mathrm{e}^{-\mathrm{i}\omega t} + A_0^*\mathrm{e}^{\mathrm{i}\omega t}]$ and $B(t) = \frac{1}{2}[B_0\mathrm{e}^{-\mathrm{i}\omega t} + B_0^*\mathrm{e}^{\mathrm{i}\omega t}]$ we find

$$\langle A(t)B(t)\rangle = \frac{1}{4\tau}\int_0^\tau \mathrm{d}t\,\Big[A_0\mathrm{e}^{-\mathrm{i}\omega t} + A_0^*\mathrm{e}^{\mathrm{i}\omega t}\Big]\Big[B_0\mathrm{e}^{-\mathrm{i}\omega t} + B_0^*\mathrm{e}^{\mathrm{i}\omega t}\Big] \tag{10.56a}$$

$$= \frac{1}{4\tau}\int_0^\tau \mathrm{d}t\,\Big[A_0B_0^* + A_0^*B_0 + A_0B_0\mathrm{e}^{-\mathrm{i}2\omega t} + A_0^*B_0^*\mathrm{e}^{\mathrm{i}2\omega t}\Big] \tag{10.56b}$$

$$= \frac{1}{4}\big[A_0B_0^* + A_0^*B_0\big] = \frac{1}{2}\,\mathrm{Re}\big[A_0B_0^*\big]. \tag{10.56c}$$

Solution 10.8
The AC DEP force on a dielectric sphere
(a) A Gauss-box argument at the surface of the sphere gives $\epsilon_1\mathbf{E}_1\cdot\mathbf{n}_1 + \epsilon_2\mathbf{E}_2\cdot\mathbf{n}_2 = q_{\mathrm{surf}}$. Here, $\mathbf{n}_1$ and $\mathbf{n}_2$ are unit vectors pointing away from the surface into the liquid and the sphere, respectively. Since the radial unit vector points from the sphere out into the liquid, we get Eq. (10.39) with the proper signs. Taking the time derivative of this equation yields $-\mathrm{i}\omega\epsilon_1E_{r,1} + \mathrm{i}\omega\epsilon_2E_{r,2} = \partial_t q_{\mathrm{surf}}$. Combining this with Eq. (10.40) gives $-\mathrm{i}\omega\epsilon_1E_{r,1} + \mathrm{i}\omega\epsilon_2E_{r,2} = \sigma_{\mathrm{el},1}E_{r,1} - \sigma_{\mathrm{el},2}E_{r,2}$. Separation of 1-terms and 2-terms followed by multiplication by i/ω leads to Eq. (10.41).

(b) The DEP force is zero if the real part of the Clausius–Mossotti factor is zero, so

$$\mathrm{Re}\left[\frac{\epsilon_2(\omega_\mathrm{c}) - \epsilon_1(\omega_\mathrm{c})}{\epsilon_2(\omega_\mathrm{c}) + 2\epsilon_1(\omega_\mathrm{c})}\right] = \mathrm{Re}\left[\frac{[\epsilon_2(\omega_\mathrm{c}) - \epsilon_1(\omega_\mathrm{c})][\epsilon_2(\omega_\mathrm{c}) + 2\epsilon_1(\omega_\mathrm{c})]^*}{[\epsilon_2(\omega_\mathrm{c}) + 2\epsilon_1(\omega_\mathrm{c})][\epsilon_2(\omega_\mathrm{c}) + 2\epsilon_1(\omega_\mathrm{c})]^*}\right] = 0. \tag{10.57}$$

Using $\epsilon(\omega_\mathrm{c}) = \epsilon - \mathrm{i}\sigma/\omega_\mathrm{c}$, the real part of the enumerator becomes,

$$(\epsilon_2 - \epsilon_1)(\epsilon_2 + 2\epsilon_1) + \frac{1}{\omega_\mathrm{c}^2}(\sigma_{\mathrm{el},2} - \sigma_{\mathrm{el},1})(\sigma_{\mathrm{el},2} + 2\sigma_{\mathrm{el},1}) = 0, \tag{10.58}$$

from which Eq. (10.45) follows.

(c) The Clausius–Mossotti function $K(\omega)$ for the parameter values corresponding to a biological cell, $\epsilon_2 = 60.0\epsilon_0$ and $\sigma_{\mathrm{el},2} = 0.1$ S/m, in water, $\epsilon_1 = 78.0\epsilon_0$ and $\sigma_{\mathrm{el},1} = 0.01$ S/m. The characteristic frequency is $\omega_\mathrm{c} = 1.88 \times 10^8$ rad/s.

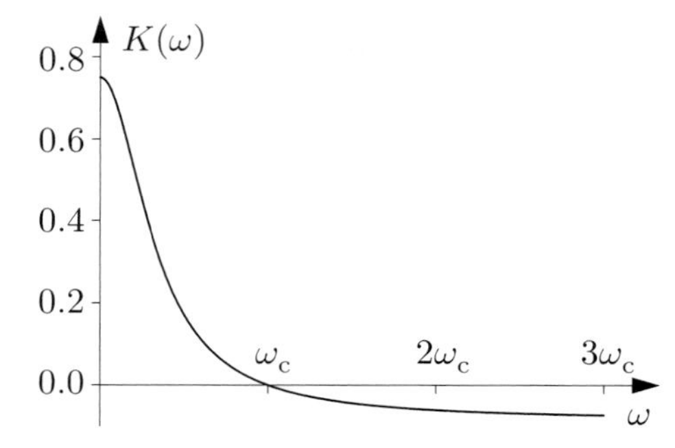

Fig. 10.6 The Clausius–Mossotii function $K(\omega)$.

11 Magnetophoresis

Magnetophoresis (MAP) is the magnetic analog to dielectrophoresis (DEP). Particles possessing either an induced or a permanent magnetization **M** can be moved around inside a microfluidic channel by applying an external, inhomogeneous magnetic field **H**. Whereas the dielectric response of virtually all materials is strong enough to allow for dielectrophoretic effects, the magnetic response is often too weak for most materials to make magnetophoresis happen. In fact, one must often carefully prepare a given sample by attaching magnetic particles before launching it into a MAP device. However, this seemingly annoying feature is actually the strength of MAP, because it ensures full control over which part of a sample is subject to MAP. In DEP devices the strong dielectric response of both target and auxiliary particles can clutter the functionality of the device and make the DEP device difficult to operate.

11.1 Magnetophoresis and bioanalysis

Magnetophoresis in microsystems is currently undergoing a rapid development and is already a strong tool especially for bioanalysis. On the one hand, most biological samples are non-magnetic and are not affected or destroyed by the relatively weak magnetic fields employed in MAP, and on the other hand, it is possible to label cells or biomolecules specifically with magnetic microbeads. This provides a versatile physical handle for manipulation and handling of biological samples.

A sketch of a typical biocoated, magnetic microbead is shown in Fig. 11.1. The main body is a non-magnetic polymer sphere, often made of polystyrene, containing a large number of magnetic nanoparticles. The diameter of the polymer sphere is of the order 1 µm to make it practical for handling, while the magnetic particles preferably have a diameter of about 10 nm. In magnetic particles of such small size all atomic magnetic moments are aligned, but the direction of this total magnetic moment can rotate freely under the influence of thermal fluctuations at room temperature, see Exercise 11.1. This is known as superparamagnetism. The advantage of using superparamagnetic microbeads is twofold. First, such particles have a vanishing average magnetic moment in the absence of an external magnetic field and they exhibit only a tiny hysteresis effect, meaning that the average magnetic moment returns to the value zero after removing any applied external magnetic field. Secondly, the superparamagnetic microbeads acquire a large magnetic moment once placed in an external magnetic field. Consequently, such particles are ideal for capturing by turning on an external magnetic field, and for releasing them by turning it off.

The advantage of using polymers as the main body is that it allows for coating the microbead with specific biomolecules. Through well-controlled biochemical processes the

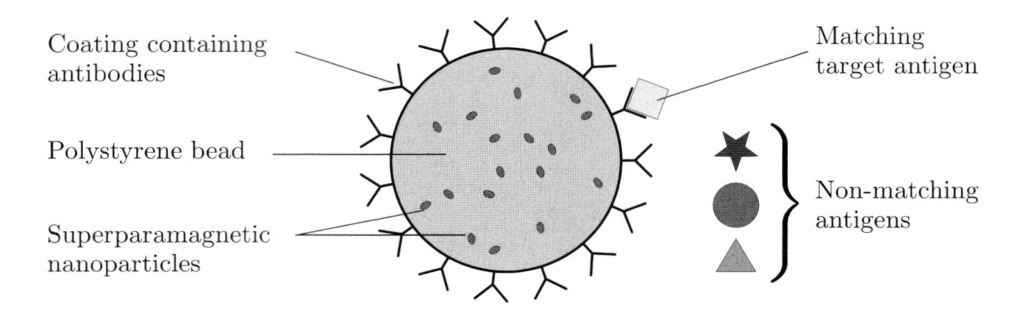

Fig. 11.1 A typical polystyrene microbead used for magnetic separation in lab-on-a-chip systems. The bead has a radius of about 1 µm and contains inclusions in the form of paramagnetic nanoparticles. The surface is coated with a specific antibody chosen to capture a given target antigen (white square) and not to interact with any other antigens (triangle, pentagram, and circle).

surface can by coated with carefully chosen antibodies, DNA strings or RNA. Once coated with such biomolecules the microbead can act as a highly specific capture probe for specific target molecules. An example of this capture selectivity is sketched in Fig. 11.1, where a surface coating of a specific antibody allows for capturing a specific target antigen, leaving all non-matching antigens untouched.

Magnetic separation of biomolecules can be implemented in lab-on-a-chip systems such as the one sketched in Fig. 11.2. Imagine a microfluidic channel with magnetic structures placed at the bottom wall. These structures can either be onchip electromagnets or magnetic material, magnetized by external electrical currents or magnetic fields, respectively. When superparamagnetic, biocoated microbeads are flushed through the channel they will be attracted by magnetophoretic forces to the magnetic structures if these are turned on. On reaching the magnetic structures, the beads are immobilized and the antibodies on their surfaces will form a layer of capture probes ready to bind with the proper antigen. If a sample containing many different antigens then is flowing through the microchannel, only the specific antigen matching the antibody on the microbeads will be captured. If the channel is flushed with a rinsing buffer after the capturing, we have achieved an upconcentration of the target antigen. This target sample can be released by turning off the magnetic field, flushing out the sample and collecting it at the outlet.

11.2 Magnetostatics

The theory of magnetophoresis has many similarities with that of dielectrophoresis. The starting point is the magnetostatic part of Maxwell's equations, i.e. assuming only stationary current densities and neglecting all time derivatives otherwise appearing. The two magnetostatic Maxwell equations are

$$\boldsymbol{\nabla}\cdot\mathbf{B} = 0, \tag{11.1a}$$

$$\boldsymbol{\nabla}\times\mathbf{B} = \mu_0\mathbf{J}_{\mathrm{tot}} = \mu_0\mathbf{J}_{\mathrm{ext}} + \mu_0\mathbf{J}_{\mathrm{mag}}, \tag{11.1b}$$

where $\mu_0 = 4\pi\times 10^{-7}$ H/m is the magnetic permeability of vacuum, and where $\mathbf{J}_{\mathrm{tot}}$, $\mathbf{J}_{\mathrm{ext}}$ and $\mathbf{J}_{\mathrm{mag}}$ all are stationary current densities; $\mathbf{J}_{\mathrm{tot}}$ is the total current density, $\mathbf{J}_{\mathrm{ext}}$ is the external transport current density running in conductors, and $\mathbf{J}_{\mathrm{mag}}$ is the current density

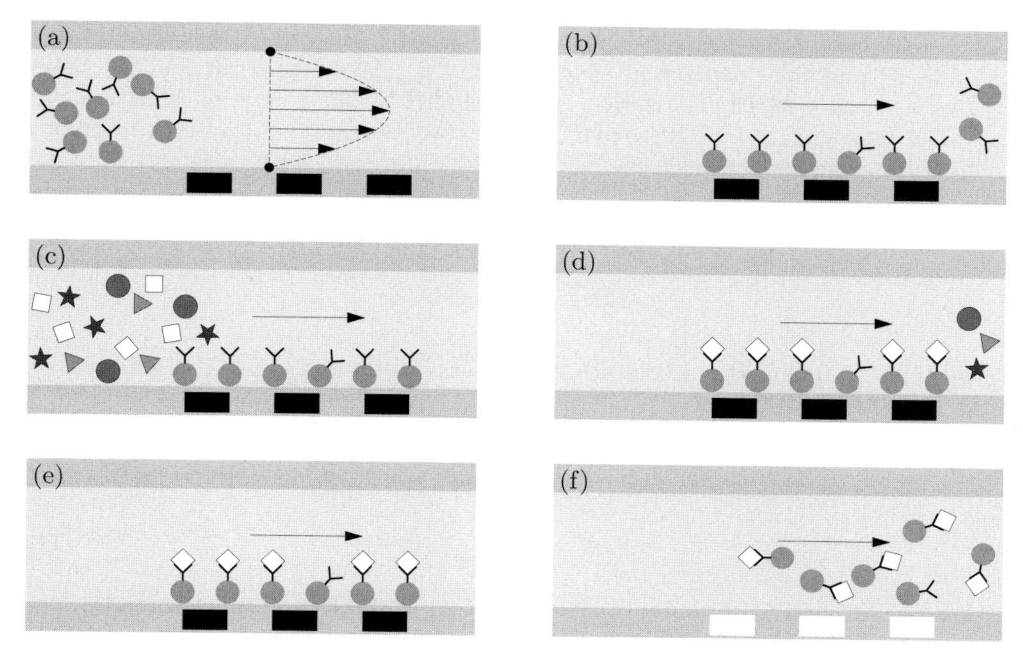

Fig. 11.2 The principle in magnetic separation for biosampling using magnetic beads flowing in a microfluidic channel. (a) A Poiseuille flow (light gray) carrying magnetic microbeads (dark circles) coated with suitable antibodies (attached Y-shapes). (b) Immobilization of the magnetic antibody beads by activating magnets (black rectangles) placed in the bottom wall (gray). (c) Introduction of sample containing the target antigen (white squares) and a number of other antigens (triangles, circles and pentagrams). (d) Capture of the target antigen by the immobilized antibody beads. (e) Thorough rinsing. (f) Release of the target sample by deactivating the magnets (now white rectangles) followed by collection at the microchannel outlet to the right.

bound to magnetic material in the form of atomic/molecular current loops and quantum-mechanical spins. This division of the total current density is analogous to the division of the total charge density into external and polarization charge densities, see Eq. (8.11).

In analogy with the dipole moment $\mathbf{p} = q\mathbf{d}$ of Eq. (8.7) the magnetic dipole moment $\mathbf{m}$ for a circulating magnetization current I_{mag} enclosing a flat area $\mathcal{A}$ with the surface normal $\mathbf{n}$ is defined by

$$\mathbf{m} = I_{\text{mag}}\mathcal{A}\,\mathbf{n}. \tag{11.2}$$

In analogy with the polarization $\mathbf{P}$ of Eq. (8.8) the magnetization $\mathbf{M}$ is defined as the magnetic moment density,

$$\mathbf{M}(\mathbf{r}_0) = \lim_{\text{Vol}(\Omega^*)\to 0}\left[\frac{1}{\text{Vol}(\Omega^*)}\int_{\Omega^*} d\mathbf{r}\,\mathbf{m}(\mathbf{r}_0+\mathbf{r})\right]. \tag{11.3}$$

A careful mathematical analysis, Jackson (1975), shows that the magnetization $\mathbf{M}$ is related to the density $\mathbf{J}_{\text{mag}}$ of the circulating magnetization current as

$$\mathbf{J}_{\text{mag}} = \boldsymbol{\nabla}\times\mathbf{M}, \tag{11.4}$$

an expression that is the magnetic analog of the relation Eq. (8.10) between the electric polarization $\mathbf{P}$ and the polarization charge density ρ_{pol}. Note that the analogy between magnetization and electric polarization is not perfect, as the source of $\mathbf{M}$ is the vector $\mathbf{J}_{\mathrm{mag}}$, while the source of $\mathbf{P}$ is the scalar ρ_{pol}. But we, nevertheless, continue to point out the similarities between the magnetic and the electric theory. Just as $\mathbf{E}$ and $\mathbf{P}$ could be combined to give the useful concept of the $\mathbf{D}$ field, it is beneficial to combine the magnetic induction $\mathbf{B}$ and the magnetization $\mathbf{M}$ in the definition of the magnetic field $\mathbf{H}$ as

$$\mathbf{H} = \frac{1}{\mu_0}\,\mathbf{B} - \mathbf{M}. \tag{11.5}$$

We note that by taking the divergence of this expression and using the Maxwell equation (11.1a), we find

$$\boldsymbol{\nabla}\cdot\mathbf{H} = -\boldsymbol{\nabla}\cdot\mathbf{M}. \tag{11.6a}$$

Finally, by inserting Eqs. (11.4) and (11.5) into Eq. (11.1b) we obtain

$$\boldsymbol{\nabla}\times\mathbf{H} = \mathbf{J}_{\mathrm{ext}}. \tag{11.6b}$$

Equations (11.6a) and (11.6b) state that the sources of the $\mathbf{H}$ field are a spatially varying magnetization and an external current density.

The last fundamental concept to be introduced here in connection with magnetostatics is the magnetic susceptibility tensor χ defined as

$$\chi \equiv \left(\frac{\partial\mathbf{M}}{\partial\mathbf{H}}\right)_{V,T}. \tag{11.7}$$

For hysteresis-free, isotropic materials $\mathbf{M}$ and $\mathbf{H}$ are parallel, and χ becomes a scalar. Furthermore, in some cases the relation between $\mathbf{M}$ and $\mathbf{H}$ is linear, and we get the simplest expressions, which are analogous to Eq. (8.12),

$$\mathbf{M} = \chi\,\mathbf{H}, \tag{11.8a}$$

$$\mathbf{B} = \mu_0(\mathbf{H}+\mathbf{M}) = \mu_0(1+\chi)\mathbf{H} \equiv \mu_0\mu_{\mathrm{r}}\mathbf{H} \equiv \mu\,\mathbf{H}. \tag{11.8b}$$

Here, the coefficients μ and μ_{r} are called the permeability and relative permeability, respectively.

11.3 Basic equations for magnetophoresis

We are going to study the magnetic forces acting on a magnetizable body when it is placed in an external magnetic field. Consider a given external magnetic field characterized by $\mathbf{H}_{\mathrm{ext}}$ before the magnetizable object enters. When the object is then placed in the magnetic field, it acquires a magnetization $\mathbf{M}$, and this in turn generates extra contributions to the magnetic field, which therefore changes from $\mathbf{H}_{\mathrm{ext}}$ to $\mathbf{H}$.

To find the magnetic force $\mathbf{F}$ on the magnetizable object is not entirely simple. We shall not go through the derivation here, but just mention that a safe way to proceed is to consider the free energy of the system (Landau, Lifshitz and Pitaevskiĭ, 1984). The differential of the free energy density is generally given by $-\mathbf{B}\cdot\mathrm{d}\mathbf{H}$. From this the change $\delta\mathcal{F}$ in total free-energy $\mathcal{F}$ upon a small spatial displacement $\delta\mathbf{r}$ in the system can be calculated. Now, free energy and force are in general related by the differential relation

$$\delta\mathcal{F} = -\mathbf{F}\cdot\delta\mathbf{r}. \tag{11.9}$$

Carrying out this analysis results in the following expression for the force $\mathbf{F}$ on the magnetizable object, valid if the fields are static, $\boldsymbol{\nabla}\times\mathbf{H}_{\mathrm{ext}} = 0$, and the body is nonconducting:

$$\mathbf{F} = \mu_0 \int_{\mathrm{body}} \mathrm{d}\mathbf{r}\,(\mathbf{M}\cdot\boldsymbol{\nabla})\mathbf{H}_{\mathrm{ext}}, \quad \text{for}\ \ \boldsymbol{\nabla}\times\mathbf{H}_{\mathrm{ext}} = 0. \tag{11.10}$$

This expression with its underlying assumptions is the basis for understanding magnetophoresis.

To be a little more specific we consider, in analogy with the dielectric treatment in Section 10.3, a homogeneous, isotropic, and nonconducting sphere of radius a and permeability μ. As $\boldsymbol{\nabla}\times\mathbf{H}_{\mathrm{ext}} = 0$ it is possible, as in Eqs. (8.2a) and (8.3a) to introduce a magnetic potential ϕ_m such that

$$\mathbf{H}_{\mathrm{ext}} = -\boldsymbol{\nabla}\phi_m. \tag{11.11}$$

The analogy to the dielectric case is obvious, and the resulting expression for the magnetization $\mathbf{M}$ of the sphere after being placed in the homogeneous field $\mathbf{H}_{\mathrm{ext}}$ is given by

$$\mathbf{M} = 3\frac{\mu-\mu_0}{\mu+2\mu_0}\,\mathbf{H}_{\mathrm{ext}}. \tag{11.12}$$

Note how closely this expression resembles Eq. (10.18) for the induced dipole moment in electrostatics. The analogy is perfect when we consider the resulting MAP force $\mathbf{F}_{\mathrm{MAP}}$,

$$\mathbf{F}_{\mathrm{MAP}}(\mathbf{r}_0) = 2\pi\mu_0\,K(\mu_0,\mu)\,a^3\nabla\Big[\mathbf{H}_{\mathrm{ext}}(\mathbf{r}_0)^2\Big], \tag{11.13}$$

where even the Clausius–Mossotti factor from Eq. (10.19) reappears. Note how the MAP force is dependent on the gradient of the square of the magnetic field, just as the DEP force.

In conclusion, we can say that force expressions applying to magnetophoresis and dielectrophoresis can be very similar or even in special cases identical. However, it is important to bear in mind that often there will be substantial deviations between the two descriptions, primarily due to the intrinsic nonlinear nature of magnetic materials.

11.4 Calculation of magnetic-bead motion

The performance of a microfluidic, magnetic-bead-capture system can be evaluated by numerical simulation. For a single bead the calculation can be simplified by neglecting the (small) influence from the bead on liquid flow field $\mathbf{v}$. The latter can therefore be calculated before adding the bead and hence be treated as a known field. Similarly, the magnetic microbead does not affect the applied external magnetic field significantly, thus $\mathbf{H}_{\mathrm{ext}}$ can also be calculated without the presence of the bead, whence it is also a known field. Moreover, from the discussion in Section 6.4 we know that the inertia of the microbead can be neglected, so at a given instant the velocity $\mathbf{u}$ of the bead is given by the force balance between the magnetophoretic force $\mathbf{F}_{\mathrm{MAP}}$, Eq. (11.13), and the viscous drag force $\mathbf{F}_{\mathrm{drag}}$,

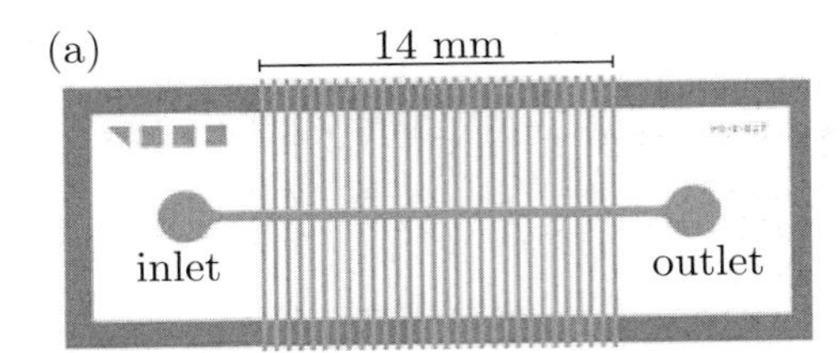

Fig. 11.3 (a) Design mask for a 400 μm wide and 14 mm long channel passing straight from the inlet to the outlet (horizontal gray line). There are placed 30 magnetic structures along each side of the channel (vertical gray lines). (b) The silicon-based (black) lab-on-a-chip system with the magnetic permalloy structures (white) at the side and a pyrex lid for sealing. The 2 mm wide inlet and outlet is seen at each end of the channel (gray). Adapted from Lund-Olsen, Bruus and Hansen (2007), courtesy of Mikkel Fougt Hansen, DTU Nanotech.

Eq. (3.128). Isolating $\mathbf{u}$ in the force balance $\mathbf{F}_{\mathrm{MAP}} + \mathbf{F}_{\mathrm{drag}} = \mathbf{0}$ leads to the following expression for the incremental change $\mathbf{dr}$ in the position $\mathbf{r}$ of the bead during the time step dt,

$$\mathbf{dr} = \mathbf{u}\, dt = \left(\mathbf{v}(\mathbf{r}) + \frac{\mu_0\, a^2}{3\,\eta}\, K(\mu_0, \mu)\, \nabla \Big[\mathbf{H}_{\mathrm{ext}}(\mathbf{r})^2 \Big] \right) \mathrm{d}t. \tag{11.14}$$

The trajectory can then be found by numerical integration using, e.g. a Runge–Kutta integration routine.

It may not be feasible to accurately simulate a many-bead system using the single-bead approach outlined above. Instead, one may model the microbeads by a continuous particle density $c(\mathbf{r}, t)$ and an associated particle current density $\mathbf{J}$. The governing equations of these two fields are the continuity equation, see Section 5.2, and a magnetic force variant of the Nernst–Planck equation, see Section 9.1,

$$\partial_t c = -\boldsymbol{\nabla}\cdot \mathbf{J}, \tag{11.15a}$$

$$\mathbf{J} = -D\, \boldsymbol{\nabla} c + c\, \mathbf{v} + c\, b\, \mathbf{F}_{\mathrm{MAP}}, \tag{11.15b}$$

where $D = k_{\mathrm{B}}T/(6\pi\eta a)$ is the Einstein diffusivity, Eq. (6.49), and $b = 1/(6\pi\eta a)$ is the Stokes mobility, Eq. (3.128). The boundary condition for the system depends on detailed interactions between the walls of the system and the beads. One possible assumption is the condition that once touching the wall any bead gets stuck, which mathematically can be formulated as

$$\mathbf{n}\cdot\mathbf{J} = \begin{cases} c\, b\, \mathbf{n}\cdot\mathbf{F}_{\mathrm{MAP}}, & \text{for } \mathbf{n}\cdot\mathbf{F}_{\mathrm{MAP}} > 0, \\ 0, & \text{for } \mathbf{n}\cdot\mathbf{F}_{\mathrm{MAP}} < 0, \end{cases} \tag{11.16}$$

where $\mathbf{n}$ is the outward-pointing surface normal. This boundary condition prevents a bead current from the wall into the system, while it allows a bead current to flow into the wall.

As shown by Mikkelsen and Bruus (2005) the continuum description also makes it possible to take the influence from the beads onto the liquid into account. Due to the force balance between the magnetophoretic force and the viscous drag, the former must be included in the Navier–Stokes equation as body force,

$$\rho\partial_t \mathbf{v} + \rho(\mathbf{v}\cdot\boldsymbol{\nabla})\mathbf{v} = -\boldsymbol{\nabla} p + \eta\nabla^2 \mathbf{v} + c\mathbf{F}_{\mathrm{MAP}}. \tag{11.17}$$

By comparing numerical simulation with and without the magnetophoretic body force in the Navier–Stokes equation, it was found that for concentrations so large that the interbead distance is below 10 µm the magnetophoretic body force on the liquid is significant for the motion of the beads. The motion of one bead due to the magnetophoretic force induces a viscous drag in the liquid, a kind of wake, which is felt by the other beads, and for high bead concentrations this hydrodynamic force cannot be neglected.

11.5 Magnetophoretic lab-on-a-chip systems

We end this short introduction to magnetophoresis by showing one explicit example of a magnetophoretic lab-on-a-chip system fabricated in the group of Fougt Hansen at MIC.

In Fig. 11.3(a) is shown the design mask for fabricating a magnetic microfluidic separation chip. The core of the design is the 14 mm long, 400 µm wide and 80 µm deep channel with 30 magnetic structures placed along each side. At each end of the channels are placed an inlet and an outlet reservoir with diameters of 2 mm.

The finished device is seen on the picture in Fig. 11.3(b). The main structure is fabricated in silicon. The magnetic structures are filled with permalloy (20% Fe and 80% Ni) by electroplating before sealing the channel with a pyrex lid.

To operate the device it is placed in the middle of the gap of an electromagnet and a syringe pump is attached by teflon tubes. When the electromagnet is turned on, the permalloy structures are magnetized and strong field gradients are created in the channel. Through an inlet valve solutions of superparamagnetic beads with a diameter of 1 µm are led through the channel in a Poiseuille flow with a maximum speed around 1 mm/s. The MAP force drags the beads to the side of the channel and captures them at the edges of the magnetic structures.

In Fig. 11.4 is seen a micrograph of the first 10 structures (top panel) and the next 10 structures (bottom channel). The captured beads are seen as a dark gray shadow at the

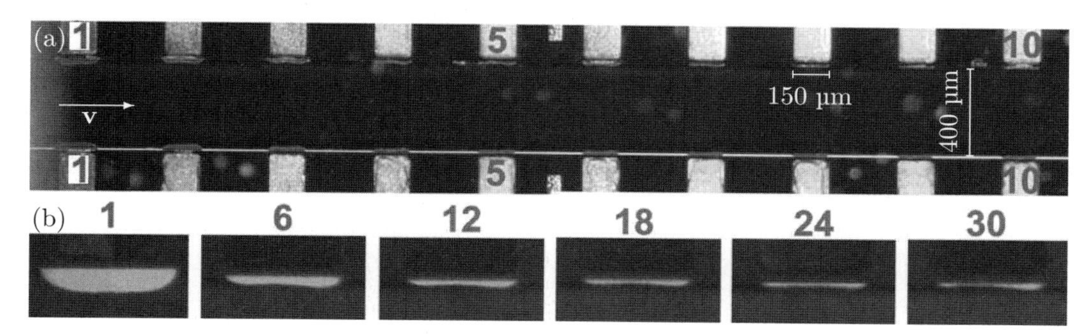

Fig. 11.4 (a) Micrograph of magnetic bead capture by magnetophoresis in a 400 µm wide and 80 µm deep silicon-based microchannel with a pyrex lid. The 150 µm wide magnetic structures (light rectangles) are made of permalloy, and the first ten pairs of these are shown. Inside the channel, in front of the magnetic structures, are seen some light gray clouds of captured 1-µm-sized beads coated with fluorescein. For an improved design see Fig. 14.3. (b) A zoom-in on the captured beads at magnetic structure no. 1, 6, 12, 18, 24 and 30. The amount of captured beads is largest at the first structure and then decreases monotonically. Micrographs courtesy of Torsten Lund-Olesen and Mikkel Fougt Hansen, DTU Nanotech.

end of the magnetic structures (light gray rectangles). The number of beads being caught decreases along the channel. The largest amount of captured beads are seen on element number 1.

So far the system has mainly been run with magnetic beads without any biocoating. The system has now been successfully tested and is ready to be applied for more biorelated investigations. For further developments involving integrated micromixers, see Section 14.4 and Lund-Olesen, Bruus and Hansen (2007).

11.6 Further reading

The theory of magnetic forces on magnetizable bodies is not straightforward and some inconsistencies even appear in the current literature. A good starting point, however, is provided by Jackson (1975) and Landau, Lifshitz and Pitaevskiĭ (1984). An introduction to the basic principles of magnetophoresis is given in the book of Jones (1995), while the review papers by Gijs (2004) and Pamme (2006) give excellent descriptions of magnetophoresis used for bioanalysis in microfluidic systems.

11.7 Exercises

Exercise 11.1
Superparamagnetic nanoparticles
In the simplest cases the relaxation time τ for the net magnetization of a superparamagnetic particle follows a thermal activation law,

$$\tau = \tau_0 \,\exp\left(\frac{K\mathcal{V}}{k_\mathrm{B}T}\right), \tag{11.18}$$

where the pre-exponential factor is $\tau_0 = 10^{-12} - 10^{-10}$ s, $\mathcal{V}$ is the volume of the particle, and $K \approx 10^4$ J/m^3 is the so-called anisotropy energy density.

Argue why particles need to have a radius of about 5 nm or smaller to exhibit superparamagnetism at room temperature.

Exercise 11.2
The torque from the magnetic moment of a square current loop
Prove that the mechanical torque $\boldsymbol{\tau}$ from the magnetic moment $\mathbf{m} = I\mathcal{A}\,\mathbf{n}$ is given by

$$\boldsymbol{\tau} = I\mathcal{A}\,\mathbf{n}\times\mathbf{B} = \mathbf{m}\times\mathbf{B}, \tag{11.19}$$

for a thin wire formed as a square closed loop of side length L in the positive direction, carrying a circulating current I, and placed in the horizontal xy plane, when a constant magnetic field $\mathbf{B}$ of arbitrary orientation relative to the surface normal $\mathbf{n}$ of the area $\mathcal{A} = L^2$ enclosed by the loop. Hint: use the fact that the force $\mathbf{F}$ on a thin wire of length L carrying a current I is given by $\mathbf{F} = IL\,\mathbf{e}_L \times \mathbf{B}$, where $\mathbf{e}_L$ is a unit vector indicating the current direction along the wire.

Exercise 11.3
The rotation of M
Prove Eq. (11.6b), $\boldsymbol{\nabla}\times\mathbf{H} = \mathbf{J}_\mathrm{ext}$, from the preceding part of Section 11.2.

Exercise 11.4
Magnetic two-length-scale bead separator
When implementing magnetic elements for magnetophoretic capturing of beads in microfluidic systems, it turns out to be difficult to apply a sufficiently strong magnetic field to generate the needed capturing force. This problem can be overcome by the use of a two-length-scale magnetic system: A large permanent magnet placed at some distance from the microchannel, and a small paramagnetic element placed directly next to the microchannel.

Consider a infinite parallel-plate microchannel placed as in Fig. 3.6 between $z = 0$ and $z = h$. Beneath the channel a cylindrical permanent magnet with magnetization M_{pm} and radius R is placed with its axis parallel to the y axis, i.e. perpendicular to the flow direction along the x axis, and crossing the z axis at $z_{\mathrm{pm}} < 0$. From magnetostatics it is known that the magnetic field $\mathbf{H}_{\mathrm{pm}}(z)$ along the z axis from such a magnet is given by

$$\mathbf{H}_{\mathrm{pm}} = C_1 \frac{R^2 M_{\mathrm{pm}}}{(z - z_{\mathrm{pm}})^2}\, \mathbf{e}_z, \tag{11.20}$$

where C_1 is a constant.

(a) Combine Eqs. (11.10) and (11.20) to derive an expression for the magnetophoretic force F_{pm} on a bead placed in the channel at position $0 < z < h$ on the z axis. Hint: the external magnetic field can be assumed constant across the bead.

(b) Now imagine a magnetic element in the form of a paramagnetic cylinder with radius $r \ll R$ placed also with its axis parallel to the y axis, but crossing the z axis at z_{me} much closer to the microchannel, $z_{\mathrm{pm}} \ll z_{\mathrm{me}} < 0$. This is possible by electroplating techniques. The magnetic element is magnetized by the permanent magnetic cylinder and acquires the magnetization M_{pm}. Derive an expression for the magnetophoretic force F_{me} from this magnetic element on a bead placed in the channel at position $0 < z < h$ on the z axis.

(c) Write down the ratio $F_{\mathrm{me}}/F_{\mathrm{pm}}$ between the two magnetophoretic forces an a bead, and estimate its value given the following parameter values typical for a microfluidic system: $R = 1$ mm, $r = 5$ µm, $M_{\mathrm{pm}} = M_{\mathrm{me}}$, $z_{\mathrm{pm}} = -1.5$ mm, and $z_{\mathrm{me}} = -10$ µm.

11.8 Solutions

Solution 11.1
Superparamagnetic nanoparticles
The critical volume $\mathcal{V}^*$ below which a particle at room temperature begins to exhibit superparamagnetic behavior can be estimated as the volume which makes the argument of the exponential function in Eq. (11.18) unity, i.e.

$$\mathcal{V}^* \equiv \frac{k_{\mathrm{B}}T}{K} = \frac{4.1 \times 10^{-21}\ \mathrm{J}}{10^4\ \mathrm{J\,m^{-3}}} = 1.0 \times 10^{-25}\ \mathrm{m}^3 = \frac{4\pi}{3}(4.6\ \mathrm{nm})^3. \tag{11.21}$$

Consequently, spherical particles with radii less than 5 nm exhibits superparamagnetism.

Solution 11.2
The torque from the magnetic moment of a square current loop
We place the $L \times L$ square current loop in the xy plane centered around the origin and parallel to the x and y axis. The current I is assumed to flow counterclockwise. Starting in the lower left corner and following the current direction, the four corners are denoted a, b, c and d. The

four sides have center of mass positions $\mathbf{r}_{ab} = -(L/2)\mathbf{e}_y$, $\mathbf{r}_{bc} = (L/2)\mathbf{e}_x$, $\mathbf{r}_{cd} = (L/2)\mathbf{e}_y$, and $\mathbf{r}_{da} = -(L/2)\mathbf{e}_x$ as well as the unit vectors $\mathbf{e}_{ab} = \mathbf{e}_x$, $\mathbf{e}_{bc} = \mathbf{e}_y$, $\mathbf{e}_{cd} = -\mathbf{e}_x$, and $\mathbf{e}_{da} = -\mathbf{e}_y$.

Using the vector identity $\mathbf{a} \times (\mathbf{b} \times \mathbf{c}) = (\mathbf{a}\cdot\mathbf{c})\mathbf{b} - (\mathbf{a}\cdot\mathbf{b})\mathbf{c}$ proved in Exercise 1.4(d), the torque $\boldsymbol{\tau}$ is then calculated straightforwardly as

$$\begin{aligned} \boldsymbol{\tau} &= \sum_{i=1}^{4} \mathbf{r}_i \times (IL\mathbf{e}_i \times \mathbf{B}) \\ &= IL\frac{L}{2}\Big[-\mathbf{e}_y \times (\mathbf{e}_x \times \mathbf{B}) + \mathbf{e}_x \times (\mathbf{e}_y \times \mathbf{B}) + \mathbf{e}_y \times (-\mathbf{e}_x \times \mathbf{B}) - \mathbf{e}_x \times (-\mathbf{e}_x \times \mathbf{B})\Big] \\ &= I\mathcal{A}\,(-B_y\mathbf{e}_x + B_x\mathbf{e}_y) = I\mathcal{A}\,\mathbf{n} \times \mathbf{B} = \mathbf{m} \times \mathbf{B}. \end{aligned} \tag{11.22}$$

Solution 11.3
The rotation of M

From Eq. (11.1b) we have $\mathbf{J}_{\text{ext}} = \boldsymbol{\nabla} \times (\frac{1}{\mu_0}\mathbf{B}) - \mathbf{J}_{\text{mag}}$. By inserting $\mathbf{J}_{\text{mag}} = \boldsymbol{\nabla} \times \mathbf{M}$ from Eq. (11.4) in this equation, we arrive at $\mathbf{J}_{\text{ext}} = \boldsymbol{\nabla} \times (\frac{1}{\mu_0}\mathbf{B} - \mathbf{M}) = \boldsymbol{\nabla} \times \mathbf{H}$.

Solution 11.4
Magnetic two-length-scale bead separator

Given the symmetry all vectors are parallel to the z axis. In the following a subscript "b" refer to the bead.

(a) From Eqs. (11.10) and (11.20) we get

$$F_{\text{pm}}(z) = \mu_0 \mathcal{V}_{\text{b}} M_{\text{b}} \partial_z \left[C_1 \frac{R^2 M_{\text{pm}}}{(z - z_{\text{pm}})^2}\right] = C_2 \frac{R^2 M_{\text{pm}}}{(z - z_{\text{pm}})^3}, \tag{11.23}$$

where C_2 is a new constant.

(b) In analogy with Eq. (11.23) we get

$$F_{\text{me}}(z) = \mu_0 \mathcal{V}_{\text{b}} M_{\text{b}} \partial_z \left[C_1 \frac{r^2 M_{\text{me}}}{(z - z_{\text{me}})^2}\right] = C_2 \frac{r^2 M_{\text{me}}}{(z - z_{\text{me}})^3}, \tag{11.24}$$

where C_2 is the same constant as before.

(c) The force ratio is derived directly from Eqs. (11.23) and (11.24)

$$\frac{F_{\text{me}}}{F_{\text{pm}}} = \frac{r^2 M_{\text{me}}}{R^2 M_{\text{pm}}} \frac{(z - z_{\text{pm}})^3}{(z - z_{\text{me}})^3} = \frac{5^2}{1000^2} \frac{1500^3}{10^3} \approx 10^2, \tag{11.25}$$

and the given lengths have been inserted all in µm. The result indicates that the introduction of the paramagnetic element increases the magnetophoretic force by nearly two orders of magnitude.

12 Thermal transfer

The theory of thermal transfer in microfluidic systems is complex. Even for the simplest case of a single fluid, all three governing equations must be taken into account leading to five coupled, second-order, partial differential equations for the scalar fields $v_x(\mathbf{r},t)$, $v_y(\mathbf{r},t)$, $v_z(\mathbf{r},t)$, $p(\mathbf{r},t)$ and $T(\mathbf{r},t)$, and the three principal parameters, viscosity η, thermal conductivity κ and density ρ all depend on temperature.

In this chapter we shall study some simple, idealized cases that allow for analytical solutions, and that lead us to physical insight into the basic thermal phenomena in microfluidics. Our analysis shall often rely on perturbation theory, introduced in Section 1.5. Perturbation theory can be applied because η, κ, and ρ have only a weak dependence on temperature, as is evident when studying the thermal perturbation coefficients of water, $(\partial_T\eta)/\eta$, $(\partial_T\kappa)/\kappa$, and $(\partial_T\rho)/\rho$, listed in Table 12.1. When multiplied by ΔT they become the useful, dimensionless perturbation parameters α_η, α_κ, and α_ρ, which even for the significant temperature difference $\Delta T = 10$ ℃ are small

$$\left.\begin{aligned} \alpha_\eta &\equiv \frac{\partial_T\eta}{\eta}\,\Delta T = -2.5\times 10^{-1} \\ \alpha_\kappa &\equiv \frac{\partial_T\kappa}{\kappa}\,\Delta T = 2.8\times 10^{-2} \\ \alpha_\rho &\equiv \frac{\partial_T\rho}{\rho}\,\Delta T = -2.1\times 10^{-3} \end{aligned}\right\},\ \text{for } T_0 = 20\ ℃ \text{ and } \Delta T = 10\ ℃. \tag{12.1}$$

We note that $|\alpha_\rho| \ll |\alpha_\kappa| \ll |\alpha_\eta| < 1$, so not only is it possible to perform well-defined perturbation calculations in the temperature, but we can also conclude that for water the smallest thermal effects are due to changes in density, followed by changes in thermal con-

Table 12.1 Temperature-dependent parameters for pure water: the viscosity η, the thermal conductivity κ and the density ρ as well as the thermal perturbation coefficients $(\partial_T\eta)/\eta$, $(\partial_T\kappa)/\kappa$, and $(\partial_T\rho)/\rho$. Data are taken from *CRC Handbook of Chemistry and Physics.*

T_0 [℃]	η [mPa s]	$(\partial_T\eta)/\eta$ [K^{-1}]	κ [W m^{-1}K^{-1}]	$(\partial_T\kappa)/\kappa$ [K^{-1}]	ρ [kg m^{-3}]	$(\partial_T\rho)/\rho$ [K^{-1}]
10	1.307	-2.9×10^{-2}	0.580	3.1×10^{-3}	999.7	-0.9×10^{-4}
20	1.002	-2.5×10^{-2}	0.597	2.8×10^{-3}	998.2	-2.1×10^{-4}
30	0.798	-2.1×10^{-2}	0.613	2.5×10^{-3}	995.6	-3.0×10^{-4}
40	0.653	-1.9×10^{-2}	0.628	2.2×10^{-3}	992.2	-3.8×10^{-4}
60	0.467	-1.5×10^{-2}	0.652	1.6×10^{-3}	983.2	-5.2×10^{-4}
80	0.355	-1.3×10^{-2}	0.670	1.1×10^{-3}	971.6	-6.4×10^{-4}

ductivity, while the temperature dependence of the viscosity gives rise to the largest thermal effects.

12.1 Thermal effects in hydrostatics

As our first example of thermal effects in microfluidics we study water at rest in the infinite parallel-plate channel introduced in Section 3.4.2. We apply constant temperatures T_0 and $T_0 + \Delta T$ on the bottom and top plates, respectively, and look for the steady-state solution

$$T = T_0 + \Delta T\,\Theta(z), \tag{12.2}$$

where the dimensionless temperature deviation $\Theta(z) \equiv [T(z) - T_0]/\Delta T$ does not depend on x and y due to symmetry. Viscosity plays no role for fluids at rest, so the most significant temperature dependence stems from the conductivity κ. According to Eq. (12.1), the temperature dependence of κ is so weak that it is well captured by a first-order Taylor expansion in ΔT around $\kappa_0 \equiv \kappa(T_0)$,

$$\kappa(T) \approx \kappa_0 + (\partial_T \kappa_0)\Delta T\,\Theta(z) = \big[1 + \alpha_\kappa \Theta(z)\big]\kappa_0. \tag{12.3}$$

Finally, we introduce the dimensionless co-ordinate

$$\zeta \equiv \frac{z}{h}, \tag{12.4}$$

and the heat-transfer equation (2.65) with boundary conditions consequently simplifies to

$$0 = \partial_\zeta\Big(\big[1 + \alpha_\kappa \Theta(z)\big]\partial_\zeta \Theta(\zeta)\Big), \text{ with } \Theta(0) = 0, \text{ and } \Theta(1) = 1. \tag{12.5}$$

We solve this equation as a perturbation problem in α_κ, so we write $\Theta(\zeta)$ as

$$\Theta(\zeta) = \Theta_0(\zeta) + \alpha_\kappa \Theta_1(\zeta) + \alpha_\kappa^2 \Theta_2(\zeta) + \cdots. \tag{12.6}$$

To first order in α_κ, Eq. (12.5) becomes

$$0 = \partial_\zeta\big[(1 + \alpha_\kappa \Theta_0)\,\partial_\zeta(\Theta_0 + \alpha_\kappa \Theta_1)\big] \approx \partial_\zeta^2 \Theta_0 + \alpha_\kappa\big[\Theta_0 \partial_\zeta^2 \Theta_0 + (\partial_\zeta \Theta_0)^2 + \partial_\zeta^2 \Theta_1\big], \tag{12.7}$$

which in analogy to Eq. (1.35) leads to the following zero- and first-order equations and corresponding boundary conditions

$$\partial_\zeta^2 \Theta_0 = 0, \qquad \text{with } \Theta_0(0) = 0, \text{ and } \Theta_0(1) = 1, \tag{12.8a}$$

$$\partial_\zeta^2 \Theta_1 = -\Theta_0 \partial_\zeta^2 \Theta_0 - (\partial_\zeta \Theta_0)^2, \text{ with } \Theta_1(0) = 0, \text{ and } \Theta_1(1) = 0. \tag{12.8b}$$

The solutions

$$\Theta_0 = \zeta, \quad \text{and} \quad \Theta_1 = \frac{1}{2}(\zeta - \zeta) \tag{12.9}$$

are easily found by integration of $\partial_z^2 \Theta_0$ and then $\partial_z^2 \Theta_1$, so to first order in α_κ the resulting temperature profile is seen to be given by

$$\Theta(\zeta) = \zeta + \frac{\alpha_\kappa}{2}(\zeta - \zeta^2), \tag{12.10a}$$

$$T(z) = T_0 + \left(\frac{z}{h} - \frac{\alpha_\kappa}{2}\left[\frac{z}{h} - \left(\frac{z}{h}\right)^2\right]\right)\Delta T. \tag{12.10b}$$

Consider now the situation with the same thermal boundary conditions, but where the space between $z = 0$ and $z = h$ is filled with N parallel slabs of different materials, say a combination of liquid layers separated by thin solid plates. The ith slab has the thermal conductivity κ_i and it is positioned between $z = z_{i-1}$ and $z = z_i$, where $0 = z_0 < z_1 < \ldots < z_{N-1} < z_N = h$. The steady-state temperature field then takes the form

$$T = T_0 + \Delta T\, \Theta_i(z), \quad \text{for } z_{i-1} < z < z_i, \quad \text{and } i = 1, 2, \ldots, N. \tag{12.11}$$

As before, Fourier's equation for each slab is $\partial_z^2 \Theta_i(z) = 0$, so the temperature profile is piecewise linear, and within each slab it is given by

$$\Theta_i(z) = A_i\, z + \Theta_i^{(0)}, \tag{12.12}$$

where the constants A_i and $\Theta_i^{(0)}$ must be determined by the boundary conditions. These are given by the requirement that in steady state heat cannot accumulate anywhere, so the heat current density J_heat must be constant. In particular it must be constant at the interfaces, $J_{\text{heat},i}(z_i) = J_{\text{heat},i+1}(z_i)$, which by Fourier's law Eq. (2.55) translates into

$$\kappa_i \partial_z \Theta_i(z_i) = \kappa_{i+1} \partial_z \Theta_{i+1}(z_i), \quad i = 1, 2, \ldots, N-1. \tag{12.13}$$

Moreover, the temperature field must be continuous across the interfaces, which yields the conditions

$$\Theta_i(z_i) = \Theta_{i+1}(z_i), \quad i = 1, 2, \ldots, N-1. \tag{12.14}$$

It is left as an exercise for the reader to show that for a two-layer system with its interface placed at $z_1 = \alpha h$, where $0 < \alpha < 1$, the resulting temperature profile is

$$T(z) = T_0 + \Delta T \times \begin{cases} \dfrac{\kappa_2}{\kappa_1 + (\kappa_2 - \kappa_1)\alpha}\, \dfrac{z}{h}, & \text{for} \quad 0 < z < \alpha h, \\ 1 + \dfrac{\kappa_1}{\kappa_1 + (\kappa_2 - \kappa_1)\alpha} \left(\dfrac{z}{h} - 1\right), & \text{for} \quad \alpha h < z < h. \end{cases} \tag{12.15}$$

With the steady-state solution at hand we move on to find the time-dependent solution to the same system of water at rest in the infinite parallel-plate channel. For times $t < 0$ the entire system has the uniform temperature T_0. At time $t = 0$ the temperature of the top plate is suddenly raised to $T_0 + \Delta T$ and kept there for $t > 0$. For simplicity, we disregard the temperature dependence of the thermal conductivity κ, and the problem therefore reduces to Fourier's equation (2.68) characterized by the thermal diffusivity $D_\text{th} = \kappa/(\rho c_\text{p})$. We use the thermal diffusion time τ_th to introduce the dimensionless time τ as

$$\tau_\text{th} \equiv \frac{h^2}{D_\text{th}} \quad (= 70 \text{ ms for water with } h = 100\ \mu\text{m}), \tag{12.16a}$$

$$\tau \equiv \frac{t}{\tau_\text{th}}, \tag{12.16b}$$

whereby the dimensionless form of Fourier's equation becomes the diffusion equation

$$\partial_\tau \Theta(\zeta, \tau) = \partial_\zeta^2 \Theta(\zeta, \tau), \tag{12.17}$$

with a diffusion constant of unity, and with the following boundary and initial conditions,

$$\Theta(0, \tau) = 0, \quad \Theta(1, \tau) = 1, \quad \Theta(\zeta, 0) = 0, \quad \Theta(\zeta, \infty) = \zeta. \tag{12.18}$$

Note that the steady-state solution at $\tau = \infty$ is given by the zero-order solution Θ_0 from Eq. (12.9), $\Theta(\zeta, \infty) \equiv \Theta_0(\zeta) = \zeta$. The thermal problem defined by Eqs. (12.17) and (12.18) is almost identical to the constant planar-source diffusion of Section 5.3.3 and the momentum diffusion in transient Poiseuille flow of Section 6.2.

For short times, $\tau \ll 1$, we can use the result for constant planar-source diffusion given in Eq. (5.41) in terms of the complementary error function $\mathrm{erfc}(s)$ and here written as

$$\Theta(\zeta,\tau) \approx \mathrm{erfc}\left[\frac{1-\zeta}{\sqrt{\tau}}\right], \qquad \text{for } \tau \ll 1, \tag{12.19a}$$

$$T(z,t) \approx T_0 + \Delta T\, \mathrm{erfc}\left[\frac{h-z}{\sqrt{D_\mathrm{th} t}}\right], \text{ for } t \ll \tau_\mathrm{th}. \tag{12.19b}$$

The high value of the temperature profile propagates from $\zeta = 1$ towards $\zeta = 0$, and for $\tau \ll 1$ the values $\Theta(0,\tau)$ are exponentially small, thus to a very good approximation ensuring the fulfillment of the boundary condition $\Theta(0,\tau) = 0$. As τ increases toward 1, the short-time solution violates this boundary condition, and another solution must be found.

For long times, $\tau \gg 1$, we utilize the similarity of the present problem and the transient Poiseuille flow problem defined by Eqs. (6.60) and (6.61). We employ the same Fourier expansion method and write the solution for $\Theta(\zeta,\tau)$ in analogy with Eq. (6.72) as

$$\Theta(\zeta,\tau) = \Theta(\zeta,\infty) - \sum_{n=1}^{\infty} A_n\, f_n(\zeta)\, \mathrm{e}^{-\lambda_n \tau} = \zeta - \sum_{n=1}^{\infty} A_n\, \sin(n\pi\zeta)\, \mathrm{e}^{-n^2\pi^2\tau}. \tag{12.20}$$

Since $\partial_\tau\big[\sin(n\pi\zeta)\mathrm{e}^{-n^2\pi^2\tau}\big] = -n^2\pi^2 \sin(n\pi\zeta)\mathrm{e}^{-n^2\pi^2 t} = \partial_\zeta^2\big[\sin(n\pi\zeta)\mathrm{e}^{-n^2\pi^2\tau}\big]$, we see that Eq. (12.17) is fulfilled. It is also straightforward to see that the first, second and fourth conditions of Eq. (12.18) are satisfied. To fulfill the third condition, $\Theta(\zeta,0) = 0$, we must demand that

$$\sum_{k=1}^{\infty} A_k \sin(k\pi\zeta) = \zeta. \tag{12.21}$$

The coefficient A_n is found by multiplying this equation by $\sin(n\pi\zeta)$, followed by integration over ζ from 0 to 1, and by use of the sine-function orthogonality relation, Eq. (6.70),

$$A_n = 2\int_0^1 \mathrm{d}\zeta\, \zeta\, \sin(n\pi\zeta) = \frac{2}{\pi}\, \frac{(-1)^{n+1}}{n}. \tag{12.22}$$

The result for $\Theta(\zeta,\tau)$ or $T(z,t)$ is then

$$\Theta(\zeta,\tau) = \zeta - \frac{2}{\pi}\sum_{n=1}^{\infty} \frac{(-1)^{n+1}}{n}\, \sin(n\pi\zeta)\, \mathrm{e}^{-n^2\pi^2\tau}, \tag{12.23a}$$

$$T(z,t) = T_0 + \Delta T\left[\frac{z}{h} - \frac{2}{\pi}\sum_{n=1}^{\infty} \frac{(-1)^{n+1}}{n}\, \sin\!\Big(n\pi\frac{z}{h}\Big)\, \exp\!\Big(-n^2\pi^2\frac{D_\mathrm{th}}{h^2}t\Big)\right]. \tag{12.23b}$$

The above Fourier sums converge poorly for $\tau \to 0$. However, for finite times the large pre-factor $n^2\pi^2$ in the argument of the exponential function ensures rapid convergence. In particular, the time dependence is well described by the $n = 1$ term alone once $\tau > \tau_0$, where τ_0 is defined as the time where the $n = 2$ term is one order of magnitude smaller than the $n = 1$ term, $\exp(-2^2\pi^2\tau_0) = 0.1\exp(-1^2\pi^2\tau_0)$, i.e. $\tau_0 = 0.078$. Hence

$$\Theta(\zeta,\tau) \approx \zeta - \frac{2}{\pi}\,\sin(\pi\zeta)\,\mathrm{e}^{-\pi^2\tau}, \qquad \text{for } \tau > 0.078, \tag{12.24a}$$

$$T(z,t) \approx T_0 + \Delta T\left[\frac{z}{h} - \frac{2}{\pi}\,\sin\Big(\pi\frac{z}{h}\Big)\,\exp\Big(-\pi^2\frac{D_\mathrm{th}}{h^2}t\Big)\right], \quad \text{for } t > 0.078\,\frac{h^2}{D_\mathrm{th}}. \tag{12.24b}$$

We note from Eqs. (12.16a) and (12.24b) that the thermal transient time for a 100 µm thick water layer at 20 °C is $h^2/(\pi^2 D_\mathrm{th}) = 7.1$ ms.

12.2 Poiseuille flow in a transverse temperature gradient

As a simple example of the more complex situation where both conduction and convection of heat are present, we now impose a Poiseuille flow in the transversely heated, infinite parallel-plate channel from the previous section. The smallness of the thermal perturbation coefficients $|(\partial_T\rho)/\rho|$ and $|(\partial_T\kappa)/\kappa|$ relative to $|(\partial_T\eta)/\eta|$, see Table 12.1, makes it possible to treat the density ρ and the thermal conductivity κ as constants and only take the temperature dependence of the viscosity η into account.

The top and bottom plates of the flat microchannel are placed parallel to the xy-plane at $z = 0$ and $z = h$, respectively. We consider a finite segment of length L of an infinitely long system, which is translationally invariant along the flow direction given by the x axis. In steady state for this particular setup the heat-transfer equation (2.66) becomes

$$v_x\partial_x T(z) = \frac{\kappa}{\rho c_\mathrm{p}}\partial_z^2 T(z) + \frac{\eta}{2\rho c_\mathrm{p}}\,(\partial_z v_x)^2. \tag{12.25}$$

The heat generated by the viscosity is minute since

$$\frac{\left|\frac{\eta}{2\rho c_\mathrm{p}}\,(\partial_z v_x)^2\right|}{\left|\frac{\kappa}{\rho c_\mathrm{p}}\partial_z^2 T(z)\right|} \approx \frac{\eta v_0^2}{2\kappa\Delta T} = \frac{10^{-3}\ \mathrm{Pa\,s}\times 10^{-6}\ \mathrm{m^2\,s^{-2}}}{1\ \mathrm{W\,m^{-1}\,K^{-1}}\times 10\ \mathrm{K}} = 10^{-10}, \tag{12.26}$$

where we have taken the characteristic velocity $v_0 = 1$ mm/s and the imposed temperature difference $\Delta T = 10$ K. The contribution from viscosity is therefore neglected, and since also the convection term $\partial_x T(z)$ is zero by symmetry, we end up with the Laplace equation $\nabla^2 T = 0$ for the temperature field. The solution is given by Eq. (12.10b) with $\alpha_\kappa = 0$,

$$T(z) = T_0 + \Delta T\frac{z}{h}. \tag{12.27}$$

By the standard symmetry arguments for Poiseuille flow, the velocity and pressure fields are seen to have the forms $\mathbf{v} = v_x(z)\mathbf{e}_x$ and $p = p_0 + (1 - x/L)\Delta p$, respectively. The Navier–Stokes equation for the flow field has the same form as in Eq. (3.28a),

$$\partial_z^2 v_x(z) = -\frac{\Delta p}{\eta\big(T(z)\big)L}, \tag{12.28}$$

but due to the z dependence of the viscosity $\eta\big(T(z)\big)$ the right-hand side of this equation is no longer a constant. To find $v_x(z)$ we employ perturbation theory in the perturbation parameter α, which in analogy with Eq. (12.3) arises from a Taylor expansion of the viscosity,

$$\eta(T) \approx \eta_0 + \partial_T \eta(T_0)\,\big(T(z) - T_0\big) = \eta_0 + \eta_0' \Delta T \frac{z}{h} = \Big(1 + \alpha_\eta \frac{z}{h}\Big)\,\eta_0. \tag{12.29}$$

In Table 12.1 we see that for $T_0 = 20$ °C the perturbation coefficient is $\partial_t \eta/\eta = -0.025\ \mathrm{K}^{-1}\Delta T$, which, as stated in Eq. (12.1), yields a perturbation parameter $\alpha_\eta = -0.25$. Note that α_η is negative: the viscosity decreases as the temperature increases, see Fig. 2.2.

To facilitate the calculation of the velocity field $v_x(z)$ we introduce dimensionless variables ζ and $\tilde{v}$ for the vertical co-ordinate and horizontal velocity as follows,

$$z \equiv h\,\zeta, \qquad \text{and} \qquad v_x(z) \equiv \frac{\Delta p h^2}{\eta_0} L\,\tilde{v}(\zeta). \tag{12.30}$$

In terms of ζ and $\tilde{v}$ the Navier–Stokes equation (12.28) becomes

$$\partial_\zeta^2 \tilde{v} = \frac{-1}{1 + \alpha_\eta \zeta} \approx -1 + \alpha_\eta \zeta - \alpha_\eta^2 \zeta^2 + \cdots. \tag{12.31}$$

Following the standard perturbation scheme we introduce an expansion in α_η of $\tilde{v}$,

$$\tilde{v} = \tilde{v}_0 + \alpha_\eta \tilde{v}_1 + \alpha_\eta^2 \tilde{v}_2 + \cdots, \tag{12.32}$$

which upon insertion into Eq. (12.31) leads to the following zero-order and first-order equations with no-slip boundary conditions,

$$\partial_\zeta^2 \tilde{v}_0 = -1, \qquad \tilde{v}_0(0) = 0, \qquad \tilde{v}_0(1) = 0, \tag{12.33a}$$

$$\partial_\zeta^2 \tilde{v}_1 = \zeta, \qquad \tilde{v}_1(0) = 0, \qquad \tilde{v}_1(1) = 0. \tag{12.33b}$$

The solutions are straightforward

$$\tilde{v}_0(\zeta) = -\frac{1}{2}(\zeta^2 - \zeta), \quad \text{and} \quad \tilde{v}_1(\zeta) = \frac{1}{6}(\zeta^3 - \zeta), \tag{12.34}$$

so we can write the dimensionless and full solutions to first order in α_η as

$$\tilde{v}(\zeta) = \frac{1}{2}(\zeta - \zeta^2) - \frac{\alpha_\eta}{6}(\zeta - \zeta^3), \tag{12.35}$$

$$v_x(z) = \frac{\Delta p}{2\eta_0 L}\Big[z(h - z) - \frac{\alpha_\eta}{3h}\, z(h^2 - z^2)\Big]. \tag{12.36}$$

As expected, the zero-order velocity field is the same as Eq. (3.29), and since $\alpha_\eta < 0$ for $\Delta T > 0$ we find an increasing flow as the temperature is increased, as shown in Fig. 12.1 for the specific value $\alpha_\eta = -0.22$ corresponding to $\Delta T = 10$ K.

Using the solution for $v_x(z)$ we can calculate the temperature dependent flow rate $Q(\Delta T)$ and hydraulic resistance $R(\Delta T)$. As shown in Exercise 12.1 we find

$$Q(\Delta T) = \int_0^w \mathrm{d}y \int_0^h \mathrm{d}z\; v_x(z) = \Big(1 - \frac{1}{2}\alpha_\eta\Big) \frac{\Delta p\, h^3 w}{12\eta_0 L}, \tag{12.37}$$

from which follows the hydraulic resistance to first order in α_η

$$R(\Delta T) = \frac{\Delta p}{Q(\Delta T)} = \Big(1 + \frac{1}{2}\alpha_\eta\Big) \frac{12\eta_0 L}{h^3 w} = \Big(1 + \frac{\eta_0'}{\eta_0}\Delta T\Big)\, R_{\mathrm{hyd}}(T_0). \tag{12.38}$$

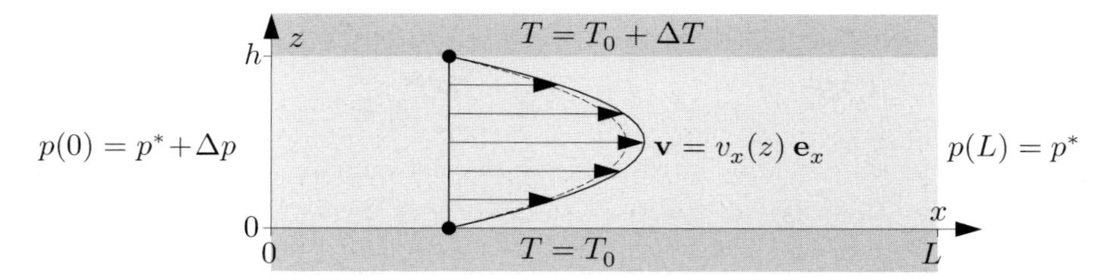

Fig. 12.1 Poiseuille flow in an infinite, parallel-plate channel of height h with fixed temperatures T_0 and $T_0 + \Delta T$ imposed on the bottom and top plate, respectively. The z dependent temperature field, $T(z) = T_0 + \Delta T \frac{h}{z}$, leads to a z-dependent viscosity $\eta(z) \approx \eta_0 + \eta_0' \Delta T \frac{h}{z}$, where $\eta_0' \equiv \partial_T \eta(T_0)$. The arrows and the full line indicate the perturbed velocity field for $\alpha = \frac{\eta_0' \Delta T}{\eta_0} = -0.22$, while the dashed line indicates the unperturbed velocity field at constant temperature T_0. Note that according to Eq. (12.36), the thermally perturbed velocity profile is no longer a symmetric parabola. However, the assymetry is not clearly visible on the figure due to the smallness of the perturbation α.

The previous example became particularly simple because the temperature field T did not depend on the x co-ordinate. We therefore relax this constraint and allow T to depend on both x and z, $T = T(x, z)$. We return to the setup shown in Fig. 12.1, but while keeping the bottom plate at the fixed temperature T_0 we now change the temperature profile of the top plate, such that $T = T_0$ for $0 < x \leq L/2 - d$, $T = T_0 + \Delta T$ for $L/2 + d \leq x < L$, and T increases monotonically from T_0 to $T_0 + \Delta T$ for $L/2 - d \leq x \leq L/2 + d$ around the midpoint of the channel. With this setup the transverse temperature is zero at the left end of the channel, while it is $\Delta T/h$ at the right end of the channel. The temperature of the top plate can be written as

$$T(x, h) \equiv T_0 + \Delta T\, f\Big(\frac{x}{d}\Big), \tag{12.39}$$

where one possible choice for f could be tangent hyperbolic,

$$f\Big(\frac{x}{d}\Big) = \frac{1}{2}\Big[1 + \tanh\Big(\frac{x - L/2}{d}\Big)\Big]. \tag{12.40}$$

If we consider the case where the length scale d for the changes in temperature along the channel is larger than the height h of the channel, $h \ll d \ll L$, we can obtain some simplifying approximations, which are necessary to provide analytical solutions. First, we notice that we can introduce the small parameter α defined by

$$\alpha \equiv \frac{h}{d} \ll 1. \tag{12.41}$$

Then, in analogy with ζ in Eq. (12.4), we introduce the dimensionless co-ordinate ξ,

$$\xi \equiv \frac{x}{d}, \tag{12.42}$$

so that the dimensionless temperature difference Θ introduced in Eq. (12.2) now becomes

$$\Theta = \Theta(\xi, \zeta). \tag{12.43}$$

In terms of the dimensionless variables ξ and ζ the heat-transfer equation Eq. (2.66) becomes

$$\frac{v_x}{d}\partial_\xi\Theta = \frac{D_{\rm th}}{d^2}\,\partial_\xi^2\Theta + \frac{D_{\rm th}}{h^2}\,\partial_\zeta^2\Theta. \tag{12.44}$$

This equation can be simplified by introducing the Péclet number $P\acute{e}_{\rm th}$ for thermal diffusion in analogy with the Péclet number Pe of Eq. (5.53) for mass diffusion,

$$P\acute{e}_{\rm th} \equiv \frac{v_x h}{D_{\rm th}} = \frac{\tau_{\rm th}}{h/v_x}. \tag{12.45}$$

The thermal Péclet number is the ratio of the thermal diffusion time $\tau_{\rm th}$ relative to the convection time h/v_x. Multiplying the heat-transfer equation (12.44) by $h^2/D_{\rm th}$ it becomes

$$\alpha P\acute{e}_{\rm th}\partial_\xi\Theta = \alpha^2\,\partial_\xi^2\Theta + \partial_\zeta^2\Theta. \tag{12.46}$$

We seek solutions Θ to this equation, which fulfills the boundary conditions

$$\Theta(0,\zeta) = 0, \quad \Theta(1,\zeta) = \zeta, \quad \Theta(\xi,0) = 0, \quad \Theta(\xi,1) = f(\xi). \tag{12.47}$$

In the case of zero convection, i.e. $P\acute{e}_{\rm th} = 0$, we see that

$$\Theta_0(\xi,\zeta) = f(\xi)\,\zeta \tag{12.48}$$

is an approximative solution to Eq. (12.46), given that $\alpha^2 f''(\xi)\zeta \ll 1$. But since $\alpha = h/d \ll 1$ this is a very good approximation. For low thermal Péclet numbers, $\alpha P\acute{e}_{\rm th} \ll 1$, we can write the next level of approximation as

$$\Theta(\xi,\zeta) \equiv f(\xi)\,\zeta + \alpha P\acute{e}_{\rm th}\Theta_1(\xi,\zeta). \tag{12.49}$$

When this is inserted into the heat-transfer equation, we find that terms without α drop out, while terms with α^2 are neglected due to their smallness, so we are left with terms containing the factor $\alpha P\acute{e}_{\rm th}$. As a result, we get the following equation for Θ_1,

$$\partial_\zeta^2\Theta_1(\xi,\zeta) = f'(\xi)\,\zeta, \tag{12.50}$$

with the solution

$$\Theta_1(\xi,\zeta) = \frac{1}{6}f'(\xi)\,(\zeta^3 - \zeta). \tag{12.51}$$

In conclusion, we have determined the temperature field to first order in $\alpha P\acute{e}_{\rm th}$,

$$\Theta(\xi,\zeta) \approx f(\xi)\,\zeta + \frac{1}{6}\,\alpha P\acute{e}_{\rm th} f'(\xi)\,(\zeta^3 - \zeta), \tag{12.52a}$$

$$T(x,z) \approx T_0 + \Delta T\left[f\Big(\frac{x}{d}\Big)\,\frac{z}{h} + \frac{v_x h^2}{6D_{\rm th}d}\,f'\Big(\frac{x}{d}\Big)\,\Big(\frac{z^3}{h^3} - \frac{z}{h}\Big)\right]. \tag{12.52b}$$

The temperature field can be used to estimate the ability to use a microfluidic flow as a cooling device. Two time scales are relevant to study,

$$\tau_{\rm conv} \equiv \frac{d}{v_x}, \qquad \tau_{\rm th} \equiv \frac{h^2}{D_{\rm th}}, \tag{12.53}$$

namely the time $\tau_{\rm conv}$ it takes to convect along the temperature step of width d along the x axis, and the time $\tau_{\rm th}$ it takes heat to diffuse across the channel of height h. In fact, we have

$\alpha P\acute{e}_{\rm th} = \tau_{\rm th}/\tau_{\rm conv}$. For a slow flow $\tau_{\rm conv} \gg \tau_{\rm th}$ and the temperature field is unaffected by the flow, while for a fast flow $\tau_{\rm conv} \ll \tau_{\rm th}$, and the temperature field changes significantly.

To estimate the improvement in cooling by turning on a flow velocity, we notice that the change $\tilde{T}$ in temperature field is given by

$$\tilde{T}(x,z) = \frac{\Delta T}{6}\,\frac{\tau_{\rm th}}{\tau_{\rm conv}}\,f'\Big(\frac{x}{d}\Big)\left(\frac{z^3}{h^3} - \frac{z}{h}\right). \tag{12.54}$$

The extra heat current density $\Delta J^{\rm cond}_{\rm heat}$ by conduction through the hot top plate is given by

$$\Delta J^{\rm cond}_{\rm heat}(x) = \kappa\partial_z\tilde{T}(x,h) = \kappa\,\frac{\Delta T}{3h}\,\frac{\tau_{\rm th}}{\tau_{\rm conv}}\,f'\Big(\frac{x}{d}\Big). \tag{12.55}$$

The extra heat current $\Delta I_{\rm th}$, in units of W, that is induced by the liquid flow in the channel can easily be calculated by a surface integral over the top plate,

$$\Delta I_{\rm th} = \int_0^w {\rm d}y \int_0^L {\rm d}x\,\Delta J^{\rm cond}_{\rm heat}(x) = \frac{1}{3}\,\frac{\tau_{\rm th}}{\tau_{\rm conv}}\,\kappa\,\frac{\Delta T}{h}\,wd. \tag{12.56}$$

For typical parameter values for water in a channel of height $h = 100$ µm, width $w = 1$ mm, and thermal step length $d = 1$ mm, we obtain for a temperature difference of $\Delta T = 50$ K (as found in PCR chips) that

$$\frac{\kappa}{3}\,\frac{\Delta T}{h}\,wd = 0.12\ \mathrm{W}. \tag{12.57}$$

In this case $\alpha = h/d = 0.1$, and if we also insist that $\alpha P\acute{e}_{\rm th} = \tau_{\rm th}/\tau_{\rm conv} = 0.1$ we find $v_x = 1.4$ mm/s. Thus for reasonable microfluidic parameters we find that the cooling power of water flow in the setup in question is a decent 12 mW.

12.3 Equivalent circuit model for heat transfer

In Section 4.7 we studied how fluid flow in microfluidic networks could be analyzed in terms of an equivalent circuit model. Likewise, we shall now see that it is also possible and useful to do the same for heat flow, where the temperature difference ΔT is analogous to the voltage drop ΔU, the amount of heat $Q_{\rm th}$ is analogous to the charge Q, and the thermal current $I_{\rm th}$, heat per unit time, is analogous to electric current I. Our starting point is the infinite parallel-plate channel of hight h filled with water at rest, see Section 12.1.

From Eq. (12.23b) in steady state, $t \to \infty$, we see that $|\boldsymbol{\nabla} T| = \Delta T/h$. Using this with Fourier's law of heat conduction, Eq. (2.55), we find the following expression for the heat current $I_{\rm th}$ streaming from the top plate to the bottom plate, both of area $\mathcal{A}$,

$$I_{\rm th} = \mathcal{A}|\mathbf{J}_{\rm th}| = \mathcal{A}\kappa|\boldsymbol{\nabla} T| = \mathcal{A}\kappa\frac{1}{h}\,\Delta T \equiv \frac{1}{R_{\rm th}}\,\Delta T, \tag{12.58}$$

where we, in analogy with Ohm's law, have defined the thermal resistance $R_{\rm th}$ by

$$R_{\rm th} \equiv \frac{\Delta T}{I_{\rm th}} = \frac{h}{\kappa\mathcal{A}}. \tag{12.59}$$

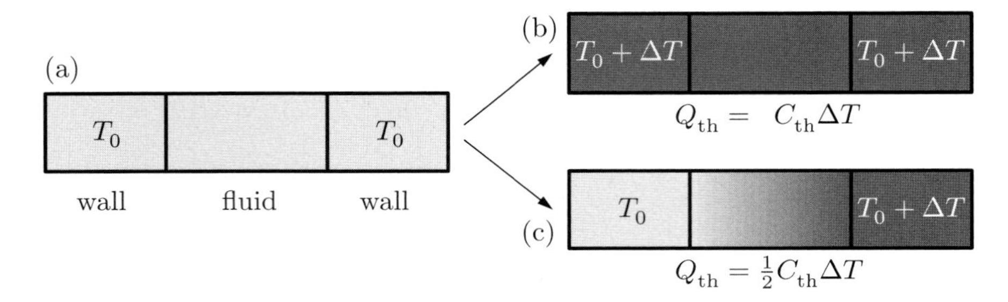

Fig. 12.2 Grayscale plot of the temperature ranging from T_0 (light gray) to $T_0 + \Delta T$ (dark gray) in a slab of homogeneous fluid between two walls. (a) Base setup, where the temperatures of both walls are kept constant at T_0. (b) The net transfer of heat, $Q_{\rm th} = C_{\rm th}\Delta T$, to the fluid when the temperature of both walls is raised to $T_0 + \Delta T$. (c) The net transfer of heat, $Q_{\rm th} = \frac{1}{2}C_{\rm th}\Delta T$, to the fluid when the temperature of one wall is kept at T_0 and the other is raised to $T_0 + \Delta T$.

We can also define a thermal capacitance $C_{\rm th}$. If the temperature of an isothermal body of volume $\mathcal{V}$ is raised from T_0 to $T_0 + \Delta T$ the amount of extra heat $Q_{\rm th}$ transferred to the body is given by

$$Q_{\rm th} = \mathcal{V}\rho c_{\rm p}\,\Delta T \equiv C_{\rm th}\,\Delta T, \quad \text{(isothermal case)}, \tag{12.60}$$

where, in analogy with the electric capacitance $C = Q/\Delta U$, we have defined the thermal capacitance $C_{\rm th}$ by

$$C_{\rm th} \equiv \frac{Q_{\rm th}}{\Delta T} = \mathcal{V}\rho c_{\rm p}. \tag{12.61}$$

If a constant gradient temperature gradient $\Delta T/h$ is imposed across a slab of material of thickness h, the average temperature of the slab is $T_0 + \Delta T/2$. In this case the amount of heat $Q_{\rm th}$ accumulated in the slab due to the temperature rise ΔT is

$$Q_{\rm th} = \mathcal{V}\rho c_{\rm p}\,\frac{\Delta T}{2} = \frac{1}{2}C_{\rm th}\,\Delta T, \quad \text{(temperature gradient)}. \tag{12.62}$$

The isothermal case is compared to the case of constant temperature gradient in Fig. 12.2.

Given the thermal resistance and the thermal capacitance we can of course also introduce a thermal RC time by

$$R_{\rm th}C_{\rm th} = \frac{h^2\rho c_{\rm p}}{\kappa} = \frac{h^2}{D_{\rm th}} = \tau_{\rm th}. \tag{12.63}$$

The thermal RC time $t_{\rm th}$ is thus equal to the thermal diffusion time $\tau_{\rm th}$ introduced in Eq. (12.16a), and using $R_{\rm th}$ and $C_{\rm th}$ elements we can in fact obtain a simple equivalent circuit that models reasonably well the heat current corresponding to the thermal transient $T(z,t)$ of Eq. (12.24b). We use the RCR circuit depicted in Fig. 12.3(a). The thermal resistance $R_{\rm th}$ defined in Eq. (12.59) is divided in two equal resistors of value $R_{\rm th}/2$ coupled in series. In this way, we can attach the heat capacitor with capacitance $C_{\rm th}$, defined in Eq. (12.61), at the midpoint of the water layer, where the temperature in steady state is $T_0 + \Delta T/2$.

Following the description in Section 12.1 we have the same temperature T_0 at both ends of the circuit for time $t < 0$. At time $t = 0$ we raise the input temperature to $T_0 + \Delta T$, and keep

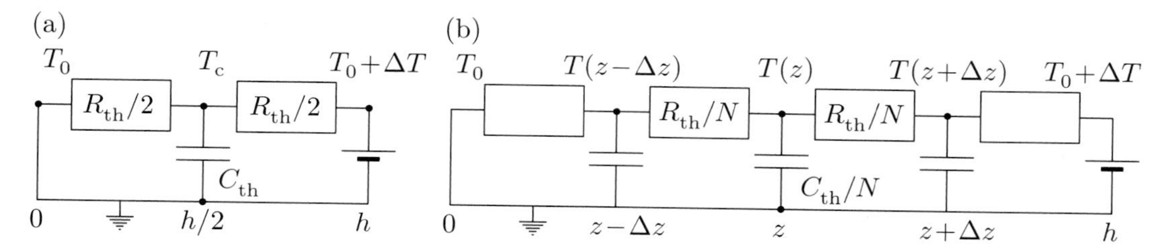

Fig. 12.3 (a) An equivalent RCR circuit model for heating of a liquid in an a infinite parallel-plate channel with temperature T_0 and $T_0+\Delta T$ at the bottom and top plate, respectively. (b) A N-stage RC transmission-line equivalent circuit model of the heat-diffusion equation.

this value for $t > 0$. The temperature at the center point is denoted $T_c(t)+T_0$, and at steady state it is $T_c(\infty)+T_0 = \Delta T/2+T_0$. The heat currents, Eq. (12.58), through the resistors are $I_1 = (\Delta T - T_c)/(R_{th}/2)$ and $I_2 = T_c/(R_{th}/2)$, while the current to the capacitor is given by the time derivative of Eq. (12.62) as $I_C = \partial_t Q_{th} = \partial_t(\frac{1}{2}C_{th}T_c) = \frac{1}{2}C_{th}\partial_t T_c$. Conservation of heat can be expressed as $I_1 = I_2 + I_C$ and from this we obtain the differential equation

$$\partial_t T_c = \frac{4}{R_{th}C_{th}}(\Delta T - 2T_c), \tag{12.64}$$

which has the solution

$$T_c(t) = \left[1 - \exp\left(-\frac{8}{R_{th}C_{th}}\,t\right)\right]\frac{\Delta T}{2}. \tag{12.65}$$

In the RCR model the heat accumulated in the water layer at time t is given by $Q_{th}^{RC}(t) = \mathcal{A}h\rho c_p T_c(t)$, which together with the expression $R_{th}C_{th} = \tau_{th}$ leads to

$$Q_{th}^{RC}(t) = \left[1 - \exp\left(-\frac{8}{\tau_{th}}\,t\right)\right]C_{th}\frac{\Delta T}{2}. \tag{12.66}$$

The accumulated heat can also be calculated from the expression Eq. (12.24b) by integration of the heat $\mathrm{d}z\,\mathcal{A}\rho c_p[T(z,t)-T_0]$ accumulated in each slab of thickness $\mathrm{d}z$,

$$\begin{aligned} Q_{th}(t) &= \int_0^h \mathrm{d}z\,\mathcal{A}\rho c_p[T(z,t) - T_0] = C_{th}\Delta T\int_0^1 \mathrm{d}\zeta\left[\zeta - \frac{2}{\pi}\sin(\pi\zeta)\mathrm{e}^{-\pi^2\tau}\right] \\ &= \left[1 - \frac{8}{\pi^2}\exp\left(-\frac{\pi^2}{\tau_{th}}\,t\right)\right]C_{th}\frac{\Delta T}{2}, \end{aligned} \tag{12.67}$$

where in the last equality we have used Eq. (12.16b) to express the dimensionless time τ in real time t. To the extent that 8 approximates $\pi^2 = 9.87$ the expressions $Q_{th}^{RC}(t)$ and $Q_{th}(t)$ for the accumulated heat in the water layer obtained by the RCR circuit model and the heat-transfer equation are in agreement. It is thus possible to gain a good insight into heat transfer in a given system by constructing the corresponding equivalent circuit model.

Whereas the RCR circuit of Fig. 12.3(a) is only approximating the actual time-dependent thermal diffusion, it turns out that in the limit of $N \to \infty$ the N-stage RC transmission

(a)

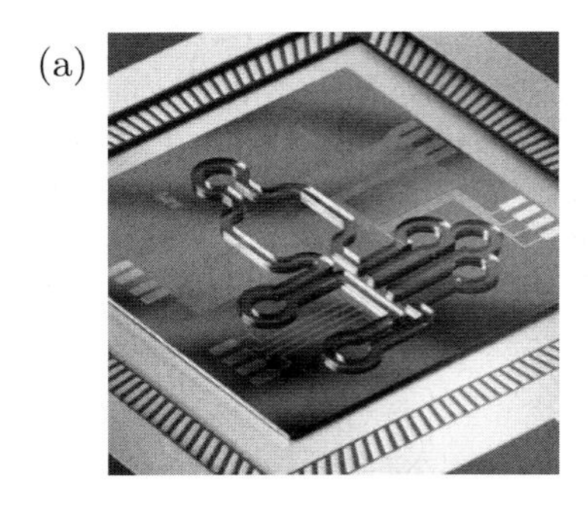

(b)

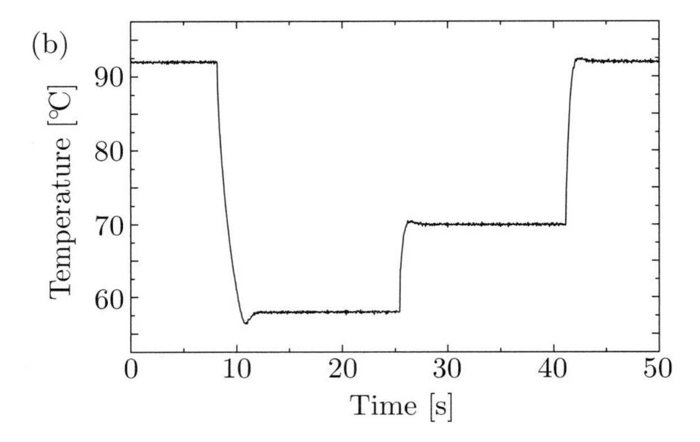

Fig. 12.4 (a) A photograph of a PCR chip with integrated DEP-based pre-treatment chamber. Courtesy of Anders Wolff, DTU Nanotech. (b) One thermal cycle measured on the PCR chip with heating and cooling rates of up to 50 and 30 ℃/s, respectively. Measurements by El-Ali *et al.*, *Sens. Actuators, A* **110**, 3 – 10 (2004), reproduced by permission of Elsevier.

line shown in Fig. 12.3(b) is an exact representation of the thermal diffusion equation $\partial_t T = D_{\text{th}}\nabla^2 T$, Eq. (2.68). This can be shown as follows. The water layer of thickness h is divided into N parallel layers each of thickness $\Delta z = h/N$. Each layer has the thermal resistance $R_{\text{th}}/N = R_{\text{th}}\Delta z/h$ and thermal capacitance $C_{\text{th}}/N = C_{\text{th}}\Delta z/h$, where R_{th} and C_{th} are the total resistance and capacitance of the water layer given by Eqs. (12.59) and (12.61), respectively. The thin layers have different temperatures, but a given layer is to a good approximation isothermal with the temperature $T(z)$. The thermal current $I_{\text{th}}(z + \Delta T)$ running through the layer at height $z + \Delta z$ into the layer at height z divides into a thermal current $I_{\text{th}}(z)$ running out of the layer down into the layer at height $z - \Delta z$ and heating current $\partial_t Q_{\text{th}}$ of the layer at z. This conservation of thermal current can be written as

$$\frac{h}{R_{\text{th}}\Delta z}\Big[T(z+\Delta z) - T(z)\Big] = \frac{h}{R_{\text{th}}\Delta z}\Big[T(z) - T(z-\Delta z)\Big] + \frac{C_{\text{th}}\Delta z}{h}\,\partial_t T(z), \tag{12.68}$$

which upon isolation of $\partial_t T$ leads to

$$\partial_t T = \frac{h^2}{R_{\text{th}}C_{\text{th}}}\,\frac{T(z+\Delta z) - 2T(z) + T(z-\Delta z)}{\Delta z^2} \underset{\Delta z\to 0}{\longrightarrow} \frac{h^2}{R_{\text{th}}C_{\text{th}}}\,\partial_z^2 T = D_{\text{th}}\partial_z^2 T, \tag{12.69}$$

and we have arrived at Fourier's equation. The approximate nature of the *RCR* model is seen to arise from truncating the exact *RC* transmission-line model containing infinitely many *RC* elements to a model of just one *RC* element with an additional *R* element added.

12.4 The PCR biochip

The invention of the polymerase chain reaction (PCR) in 1983 by Kary Mullis is a major achievement in the development of powerful analysis tools for genetical studies. The discovery led to the Nobel Prize for Chemistry ten years later in 1993. One of the key components in the PCR technology is the cyclic switching between three temperature levels of 55 ℃, 72 ℃, and 95 ℃. The small thermal times, $\tau_{\text{th}} \approx 70$ ms for $h \approx 100$ µm, makes lab-on-a-chip systems ideal for implementation of fast and reliable PCR devices.

Table 12.2 Material parameters relevant for describing the thermal properties of materials used in lab-on-a-chip devices.

material	κ $[\mathrm{W\,m^{-1}\,K^{-1}}]$	ρ $[\mathrm{kg\,m^{-3}}]$	c_p $[\mathrm{J\,kg^{-1}\,K^{-1}}]$	D_th $[\mathrm{m^2\,s^{-1}}]$
Water	6.0×10^{-1}	1.0×10^{3}	4.2×10^{3}	1.4×10^{-7}
Pyrex	1.1×10^{0}	2.2×10^{3}	7.8×10^{2}	6.3×10^{-7}
Silicon	$1.5 \times 10^{+2}$	2.3×10^{3}	7.1×10^{2}	9.2×10^{-5}
Topas	1.3×10^{-1}	1.0×10^{3}	1.5×10^{3}	8.5×10^{-8}
PMMA	1.7×10^{-1}	1.2×10^{3}	1.4×10^{3}	1.0×10^{-7}
SU-8	2.0×10^{-1}	1.2×10^{3}	1.5×10^{3}	1.1×10^{-7}

In Fig. 12.4 is shown an example of a PCR chip designed, fabricated and tested at MIC. The main thermal properties of this lab-on-a-chip device can be analyzed using the equivalent circuit theory introduced in Section 12.3. This is illustrated in Fig. 12.5. The main simplification arises from the planar structure of the typical lab-on-a-chip device. The wide extension in the xy-plane makes the structure almost 1D in the z direction, and hence the thermal resistance R_th and the thermal capacitance can be used as defined in the previous section.

Note that in the equivalent circuit diagram in Fig. 12.5(b) each layer is modelled by a single RC element and not an RCR element as in Fig. 12.3(a). This is possible as no short circuits to ground are present, hence there is no need to bury the thermal capacitor between two half-resistors. Note also that the heat convection by air from the lid of the structure is modelled as a resistor R_a. Its value is determined as

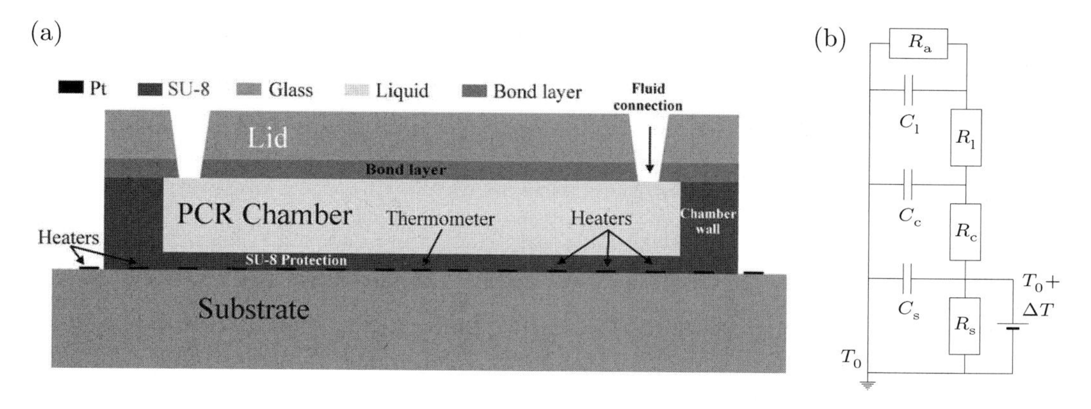

Fig. 12.5 (a) Sketch of the structure of the PCR chip showing the substrate, the chamber, and the lid. Moreover, the heaters and the thermometer placed on the substrate under the chamber are shown. Diagram by El-Ali *et al.*, *Sens. Actuators, A* **110**, 3 – 10 (2004), reproduced by permission of Elsevier. (b) The equivalent circuit model for the thermal properties of the PCR chip. Three RC elements with subscripts "s","c", and "l" model the thermal resistance and capacitance of the substrate, the chamber and the lid, respectively. The convection through the air is modelled by the resistor R_a, and the heaters are modelled by the battery providing a temperature difference ΔT.

$$R_{\rm a} \equiv \frac{1}{K\mathcal{A}}, \quad \text{heat resistance due to convection in air,} \tag{12.70}$$

where the heat-transfer coefficient K is given by

$$K = 7.5\ \mathrm{W\,m^{-2}\,K^{-1}}. \tag{12.71}$$

In Table 12.2 are listed a number of material parameters relevant for describing thermal properties of common lab-on-a-chip systems.

12.5 Exercises

Exercise 12.1
The temperature-dependent Poiseuille flow rate
Show that to first order in the perturbation parameter α_η, the temperature-dependent velocity field, Eq. (12.36), leads to the flow rate $Q(\Delta T)$ given in Eq. (12.37).

Exercise 12.2
Steady-state temperature profile in a two-layer system
Verify that the expression given in Eq. (12.15) for the temperature profile in a two-layer system is correct.

Exercise 12.3
The thermal diffusion time for water
Calculate the thermal diffusion time for one layer of water 10 µm thick and another one 10 mm thick. Discuss the result.

Exercise 12.4
Thermal diffusion at short time scales
Consider the evolution at short of the dimensionless temperature difference $\Theta(\zeta,\tau)$ after the sudden temperature increase from 0 to 1 has been applied at $\zeta = 1$.

(a) Calculate the values of the dimensionless temperature $\Theta(0,\tau)$ at the cold wall $\zeta = 0$ for the times $\tau = 0.1$, 0.2 and 0.3 using the approximate solution given by Eq. (12.19a) valid for $\tau \ll 1$.

(b) Given the results obtained in (a), discuss at which time τ it is appropriate to switch from the short time scale expression Eq. (12.19a) for Θ to the long time scale expression Eq. (12.24a).

Exercise 12.5
The general expression for 1D thermal diffusion
Verify that the general expression Eq. (12.23b) for 1D thermal diffusion indeed does fulfill Fourier's equation and the appropriate boundary conditions.

Exercise 12.6
Poiseuille flow in a thermal gradient
Consider the non-constant temperature profile of the top plate in the infinite parallel-plate channel given in Eq. (12.40). Verify the correctness to first order in α of the expression for Θ given by Eq. (12.52a).

Exercise 12.7
The equivalent circuit model for heat transfer
Consider the equivalent circuit model for heat transfer in the PCR chip defined in Fig. 12.5 and the table of material parameters Table 12.2. Estimate the values of the resistors and capacitors in the equivalent circuit model.

12.6 Solutions

Solution 12.1
The temperature-dependent Poiseuille flow rate
By definition the flow rate is given by $Q = \int_0^w \mathrm{d}y \int_0^h \mathrm{d}z\, v_x(z) = whv_0 \int_0^1 \mathrm{d}\zeta \tilde{v}(\zeta)$, which gives $Q = whv_0 \int_0^1 \mathrm{d}\zeta \big[\frac{1}{2}(\zeta-\zeta^2) - \frac{\alpha_\eta}{6}(\zeta-\zeta^3)\big] = \frac{whv_0}{2}\big[(\frac{1}{2}-\frac{1}{3}) - \frac{\alpha_\eta}{3}(\frac{1}{2}-\frac{1}{4})\big] = \frac{whv_0}{12}(1-\frac{1}{2}\alpha_\eta)$.

Solution 12.2
Steady-state temperature profile in a two-layer system
First, for a 1D system, the steady-state Fourier law is $\partial_z^2 T(z) = 0$, which is fulfilled as Eq. (12.15) is piecewise linear. Secondly, the boundary conditions $T(0) = T_0$ and $T(h) = T_0 + \Delta T$ are checked by direct insertion. Thirdly, the heat current densities $J_{\mathrm{heat}} = \kappa\, \partial_z T$ in the two layers are seen to be identical, $\kappa_1 \partial_z T_1(z) = \Delta T \kappa_1 [\kappa_2/h]/[\kappa_1 + (\kappa_2 - \kappa_1)\alpha]$ and $\kappa_2 \partial_z T_2(z) = \Delta T \kappa_2 [\kappa_1/h]/[\kappa_1 + (\kappa_2 - \kappa_1)\alpha]$, so in particular the heat current is continuous at the interface. Finally, the temperature is continuous at the interface since $T_1(\alpha h) = T_0 + \Delta T \alpha \kappa_2/[\kappa_1 + (\kappa_2 - \kappa_1)\alpha]$ and $T_2(\alpha h) = T_0 + \Delta T\big[1 + (\alpha - 1)\kappa_1/[\kappa_1 + (\kappa_2 - \kappa_1)\alpha\big] = T_0 + \Delta T \alpha \kappa_2/[\kappa_1 + (\kappa_2 - \kappa_1)\alpha]$.

It has thus been proven that the temperature profile Eq. (12.15) fulfills all four conditions for the two-layer system, and this implies that the profile is correct.

Solution 12.3
The thermal diffusion time for water
The thermal diffusion time for a water layer of thickness h is denoted $\tau_{\mathrm{th}}(h) = h^2/D_{\mathrm{th}}$. From Eq. (12.16a) we have $\tau_{\mathrm{th}}(100\ \mathrm{\mu m}) = 70$ ms. From this we get $\tau_{\mathrm{th}}(10\ \mathrm{\mu m}) = (10^{-5}/10^{-4})^2 \times 70\ \mathrm{ms} = 10^{-2} \times 70\ \mathrm{ms} = 0.7$ ms and $\tau_{\mathrm{th}}(10\ \mathrm{mm}) = (10^{-2}/10^{-4})^2 \times 70\ \mathrm{ms} = 10^4 \times 70\ \mathrm{ms} = 700$ s. A dramatic increase in the thermal diffusion time by a factor 10^6 results by the change from µm to mm scale. In the latter case the thermal time scale may be prohibitively long and necessitate the use of microsystems.

Solution 12.4
Thermal diffusion at short time scales
At the cold wall $\zeta = 0$ the dimensionless temperature Eq. (12.19a) becomes $\Theta(0,\tau) = \mathrm{erfc}\big(1/\sqrt{\tau}\big)$ for $\tau \ll 1$.

(a) Mathematical tables for the complementary error function give $\Theta(0, 0.1) = 7.7\times10^{-6}$, $\Theta(0, 0.2) = 1.6 \times 10^{-3}$, and $\Theta(0, 0.3) = 9.8 \times 10^{-3}$.

(b) Ideally, the temperature rise Θ at the cold wall $\zeta = 0$ should be zero, but values much smaller than the temperature rise of unity at the hot wall $\zeta = 1$ would be acceptable for approximative solutions. In (a) it was found that $|\Theta(0,\tau)| < 0.002$ for $\tau < 0.2$, and in Eq. (12.24a) an approximation valid for $\tau > 0.07$ was given. Thus, good approximative results are obtained for the entire time range when switching at $\tau = 0.2$ from the short- to the long-time approximation.

Solution 12.5
The general expression for 1D thermal diffusion

Fourier's equation for the temperature field $T(z,t)$ reads $\partial_t T = D_{\mathrm{th}} \partial_z^2 T$. The solution Eq. (12.23b) for T to be checked has the form $T(z,t) = T_0 + \Delta T\big[z/h - \sum_{n=1}^{\infty} A_n f_n(z) g_n(t)\big]$, where $f_n(z) = \sin(n\pi z/h)$ and $g_n(t) = \exp(-n^2\pi^2 D_{\mathrm{th}} t/h^2)$.

It is straightforward to see that the term proportional to z vanishes when acted on by ∂_t and ∂_z^2, while for the Fourier terms we get $\partial_t[f_n(z)g_n(t)] = (-n^2\pi^2 D_{\mathrm{th}}/h^2)\, f_n(z)g_n(t)$ and $\partial_z^2[f_n(z)g_n(t)] = (-n^2\pi^2/h^2)\, f_n(z)g_n(t)$. Thus, Fourier's equation is seen to be satisfied.

The boundary conditions are checked as follows. By definition of the Fourier coefficients A_n we have $\sum_{n=1}^{\infty} A_n f_n(z) \equiv z/h$ so for $t=0$, as $g_n(0)=1$, we get $T(z,0) = T_0 + \Delta T\big[z/h - \sum_{n=1}^{\infty} A_n\, f_n(z)\big] = T_0$. For $t=\infty$ we have $g_n(\infty)=0$ and thus $T(z,\infty) = T_0 + \Delta T\, z/h$. Finally, for $z=0$ and $z=h$, we get that $f_n(0) = f_n(h) = 0$ and thus $T(0,t) = T_0$ and $T(h,t) = T_0 + \Delta T$.

Solution 12.6
Poiseuille flow in a thermal gradient

To first order in α the heat-transfer equation (12.46) becomes $\alpha P\acute{e}_{\mathrm{th}} \partial_\xi \Theta = \partial_\zeta^2 \Theta$, and the task is to verify that $\Theta(\xi,\zeta) \approx f(\xi)\,\zeta + \frac{1}{6}\,\alpha P\acute{e}_{\mathrm{th}} f'(\xi)\,(\zeta^3 - \zeta)$ is a solution.

The left-hand side of the heat-transfer equation already contains a factor α, so here it suffices to insert the zero-order expression $\Theta(\xi,\zeta) = f(\xi)\,\zeta$ giving $\alpha P\acute{e}_{\mathrm{th}} \partial_\xi \Theta \approx \alpha P\acute{e}_{\mathrm{th}} f'(\xi)\,\zeta$. On the right-hand side the second derivative ∂_ζ^2 removes the term $f(\xi)\zeta$ from Θ leaving only the term proportional to α. Insertion of this term gives $\partial_\zeta^2 \Theta \approx \frac{1}{6}\,\alpha P\acute{e}_{\mathrm{th}} f'(\xi)\,(3 \times 2 \times \zeta - 0) = \alpha P\acute{e}_{\mathrm{th}} f'(\xi)\,\zeta$. Consequently, the heat-transfer equation is fulfilled to first order in α.

The boundary conditions are satisfied: $\Theta(\xi,0) = 0+0 = 0$ and $\Theta(\xi,1) = f(\xi)+0 = f(\xi)$.

Solution 12.7
The equivalent circuit model for heat transfer

We consider a 1D sandwich structure consisting of a substrate (Pyrex of thickness $h_{\mathrm{s}} = 1$ mm), a chamber (water of thickness $h_{\mathrm{c}} = 0.4$ mm), and a lid (Pyrex of thickness $h_{\mathrm{l}} = 0.5$ mm). The area of the system is $\mathcal{A} = 7\ \mathrm{mm} \times 7\ \mathrm{mm} = 4.9 \times 10^{-5}\ \mathrm{m}^2$. The layers have the thermal resistances, $R_{\mathrm{th}} = h/(\kappa\mathcal{A})$, and capacitances, $C_{\mathrm{th}} = \frac{1}{2}\,\mathcal{A}h\rho c_{\mathrm{p}}$. We note that for Pyrex $\kappa\mathcal{A} = 5.39 \times 10^{-5}$ W m/K, while for water it is $\kappa\mathcal{A} = 2.94 \times 10^{-5}$ W m/K. Likewise, for Pyrex we have $\frac{1}{2}\mathcal{A}\rho c_{\mathrm{p}} = 42.0$ J/(K m) and for water $\frac{1}{2}\mathcal{A}\rho c_{\mathrm{p}} = 103.0$ J/(K m).

$$R_{\mathrm{a}} = \frac{1}{7.5 \frac{\mathrm{W}}{\mathrm{m^2\,K}}\ 4.9 \times 10^{-5}\ \mathrm{m}^2} = 2.7 \times 10^3\ \frac{\mathrm{K}}{\mathrm{W}}, \tag{12.72}$$

$$R_{\mathrm{l}} = \frac{5.0 \times 10^{-4}\ \mathrm{m}}{5.39 \times 10^{-5}\ \frac{\mathrm{W\,m}}{\mathrm{K}}} = 9.3\ \frac{\mathrm{K}}{\mathrm{W}}, \qquad C_{\mathrm{l}} = 5.0 \times 10^{-4}\ \mathrm{m}\ 42.0\ \frac{\mathrm{J}}{\mathrm{K\,m}} = 21.0\ \frac{\mathrm{mJ}}{\mathrm{K}},$$

$$R_{\mathrm{c}} = \frac{4.0 \times 10^{-4}\ \mathrm{m}}{2.94 \times 10^{-5}\ \frac{\mathrm{W\,m}}{\mathrm{K}}} = 13.6\ \frac{\mathrm{K}}{\mathrm{W}}, \qquad C_{\mathrm{c}} = 4.0 \times 10^{-4}\ \mathrm{m}\ 103.0\ \frac{\mathrm{J}}{\mathrm{K\,m}} = 41.2\ \frac{\mathrm{mJ}}{\mathrm{K}},$$

$$R_{\mathrm{s}} = \frac{1.0 \times 10^{-3}\ \mathrm{m}}{5.39 \times 10^{-5}\ \frac{\mathrm{W\,m}}{\mathrm{K}}} = 18.6\ \frac{\mathrm{K}}{\mathrm{W}} \qquad C_{\mathrm{s}} = 1.0 \times 10^{-3}\ \mathrm{m}\ 42.0\ \frac{\mathrm{J}}{\mathrm{K\,m}} = 42.0\ \frac{\mathrm{mJ}}{\mathrm{K}}.$$

These components can then be used to determine the time response of the system using standard circuit analysis.

13
Two-phase flow

In the previous chapters we have been concerned with the microfluidic properties of a single liquid. In Section 5.2 we went beyond the case of a pure liquid and studied the convection-diffusion equation for solutes in a solvent. This theme was extended in Section 8.3, where we analyzed the influence of dissolved ions on the electrohydrodynamic properties of electrolytes. The topic of the present chapter is the behavior of two different fluids flowing simultaneously in a microchannel; a situation denoted two-phase flow. We will treat examples of flow involving either one liquid and one gas phase, or two different liquid phases.

Two-phase flow is very important in lab-on-a-chip systems, where it is often desirable to bring together two liquids to prepare for further treatment, to obtain certain chemical reactions, or to perform chemical analysis. The presence of gas bubbles in microchannels is another major two-phase flow issue in microfluidics, either because the mechanical properties of the bubbles are used for some functionality of the device, or because unwanted bubbles appear as a consequence of the introduction of some liquid or as the result of electrolysis.

In the following we will study some basic theoretical aspects of simple two-phase flow in microfluidic systems. By keeping the examples simple we can gain some insight in a topic for which the mathematical formalism otherwise easily can grow to an almost intractable level of complexity.

13.1 Two-phase Poiseuille flow

Consider the infinite parallel-plate geometry analyzed in Section 3.4.2 for basic Poiseuille flow, but assume now that we have managed to establish a steady-state flow of two different liquids with a flat interface situated at $z = h^*$ as shown in Fig. 13.1. The flow is driven by a pressure drop Δp over the distance L along the x axis. The bottom layer $0 < z < h^*$ is liquid 1 with viscosity η_1, while the top layer $h^* < z < h$ is liquid 2 with viscosity η_2. As in Section 3.4.2 the system is translationally invariant along the x and the y axis, so as before the velocity field $\mathbf{v}$ and the pressure field p must be of the form

$$\mathbf{v} = v_x(z)\,\mathbf{e}_x, \quad 0 < z < h, \tag{13.1a}$$

$$p(x) = p^* + \left(1 - \frac{x}{L}\right)\Delta p. \tag{13.1b}$$

Note that because we have assumed a flat liquid/liquid interface no Young–Laplace pressure arises and the pressure field is equal to that of a single-phase Poiseuille flow.

It is natural to piece the full velocity field $v_x(z)$ together from two piecewise differentiable fields $v_{1,x}(z)$ and $v_{2,x}(z)$:

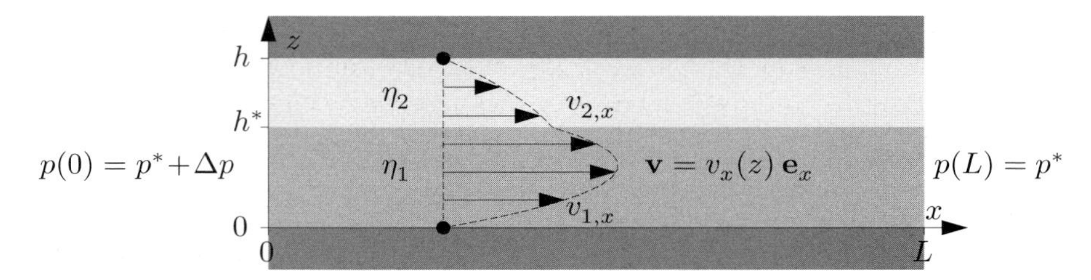

Fig. 13.1 An example of two-phase Poiseuille flow. Fluid 1 of viscosity η_1 (light gray) and fluid 2 of viscosity η_2 (gray) occupying the space of height h between two horizontally placed, parallel, infinite planar plates (dark gray). The interface situated at $z = h^*$ is assumed to be planar and stable. A constant pressure drop Δp along the x axis drives a Poiseuille flow $\mathbf{v} = v_x(z)\,\mathbf{e}_x$. The boundary conditions are no slip at the walls and continuous velocity v_x and shear stress $\sigma_{xz} = \eta\,\partial_z v_x$ at the interface $z = h^*$. Note the discontinuity in the derivative of the velocity at the interface. Here, we have used the parameter values $\eta_2 = 4\eta_1$ and $h^* = 0.6h$.

$$v_x(z) = \begin{cases} v_{2,x}(z), & \text{for} \quad h^* < z < h, \\ v_{1,x}(z), & \text{for} \quad 0 < z < h^*. \end{cases} \tag{13.2}$$

The boundary conditions for the velocity field are no slip at the walls and continuous velocity v_x and shear stress $\sigma_{xz} = \eta\,\partial_z v_x$ at the interface $z = h^*$,

$$v_{2,x}(h) = 0, \tag{13.3a}$$
$$v_{1,x}(0) = 0, \tag{13.3b}$$
$$v_{1,x}(h^*) = v_{2,x}(h^*), \tag{13.3c}$$
$$\eta_1\,\partial_z v_{1,x}(h^*) = \eta_2\,\partial_z v_{2,x}(h^*). \tag{13.3d}$$

If the shear stress were not continuous, infinite and thus unphysical forces would be present at the interface. A straightforward generalization of Eq. (3.29) for the single-phase Poiseuille flow velocity field yields the following expressions for $v_{1,x}(z)$ and $v_{1,x}(z)$ that satisfies the first two boundary conditions Eqs. (13.3b) and (13.3a),

$$v_{2,x}(z) = \frac{\Delta p}{2\eta_2 L}\,(h - z)(z - h_2), \tag{13.4a}$$
$$v_{1,x}(z) = \frac{\Delta p}{2\eta_1 L}\,(h_1 - z)z. \tag{13.4b}$$

Here, h_1 and h_2 are two constants to be determined by use of the last two boundary conditions Eqs. (13.3c) and (13.3d). Using the fact that in the setup with the given symmetries $\sigma_{xz}(z) = \eta\partial_z v_x(z)$, the two boundary conditions results in two linear equations for the two unknowns h_1 and h_2, which are readily solved to yield

$$h_2 = \frac{\left(\frac{\eta_1}{\eta_2} - 1\right)\left(1 - \frac{h^*}{h}\right)}{\frac{\eta_1}{\eta_2}\left(1 - \frac{h^*}{h}\right) + \frac{h^*}{h}}\,h^*, \tag{13.5a}$$

$$h_1 = h + h_2. \tag{13.5b}$$

In Fig. 13.1 is shown an explicit example of a velocity profile in a two-phase Poiseuille flow with a flat interface. Note the discontinuity in the derivative of the velocity field at the interface $z = h^*$. This discontinuity arises from the different values of the viscosities η_1 and η_2 and the demand for a continuous shear stress at the interface.

13.2 Capillary and gravity waves

It might be of interest to study under which conditions the interface of a two-phase Poiseuille flow is stable. The full analysis of this simple question turns out to be very difficult and is beyond the scope of this chapter. Therefore, to shed some light on the issue we are forced to consider some simplified cases, unfortunately not entirely relevant for microfluidics, namely the so-called capillary and gravity waves where the effect of viscosity can be neglected.

13.2.1 Gravity waves of short wavelength

Gravity waves of small amplitude and short wavelengths on a water/air interface is arguably the simplest example of interface waves where viscosity can be neglected. This is also known as gravity waves on an inviscid liquid with a free surface. During the calculations we make certain assumptions to progress at ease; at the end of the calculation we check the obtained solution for consistency with these assumptions. If viscosity indeed can be neglected then according to Kelvin's circulation theorem, see Exercise 13.2, the flow of the incompressible water is a potential flow,

$$\mathbf{v} = \boldsymbol{\nabla}\phi, \tag{13.6}$$

where ϕ is a scalar potential function to be determined. Assume that the body of water is infinite in the xy-plane and sustained by a solid base plane placed at $z = -h$. When at rest the water/air interface is flat and given by $z = \zeta(x, y, t) \equiv 0$. Starting from rest a small-amplitude plane wave running in the x direction is gradually established. Under these circumstances the potential function does not depend on y, thus $\phi = \phi(x, z, t)$. Neglecting the non-linear term due to the smallness of $\mathbf{v}$, the incompressible Navier–Stokes equation takes the form $\rho\partial_t\mathbf{v} = -\boldsymbol{\nabla}p + \rho\mathbf{g}$ and thus,

$$\rho\boldsymbol{\nabla}(\partial_t\phi) = -\boldsymbol{\nabla}(p + \rho g z). \tag{13.7}$$

We evaluate this expression at $z = \zeta$, which is assumed to be small. The pressure of air at the interface is the constant p^*. Moreover, keeping only first order terms in ϕ and ζ, we find that $\phi(\zeta) \approx \phi(0) + \partial_z\phi(0)\zeta \approx \phi(0)$, thus ϕ needs only to be evaluated at the equilibrium position $z = 0$. Equation (13.7) is now easily integrated and evaluated at $z = \zeta$ to give

$$\rho\big(\partial_t\phi + g\,\zeta\big) = p^*, \quad z = \zeta \approx 0, \tag{13.8}$$

which can be interpreted as continuity of the pressure going from the liquid to air across the interface. Differentiating this equation with respect to time yields

$$\partial_t^2\phi + g\,\partial_t\zeta = 0, \quad z = \zeta \approx 0. \tag{13.9}$$

We now note that on the one hand $v_z = \partial_z\phi$ and on the other hand, since $\mathbf{v}$ and ζ are both small, $v_z = \mathrm{d}\zeta/\mathrm{d}t = \partial_t\zeta + (\mathbf{v}\cdot\boldsymbol{\nabla})\zeta \approx \partial_t\zeta$. Thus, inserting $\partial_t\zeta = v_z = \partial_z\phi$ into Eq. (13.9) and rearranging the two terms leads to

$$-\partial_t^2\phi = g\,\partial_z\phi, \quad z=\zeta\approx 0. \tag{13.10}$$

The continuity equation $\boldsymbol{\nabla}\cdot\mathbf{v}=0$ must also be fulfilled. Combining this equation with the potential form $\mathbf{v}=\boldsymbol{\nabla}\phi$ yields the Laplace equation for ϕ,

$$\nabla^2\phi = \boldsymbol{\nabla}\cdot(\boldsymbol{\nabla}\phi) = \boldsymbol{\nabla}\cdot\mathbf{v} = 0. \tag{13.11}$$

Consequently, it is natural to seek plane-wave solutions of the form

$$\phi(x,z,t) = f(z)\,\cos(kx-\omega t), \tag{13.12}$$

where $k=2\pi/\lambda$ is the wave number and ω the angular frequency, see Appendix E. Inserting this into the continuity equation (13.11) yields, after removal of the common cosine factor

$$\partial_z^2 f - k^2 f = 0, \tag{13.13}$$

which have the solution

$$f(z) = \frac{A}{\sinh(kh)}\,\cosh\big[k(z+h)\big], \tag{13.14}$$

where the argument of the cosine hyperbolic function is chosen to ensure $v_z=\partial_z\phi=0$ for $z=-h$, so that no liquid is allowed to flow through the solid base plane at $z=-h$. In conclusion, the potential function ϕ and the velocity field $\mathbf{v}$ becomes

$$\phi(x,z,t) = \frac{A}{\sinh(kh)}\,\cosh\big[k(z+h)\big]\,\cos(kx-\omega t), \tag{13.15a}$$

$$\mathbf{v}(x,z,t) = \frac{A}{\sinh(kh)}\begin{pmatrix} -k\cosh\big[k(z+h)\big]\,\sin(kx-\omega t)\\ 0\\ +k\sinh\big[k(z+h)\big]\,\cos(kx-\omega t)\end{pmatrix}. \tag{13.15b}$$

At the interface, $z=0$, we obtain the velocity

$$\mathbf{v}(x,0,t) = A\begin{pmatrix} -k\tanh\big[kh\big]\,\sin(kx-\omega t)\\ 0\\ +k\cos(kx-\omega t)\end{pmatrix} \underset{kh\to\infty}{\longrightarrow} A\begin{pmatrix} -k\sin(kx-\omega t)\\ 0\\ +k\cos(kx-\omega t)\end{pmatrix}, \tag{13.16}$$

while at the base plane, $z=-h$, we get

$$\mathbf{v}(x,-h,t) = \frac{A}{\sinh(kh)}\begin{pmatrix} -k\sin(kx-\omega t)\\ 0\\ 0\end{pmatrix} \underset{kh\to\infty}{\longrightarrow} \begin{pmatrix} 0\\ 0\\ 0\end{pmatrix}. \tag{13.17}$$

Note that in general the no-slip boundary condition is not fulfilled by the velocity field given in Eq. (13.17). This is not a big surprise; as we have neglected viscosity there is no reason for $v_x=0$ at the solid wall. We can only insist that no liquid can penetrate the wall, i.e. $v_z=0$ for $z=-h$. The free-surface gravity wave is sketched in Fig. 13.2(a).

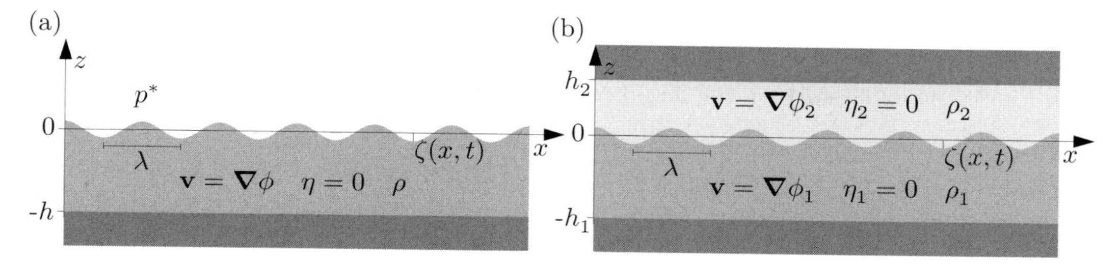

Fig. 13.2 (a) A free-surface gravity wave on an inviscid liquid with a density ρ and of depth h. The surface is displaced by the amount $\zeta(x,t)$ from the equilibrium position given by $z \equiv 0$, and the resulting wave has the wavelength λ. (b) A gravity wave at the interface between two confined liquid layers of densities ρ_1 and ρ_2 and of layer depths h_1 and h_2, respectively.

With the solution for $\phi(x,z,t)$ at hand we can use Eq. (13.10) to obtain the dispersion relation $\omega = \omega(k)$ for the gravity waves. Inserting Eq. (13.15a) we arrive at the following after removal of the common function

$$\omega^2 = gk \quad \Rightarrow \quad \omega = \sqrt{gk} = \sqrt{2\pi\,\frac{g}{\lambda}}, \tag{13.18}$$

thus the frequency increases as the wavelength decreases.

We end this treatment by assessing the validity of omitting the viscosity. To establish when $|\eta\nabla^2\mathbf{v}| \ll |\rho\partial_t\mathbf{v}|$ we note that $\nabla \to k$ and $\partial_t \to \omega = \sqrt{gk}$. So, we get

$$|\eta\nabla^2\mathbf{v}| \ll |\rho\partial_t\mathbf{v}| \quad \Rightarrow \quad \eta k^2 \ll \rho\sqrt{gk} \quad \Rightarrow \quad k^3 \ll \frac{\rho^2 g}{\eta^2} \quad \Rightarrow \quad \lambda \gg 2\pi\left[\frac{\eta^2}{\rho^2 g}\right]^{\frac{1}{3}}. \tag{13.19}$$

Inserting the values for water we obtain $\lambda \gg 60$ µm. Likewise, to establish when the non-linear term can be neglected we again use $\nabla \to k$, $\partial_t \to \omega$, and $\mathbf{v} \to \omega\zeta$:

$$|\rho(\mathbf{v}\cdot\nabla)\mathbf{v}| \ll |\rho\partial_t\mathbf{v}| \quad \Rightarrow \quad k\omega\zeta \ll \omega \quad \Rightarrow \quad \lambda \gg 2\pi\zeta. \tag{13.20}$$

Finally, the no-slip condition is only approximately fulfilled for $kh \gg 1$ or $\lambda \ll 2\pi\,h$, so the range of validity of the treatment is given by

$$\max\left\{60\ \mu\text{m},\ 2\pi\,\zeta\right\} \ll \lambda \ll 2\pi\,h. \tag{13.21}$$

Based on the solution for free-surface gravity waves we can easily analyze the gravity waves at the interface between two confined liquid layers 1 and 2 as sketched in Fig. 13.2(b). We simply try to match solutions for the potential function ϕ_1 and ϕ_2, for each layer, of the form given by ϕ in Eq. (13.15a):

$$\phi(x,z,t) = \begin{cases} \phi_2(x,z,t) = A_2\cosh\left[k(z-h_2)\right]\,\cos(kx-\omega t), & \text{for} \quad 0 < z < h_2, \\ \phi_1(x,z,t) = A_1\cosh\left[k(z+h_1)\right]\,\cos(kx-\omega t), & \text{for} \quad -h_1 < z < 0. \end{cases} \tag{13.22}$$

Here, A_1 and A_2 are constants to be determined by the boundary conditions at the interface, $z = 0$. One boundary condition is the continuity of the pressure at the interface. In analogy with Eq. (13.8), but now with the liquid pressures on both sides of the equation we obtain

$$\rho_1\big(\partial_t\phi_1 + g\,\zeta\big) = \rho_2\big(\partial_t\phi_2 + g\,\zeta\big), \quad \text{for } z = 0, \tag{13.23}$$

which can be solved with respect to ζ to yield

$$\zeta = \frac{1}{(\rho_1 - \rho_2)g}\Big(\rho_2\partial_t\phi_2 - \rho_1\partial_t\phi_1\Big), \quad \text{for } z = 0. \tag{13.24}$$

The other boundary condition simply states that the vertical velocity component $v_z = \partial_z\phi$ must be the same on either side of the interface,

$$\partial_z\phi_1 = \partial_z\phi_2, \quad \text{for } z = 0. \tag{13.25}$$

Now, by differentiation of Eq. (13.24) with respect to time and utilizing $v_z = \partial_z\phi = \partial_t\zeta$ we arrive at

$$(\rho_1 - \rho_2)g\partial_z\phi_1 = \rho_2\partial_t^2\phi_2 - \rho_1\partial_t^2\phi_1, \quad \text{for } z = 0. \tag{13.26}$$

Insertion of the wave functions Eq. (13.22) leads to the following forms of Eqs. (13.25) and (13.26), after appropriate algebraic reductions,

$$A_1\sinh\big(kh_1\big) = -A_2\sinh\big(kh_2\big), \tag{13.27a}$$

$$(\rho_1 - \rho_2)gkA_1\sinh\big(kh_1\big) = \Big[\rho_1 A_1\cosh\big(kh_1\big) - A_2\cosh\big(kh_2\big)\Big]\,\omega^2. \tag{13.27b}$$

Solving for ω we find the dispersion relation

$$\omega = \sqrt{\frac{(\rho_1 - \rho_2)gk}{\rho_1\coth\big(kh_1\big) + \rho_2\coth\big(kh_2\big)}}. \tag{13.28}$$

When both liquids are very deep, $kh_1 \gg 1$ and $kh_2 \gg 1$ the result is simply

$$\omega = \sqrt{gk\,\frac{\rho_1 - \rho_2}{\rho_1 + \rho_2}}. \tag{13.29}$$

The physical interpretation of this expression for $\rho_1 > \rho_2$ and for $\rho_1 < \rho_2$ is discussed in Exercise 13.3.

13.2.2 Capillary waves

So far the role of surface tension between the fluids in the two-phase flow has been neglected, but to add it is not too difficult, at least conceptually. We have already in connection with Eq. (7.9) discussed how surface tension must be included as a Young–Laplace pressure-drop discontinuity in the boundary condition at the interface for the stress tensor projected along the surface normal. Since the Young–Laplace pressure drop Δp_{surf} is given by the surface tension γ times the curvature κ of the interface,

$$\Delta p_{\text{surf}} = \gamma\,\kappa, \tag{13.30}$$

we choose to study the influence of surface tension for the gravity waves in the limit of small wavelengths, see Section 13.2.1, as the curvature increases as the wavelength decreases.

From Eq. (7.10) we have that for a curve $\mathbf{r}(s)$ described by the parameter s, the curvature $\kappa(s)$ is given by

$$\kappa(s) = \frac{|\mathbf{r}'(s) \times \mathbf{r}''(s)|}{|\mathbf{r}'(s)|^3}, \tag{13.31}$$

If we, as in Section 13.2.1, study waves only propagating along the x direction, then the x co-ordinate can serve as a parameter to define the shape $\mathbf{r}(x)$ of the interface at any given time t and y co-ordinate,

$$\mathbf{r} = \begin{pmatrix} x \\ y \\ \zeta(x) \end{pmatrix}, \quad \mathbf{r}' = \begin{pmatrix} 1 \\ 0 \\ \zeta'(x) \end{pmatrix}, \quad \mathbf{r}'' = \begin{pmatrix} 0 \\ 0 \\ \zeta''(x) \end{pmatrix}, \tag{13.32}$$

where ζ is the displacement along the z axis of the interface away from the equilibrium position at $z = 0$. Keeping terms of linear order in ζ gives the curvature

$$\kappa(x) = \zeta''(x) + \mathcal{O}(\zeta^2). \tag{13.33}$$

Using the potential description of the velocity field, $\mathbf{v} = \boldsymbol{\nabla}\phi$ from Section 13.2.1 we obtain by combining Eqs. (7.9) and (13.8) the following partial differential equation for ϕ and ζ:

$$\rho(\partial_t \phi + g\,\zeta) = p^* + \gamma \partial_x^2 \zeta. \tag{13.34}$$

In analogy with the derivation of Eq. (13.10), we differentiate Eq. (13.34) with respect to time and utilize $\partial_t \zeta = \partial_z \phi$ to obtain

$$-\partial_t^2 \phi = g\,\partial_z \phi - \frac{\gamma}{\rho}\,\partial_x^2 \partial_z \phi, \text{for } z = 0. \tag{13.35}$$

Moreover, the incompressibility condition $\boldsymbol{\nabla}\cdot\mathbf{v} = \nabla^2\phi$ still holds, so ϕ still takes the form of Eq. (13.15a),

$$\phi(x, z, t) = A \cosh\big[k(z + h)\big] \cos(kx - \omega t). \tag{13.36}$$

Inserting this into Eq. (13.35) leads, in analogy with Eq. (13.18), to the dispersion relation $\omega(k)$ for capillary waves,

$$\omega^2 = gk + \frac{\gamma}{\rho}\,k^3 \quad \Rightarrow \quad \omega = \sqrt{gk + \frac{\gamma}{\rho}\,k^3}. \tag{13.37}$$

As expected, the effect of surface tension becomes more dominant at shorter wavelengths where the wave number is big. Surface tension dominates in the square root of the dispersion relation when

$$\frac{\gamma}{\rho}\,k^3 \gg gk \quad \Rightarrow \quad k \gg \sqrt{\frac{\rho g}{\gamma}} = \frac{1}{\ell_\text{cap}} \quad \Rightarrow \quad \lambda \ll 2\pi\ell_\text{cap}, \tag{13.38}$$

where the capillary length ℓ_cap, introduced in Eq. (7.16) for characterizing capillary rise, now reappears.

13.3 Gas bubbles in microfluidic channels

Many microfluidic networks in lab-on-a-chip devices contain channel contractions. These tend to become problematic if, as is often the case, gas bubbles are introduced into the liquid at the inlets or by electrochemical processes. Due to the small channel dimensions gas bubbles can easily be big enough to span the entire channel cross-section. Such "large" bubbles are prone to get stuck at the channel contraction, whereby they can clog the flow and disturb measurements or functionality of the system in an uncontrolled manner. To clear the clogged channel an external pressure, the so-called clogging pressure, has to be applied to push the clogging bubble out of the system.

A complete analysis of the motion of a large bubble through a microchannel contraction involves many different physical effects, some of which are not completely understood. Any comprehensive analysis would at least require detailed modelling of the liquid/gas, liquid/solid, and solid/gas interfaces, see Section 7.1, as well as the dynamics in the bulk fluids. But also more complicated processes near the contact lines need to be addressed, e.g. wetting, contact line pinning and hysteresis, dynamic contact angles and contact lines, and static and dynamic friction.

Following Jensen, Goranović, and Bruus (2004) we will study the passage of a single gas bubble through a liquid-filled microchannel constriction. To obtain a simple tractable model, we will restrict our analysis to the simple quasi-static motion of bubbles. By this we mean that the velocity of the bubble is nearly zero and that the entire model system remains arbitrarily close to equilibrium for all bubble positions. All dynamic aspects are thus neglected, and basically the model involves only the free energy of the internal interfaces of the system, external pressures and the geometry of the system. We also choose to work with axisymmetric channels of smooth (but otherwise arbitrary) contraction geometries free from any sharp corners and other singularities. With these simplifications the forces or pressures needed to push a bubble through the system can be calculated accurately without losing the essential physics of the problem. This in turn enables us to formulate design rules for microchannel contractions to prevent or reduce clogging.

Consider a hydrophilic microfluidic channel, such as the one depicted in Fig. 13.3. For simplicity, it is chosen to be axisymmetric about the x axis with a position-dependent channel radius $a(x)$. The local tapering angle $\theta_\mathrm{t}(x)$ of the channel wall can therefore be defined as

$$\theta_\mathrm{t}(x) \equiv \partial_x a(x). \tag{13.39}$$

The channel is filled with a liquid (light gray). A large bubble (white) of some other fluid, we think mainly of a gas such as air, is present in the liquid. By large we mean that the volume of the bubble is larger than the volume $V_\mathrm{sph}^\mathrm{max}$ of the largest inscribed sphere that can be placed anywhere in the microchannel. A large bubble divides the liquid in two disconnected parts, left and right of the bubble, respectively. The bubble itself consists of a bulk part in direct contact with the walls of the channel and of two menisci, in contact with the liquid, capping the ends of the bubble.

The bubble is assumed to be in quasi-static equilibrium. In that case it is relatively simple to combine mass conservation with geometric constraints to determine, as a function of the bubble position, the pressure drops over the two menisci needed to maintain this equilibrium. We define our central concept, the clogging pressure, as the maximum of the position-dependent pressure drop across the bubble, i.e. the minimal external pressure that must be supplied to push the bubble through the microchannel.

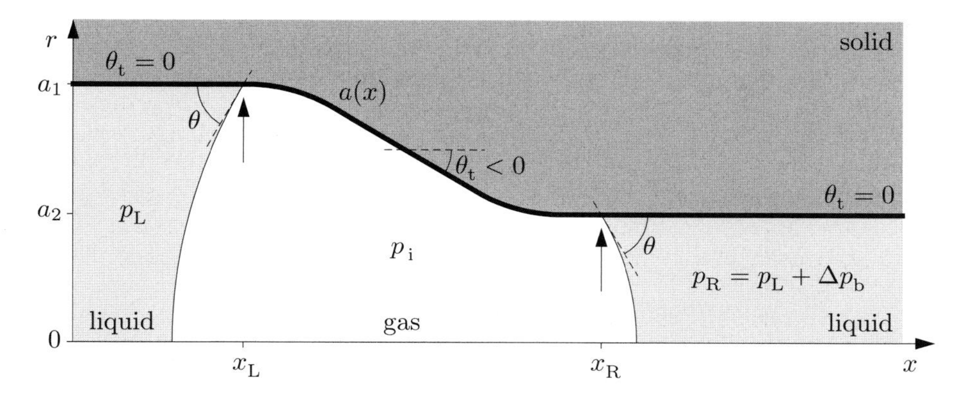

Fig. 13.3 A side view of a gas bubble (white) with internal pressure p_i surrounded by a liquid (light gray) inside a hydrophilic axisymmetric channel (dark gray) of varying radius $a(x)$. The left (right) contact line (little arrows) has the co-ordinate x_L (x_R) and a contact angle θ. The channel is contracting from a straight part of radius a_1 to one of smaller radius a_2. Throughout this chapter we have chosen $a(x)$ to be a sloped straight line joined to the straight parts by two circular arcs. The tapering angle θ_t is given by $\tan\theta_t = \partial_x a(x)$. The pressure left (right) of the bubble is denoted p_L (p_R) and the pressure difference across the bubble is Δp_b.

The essential physical parameters are the three surface tensions, see Eq. (7.2), γ_{lg}, γ_{sl}, and γ_{sg} for the liquid/gas, solid/liquid, and solid/gas interfaces, respectively. In equilibrium the contact angle θ is determined by the surface tensions through the Young equation (7.14). To sustain a curved interface, say the left one at $x = x_L$, with the main radii of curvature R_1^c and R_2^c between a gas of pressure p_i and a liquid of pressure p_L, the pressure difference $\Delta p = p_i - p_L$ must obey the Young–Laplace equation (7.8),

$$\Delta p(x_L) = \gamma_{lg}\left(\frac{1}{R_1^c} + \frac{1}{R_2^c}\right) = 2\,\gamma_{lg}\,\frac{\cos\left[\theta + \theta_t(x_L)\right]}{a(x_L)}, \tag{13.40}$$

where the last equation is applicable for a circular cross-section of radius $a(x)$ and a tapering angle $\theta_t(x)$. We use the standard convention that the radius of curvature is taken as positive if the interface is concave when seen from the gas.

In the rest of the analysis we consider a "large" bubble having the initial position "1" in the widest part of the channel. The initial volume is $V_1 = \gamma V_{sph}^{max}$, where $\gamma > 1$ and $V_{sph}^{max} = 4\pi a_1^3/3$, and the corresponding internal pressure is $p_{i,1}$. At a later stage the bubble is moved to a position "2", where the volume is V_2 and the internal pressure $p_{i,2}$. In the quasi-static case the bubble motion is isothermal and hence the compressibility condition applies,

$$p_{i,1}V_1 = p_{i,2}V_2. \tag{13.41}$$

The pressure p_i within the bubble is given as the ambient pressure p^* plus the pressure change Δp across the curved interface, given by Eq. (13.40).

The most extreme compression is obtained by pressing a large bubble, which floats without geometrical constraints in a bulk liquid of pressure p^*, into a narrow circular channel of radius a_2. Combining Eqs. (13.40) and (13.41) yields

$$\frac{V_1}{V_2} = \frac{p_{\mathrm{i},2}}{p_{\mathrm{i},1}} \approx \frac{p_{\mathrm{i},2}}{p^*} = 1 + \frac{2\gamma_{\mathrm{lg}}\cos\theta}{a_2 p^*}. \tag{13.42}$$

For example, moving a large spherical air bubble in water ($\gamma_{\mathrm{lg}} = 0.0725\ \mathrm{J\,m^{-2}}$) at the ambient pressure $p^* = 10^5$ Pa into a channel of radius $a_2 = 25$ µm leads to $V_1/V_2 \approx 1.06$, i.e. a volume compression of 6%. Moving a bubble from a 300 µm to a 190 µm wide channel yields a compression of about 0.2%.

For a bubble positioned in a microchannel contraction the total internal energy E_{tot} is the sum of the surface free energy, gravitational energy, kinetic energy, and frictional energy. We regard the surrounding pressures as external energy. By our definition quasi-static motion of an incompressible bubble implies that the kinetic energy is zero and that friction is zero because of hydrostatic and thermodynamic equilibrium. Finally, we treat channels of characteristic dimensions $2a_2$ less than 300 µm, which is significantly smaller than the capillary length of water, $\ell_{\mathrm{cap}} \approx 2.7$ mm, Eq. (7.16). So, the gravitational energy can also be neglected, which ensures that the menisci may be approximated by spherical caps. The total internal energy E_{tot} of the microchannel containing a quasi-statically moving bubble is given only by the surface free energy, i.e. the sum of interfacial energies γ_i times interfacial areas A_i,

$$E_{\mathrm{tot}} = \sum_i \gamma_i A_i = \gamma_{\mathrm{lg}} A_{\mathrm{lg}} + \gamma_{\mathrm{sg}} A_{\mathrm{sg}} + \gamma_{\mathrm{sl}} A_{\mathrm{sl}}. \tag{13.43}$$

The Young–Laplace pressure drops, Eq. (13.40), at the menisci are given by,

$$\Delta p_{\mathrm{L}} = p_{\mathrm{i}} - p_{\mathrm{L}}, \tag{13.44a}$$

$$\Delta p_{\mathrm{R}} = p_{\mathrm{i}} - p_{\mathrm{R}}. \tag{13.44b}$$

The total pressure drop $\Delta p_{\mathrm{b}}(x_{\mathrm{cm}})$ over the bubble as a function of its center of mass x_{cm} is given by

$$\Delta p_{\mathrm{b}}(x_{\mathrm{cm}}) = p_{\mathrm{R}} - p_{\mathrm{L}} = \Delta p_L(x_{\mathrm{cm}}) - \Delta p_R(x_{\mathrm{cm}}). \tag{13.45}$$

The clogging pressure p_{clog} is defined as the maximal position-dependent pressure drop across the bubble,

$$p_{\mathrm{clog}} = \max\big\{-\,\Delta p_{\mathrm{b}}(x_{\mathrm{cm}})\big\}. \tag{13.46}$$

The clogging pressure expresses the minimal amount by which the left-side pressure p_{L} must exceed the right-side pressure p_{R} to push the bubble through the contraction quasi-statically from left to right. Combining the geometry defined in Fig. 13.3 with Eqs. (13.40) and (13.45) the central expression of our analysis is easily derived,

$$\Delta p_{\mathrm{b}} = 2\gamma_{\mathrm{lg}}\left(\frac{\cos[\theta + \theta_{\mathrm{t}}(x_L)]}{a(x_{\mathrm{L}})} - \frac{\cos[\theta - \theta_{\mathrm{t}}(x_R)]}{a(x_{\mathrm{R}})}\right). \tag{13.47}$$

From the previous discussion it follows that if $\Delta p_{\mathrm{b}} < 0$ then the contraction causes bubble clogging, whereas for $\Delta p_{\mathrm{b}} > 0$ the bubble tends to move spontaneously through the contraction towards the narrow part.

Instead of the pressure we can also calculate the change in energy resulting from moving a bubble inside a channel of radius a_1 into a smaller channel of radius $a_2 < a_1$, e.g. by moving

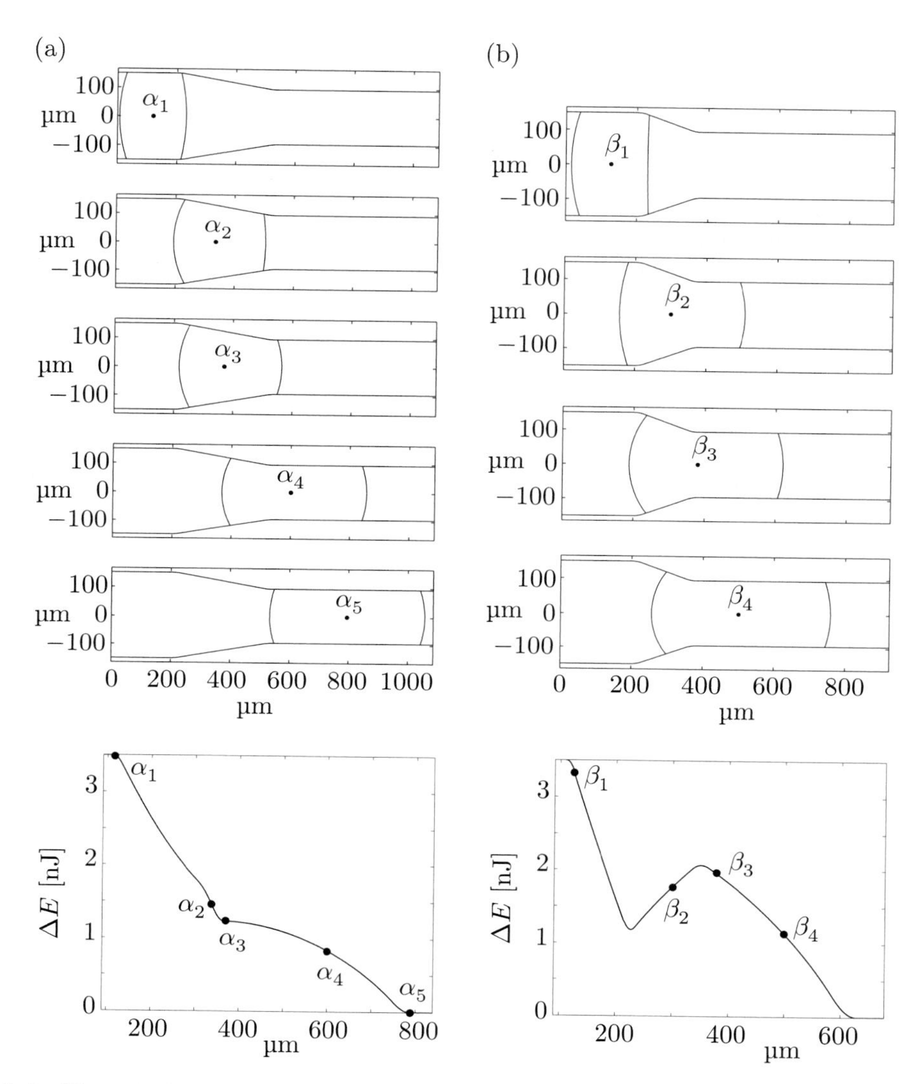

Fig. 13.4 Numerical analysis of quasi-static bubble motion in a microchannel contraction. (a) Five different positions of a large bubble with $\gamma = 1.02$ inside a 1000 µm long hydrophilic channel with contact angle $\theta = 72°$ and a maximum tapering angle $\theta_t = -10°$. The energy difference ΔE as a function of position. The monotonically decreasing graph implies that the bubble moves spontaneously from the large channel at the left into the narrow channel at the right. (b) Similar setup except for a larger maximum tapering angle of $-20°$. The energy difference ΔE as a function of position shows a clogging effect. For positions between 210 µm and 350 µm the energy increases and consequently energy needs to be supplied to press the bubble into the narrow channel at the right, say from position β_2 to β_3. Adapted from Jensen, Goranović, and Bruus (2004), courtesy of Mads Jakob Jensen, DTU Nanotech.

it from left to right in the channel depicted in Fig. 13.3. Intuitively, we would expect the energy to increase as a result of the move. In most cases this intuition is correct, however, we shall see that in some cases the system gains energy by the move, solely due to geometric conditions. With the above-mentioned large initial volume $V_1 = \gamma V_{\mathrm{sph}}^{\mathrm{max}}$, the bubble is forced to touch the walls regardless of its position. According to Eqs. (13.40) and (13.44b) the internal pressure of the bubble is

$$p_{\mathrm{i},1} = p_{\mathrm{R}} + 2\gamma_{\mathrm{lg}} \frac{\cos\theta}{a_1}. \tag{13.48}$$

The volume of the bubble is the sum of two spherical cap volumes and the volume of a cylinder of initial length L. Once the length L is known, the relevant interfacial areas A_{lg} and A_{sg} may be found.

The gas bubble is now moved to the narrow part of the channel with radius a_2, and according to Eqs. (13.40), (13.41), and (13.44b) the bubble pressure $p_{\mathrm{i},2}$ and volume V_2 are,

$$p_{\mathrm{i},2} = p_{\mathrm{R}} + 2\gamma_{\mathrm{lg}} \frac{\cos\theta}{a_2}, \tag{13.49}$$

$$V_2 = \frac{p_{\mathrm{i},1}}{p_{\mathrm{i},2}}\, V_1. \tag{13.50}$$

By solving Eq. (13.50) it is straightforward to find the change in total free-surface energy,

$$\Delta E_{\mathrm{tot}} = E_{\mathrm{tot},2} - E_{\mathrm{tot},1} \tag{13.51}$$

$$= \gamma_{\mathrm{lg}}\big(A_{\mathrm{lg},2} - A_{\mathrm{lg},1}\big) + \gamma_{\mathrm{lg}} 2\pi (a_2 \ell - a_1 L)\cos\theta, \tag{13.52}$$

where ℓ is the length of the bubble in the channel of radius $a_2 < a_1$, and where Young's relation Eq. (7.14) has been used to eliminate the solid/liquid and solid/gas interfacial energies. In Fig. 13.4 is shown the result of two calculations of the energy as a function of position. One of the cases, panel (a), demonstrates the possibility of having a microchannel constriction in which the gas bubble moves spontaneously into the narrow section. The other case, panel (b), is more common. An energy barrier prevents the bubble from moving into the narrow part without an external pressure pushing it through.

For experimental studies of gas bubbles moving in microchannel constrictions see Chio *et al.* (2006).

13.4 Droplets in microfluidic junctions and digital fluidics

Microfluidics offers new opportunities to study the long-standing, complex and fundamental hydrodynamic problem of droplet breakup, when droplets of one fluid move immersed in another fluid. Moreover, droplet microfluidics is important for a growing number of applications ranging from emulsion generation and microparticle production, via chemical screening to protein-crystal encapsulation and immunoassay. The fabrication and manipulation of microdroplets are now being termed "digital fluidics".

It is therefore of considerable interest to quantify the droplet-breakup processes. A geometrical condition for droplet breakup has been found by Ménétrier-Deremble and Tabeling (2006). They studied droplets of an aqueous solution of fluorescein and glycerin moving in a continuous phase of hexadecane, see Fig. 13.5. The device contains a droplet formation

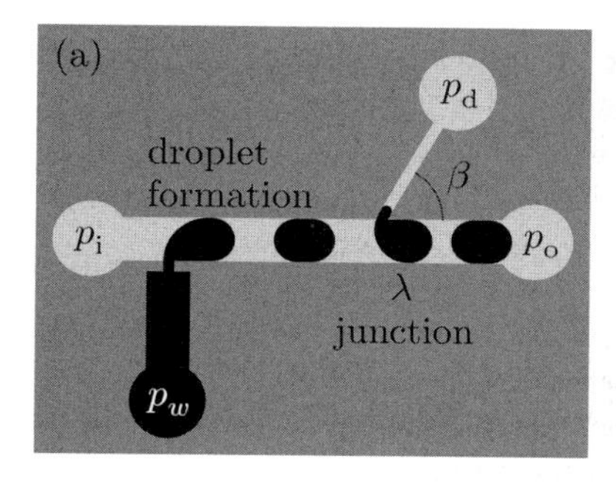

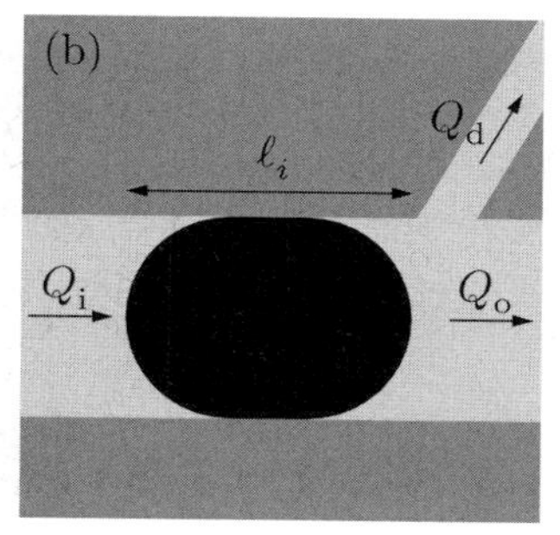

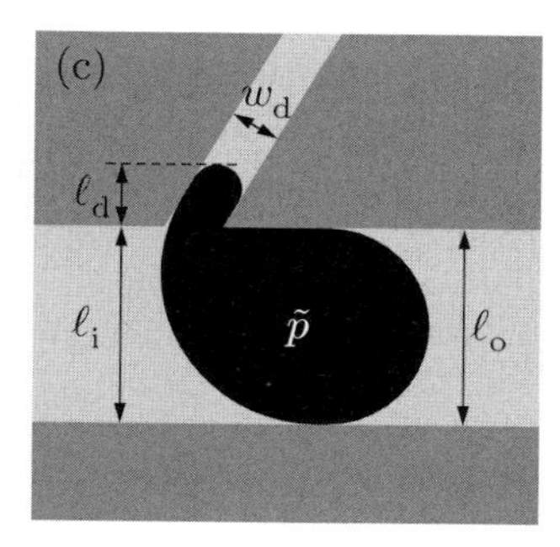

Fig. 13.5 (a) Sketch of a droplet generator and λ junction adapted from Ménétrier-Deremble and Tabeling (2006). Droplets of an aqueous solution of fluorescein and glycerin (black) move in a continuous phase of hexadecane (white). Four pressures are indicated: the inlet pressure p_i, the water pressure p_w, the outlet pressure p_o and the daughter channel pressure p_d. (b) A closeup of a droplet of length ℓ_i approaching the λ junction. The flow rate of the inlet, outlet, and daughter channel are indicated, respectively, as Q_i, Q_o and Q_d. (c) A droplet with the internal pressure $\tilde{p}$ in the λ junction. The length ℓ_d of the drop-finger inside the daughter channel is indicated as well as the width w_d of the daughter channel.

segment and a so-called λ junction named after its geometrical λ-like shape. The main liquid flow is in the center channel leading straight from the inlet to the outlet. However, at the λ junction liquid can leave the main channel through a daughter channel. Depending on the pressures p_i, p_o and p_d of the inlet, outlet and daughter channel, respectively, a droplet will either break up directly into two pieces at the λ junction, or pass without breaking up, or exhibit a retarded breakup, where the part of the droplet entering the daughter channel, the so-called finger, begins to retract prior to the breakup.

For the full treatment of this problem the reader is referred to the original paper, Ménétrier-Deremble and Tabeling (2006), here we will just briefly discuss the criterion for direct breakup. As seen in Fig. 13.5(c), the backside of the droplet forms a circle arc with tangents along the outlet and daughter channel. Breakup occurs through a Rayleigh–Plateau instability, if the thickness of the droplet between its backside and the sharp top-right corner becomes small enough. This will happen if the volume of the droplet and the geometry of the channels allow the finger part of the droplet to advance sufficiently into the daughter channel, the distance ℓ_d in Fig. 13.5(c), without mass conservation forcing the opening of a tunnel for the continuous phase to pass around the droplet and entering the daughter channel.

Three equations can be stated that eventually establish the criterion for direct breakup. The first is mass conservation for the fluid flow,

$$Q_i = Q_o + Q_i. \tag{13.53}$$

The second equation is also related to mass conservation, as it relates the length ℓ_d of the droplet finger in the daughter channel,

$$\ell_d \approx \frac{wQ_d}{w_d Q_i}\,\ell_i. \tag{13.54}$$

The third equation involves the pressure drops experienced from any of the inlets or outlets to the interior of the droplet, where the pressure is denoted $\tilde{p}$. The pressure drops comprise Hagen–Poiseuille pressure drops along the channels and Young–Laplace pressure drops across the droplet interface. From Eqs. (4.1) and (7.8), see Exercise 13.6, we get

$$\tilde{p} = p_i - R_i Q_i + \gamma\kappa_i, \tag{13.55a}$$

$$\tilde{p} = p_o + R_o Q_o + \gamma\kappa_o, \tag{13.55b}$$

$$\tilde{p} = p_d + R_d Q_d + \gamma\kappa_d, \tag{13.55c}$$

where R_x and κ_x are the hydraulic resistance and the mean curvature of the interface in channel x. If each of the three equations are divided by the respective channel resistance followed by addition of them, we arrive at

$$\tilde{p}\left(\frac{1}{R_i} + \frac{1}{R_o} + \frac{1}{R_d}\right) = \left(\frac{p_i}{R_i} + \frac{p_o}{R_o} + \frac{p_d}{R_d}\right) + \gamma\left(\frac{\kappa_i}{R_i} + \frac{\kappa_o}{R_o} + \frac{\kappa_d}{R_d}\right). \tag{13.56}$$

By further manipulation the criterion for direct droplet breakup is found to be

$$Q_d^* = \frac{Q_\mathrm{cap}}{1 - \ell_c/\ell_d^*}, \tag{13.57}$$

where the quantities appearing are defined by

$$Q_d^* \equiv \frac{p - p_d}{R_d}, \qquad Q_\mathrm{cap} \equiv \frac{\gamma R}{R_d}\left(\frac{\kappa_i}{R_i} + \frac{\kappa_o}{R_o} + \frac{\kappa_d}{R_d}\right), \tag{13.58a}$$

$$\ell_d^* \equiv \ell_i \frac{Q_d^* \, w}{Q_i \, w_d}, \qquad R \equiv \left(\frac{1}{R_i} + \frac{1}{R_o} + \frac{1}{R_d}\right)^{-1}, \tag{13.58b}$$

$$Q_i = \frac{p_i - p}{R_i}, \qquad p \equiv R\left(\frac{p_i}{R_i} + \frac{p_o}{R_o} + \frac{p_d}{R_d}\right). \tag{13.58c}$$

Given the otherwise complex hydrodynamics at breakup, it is remarkable that the criterion Eq. (13.57) can be established based mainly on pure geometrical considerations.

13.5 Further reading

There exists a growing literature on the broad subject of digital fluidics and bubbles/droplet formation in microfluidics. The following papers, together and the many references they contain, cover a variety of topics in the field. They could provide a good starting point for the reader who wants to know more: Garstecki, Stone, and Whitesides (2005), Garstecki, Fuerstman, and Whitesides (2005), Utada *et al.* (2005), Willaime *et al.*(2006), Jensen, Stone and Bruus (2006), as well as Kim *et al.* (2007).

13.6 Exercises

Exercise 13.1
Two-phase Poiseuille flow
Consider the two-phase Poiseuille flow defined in Fig. 13.1.

(a) Based on basic physical considerations, explain qualitatively why the velocity profile shown in the figure is correct.

(b) Derive the expressions Eq. (13.5) for the two constants h_1 and h_2 that determines the velocity field in the two-phase Poiseuille flow.

Exercise 13.2
Kelvin's circulation theorem and potential flow

The velocity circulation Γ is defined as the closed contour integral $\Gamma \equiv \oint_{\mathcal{C}} \mathbf{v}\cdot\delta\mathbf{r}$, where the contour $\mathcal{C}$ is defined by a string of connected fluid elements forming a closed loop, i.e. a Lagrangean approach. As a function of time each fluid element on the loop moves, and consequently the contour they define is deformed. To avoid confusion between spatial and temporal derivatives we use the symbol δ for the former and d for the latter. The infinitesimal curve length $\delta\mathbf{r}$ appearing in the integral is defined as the difference of the position vectors $\mathbf{r}$ of neighboring fluid elements defining the contour $\mathcal{C}$.

(a) Argue why the time derivative of the circulation is given by

$$\frac{\mathrm{d}\Gamma}{\mathrm{d}t} = \frac{\mathrm{d}}{\mathrm{d}t}\oint_{\mathcal{C}} \mathbf{v}\cdot\delta\mathbf{r} = \oint_{\mathcal{C}} \frac{\mathrm{d}\mathbf{v}}{\mathrm{d}t}\cdot\delta\mathbf{r} + \oint_{\mathcal{C}} \mathbf{v}\cdot\frac{\mathrm{d}\delta\mathbf{r}}{\mathrm{d}t}. \tag{13.59}$$

(b) Show the following differential relation between small distance differences $\delta\mathbf{r}$ and the associated small velocity differences $\delta\mathbf{v}$,

$$\mathbf{v}\cdot\frac{\mathrm{d}(\delta\mathbf{r})}{\mathrm{d}t} = \mathbf{v}\cdot\delta\mathbf{v} = \delta\big(\tfrac{1}{2}v^2\big). \tag{13.60}$$

(c) Insert the Navier–Stokes equation for an incompressible and inviscid fluid as well as Eq. (13.60) into Eq. (13.59) and show that the time-derivative $\mathrm{d}_t\Gamma$ of the velocity circulation Γ is zero.

(d) Consider a situation where the velocity field $\mathbf{v}(t)$ at the present time is slowly built up from a state of rest in the distant past, $\mathbf{v}(t=-\infty)\equiv\mathbf{0}$. Argue that in this case $\Gamma(t)\equiv 0$ and $\boldsymbol{\nabla}\times\mathbf{v}=0$ and thus a scalar potential function ϕ exists such that

$$\mathbf{v} = \boldsymbol{\nabla}\phi. \tag{13.61}$$

When this is the case, the flow is for obvious reasons denoted a potential flow. The result can be extended to isentropic, compressible fluids using Eq. (15.4) in which $(1/\rho)\boldsymbol{\nabla}p$ is written as a gradient of a scalar function.

Exercise 13.3
Stability of gravity waves

The dispersion relation $\omega(k)$ for gravity waves is given by Eq. (13.29). Discuss the physical interpretation of this expression for $\rho_1 > \rho_2$ and for $\rho_1 < \rho_2$. Hints: consider the time evolution using the complex notation $\mathrm{e}^{-\mathrm{i}\omega t}$ for the time dependence.

Exercise 13.4
Capillary waves

The basic theory for capillary waves was given in Section 13.2.2.

(a) In analogy with the argument leading from Eq. (13.9) to Eq. (13.10), show that the wave equation Eq. (13.35) follows from Eq. (13.34).

(b) Verify the dispersion relation Eq. (13.37).

Exercise 13.5
The pressure drop across a gas bubble in a deformed microchannel
Derive the expression given in Eq. (13.47) for the pressure drop Δp_b across a bubble in a cylindrical microfluidic channel with position-dependent radius $a(x)$.

Exercise 13.6
The criterion for direct droplet breakup
Derive the expression Eq. (13.55) for the pressure $\tilde{p}$ inside the gas bubble by considering each of the three pressures p_i, p_o and p_d for the the inlet and the two outlets.

13.7 Solutions

Solution 13.1
Two-phase Poiseuille flow

(a) In Fig. 13.1 $\eta_2 = 4\,\eta_1$. In the limit $\eta_2 \to \infty$ fluid 2 is appearing as a solid wall, and thus velocity field 1 should approach the no-slip condition $v_1 = 0$ at the interface $z = h^*$ and appear as a full parabola between $z = 0$ and $z = h^*$.

In the other limit $\eta_1 \to 0$ the presence of fluid 1 does not lead to any shear stress in fluid 2 at the interface, thus velocity field v_2 should approach the no-stress condition $\partial_z v_2 = 0$ and appear as a half-parabola with no-slip at $z = h$ and maximum value at the interface $z = h^*$.

The actual case with $\eta_2 = 4\eta_1$ is in between the two limiting cases, and we observe in Fig. 13.1 that the parabolic velocity field v_1 does not quite go to zero at the interface, while the stress or velocity field gradient in fluid 2 at the interface is not quite zero.

(b) Given the proposed solution Eq. (13.4), the no-stress condition $\eta_1 v_{1,x} = \eta_2 v_{2,x}$ at $z = h^*$ becomes $-2z + h_1 = -2z + h + h_2$. From this we immediately get $h_1 = h + h_2$ as stated in Eq. (13.5b).

The continuity of the velocity field $v_{1,x}(h^*) = v_{2,x}(h^*)$ leads to $\frac{1}{\eta_1}(h + h_2 - h^*)h^* = \frac{1}{\eta_2}(h - h^*)(h^* - h_2)$, where on the left-hand side we have used the previous result $h_1 = h + h_2$. Collecting terms proportional to h_2 gives $\left[1 + \frac{\eta_1}{\eta_2}\left(\frac{h}{h^*} - 1\right)\right] h_2 = \left[\frac{\eta_1}{\eta_2} - 1\right](h - h^*)$. From this Eq. (13.5a) is readily obtained.

Solution 13.2
Kelvin's circulation theorem and potential flow

(a) Since the contour itself is defined by moving fluid particles, both the velocity field $\mathbf{v}$ and the contour line element $\delta\mathbf{r}$ in the integrand of $\mathrm{d}_t\Gamma$ in Eq. (13.59) depend on time. Consequently, the time derivative of Γ involves the differentiation of a scalar product of two time-dependent vector functions resulting in the two final terms of Eq. (13.59).

(b) First, we note that $\mathbf{v} = \mathrm{d}_t\mathbf{r}$, since $\mathbf{r}$ denotes the position of a fluid particle. Then, since the temporal and spatial derivatives d and δ commute we get $\mathrm{d}_t(\delta\mathbf{r}) = \delta(\mathrm{d}_t\mathbf{r}) = \delta\mathbf{v}$, which proves the first equality in Eq. (13.60). The second equality follows from the basic rule of differentiation of a product.

(c) If the viscosity term can be neglected in the Navier–Stokes equation we have $\mathrm{d}\mathbf{v}/\mathrm{d}t = \boldsymbol{\nabla}(-p/\rho - gz)$. The first term in Eq. (13.59) is thus a closed-contour integral of a gradient, which integrates to zero. Likewise, the second term is the closed-contour integral of the total differential $\delta\left(\frac{1}{2}v^2\right)$, which also integrates to zero. We conclude that for the given assumptions $\mathrm{d}_t\Gamma = 0$ and thus that Γ is time independent.

(d) Let the velocity field $\mathbf{v}(t)$ at present be built up infinitely slowly from a state of rest in the distant past at $t = -\infty$, $\mathbf{v}(t = -\infty) \equiv \mathbf{0}$, by an adiabatic turn-on of the pressure and gravity, $p(t) = p\,\mathrm{e}^{\gamma t}$ and $g(t) = g\,\mathrm{e}^{\gamma t}$. By construction $\Gamma(-\infty) \equiv 0$, and since Γ is time independent we conclude that $\Gamma(t) = \Gamma(-\infty) = 0$. By Stokes' theorem $\Gamma = 0$ implies that $\boldsymbol{\nabla} \times \mathbf{v} = 0$ and thus a scalar potential function ϕ exists such that $\mathbf{v} = \boldsymbol{\nabla}\phi$.

Solution 13.3
Stability of gravity waves

The dispersion relation $\omega(k)$ for gravity waves given by Eq. (13.29) may acquire a non-zero imaginary part,

$$\mathrm{Im}\big[\omega(k)\big] = \begin{cases} 0 & , \text{ for } \rho_1 > \rho_2, \\ \sqrt{gk\,\dfrac{\rho_2 - \rho_1}{\rho_1 + \rho_2}} & , \text{ for } \rho_1 < \rho_2. \end{cases} \tag{13.62}$$

In complex notation the time dependence can be written as $\exp(-\mathrm{i}\omega t) = \exp(-\mathrm{i}\{\mathrm{Re}[\omega] + \mathrm{i}\mathrm{Im}[\omega]\}t) = \exp(-\mathrm{i}\mathrm{Re}[\omega]t)\,\exp(\mathrm{Im}[\omega]t)$. Given the result in Eq. (13.62) we see that for $\rho_1 > \rho_2$, i.e. when the light fluid is on top, the two-phase system cannot acquire a growing amplitude. Conversely, for $\rho_1 < \rho_2$, i.e. when the heavy fluid is on top, the system has a diverging amplitude factor $\exp(\mathrm{Im}[\omega]t)$, which means that the system is unstable.

Solution 13.4
Capillary waves

(a) The time derivative of Eq. (13.34) is $\rho\big(\partial_t^2\phi + g\,\partial_t\zeta\big) = \gamma\partial_x^2\partial_t\zeta$, and by insertion of $\partial_t\zeta = \partial_z\phi$ and isolation of $-\partial_t^2\phi$ we arrive at Eq. (13.35).

(b) The proposed solution Eq. (13.36) is inserted into Eq. (13.35). The derivatives are elementary, and upon evaluation of the result at $z = 0$ as well as removal of common factors we get $-\omega^2/\tanh(kh) = -gk - (\gamma/\rho)k^3$, which in the limit $kh \gg 1$ leads to Eq. (13.37).

Solution 13.5
The pressure drop across a gas bubble in a deformed microchannel

As in Eq. (7.18) the relation between the radius of curvature $R(x)$ and the local radius $a(x)$ of the cylindrical microchannel is given by $a(x) = R(x)\,\cos[\phi(x)]$, where $\phi(x)$ is defined in Fig. 13.3 as the angle between the horizontal direction and the tangent to the liquid/gas interface. For a tapered microchannel this angle is the sum of the angle between the interface tangent and the wall tangent, i.e. the contact angle θ, and the angle between the wall tangent and horizontal, i.e. the tapering angle θ_t,

$$\phi(x) = \theta + \theta_t(x). \tag{13.63}$$

Thus, for the left meniscus of the gas bubble at $x = x_L$ shown in Fig. 13.3 we have the Young–Laplace pressure drop $\Delta p_{\mathrm{surf}}(x_L) = 2\gamma_{\mathrm{lg}}\,\cos[\theta + \theta_t(x_L)]/a(x_L)$. For the right meniscus at $x = x_L$ the geometry appears as the mirror image in the yz-plane of the left meniscus, and consequently the expression for the right side includes a sign change in both the $\theta_t(x_R)$-term and in the radius of curvature, $\Delta p_{\mathrm{surf}}(x_R) = 2\gamma_{\mathrm{lg}}\,\cos[\theta - \theta_t(x_R)]/[-a(x_L)]$. In conclusion

$$\Delta p_b = \Delta p_{\mathrm{surf}}(x_L) + \Delta p_{\mathrm{surf}}(x_R) = 2\gamma_{\mathrm{lg}}\left[\frac{\cos[\theta + \theta_t(x_L)]}{a(x_L)} - \frac{\cos[\theta - \theta_t(x_R)]}{a(x_R)}\right]. \tag{13.64}$$

Solution 13.6
The criterion for direct droplet breakup
Consider first the path leading from the inlet with pressure p_i to the interior of the gas bubble with pressure $\tilde{p}$. Since the flow is in the same direction as the considered path, there is a downstream pressure decrease of $-R_i Q_i$ from the inlet to just outside the bubble. Due to the geometry of the curved interface there is a pressure increase when passing from just outside to just inside the bubble of $+\gamma\kappa_i$. Thus, for the final pressure we have $\tilde{p} = p_i - R_i Q_i + \gamma\kappa_i$.

Consider likewise the path leading from the outlet with pressure p_o to the interior of the gas bubble with pressure $\tilde{p}$. In this case the path is against the flow direction, so there is an upstream pressure increase of $+R_o Q_o$ from the outlet to just outside the bubble. The rest of the argument is identical with the former case, and we arrive at $\tilde{p} = p_o + R_o Q_o + \gamma\kappa_o$.

The second outlet with pressure p_d is identical in form to the first outlet. We therefore get $\tilde{p} = p_d + R_d Q_d + \gamma\kappa_d$.

These pressure considerations then lead to the criterion for direct droplet breakup.

14

Complex flow patterns

Viscous forces dominate in microfluidics and tend to favor laminar flow at the expense of turbulence. Most laminar flow patterns are simple, and in the case of creeping flow they even closely follow the geometry of the enclosing channel. Nevertheless, it is possible by careful design to create complex flow patterns in microfluidic lab-on-a-chip systems. In this chapter we will study two examples of such an increasing level of complexity in microflows.

14.1 Pressure-driven flow in shape-perturbed microchannels

In Section 3.5 we studied the effects of a shape-perturbed, but constant cross-section, $\partial\mathcal{C} = \partial\mathcal{C}(y,z)$, in Poiseuille flow through straight microchannels along the x axis. Now we are going to raise the level of complexity by allowing for variation of the cross-section also in the x direction along the channel in pressure-driven flows. We restrict the treatment to the 2D case, where the top plate in an infinite parallel-plate channel is perturbed from being planar at position $z = h$ to have an arbitrary shape $h(x)$, rendering the channel an average height h_0 as illustrated in Fig. 14.1(a).

We wish to calculate the velocity field in the shape-perturbed channel given an applied pressure $p(0) = p^* + \Delta p$ at $x = 0$ and $p(L) = p^*$ at $x = L$. The unperturbed problem with a constant channel height $h = h_0$ is solved in Section 3.4.2, and here the characteristic scales are set for the pressure P_0, the velocity V_0, the time T_0, the Reynolds number Re, and the aspect ratio ϵ,

$$P_0 \equiv \Delta p = 12\,\frac{\eta L V_0}{h_0^2}, \tag{14.1a}$$

$$V_0 \equiv \frac{Q_0}{w h_0} = \frac{1}{w h_0}\,\frac{\Delta p}{R_\mathrm{hyd}} = \frac{h_0^2 \Delta p}{12 \eta L}, \tag{14.1b}$$

$$T_0 \equiv \frac{L}{V_0}, \tag{14.1c}$$

$$Re \equiv \frac{\rho V_0 h_0}{\eta}, \tag{14.1d}$$

$$\epsilon \equiv \frac{h_0}{L}. \tag{14.1e}$$

To calculate the velocity field in the shape-perturbed case we apply perturbation theory to the unperturbed infinite parallel-plate channel. The perturbation parameter α is introduced as the dimensionless amplitude of the deviation $h(x) - h_0$ of the channel height $h(x)$ from its unperturbed value h_0,

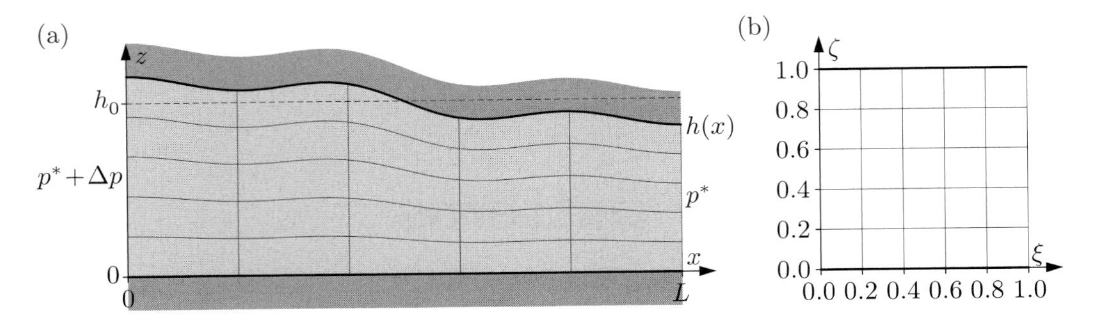

Fig. 14.1 (a) An infinite nearly-parallel-plate channel of length L with a shape-perturbed top plate (top dark gray) defined by $z = h(x)$ (thick top line) with the average value h_0 (dashed line). The bottom plate remains planar (bottom dark gray), and the liquid (light gray) is driven by the overpressure Δp at $x = 0$. The thin lines indicate iso-ξ and iso-ζ lines. (b) The dimensionless variables ξ and ζ, defined in the unit square, are introduced to parameterize the physical co-ordinates according to the prescriptions $x(\xi,\zeta) = L\,\xi$ and $z(\xi,\zeta) = h(L\xi)\,\zeta$. The bottom and top walls are given by $\zeta = 0$ and $\zeta = 1$, respectively.

$$h(x) \equiv \Big[1 + \alpha\,\lambda\big(\tfrac{x}{L}\big)\Big]\,h_0, \tag{14.2}$$

where λ is a dimensionless shape function of magnitude unity, $|\lambda| \lesssim 1$, which must have a zero average $\langle\lambda\rangle$ to ensure $\langle h(x)\rangle = h_0$,

$$\langle\lambda\rangle \equiv \frac{1}{L}\int_0^L \mathrm{d}x\,\lambda\big(\tfrac{x}{L}\big) = 0, \tag{14.3a}$$

$$\langle h(x)\rangle \equiv \frac{1}{L}\int_0^L \mathrm{d}x\,h(x) = h_0. \tag{14.3b}$$

In the following perturbation calculation, we only keep zero- and first-order terms in α.

Since the position $z = h(x)$ of the top wall is no longer constant in space, it is, as in Section 3.5.1, advantageous to ensure the no-slip boundary condition by introducing the dimensionless parameters ξ and ζ defined in the unit square as

$$\xi(x,z) \equiv \frac{x}{L}, \tag{14.4a}$$

$$\zeta(x,z) \equiv \frac{z}{h(x)} = \frac{1}{1+\alpha\lambda\big(\frac{x}{L}\big)}\,\frac{z}{h_0} \approx \Big[1 - \alpha\lambda\big(\tfrac{x}{L}\big)\Big]\,\frac{z}{h_0}. \tag{14.4b}$$

To first order in α, the inverse co-ordinate transformations are given by

$$x(\xi,\zeta) = \xi\,L, \tag{14.5a}$$

$$z(\xi,\zeta) = \big[1 + \alpha\lambda(\xi)\big]\,\zeta\,h_0, \tag{14.5b}$$

and we see, as illustrated in Fig. 14.1(b), that the top and bottom walls are given by the $\zeta = 1$ and $\zeta = 0$ isolines $z(\xi,1)$ and $z(\xi,0)$, respectively. Using these dimensionless co-ordinates together with the above-mentioned velocity scale V_0 we can introduce the dimensionless velocities $\tilde{v}_x$ and $\tilde{v}_z$. However, as already stated in Eq. (4.36), we do not expect the two velocity components to have the same amplitude. In the time interval T_0 the convection distance along the x direction is approximately $L = T_0 V_0$, but in the z direction it is only expected to be $h_0 = \epsilon L = T_0 \epsilon V_0$. It is therefore reasonable to use the amplitude V_0 for v_x and ϵV_0 for v_z. The dimensionless velocity components are therefore defined as

$$\tilde{v}_x(\xi,\zeta) \equiv \frac{1}{V_0}\, v_x\big(x(\xi,\zeta), z(\xi,\zeta)\big), \tag{14.6a}$$

$$\tilde{v}_z(\xi,\zeta) \equiv \frac{1}{\epsilon V_0}\, v_z\big(x(\xi,\zeta), z(\xi,\zeta)\big), \tag{14.6b}$$

which fulfill the no-slip boundary conditions

$$\begin{pmatrix} \tilde{v}_x(\xi,0) \\ \tilde{v}_z(\xi,0) \end{pmatrix} = \mathbf{0} \qquad \text{and} \qquad \begin{pmatrix} \tilde{v}_x(\xi,1) \\ \tilde{v}_z(\xi,1) \end{pmatrix} = \mathbf{0}. \tag{14.7}$$

In a similar manner we introduce the dimensionless pressure $\tilde{p}(\xi,\zeta)$ by

$$\tilde{p}(\xi,\zeta) \equiv \frac{1}{P_0}\Big[p\big(x(\xi,\zeta), z(\xi,\zeta)\big) - p^*\Big], \tag{14.8}$$

with the boundary conditions

$$\tilde{p}(0,\zeta) = 1 \qquad \text{and} \qquad \tilde{p}(1,\zeta) = 0. \tag{14.9}$$

The next step in the perturbation calculation is to rewrite the gradient and Laplace operator in terms of the (ξ,ζ) co-ordinates keeping only the zero and first-order terms in both α and the aspect ratio ϵ. When this is done we can formulate the continuity and Navier–Stokes equation in terms of ξ and ζ. Based on Eq. (14.4) we first calculate the Jacobian matrix containing the first derivatives,

$$\begin{pmatrix} \partial_x \xi & \partial_x \zeta \\ \partial_z \xi & \partial_z \zeta \end{pmatrix} = \frac{1}{h_0}\begin{pmatrix} \epsilon & -\alpha\epsilon\lambda'(\xi)\,\zeta \\ 0 & 1-\alpha\lambda(\xi) \end{pmatrix}. \tag{14.10}$$

From this and the chain rule, e.g. $\partial_x = (\partial_x\xi)\partial_\xi + (\partial_x\zeta)\partial_\zeta$, it is straightforward to calculate the gradient operator to first order in α and ϵ,

$$\boldsymbol{\nabla} \approx \frac{1}{h_0}\begin{pmatrix} \epsilon\,\partial_\xi \\ \partial_\zeta \end{pmatrix} + \frac{\alpha}{h_0}\begin{pmatrix} \epsilon\lambda'(\xi)\,\zeta\,\partial_\zeta \\ \lambda(\xi)\,\partial_\zeta \end{pmatrix}. \tag{14.11}$$

Note that the α term in $\boldsymbol{\nabla}$ does not contain ∂_ξ but only ∂_ζ, and that the x component in contrast to the z component is scaled by the aspect ratio ϵ. To first order in α and ϵ, the Laplacian operator ∇^2 can be obtained from the gradient operator $\boldsymbol{\nabla}$,

$$\nabla^2 = \boldsymbol{\nabla}^2 \approx \frac{1}{h_0^2}\,\partial_\zeta^2 - \frac{\alpha}{h_0^2}\,\big[2\lambda(\xi)\,\partial_\zeta^2\big], \tag{14.12}$$

and we note that both the zero- and first-order terms in alpha are proportional to ∂_ζ^2. See Exercise 14.1 for more details concerning the transformation of $\boldsymbol{\nabla}$ and ∇^2 from (x,z) co-ordinates to (ξ,ζ) co-ordinates.

The continuity equation can be written in the (ξ,ζ) co-ordinates by insertion of Eqs. (14.6) and (14.11) into $\boldsymbol{\nabla}\cdot\mathbf{v}=0$. After multiplication by $h_0/(\epsilon V_0)$ and sorting after α we get

$$\big[\partial_\xi \tilde{v}_x + \partial_\zeta \tilde{v}_z\big] + \alpha\,\big[\lambda'(\xi)\zeta\partial_\zeta\,\tilde{v}_x + \lambda(\xi)\,\partial_\zeta \tilde{v}_z\big] = 0. \tag{14.13}$$

The co-ordinate transformation of the Navier–Stokes is a little more involved, but as shown in Exercise 14.2 the inertial terms are proportional to ϵRe and $\epsilon^3 Re$ for the x and z component, respectively. It is noteworthy that these effective Reynolds numbers appear instead of the ordinary Reynolds number. In flat microfluidic channels these effective Reynolds numbers are minuscule and can safely be neglected, as we already discussed briefly in Section 4.4. To first order in α and ϵ the remaining pressure and viscosity terms lead to the following form of the x and z components of the Navier–Stokes equation sorted in powers of α,

$$\big[\partial_\zeta^2 \tilde{v}_x - \partial_\xi \tilde{p}\big] - \alpha\,\big[2\lambda(\xi)\partial_\zeta^2 \tilde{v}_x - \lambda'(\xi)\zeta\,\partial_\zeta \tilde{p}\big] = 0, \tag{14.14a}$$

$$-\partial_\zeta \tilde{p} + \alpha\,\lambda(\xi)\,\partial_\zeta \tilde{p} = 0. \tag{14.14b}$$

We remark that only the pressure and no velocity components appear in the z component of the Navier–Stokes equation when including only terms up to first order in α and ϵ.

We now write the first-order perturbation expansion for the fields, which we then have to insert in the governing equations (14.13) and (14.14)

$$\tilde{v}_x = \tilde{v}_x^{(0)} + \alpha\,\tilde{v}_x^{(1)}, \tag{14.15a}$$

$$\tilde{v}_z = \tilde{v}_z^{(0)} + \alpha\,\tilde{v}_z^{(1)}, \tag{14.15b}$$

$$\tilde{p} = \tilde{p}^{(0)} + \alpha\,\tilde{p}^{(1)}. \tag{14.15c}$$

The dimensionless zero-order continuity and Navier–Stokes equations become

$$\partial_\xi \tilde{v}_x^{(0)} + \partial_\zeta \tilde{v}_z^{(0)} = 0, \tag{14.16a}$$

$$\partial_\zeta^2\,\tilde{v}_z^{(0)} = 12\,\partial_\xi \tilde{p}^{(0)}, \tag{14.16b}$$

$$\partial_\zeta \tilde{p}^{(0)} = 0, \tag{14.16c}$$

with the well-known solution from Section 3.4.2,

$$\tilde{v}_x^{(0)}(\xi,\zeta) = 6\,\zeta(1-\zeta), \tag{14.17a}$$

$$\tilde{v}_z^{(0)}(\xi,\zeta) = 0, \tag{14.17b}$$

$$\tilde{p}^{(0)}(\xi,\zeta) = 1-\xi. \tag{14.17c}$$

The dimensionless first-order continuity and Navier–Stokes equations become

$$\partial_\xi \tilde{v}_x^{(1)} + \partial_\zeta \tilde{v}_z^{(1)} = \lambda'(\xi)\zeta\,\partial_\zeta \tilde{v}_x^{(0)} + \lambda\partial_\zeta \tilde{v}_\zeta^{(0)} = 6\lambda'(\xi)\,\zeta(1-2\zeta), \tag{14.18a}$$

$$\partial_\zeta^2\,\tilde{v}_x^{(1)} = 12\,\partial_\xi \tilde{p}^{(1)} + 2\lambda(\xi)\partial_\zeta^2 \tilde{v}_x^{(0)} = 12\,\partial_\xi \tilde{p}^{(1)} - 24\lambda(\xi), \tag{14.18b}$$

$$\partial_\zeta \tilde{p}^{(1)} = 0. \tag{14.18c}$$

In Exercise 14.3 the solution to these equations is found to be

away from the straightforward direction towards a direction more parallel to the grooves. There is less resistance to a flow along the grooves than in the perpendicular direction. This flow pattern is in accordance with the shape-perturbation flow presented above. In microfluidics the flow is directed by the geometrical shape of the walls.

It is clear that for a fluid particle being turned an angle, say to the left, by the grooves in the bottom wall of the channel, it must, due to incompressibility, rise once it is pressed against the left-side wall and acquire a vertical velocity component adding to the general horizontal forward component. Later, when it is pressed against the top wall, it will acquire a transverse velocity component directing it towards the side wall to the right. Finally, when the fluid particle is pressed against the right-side wall, it will acquire a downward pointing velocity component, *etc.* This effect is similar to rifling in a gun barrel. The viscous liquid is forced to move in a helical pattern along the channel axis.

A more complicated flow pattern arises if one places groups of grooves with different tilt angles. One famous example is the so-called staggered herring-bone mixer presented in three papers from 2002 by Ajdari (2002), Stroock *et al.* (2002a), and Stroock *et al.* (2002b). This mixer contains two sets of grooves oriented in two different directions that meet in a cusp somewhere inside the microfluidic channel. Instead of just one helical motion inside the channel, the herring-bone mixer generates several pairs of counterrotating helical flows, which break up and mix each time they pass a transition from one groove pattern to the next. The resulting flow has proven to be very efficient for mixing, moreover, the design is very robust as it does not contain any moving mechanical parts.

This design idea has been taken up by several research groups. In Fig. 14.3 is shown a staggered herring-bone device fabricated at DTU Nanotech aiming at enhancing the capture of magnetic microbeads, as described in Section 11.5. The flow is visualized using fluorescein, a molecule that can be made fluorescent and hence visible. Panel (a) is a micrograph of a 200 µm mixer with a set of staggered herring-bone grooves at the bottom of the channel. Water is injected from the left with fluorescein in the top quarter of the stream. The grooves are grouped in sets of five, and it is seen that already after passing two of these groups the fluorescein is spreading widely in the channel. Panel (b) of Fig. 14.3 is a COMSOL

Fig. 14.3 (a) A micrograph of a staggered herring-bone mixer in a 200 µm wide channel with 150 µm wide magnetic elements at the sides for capture of magnetic microbeads. Water is entering from the left containing fluorescein in the top quarter of the stream. The design is improving the capture of magnetic microbeads compared to that of the straight, smooth channel shown in Fig. 11.4. (b) A COMSOL simulation of the mixer showing good agreement with the experiment. Adapted from Lund-Olesen, Bruus and Hansen (2007), courtesy of Mikkel Fougt Hansen, DTU Nanotech.

simulation of the staggered herring-bone mixer flow. A good agreement is shown between experiment and simulation. It was shown by Lund-Olesen, Bruus and Hansen (2007) that indeed the capture of magnetic beads in channels with the staggered herring-bone mixer, see Fig. 14.3, is significantly better than in channels without, see Fig. 11.4.

A more complete theoretical treatment of the complex flow in the staggered herring-bone mixer can be done by making a perturbation expansion in powers of the amplitude of the grooves relative to the total channel height. This perturbation calculation is more complicated than the one presented in Section 14.1, because of its full 3D nature.

With this remark we end this short presentation of the pressure-driven flow in patterned microchannels and move on to discuss flow patterns generated by the induced-charge method.

14.5 Induced-charge electrolytic flow

The charged Debye layer in electrolytes provides a means to manipulate the flow pattern of the electrolytes by externally applied potentials. The basic physics of the Debye layer was treated in Section 8.3, and in Chapter 9 we studied electro-osmosis, where an electrolyte is brought into motion by sending an electrical current through it. In the following, we will study an example of the so-called induced-charge electrolytic flow. In contrast to electro-osmosis, where charge is exchanged between the electrolyte and the electrodes providing the external potential, no such charge transfer takes place in an induced-charge flow. Instead, the electric coupling is purely capacitive and motion is brought about by applying AC potentials on electrodes situated at the wall of the microchannels but electrically insulated from the electrolyte. An example of an actual induced-charge flow device is shown in Fig. 14.4.

In the following, we study a simple example of an induced-charge system with spatially symmetric electrodes. Although the problem involves the coupling between the velocity field

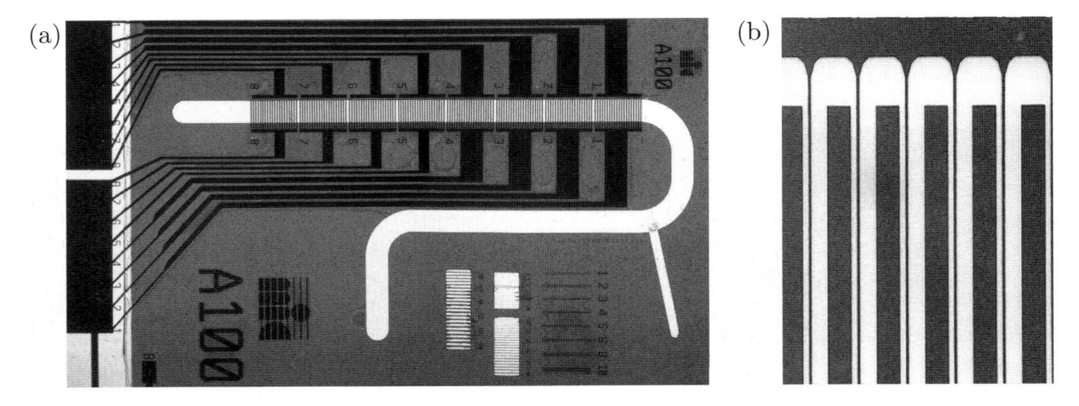

Fig. 14.4 (a) A picture of a glass-based 33 µm deep and 1 mm wide microfluidic channel (white U-shape) with an asymmetric array of interdigitated Ti/Pt-electrodes (black) placed in the bottom of the first half of the channel. When filled with an electrolyte and applying an AC-voltage of a few volt and a frequency of $\simeq$ 10 kHz on the electrode pads (the two black rectangles to the left), the device functions as a pump. (b) A close-up of the interdigitated electrode array showing the geometrical asymmetry: one electrode set is only 4 µm wide, while the other is 26 µm wide. Courtesy of Misha Marie Gregersen, DTU Nanotech.

$\mathbf{v}$, the pressure field $\mathbf{p}$, the two ionic density fields $c_\pm$, and the electrical potential ϕ, the relative simplicity does allow us to find an analytical solution. In such spatially symmetric systems the time average of the AC motion is zero, but if the spatial symmetry is broken it is possible to generate a non-zero average motion. For systems with a zero average flow, the AC induced-charge method can be used to create mixers of electrolytes, while a non-zero average can by exploited to design micropumps, admittedly with a low capacity and efficiency. The following example is a simplified version of a more complete treatment by Mortensen *et al.*, Phys. Rev. E **71**, 056306 (2005).

14.5.1 The microchannel with surface electrodes

We reconsider the binary electrolyte of Section 8.3 containing ions with charges $+Ze$ and $-Ze$, respectively. Like in Fig. 8.3 the electrolyte is confined to the semi-infinite space $z > 0$ by an impenetrable, homogeneous and planar wall. However, now the wall is an insulating layer with dielectric constant ϵ_s placed at $-d < z < 0$, see Fig. 14.5. The metallic electrode is attached at the backside of the insulator at $z < -d$, and it is biased at the surface $z = -d$ by a spatially modulated, external AC potential $V_\mathrm{ext}(x,t)$ given by

$$V_\mathrm{ext}(x,t) = V_0 \cos(qx)\, \mathrm{e}^{\mathrm{i}\omega t}, \tag{14.41}$$

where V_0 is the amplitude, q the wave number of the spatial modulation, and ω the driving angular frequency using complex notation as in Section 10.6.

There is complete translation invariance along the y axis, so the y co-ordinate drops out of our analysis, and all positions $\mathbf{r} = x\,\mathbf{e}_x + z\,\mathbf{e}_z$ in the following are therefore just referring to the xz-plane.

14.5.2 Non-equilibrium description

The following analysis combines the methods from Section 8.3 on the electrostatic Debye layer and Section 9.1 on the electrohydrodynamic Nernst–Planck transport equation. We

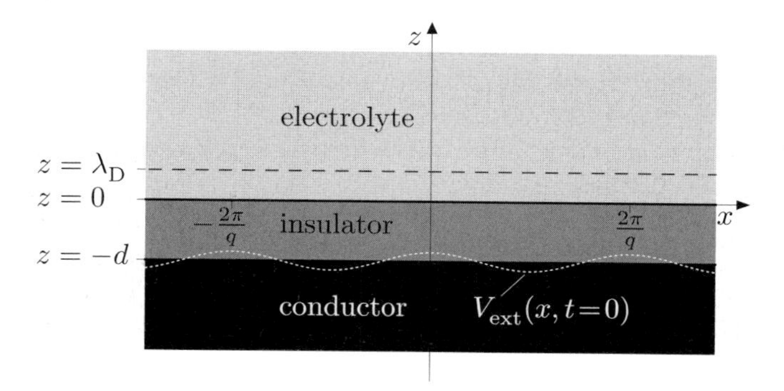

Fig. 14.5 A sketch of the induced-charge system under study. The binary electrolyte (light gray) is situated in the half-space $z > 0$. Below it, for $-d < z < 0$, is a planar wall consisting of an insulating dielectric slab of thickness d (dark gray), and below that, for $z < -d$, is a semi-infinite conductor (black). The top surface, $z = -d$, of the conductor is biased by a periodically modulated potential $V_\mathrm{ext}(x,t)$ of period $2\pi/q$ (white dotted line), which gives rise to the formation of a Debye screening layer of thickness λ_D in the electrolyte (black dashed line).

imagine that the intrinsic zeta-potential due to unpassivated surface charges on the insulator/electrolyte interface has been compensated by a corresponding DC shift, so that any non-zero potential at the $z = 0$ wall is entirely due to the applied AC potential. To simplify further, we treat the insulating layer as a simple capacitor with the surface capacitance $C_\mathrm{s} = \epsilon_\mathrm{s}/d$ per area assumed to be much larger that the Debye capacitance $C_\mathrm{D} = \epsilon/\lambda_\mathrm{D}$ of Eq. (8.40),

$$\frac{C_\mathrm{D}}{C_\mathrm{s}} = \frac{\epsilon}{\epsilon_\mathrm{s}} \frac{d}{\lambda_\mathrm{D}} \ll 1. \tag{14.42}$$

In the liquid electrolyte we consider the two ionic densities $c_\pm(\mathbf{r},t)$, the potential $\phi(\mathbf{r},t)$, the two ionic current densities (the ionic flux densities) $\mathbf{J}_\pm(\mathbf{r},t)$, the velocity field $\mathbf{v}(\mathbf{r},t)$ of the electrolyte, and the pressure $p(\mathbf{r},t)$. In the following, we suppress $(\mathbf{r},t)$ unless needed for clarity.

The number densities of the ions couple to the potential via Poisson's equation,

$$\nabla^2 \phi = -\frac{Ze}{\epsilon}\big(c_+ - c_-\big). \tag{14.43a}$$

The ionic current densities are coupled to the ionic densities by a continuity equation, which in the absence of any chemical reactions in the system is

$$\partial_t c_\pm = -\boldsymbol{\nabla} \cdot \mathbf{J}_\pm. \tag{14.43b}$$

The presence of convection or of gradients in the densities $c_\pm$ and the electric potential ϕ will generate ionic current densities $\mathbf{J}_\pm$. The Nernst–Planck equation gives these currents

$$\mathbf{J}_\pm = -D\boldsymbol{\nabla} c_\pm + c_\pm \mathbf{v} \mp \mu c_\pm \boldsymbol{\nabla}\phi, \tag{14.43c}$$

where, for simplicity, we have assumed that the two types of ions have the same diffusivity D and the same mobility μ. Note that both the diffusivity D and the electric conductivity σ are linked to the mobility μ via the Einstein relation $D = (k_\mathrm{B}T/Ze)\mu$ and $\sigma^\pm = Zec_\pm\mu$, see Exercise 5.5 and Section 8.2.2.

Finally, the velocity field and pressure of the liquid are coupled to the potential and ionic densities by the Navier–Stokes equation

$$\rho\big[\partial_t \mathbf{v} + (\mathbf{v}\cdot\boldsymbol{\nabla})\mathbf{v}\big] = -\boldsymbol{\nabla} p + \eta\nabla^2\mathbf{v} - Ze\big[c_+ - c_-\big]\boldsymbol{\nabla}\phi, \tag{14.43d}$$

where ρ is the mass density, η is the viscosity of the liquid, and p is the pressure. Furthermore, treating the electrolyte as an incompressible fluid we have

$$\boldsymbol{\nabla}\cdot\mathbf{v} = 0. \tag{14.43e}$$

The coupled field equations, Eqs. (14.43a) to (14.43e), fully govern the physical fields ϕ, $c_\pm$, $\mathbf{J}_\pm$, $\mathbf{v}$, and p.

We now turn to the boundary conditions of the fields, beginning with the potential. Assuming a vanishing intrinsic zeta-potential and noting that in the extreme limit $C_\mathrm{D}/C_\mathrm{s} \to 0$

For $q\lambda_D \ll 1$ this means that $\omega_c = \frac{\eta}{\rho}\, q^2$. In this way, for small Reynolds numbers, we get the linearized Navier–Stokes equation

$$\mathbf{0} = -\boldsymbol{\nabla} p + \eta\nabla^2\mathbf{v} + \mathbf{f}, \quad \omega \ll \omega_c, \tag{14.65}$$

which is the resulting quasi-steady flow problem, linear in the velocity field. Equation (14.65) with Eq. (14.63a) can be solved exactly, but in the following we restrict ourselves to an approximate solution based on the separation of length scales, $\lambda_D \ll q^{-1}$.

From Eq. (14.63a) we note that the electrical body force $\mathbf{f}$ Eq. (14.63a) decays in the z direction over the short length scale λ_D, while along the x direction it varies on the much longer length scale q^{-1}. Moreover, the no-slip condition forces the velocity $\mathbf{v}$ to be zero at the wall, but due to shear flow it is much easier for the parallel component v_x to acquire a significant value within the Debye layer than for the perpendicular component v_z. Thus, $v_z \approx 0$ for $0 < z < 3\lambda_D$, and the z component of the Navier–Stokes equation becomes $-\partial_z p + f_z = 0$ or

$$p(x,z) = \lambda_D F_0\, 2\cos^2(qx)\, e^{-z/\lambda_D}. \tag{14.66}$$

From this we easily find the x component of $\boldsymbol{\nabla} p$,

$$\partial_x p = q\lambda_D F_0\, 4\cos(qx)\sin(qx)\, e^{-z/\lambda_D} = 2\lambda_D q\, f_x \ll f_x. \tag{14.67}$$

So the x component of the Navier–Stokes equation can be approximated by

$$\eta(\partial_x^2 + \partial_z^2)v_x = -f_x = -F_0\, \sin(2qx)\, e^{-z/\lambda_D}. \tag{14.68}$$

From the functional dependencies we see that $|\partial_z^2 v_x| \approx |v_x|/\lambda_D^2$ and $|\partial_z^2 v_x| \approx |v_x|\, q^2$, and since $q\lambda_D \ll 1$ the $\partial_x^2 v_x$ term can be neglected. So by straightforward integration twice after z and taking the boundary condition $v_x(x,0,t) = 0$ into account we arrive at

$$v_x(x,z) = \frac{\lambda_D^2}{\eta}\, F_0\, \sin(2qx)\left(1 - e^{-z/\lambda_D}\right), \quad \text{for } 0 < z \lesssim 3\lambda_D. \tag{14.69}$$

In analogy with the EO slip velocity introduced in Eqs. (9.10) and (9.11), we now define the slip velocity $v_s(x,t)$ for the induced-charge flow as the limit at infinity of $v_x(x,\infty,t)$,

$$v_s(x,t) \equiv v_1\, \frac{\cos(2\,\omega\, t + \varphi)}{\frac{\omega}{\omega^*} + \frac{\omega^*}{\omega}}\, \sin(2\, q\, x). \tag{14.70}$$

We now leave the narrow Debye layer and study the flow in the bulk beyond the $3\lambda_D$-layer. Here, according to Eq. (14.63a) the body force vanishes, and the Navier–Stokes equation (14.65) becomes the Stokes equation (2.41) containing only the pressure and viscosity term. The system of equations to be solved can be summarized as

$$(\partial_x^2 + \partial_z^2)\mathbf{v} = \frac{1}{\eta}\, \boldsymbol{\nabla} p, \tag{14.71a}$$

$$v_x(x,0,t) = v_1\, \frac{\cos(2\,\omega\, t + \varphi)}{\frac{\omega}{\omega^*} + \frac{\omega^*}{\omega}}\, \sin(2\, q\, x), \tag{14.71b}$$

$$\boldsymbol{\nabla}\cdot\mathbf{v} = 0. \tag{14.71c}$$

Taking the divergence of the Stokes equation (14.71a) and using the incompressibility condition $\boldsymbol{\nabla}\cdot\mathbf{v} = 0$ we arrive at a Laplace equation for the pressure, as in Eq. (2.45),

$$\nabla^2 p = 0. \tag{14.72}$$

The boundary condition Eq. (14.71b) for v_x contains a $\sin(2qx)$ factor, so it is natural to seek a solution for p proportional to $\cos(2qx)$ that has the right x derivative. The Laplace equation Eq. (14.72) then forces the z dependence e^{-2qz}, so $p(x,z) = P_0\,\mathrm{e}^{-2qz}\cos(2qx)$ and thus

$$(\partial_x^2 + \partial_z^2)\mathbf{v} = \frac{-2qP_0}{\eta}\,\mathrm{e}^{-2qz}\Big(\sin(2qx)\,\mathbf{e}_x + \cos(2qx)\,\mathbf{e}_z\Big). \tag{14.73}$$

Finally, using the trial solutions $v_z = G(z)\mathrm{e}^{-2qz}\cos(2qx)$ and $v_x = H(z)\mathrm{e}^{-2qz}\sin(2qx)$ we find the solutions for v_x, v_z, and p,

$$v_x(x,z,t) = +v_1\,\frac{\cos(2\,\omega\,t+\varphi)}{\frac{\omega}{\omega^*}+\frac{\omega^*}{\omega}}\,\mathrm{e}^{-2qz}\,(1-2qz)\sin(2qx), \tag{14.74a}$$

$$v_z(x,z,t) = -v_1\,\frac{\cos(2\,\omega\,t+\varphi)}{\frac{\omega}{\omega^*}+\frac{\omega^*}{\omega}}\,\mathrm{e}^{-2qz}\,2qz\cos(2qx), \tag{14.74b}$$

$$p(x,z,t) = -4q\eta v_1\,\frac{\cos(2\omega t+\varphi)}{\frac{\omega}{\omega^*}+\frac{\omega^*}{\omega}}\,\mathrm{e}^{-2qz}\cos(2qx). \tag{14.74c}$$

If we now substitute into Eq. (14.43d) we get that the difference between the right- and left-hand term is proportional to $\mathrm{e}^{-z/\lambda_D} + \mathcal{O}(\omega/\omega_\mathrm{D}) + \mathcal{O}([q\lambda_D]^2)$. This shows that Eq. (14.74) is indeed an excellent approximation to the full solution of the non-linear time-dependent Navier–Stokes equation (14.43d) for $z \gtrsim 3\lambda_D$. Concerning the incompressibility constraint given by Eq. (14.43e), our solution gives $\boldsymbol{\nabla}\cdot\mathbf{v} = \mathcal{O}([q\lambda_D]^2)$.

In Fig. 14.6(b)–(d) we show plots of the pressure and velocity fields, Eq. (14.74). It is seen how a complex flow pattern involving counterrotating rolls appears, and how this pattern dies out at a distance $\simeq 1/q$ from the wall at $z = 0$ corresponding the wavelength along the x axis of the applied potential.

14.6 Exercises

Exercise 14.1
Co-ordinate transformation of differential operators
Derive the expressions in (ξ,ζ) co-ordinates for $\boldsymbol{\nabla}$ and ∇^2 given to first order in the perturbation parameter α and the aspect ratio $\epsilon = h_0/L$ in Eqs. (14.11) and (14.12), respectively.

Exercise 14.2
The co-ordinate transformation of the Navier–Stokes equation
Including terms to any order in ϵ but only to zero-order in α, derive the Navier–Stokes equation in (ξ,ζ) co-ordinates for the dimensionless velocity $(\tilde{v}_x,\tilde{v}_z)$ and pressure $\tilde{p}$ given in Eqs. (14.6) and (14.8), respectively.

Exercise 14.3
First-order solution to the shape-perturbation problem
Solve the dimensionless first-order continuity and Navier–Stokes equations Eq. (14.18), and show that the solution is given by Eq. (14.19).

Exercise 14.4
The flow rate in the first-order shape-perturbation problem
Use the expression given in Eq. (14.20a) to calculate the flow rate $Q(x)$ for the velocity field $v_x(x,z)$ at position x, and show that it is independent of x to first order in the perturbation parameter α.

Exercise 14.5
The diffusion equation for the concentration in the electrokinetic pump
In Section 14.5.3 a diffusion equation is set up for the ionic density difference $\nu = c_+ - c_-$.

(a) Derive the diffusion equation for the two densities c_+ and c_-, Eq. (14.48).

(b) Based on this and the Debye–Hückel approximation derive the diffusion equation for the density difference ν, Eq. (14.51).

(c) Insert the trial solution Eq. (14.52a) for ν into the diffusion equation Eq. (14.51) and show that the decay parameter κ is given by Eq. (14.52b).

Exercise 14.6
The solution for the electrical potential in the electrokinetic pump
In Section 14.5.3 the solution is presented for the electrical potential ϕ in the linearized regime.

(a) Why are there two decaying terms present in the solution Eq. (14.54)?

(b) Show, based on Eq. (14.53), that the solution for the potential ϕ is given by Eq. (14.54).

(c) Derive the expression given in Eq. (14.56) for the coefficient $\mathcal{C}_2$ by inserting Eqs. (14.52a) and (14.54) into Eq. (14.55).

(d) Derive the expression (14.57) for the coefficient $\mathcal{C}_1$ by combining Eqs. (14.44a), (14.54), and (14.56).

Exercise 14.7
The solution for the velocity field in the electrokinetic pump
In Section 14.5.3 the solution is presented for the velocity field $\mathbf{v} = (v_x, v_z)$ in the linearized regime. Determine the functions $G(z)$ and $H(z)$ such that the trial solutions $v_z = G(z)\mathrm{e}^{-2qz}\cos(2qx)$ and $v_x = H(z)\mathrm{e}^{-2qz}\sin(2qx)$ indeed solves both the continuity equation (14.71c) and the Navier–Stokes equation (14.73). Hint: insert the trial solutions in the equations and obtain simple ordinary differential equations for $G(z)$ and $H(z)$.

Exercise 14.8
Physical interpretation of the flow rolls in the electrokinetic pump
Discuss the physical mechanisms that lead from the applied harmonically varying potential $V_\mathrm{ext}(x,t)$ to the final induced-charge flow and pressure field depicted in Fig. 14.6.

14.7 Solutions

Solution 14.1
Co-ordinate transformation of differential operators
Applying the chain rule on the co-ordinate transformations $\xi = x/L$ and $\zeta = (1-\alpha\lambda)\,z/h_0$ given in Eq. (14.4), we obtain $\boldsymbol{\nabla}$ to first order in α,

$$\partial_x = (\partial_x\xi)\partial_\xi + (\partial_x\zeta)\partial_\zeta = (1/L)\partial_\xi - (\alpha\lambda'/L)(z/h_0)\partial_\zeta \approx (\epsilon/h_0)\,(\partial_\xi - \alpha\lambda'\zeta\,\partial_\zeta), \quad (14.75a)$$

$$\partial_z = (\partial_z\xi)\partial_\xi + (\partial_z\zeta)\partial_\zeta = 0 + (1-\alpha\lambda)(1/h_0)\partial_\zeta = (1/h_0)\,(\partial_\zeta - \alpha\lambda\,\partial_\zeta), \quad (14.75b)$$

where we have used that $\alpha z/h_0 \approx \alpha\zeta$ to first order in α, $\partial_x\lambda = \partial_\xi\lambda/L \equiv \lambda'/L$, and $\epsilon \equiv h_0/L$.

The Laplacian to first order in α and ϵ is found from the square of the nabla-operator,

$$\begin{aligned}\nabla^2 &= (\epsilon/h_0)^2(\partial_\xi - \alpha\lambda'\zeta\,\partial_\zeta)(\partial_\xi - \alpha\lambda'\zeta\,\partial_\zeta) + (1/h_0^2)\,(\partial_\zeta - \alpha\lambda\,\partial_\zeta)(\partial_\zeta - \alpha\lambda\,\partial_\zeta)\\ &\approx (1/h_0^2)\,(1-\alpha 2\lambda)\partial_\zeta^2,\end{aligned} \tag{14.76}$$

where the terms involving ∂_ξ and λ' drop out as they are second order in ϵ.

Solution 14.2

The co-ordinate transformation of the Navier–Stokes equation

To zeroth order in α and any order in ϵ the transformed differential operators are $\boldsymbol{\nabla} = (\partial_x, \partial_z) = (1/h_0)(\epsilon\partial_\xi, \partial_\zeta)$ and $\nabla^2 = (\epsilon^2/h_0^2)\partial_\xi^2 + (1/h_0^2)\partial_\zeta^2$. If we, furthermore, in analogy with ξ and ζ introduce the dimensionless time $\tau \equiv t/T_0$, the inertial terms in the co-ordinate-transformed Navier–Stokes equation to zeroth order in α become

$$\begin{aligned}\rho\big[\partial_t\mathbf{v} + (\mathbf{v}\cdot\boldsymbol{\nabla})\mathbf{v}\big] &= \rho\frac{V_0^2}{L}\begin{pmatrix}\partial_\tau\tilde{v}_x\\ \epsilon\partial_\tau\tilde{v}_z\end{pmatrix} + \rho\frac{V_0^2}{h_0}\big[\epsilon\partial_\xi\tilde{v}_x + \partial_\zeta(\epsilon\tilde{v}_z)\big]\begin{pmatrix}\tilde{v}_x\\ \epsilon\tilde{v}_z\end{pmatrix}\\ &= \rho\frac{V_0^2}{h_0}\begin{pmatrix}\epsilon\big[\partial_\tau\tilde{v}_x + (\partial_\xi\tilde{v}_x + \partial_\zeta\tilde{v}_z)\tilde{v}_x\big]\\ \epsilon^2\big[\partial_\tau\tilde{v}_z + (\partial_\xi\tilde{v}_x + \partial_\zeta\tilde{v}_z)\tilde{v}_z\big]\end{pmatrix}.\end{aligned} \tag{14.77}$$

Similarly, using $P_0 = \Delta p = 12\eta L V_0/h_0^2$ the force term in the Navier–Stokes equation becomes

$$\begin{aligned}-\boldsymbol{\nabla}p + \eta\nabla^2\mathbf{v} &= -12\frac{\eta L V_0}{h_0^2}\frac{1}{h_0}\begin{pmatrix}\epsilon\partial_\xi\tilde{p}\\ \partial_\zeta\tilde{p}\end{pmatrix} + \frac{\eta V_0}{h_0^2}\begin{pmatrix}(\epsilon^2\partial_\xi^2 + \partial_\zeta^2)\tilde{v}_x\\ (\epsilon^2\partial_\xi^2 + \partial_\zeta^2)(\epsilon\tilde{v}_z)\end{pmatrix}\\ &= \frac{\eta V_0}{h_0^2}\begin{pmatrix}-12\,\partial_\xi\tilde{p} + (\epsilon^2\partial_\xi^2 + \partial_\zeta^2)\tilde{v}_x\\ -12\,\frac{1}{\epsilon}\partial_\zeta\tilde{p} + \epsilon(\epsilon^2\partial_\xi^2 + \partial_\zeta^2)\tilde{v}_z\end{pmatrix}.\end{aligned} \tag{14.78}$$

Equating these two expressions and introducing the Reynolds number $Re \equiv \rho V_0 h_0/\eta$ lead to the expression for the co-ordinate-transformed Navier–Stokes equation to zeroth order in α and any order in ϵ,

$$\epsilon Re\begin{pmatrix}\big[\partial_\tau\tilde{v}_x + (\partial_\xi\tilde{v}_x + \partial_\zeta\tilde{v}_z)\tilde{v}_x\big]\\ \epsilon^2\big[\partial_\tau\tilde{v}_z + (\partial_\xi\tilde{v}_x + \partial_\zeta\tilde{v}_z)\tilde{v}_z\big]\end{pmatrix} = \begin{pmatrix}-12\,\partial_\xi\tilde{p} + (\epsilon^2\partial_\xi^2 + \partial_\zeta^2)\tilde{v}_x\\ -12\,\partial_\zeta\tilde{p} + (\epsilon^4\partial_\xi^2 + \epsilon^2\partial_\zeta^2)\tilde{v}_z\end{pmatrix}. \tag{14.79}$$

Note how the Reynolds number appears with at least one factor of ϵ in agreement with the less rigorous analysis in Section 4.4. We also note that the leading term governing the ξ-dependence of the pressure is ϵ^0, while for the ζ-dependence is suppressed by a factor ϵ^2, so the pressure depends only weakly on ζ for low aspect ratios.

Solution 14.3

First-order solution to the shape-perturbation problem

To solve Eq. (14.18) we begin by noticing that Eq. (14.18c) implies

$$\tilde{p}^{(1)} = \tilde{p}^{(1)}(\xi), \tag{14.80}$$

i.e. only a ξ-dependence. Inserting this result into Eq. (14.18b) leads to $\partial_\zeta^2 \tilde{v}_x^{(1)} = 12\partial_\xi \tilde{p}^{(1)}(\xi) - 24\lambda(\xi)$. So, by introducing the auxiliary function $f(\xi)$ defined by

$$f(\xi) \equiv 12\,\partial_\xi \tilde{p}^{(1)}(\xi) - 24\,\lambda(\xi), \tag{14.81}$$

we find $\tilde{v}_x^{(1)}(\xi)$ by integrating $\partial_\zeta^2 \tilde{v}_x^{(1)} = f(\xi)$,

$$\tilde{v}_x^{(1)}(\xi) = -\frac{1}{2}\, f(\xi)\, \zeta(1-\zeta), \tag{14.82}$$

which obeys the no-slip conditions at $\zeta = 0$ and $\zeta = 1$. Utilizing these results in Eq. (14.18a) we obtain an equation for $\tilde{v}_z^{(1)}$,

$$\partial_\zeta \tilde{v}_z^{(1)} = \frac{1}{2}\, f(\xi)\, \zeta(1-\zeta) + 6\lambda'(\xi)\, \zeta(1-2\zeta). \tag{14.83}$$

This equations simplifies significantly with the following trial solution for $f(\xi)$,

$$f(\xi) = 2A\,\lambda(\xi), \tag{14.84}$$

where A is a constant to be determined. With this Eq. (14.83) becomes

$$\partial_\zeta \tilde{v}_z^{(1)} = A\,\lambda'(\xi)\,\zeta(1-\zeta) + 6\lambda'(\xi)\,\zeta(1-2\zeta) = \big[(6+A)\zeta - (12+A)\zeta^2\big]\,\lambda'(\xi). \tag{14.85}$$

By integration and ensuring the no-slip condition $\tilde{v}_z^{(1)}(\xi, 0) = 0$ we get

$$\tilde{v}_z^{(1)}(\xi,\zeta) = \left[\frac{1}{2}\,(6+A)\zeta^2 - \frac{1}{3}\,(12+A)\zeta^3\right]\lambda'(\xi). \tag{14.86}$$

The second no-slip condition $\tilde{v}_z^{(1)}(\xi, 1) = 0$ is fulfilled if $\frac{1}{2}\,(6+A) = \frac{1}{3}\,(12+A)$ or $A = 6$. Thus, it follows from Eqs. (14.84), (14.82) and (14.86)

$$f(\xi) = 12\lambda(\xi), \tag{14.87a}$$

$$\tilde{v}_x^{(1)}(\xi,\zeta) = -6\,\lambda(\xi)\,\zeta(1-\zeta), \tag{14.87b}$$

$$\tilde{v}_z^{(1)}(\xi,\zeta) = 6\,\lambda'(\xi)\,\zeta^2(1-\zeta). \tag{14.87c}$$

The pressure follows from Eq. (14.81), $12\,\partial_\xi \tilde{p} = 36\,\lambda(\xi)$, or

$$\tilde{p}^{(1)}(\xi,\zeta) = 3\int_0^\xi \mathrm{d}s\,\lambda(s). \tag{14.87d}$$

Solution 14.4
The flow rate in the first-order shape-perturbation problem
From Eq. (14.20a) follows directly

$$\begin{aligned} Q(x) \equiv w\int_0^{h(x)} \mathrm{d}z\, v_x(x,z) &= \frac{w h_0^2 \Delta p}{2\eta L}\left[1 - \alpha\lambda\big(\tfrac{x}{L}\big)\right] h(x) \int_0^{h(x)} \frac{\mathrm{d}z}{h(x)}\,\frac{z}{h(x)}\left[1 - \frac{z}{h(x)}\right] \\ &= \frac{w h_0^2 \Delta p}{2\eta L}\left[1 - \alpha\lambda\big(\tfrac{x}{L}\big)\right] h_0\left[1 + \alpha\lambda\big(\tfrac{x}{L}\big)\right]\frac{1}{6} = \frac{w h_0^3 \Delta p}{12\eta L} + \mathcal{O}(\alpha^2). \end{aligned} \tag{14.88}$$

Solution 14.5
The diffusion equation for the concentration in the electrokinetic pump

(a) Combining the continuity and Nernst–Planck equations (14.43b) and (14.43c) gives

$$\begin{aligned}\partial_t c_\pm &= -\boldsymbol{\nabla}\cdot(-D\boldsymbol{\nabla}c_\pm + c_\pm\mathbf{v} \mp \mu c_\pm \boldsymbol{\nabla}\phi)\\ &= D\nabla^2 c_\pm - (\boldsymbol{\nabla}c_\pm)\cdot\mathbf{v} \pm \mu\boldsymbol{\nabla}\cdot(c_\pm\boldsymbol{\nabla}\phi),\end{aligned} \tag{14.89}$$

where we have utilized the incompressibility condition $\boldsymbol{\nabla}\cdot\mathbf{v} = 0$ to cancel the term $c_\pm\boldsymbol{\nabla}\cdot\mathbf{v}$.

(b) In the Debye–Hückel approximation the driving force from the potential ϕ is small and so are the responses in velocity $\mathbf{v}$ and concentration variations $c_\pm - c_0$. When discarding quadratic terms in these small quantities, Eq. (14.89) becomes $\partial_t c_\pm \approx D\nabla^2 c_\pm \pm \mu c_0 \nabla^2\phi$. Utilizing the Poisson equation (14.43a) to write $\nabla^2\phi = -Ze\nu/\epsilon$ and the Einstein relation to rewrite the mobility in terms of the diffusivity, $\mu = ZeD/k_\mathrm{B}T$ we get

$$\partial_t c_\pm = D\nabla^2 c_\pm \mp \frac{ZeD}{k_\mathrm{B}T}c_0\,\frac{Ze}{\epsilon}\,\nu. \tag{14.90}$$

Subtracting the c_--equation from the c_+-equation yields

$$\partial_t \nu = D\nabla^2\nu - D\,\frac{2(Ze)^2 c_0}{\epsilon k_\mathrm{B}T}\,\nu = D\left[\nabla^2\nu - \frac{1}{\lambda_\mathrm{D}^2}\right]\nu. \tag{14.91}$$

(c) When inserting the trial solution Eq. (14.52a), $\nu = \mathcal{C}_1 \mathrm{e}^{-\kappa z}\cos(qx)\,\mathrm{e}^{\mathrm{i}\omega t}$ into the diffusion equation (14.91), the differential operators leave the functions invariant, and we are left with an algebraic equation for the parameters, $\mathrm{i}\omega = D[(\kappa^2 - q^2) - 1/\lambda_\mathrm{D}^2]$. From this we find $(\lambda_\mathrm{D}\kappa)^2 = \mathrm{i}\lambda_\mathrm{D}^2\omega/D + (\lambda_\mathrm{D}q)^2 + 1$ or $\kappa = (1/\lambda_\mathrm{D})\sqrt{1 + (\lambda_\mathrm{D}q)^2 + \mathrm{i}\omega/\omega_\mathrm{D}}$, where $\omega_\mathrm{D} \equiv D/\lambda_\mathrm{D}^2$ is the Debye frequency.

Solution 14.6
The solution for the electrical potential in the electrokinetic pump

In Section 14.5.3 the solution is presented for the electrical potential ϕ in the linearized regime.

(a) The potential Eq. (14.54) must obey the Poisson equation, which is an inhomogeneous partial differential equation. The complete solution is the sum of a particular solution to the full equation and general solutions to the corresponding homogeneous equation, *in casu* the Laplace equation. The $\mathrm{e}^{-\kappa z}\cos(qx)$ term is a particular solution to the Poisson equation, while the $\mathrm{e}^{-qz}\cos(qx)$ term is the general solution to the Laplace equation.

(b) When acting on Eq. (14.54) with the Laplace operator we get

$$\begin{aligned}\nabla^2\phi &= \frac{Ze/\epsilon}{q^2-\kappa^2}\nabla^2\left[\left(\mathcal{C}_1\mathrm{e}^{-\kappa z} + \mathcal{C}_2\mathrm{e}^{-qz}\right)\cos(qx)\right]\mathrm{e}^{\mathrm{i}\omega t}\\ &= \frac{Ze/\epsilon}{q^2-\kappa^2}\left[(\kappa^2 - q^2)\mathcal{C}_1\mathrm{e}^{-\kappa z} + 0\right]\cos(qx)\,\mathrm{e}^{\mathrm{i}\omega t} = -\frac{Ze}{\epsilon}\mathcal{C}_1\mathrm{e}^{-\kappa z}\cos(qx)\,\mathrm{e}^{\mathrm{i}\omega t}.\end{aligned} \tag{14.92}$$

which is identical to Eq. (14.53).

(c) In the following, we omit the z-independent factor $\cos(qx)\,\mathrm{e}^{\mathrm{i}\omega t}$ that appears in all terms. From Eq. (14.52a) we get by differentiation with respect to z that $\partial_z\nu(0) = -\kappa\mathcal{C}_1$.

Likewise, Eq. (14.54) gives $(\epsilon/[Ze\lambda_{\rm D}^2])\partial_z\phi(0) = [-\kappa\mathcal{C}_1 - q\mathcal{C}_2]/[(\lambda_{\rm D}q)^2 - (\lambda_{\rm D}\kappa)^2]$. Inserting this in the zero-current boundary condition Eq. (14.55) we obtain

$$-\kappa\mathcal{C}_1 - \frac{\kappa\mathcal{C}_1 + q\mathcal{C}_2}{(\lambda_{\rm D}q)^2 - (\lambda_{\rm D}\kappa)^2} = 0. \tag{14.93}$$

Multiplying this equation by $(\lambda_{\rm D}q)^2 - (\lambda_{\rm D}\kappa)^2$ and utilizing $(\lambda_{\rm D}q)^2 - (\lambda_{\rm D}\kappa)^2 - 1 = {\rm i}\omega/\omega_{\rm D}$ from Eq. (14.52b), we arrive at Eq. (14.56).

(d) Inserting the the potential Eq. (14.54) into the boundary condition Eq. (14.44a) leads to

$$\frac{Ze/\epsilon}{q^2-\kappa^2}\,\big[\mathcal{C}_1 + \mathcal{C}_2\big] = V_0. \tag{14.94}$$

Insertion of $\mathcal{C}_2$ from Eq. (14.56) in this expression, and again utilizing $(\lambda_{\rm D}q)^2 - (\lambda_{\rm D}\kappa)^2 = 1 + {\rm i}\omega/\omega_{\rm D}$, results in the explicit expression for $\mathcal{C}_1$, Eq. (14.57).

Solution 14.7
The solution for the velocity field in the electrokinetic pump
Insertion of the trial functions $v_z = G(z){\rm e}^{-2qz}\cos(2qx)$ and $v_x = H(z){\rm e}^{-2qz}\sin(2qx)$ in the continuity equation and two-component Navier–Stokes equation results in

$$[G'(z) - 2q\,G(z)] + 2q\,H(z) = 0, \tag{14.95a}$$

$$G''(z) - 4q\,G'(z) = -\frac{2qP_0}{\eta}, \tag{14.95b}$$

$$H''(z) - 4q\,H'(z) = -\frac{2qP_0}{\eta}, \tag{14.95c}$$

after dividing the common factors out. The last two equations implies that $G(z)$ and $H(z)$ are linear in z and that they have slope $a = -P_0/(2\eta)$, but P_0 is yet to be determined. Moreover, the boundary condition for $v_x(x,z,t)$ at $z=0$, Eq. (14.70), leads to $H(0) \equiv v_s(t)$, without the $\sin(2qx)$ factor, while the vanishing of v_z at $z=0$ gives us $G(0)=0$. Thus, $H(z) = v_s(t)a\,z$ and $G(z) = -a\,z$. Inserting these boundary conditions in the continuity equation (14.95a) leads to $-a + 0 + 2q\,v_s = 0$, and consequently

$$G(z) = -v_s(t)\,2qz, \tag{14.96a}$$

$$H(z) = v_s(t)\,(1 - 2q\,z), \tag{14.96b}$$

$$P_0 = -4q\eta\,v_s(t), \tag{14.96c}$$

as is seen in the final expressions in Eq. (14.74) for the velocity field $\mathbf{v}$ and the pressure p.

Solution 14.8
Physical interpretation of the flow rolls in the electrokinetic pump
According to Eq. (14.41) the applied external potential $V_{\rm ext}(x,t)$ is harmonic with a spatial period of length $2\pi/q$, see Fig. 14.6(a). As the driving force in the electrokinetic pump is the product of the charge density and the gradient of the potential, and as both these fields are created by the applied field, the resulting flow is a second-order effect that therefore must experience a period doubling with a resulting period of length π/q.

The physical origin of the phenomena is the fact that positive ions are driven away from regions where the potential is positive, and likewise, negative ions are driven away from regions with a negative potential. The ionic motion is thus always directed from the potential antinodes at $x = n\pi/q$ towards the potential nodes located at $x = (n+1/2)\pi/q$. By viscous drag, these ions pull the liquid with them along the edge, as shown in Fig. 14.6(d).

This action, analogous to a conveyer belt, results in high pressures at the potential nodes where liquid is pushed towards and low pressures at the potential antinodes, where liquid flow away from, as shown in Fig. 14.6(b). The resulting flow pattern is therefore the oscillating flow rolls shown in Fig. 14.6(c) driven by the conveyer-belt action in the very thin Debye layer close to the wall at $x = 0$.

15 Acoustofluidics

In the previous chapters we have completely neglected the small but non-zero compressibility of liquids. However, in the study of acoustic effects, compressibility is the main feature. We use the term acoustofluidics to refer to the application of acoustic pressure fields in microfluidic systems. As the speed of sound in water at room temperature is $c_\mathrm{a} \approx 1.5 \times 10^3$ m/s, the application of ultrasound frequencies $f \gtrsim 1.5$ MHz will lead to wavelengths $\lambda \lesssim 1$ mm, which will fit into the submillimeter-sized channels and cavities in microfluidic systems. The ultrasound is typically generated by onchip, AC-biased piezo-actuators.

When the acoustic waves or radiation are propagating in liquids, the associated fast-moving and rapidly oscillating pressure and velocity fields can impart a slow non-oscillating velocity component to the liquid or to small particles suspended in the liquid. In microfluidic systems these normally quite minute effects can be of significance. Interestingly, the origin of these effects can be traced back to two hydrodynamic properties largely ignored in the preceding chapters, namely the non-linearity of the Navier–Stokes equation and the small but non-zero compressibility of ordinary liquids.

After establishing the basic equations of motion of the acoustic fields, we shall in particular study two examples of significant acoustic effects in microfluidics: acoustic streaming, where the velocity field of the entire liquid acquires an extra slowly varying component induced by the incoming acoustic waves, and acoustic radiation force, where small particles suspended in a liquid are moved by the momentum transfer from sound waves propagating in the liquid.

In contrast to electromagnetism, where the wave equation of the electromagnetic field follows directly from the basic Maxwell equations, see Section 16.1, the wave equation for acoustics is only an approximate equation derived by combining the thermodynamic equation of state expressing pressure in terms of density, the kinematic continuity equation (2.7), and the dynamic Navier–Stokes equation (2.29b). Discarding all external fields such as gravitation and electromagnetism, as well as considering only the isothermal case, these three equations form the starting point for the theory of acoustics or sound,

$$p = p(\rho), \tag{15.1a}$$

$$\partial_t \rho = -\boldsymbol{\nabla}\cdot(\rho \mathbf{v}), \tag{15.1b}$$

$$\rho\partial_t \mathbf{v} = -\boldsymbol{\nabla} p - \rho(\mathbf{v}\cdot\boldsymbol{\nabla})\mathbf{v} + \eta\nabla^2\mathbf{v} + \beta\eta\,\boldsymbol{\nabla}(\boldsymbol{\nabla}\cdot\mathbf{v}). \tag{15.1c}$$

This set of coupled non-linear, partial differential equations is notoriously difficult to solve numerically, not to mention analytically.

15.1 The acoustic-wave equation for zero viscosity

When perturbing the equilibrium pressure in a compressible fluid, pressure waves or sound can be generated. The goal of this section is to establish the wave equation for these acoustic waves. The basic concepts of wave equations are presented in Appendix E. In most cases of importance for acoustofluidics, the acoustic contributions to the pressure, density, and velocity fields, can be considered as small perturbations. One systematic way of solving the equation of motion for acoustics is therefore to employ perturbation theory. As the unperturbed state before applying any acoustic fields we take a homogeneous liquid at rest in thermal equilibrium. The zero-order terms therefore take the constant values

$$\mathbf{v}_0 = \mathbf{0}, \tag{15.2a}$$

$$\rho_0 = \rho^*, \tag{15.2b}$$

$$p_0 = p^*. \tag{15.2c}$$

According to Kelvin's circulation theorem Exercise 13.2, the velocity potential can be introduced whenever viscosity is negligible, and this is, to a good approximation, the case in acoustofluidics. Following the convention of Landau and Lifshitz (1993) and of Lighthill (2003) we introduce the velocity potential ϕ as

$$\mathbf{v} = \boldsymbol{\nabla}\phi, \quad \text{for } \eta = 0. \tag{15.3}$$

Using the inverse function $\rho(p)$ of Eq. (15.1a) the pressure gradient term can be written as

$$\frac{1}{\rho}\boldsymbol{\nabla}p = \boldsymbol{\nabla}\left[\int_{p^*}^{p} \mathrm{d}\tilde{p}\,\frac{1}{\rho(\tilde{p})}\right], \tag{15.4}$$

and since $(\mathbf{v}\cdot\boldsymbol{\nabla})\mathbf{v} = \frac{1}{2}\boldsymbol{\nabla}\big(|\boldsymbol{\nabla}\phi|^2\big)$, see Exercise 15.1, all terms in the Navier–Stokes equation (15.1c) can be written as gradients, which after integration yield the expression

$$\partial_t\phi + \frac{1}{2}\big|\boldsymbol{\nabla}\phi\big|^2 + \int_{p^*}^{p} \mathrm{d}\tilde{p}\,\frac{1}{\rho(\tilde{p})} = 0, \tag{15.5}$$

where the time-dependent integration "constant" $f(t)$ has been absorbed into the potential ϕ by the gauge transformation $\phi \to \phi - \int \mathrm{d}t\, f(t)$, which according to the definition Eq. (15.3) leaves the velocity $\mathbf{v}$ unchanged. Eq. (15.5) can be regarded as Bernoulli's theorem for unsteady potential flow of compressible fluids relating kinetic and potential pressure energy.

The acoustic-wave equation is derived by differentiating the Bernoulli equation with respect to time,

$$\partial_t^2\phi + \frac{1}{2}\partial_t\big|\boldsymbol{\nabla}\phi\big|^2 + \frac{1}{\rho}\,(\partial_\rho p)\,\partial_t\rho = 0, \tag{15.6}$$

and expressing the pressure-density term by ϕ through the continuity equation (15.1b). Inserting $\mathbf{v} = \boldsymbol{\nabla}\phi$ the latter becomes

$$\partial_t\rho = -(\boldsymbol{\nabla}\rho)\cdot(\boldsymbol{\nabla}\phi) - \rho\nabla^2\phi, \tag{15.7}$$

which upon substitution into Eq. (15.6) yields

$$\partial_t^2 \phi + \frac{1}{2}\partial_t \big|\boldsymbol{\nabla}\phi\big|^2 - \frac{1}{\rho}\,(\partial_\rho p)\,(\boldsymbol{\nabla}\rho)\cdot(\boldsymbol{\nabla}\phi) - (\partial_\rho p)\,\nabla^2\phi = 0. \tag{15.8}$$

The third term on the left-hand side can be rewritten using the Navier–Stokes equation (15.1c) as follows,

$$-\frac{1}{\rho}\,(\partial_\rho p)\,(\boldsymbol{\nabla}\rho) = -\frac{1}{\rho}\,(\boldsymbol{\nabla}p) = \partial_t \mathbf{v} + (\mathbf{v}\cdot\boldsymbol{\nabla})\mathbf{v} = \partial_t(\boldsymbol{\nabla}\phi) + \frac{1}{2}\boldsymbol{\nabla}\big|\boldsymbol{\nabla}\phi\big|^2. \tag{15.9}$$

We note that to lowest (first) order in the acoustic fields the pre-factor $(\partial_\rho p)/\rho$ is replaced by the unperturbed value $(\partial_\rho p)_0/\rho_0$, while the second-order term $|\nabla\phi|^2$ is neglected and Eq. (15.9) integrates to

$$\frac{1}{\rho_0}\,(\partial_\rho p)_0\,(\rho - \rho_0) \approx -\partial_t\phi. \tag{15.10}$$

With this we can rewrite the fourth term on the left-hand side in Eq. (15.8) by a Taylor expansion around the unperturbed state,

$$(\partial_\rho p)\nabla^2\phi \approx \Big[(\partial_\rho p)_0 + (\partial_\rho^2 p)_0\,(\rho - \rho_0)\Big]\nabla^2\phi \approx \left[(\partial_\rho p)_0 - \frac{\rho_0(\partial_\rho^2 p)_0}{(\partial_\rho p)_0}\,\partial_t\phi\right]\nabla^2\phi. \tag{15.11}$$

Before writing down the final wave equation we introduce some nomenclature. The first-order derivative $(\partial_\rho p)_0$ has the dimension of a velocity c_a squared,

$$c_\mathrm{a}^2 \equiv (\partial_\rho p)_0 = \left(\frac{\partial p}{\partial \rho}\right)_{s,\rho=\rho^*}. \tag{15.12}$$

This velocity, which in a more thorough thermodynamic treatment turns out to be the isentropic derivative of the pressure hence the subscript "s", will shortly be shown to be the speed of sound in the liquid; the subscript "a" refers to "acoustics". The dimensionless pre-factor $\rho_0(\partial_\rho^2 p)_0/(\partial_\rho p)_0$ is known as the non-linear parameter $\gamma^* - 1$ of the fluid,

$$\frac{\rho_0(\partial_\rho^2 p)_0}{(\partial_\rho p)_0} = \frac{\rho_0\,\partial_\rho(c_\mathrm{a}^2)}{c_\mathrm{a}^2} \equiv \gamma^* - 1. \tag{15.13}$$

In the special case of the ideal gas, γ^* equals the ratio $\gamma = c_\mathrm{p}/c_\mathrm{v}$ of specific heats. This is seen by using the well-known adiabatic pressure-density relation, $p = p_0\rho^\gamma$, which gives $c_\mathrm{a}^2 = \gamma p_0/\rho_0$ and $\rho_0(\partial_\rho^2 p)_0/(\partial_\rho p)_0 = \gamma - 1$.

With the results obtained in Eqs. (15.9)–(15.13) we can rewrite the wave equation Eq. (15.8) and obtain the final acoustic-wave equation for inviscid fluids correct to second order in the velocity potential ϕ,

$$c_\mathrm{a}^2\,\nabla^2\phi - \partial_t^2\phi = \partial_t\big|\boldsymbol{\nabla}\phi\big|^2 + (\gamma^* - 1)\,(\partial_t\phi)\,\nabla^2\phi. \tag{15.14}$$

This non-linear wave equation is difficult to solve analytically, but as mentioned above, the acoustic fields in microfluidics can often be treated as perturbations. We therefore write the velocity potential as a perturbation series, Eq. (1.37), with an implicit perturbation parameter α, to be determined later, relating to the amplitude of the acoustic field,

$$\phi = 0 + \phi_1 + \phi_2. \tag{15.15}$$

It will be shown later that the perturbation parameter is the ratio between the oscillation speed of the walls and the speed of sound, $\alpha = \omega\ell/c_\mathrm{a}$, see Eq. (15.53). The first- and second-order equations corresponding to Eq. (15.14) are easily derived,

$$c_\mathrm{a}^2\,\nabla^2\phi_1 - \partial_t^2\phi_1 = 0, \tag{15.16a}$$

$$c_\mathrm{a}^2\,\nabla^2\phi_2 - \partial_t^2\phi_2 = \partial_t\left[\left|\boldsymbol{\nabla}\phi_1\right|^2 + \frac{\gamma^* - 1}{2c_\mathrm{a}^2}\left(\partial_t\phi_1\right)^2\right], \tag{15.16b}$$

where in the last equation second-order terms have been rewritten using the first-order expression Eq. (15.16a). The particular form of Eq. (15.16a) is recognized as a standard wave equation. This class of equations are briefly treated in Appendix E, and it follows from Eqs. (E.1) and (E.2) that c_a must be interpreted as the propagation velocity of acoustic fields, or simply: c_a is the speed of sound.

15.2 Acoustic waves in first-order perturbation theory

In the following, we study some of the basic features of acoustic waves in first-order perturbation theory, initially for non-viscous fluids and later for real fluids having a non-zero viscosity. The basic quantity is the velocity potential ϕ, which to first order is written as $\phi = \phi_1$ and governed by the simple wave equation (15.16a). The related velocity, density and pressure fields are given by Eqs. (15.3) and (15.10)

$$\mathbf{v} = \mathbf{0} + \mathbf{v}_1 = \boldsymbol{\nabla}\phi_1, \tag{15.17a}$$

$$\rho = \rho_0 + \rho_1 = \rho_0 - \frac{\rho_0}{c_\mathrm{a}^2}\,\partial_t\phi_1, \tag{15.17b}$$

$$p = p_0 + p_1 = p_0 + c_\mathrm{a}^2\rho_1 = p_0 - \rho_0\,\partial_t\phi_1, \tag{15.17c}$$

while the corresponding first-order part of the continuity equation (15.1b) and the Navier–Stokes equation (15.1c) for zero viscosity are

$$\partial_t\rho_1 = -\rho_0\boldsymbol{\nabla}\cdot\mathbf{v}_1, \tag{15.18a}$$

$$\rho_0\partial_t\mathbf{v}_1 = -c_\mathrm{a}^2\boldsymbol{\nabla}\rho_1. \tag{15.18b}$$

One class of simple solutions to the wave equation are plane waves of constant amplitude ϕ_0 propagating along a given constant wavevector $\mathbf{k} = k\,\mathbf{e}_k$ with angular frequency ω,

$$\phi_1(\mathbf{r}, t) = \phi_0\,\mathrm{e}^{\mathrm{i}(\mathbf{k}\cdot\mathbf{r} - \omega t)}. \tag{15.19}$$

By direct substitution into Eq. (15.16a) we see that this is a solution to the wave equation if $\omega^2 = c_\mathrm{a}^2 k^2$, or in other words the linear acoustic dispersion relation is fulfilled,

$$\omega = c_\mathrm{a}k. \tag{15.20}$$

In terms of frequency $f = \omega/(2\pi)$ and wavelength $\lambda = 2\pi/k$ we have

$$f = c_\mathrm{a}\,\frac{1}{\lambda}. \tag{15.21}$$

Another class of simple solutions to the wave equation are the standing waves,

$$\phi(\mathbf{r},t) = \phi_k(\mathbf{r})\,\mathrm{e}^{-\mathrm{i}\omega t}. \tag{15.22}$$

Inserting this in the wave equation (15.16a) leads to the Helmholtz equation,

$$\nabla^2\phi_k(\mathbf{r}) = -\frac{\omega^2}{c_\mathrm{a}^2}\,\phi_k(\mathbf{r}) = -k^2\phi_k(\mathbf{r}), \tag{15.23}$$

which is an eigenvalue problem that for given boundary conditions allows only certain values of the wavevector k or frequency ω, and that results in the so-called eigenmodes or resonance modes $\phi_k(\mathbf{r})$.

The acoustic waves carry energy, intensity and momentum. The total energy density E_ac associated with the sound wave can be divided into a kinetic and a potential energy density, E_kin and E_pot, respectively. To lowest order, which in fact is second order, of the acoustic fields, we have

$$E_\mathrm{kin} = \frac{1}{2}\rho_0 v_1^2 = \frac{1}{2}\rho_0\big(\boldsymbol{\nabla}\phi\big)^2. \tag{15.24}$$

To calculate the potential energy we note that when changing the volume by $\mathrm{d}V$ the acoustic pressure work $p_1\mathrm{d}V$ is stored as the potential energy density $\mathrm{d}E_\mathrm{pot} = -(p_1\mathrm{d}V)/V$. By integration of these contributions we therefore get to lowest (second) order,

$$E_\mathrm{pot} = -\int_{V_0}^{V}\mathrm{d}V\,\frac{p_1}{V} = -\int_{\rho_0}^{\rho}\mathrm{d}(\rho^{-1})\,\frac{c_\mathrm{a}^2\rho_1}{\rho^{-1}} = \int_0^{\rho_1}\mathrm{d}(\rho_1)\,\frac{c_\mathrm{a}^2\rho_1}{\rho} \approx \frac{1}{2}\,\frac{c_\mathrm{a}^2}{\rho_0}\rho_1^2 = \frac{1}{2}\rho_0\Big(\frac{1}{c_\mathrm{a}}\,\partial_t\phi\Big)^2. \tag{15.25}$$

The total energy density in the acoustic field is therefore

$$E_\mathrm{ac} = E_\mathrm{kin} + E_\mathrm{pot} = \frac{1}{2}\rho_0\Big[\big(\boldsymbol{\nabla}\phi\big)^2 + \Big(\frac{1}{c_\mathrm{a}}\,\partial_t\phi\Big)^2\Big]. \tag{15.26}$$

Considering non-viscous fluids the acoustic waves are undamped, and the energy density must therefore fulfill a continuity equation,

$$\partial_t E_\mathrm{ac} = -\boldsymbol{\nabla}\cdot\mathbf{J}_E. \tag{15.27}$$

This expression can be used to determine the acoustic energy current density $\mathbf{J}_E$, also known as the acoustic intensity I_ac. Taking the time derivative of E_ac and using the wave equation for ϕ to establish the second equality lead to

$$-\boldsymbol{\nabla}\cdot\mathbf{J}_E = \rho_0\Big[\boldsymbol{\nabla}\phi\cdot\boldsymbol{\nabla}(\partial_t\phi) + \frac{1}{c_\mathrm{a}^2}\partial_t\phi\,\partial_t^2\phi\Big] = \rho_0\Big[\boldsymbol{\nabla}\phi\cdot\boldsymbol{\nabla}(\partial_t\phi) + \partial_t\phi(\nabla^2\phi)\Big] = \rho_0\boldsymbol{\nabla}\cdot\Big[(\boldsymbol{\nabla}\phi)(\partial_t\phi)\Big], \tag{15.28}$$

so the acoustic energy current density $\mathbf{J}_E$ becomes

$$\mathbf{J}_E = -\rho_0\,(\boldsymbol{\nabla}\phi)(\partial_t\phi). \tag{15.29}$$

Likewise, we can determine the acoustic momentum current density tensor $\mathbb{J}_m$ from the acoustic momentum density $\mathbf{m}$, which to lowest (first) order is given by

$$\mathbf{m} = \rho_0 \mathbf{v}_1. \tag{15.30}$$

The continuity equation for the momentum density tensor becomes

$$\boldsymbol{\nabla} \cdot \mathbb{J}_m = -\rho_0 \partial_t \mathbf{v}_1 = -c_\mathrm{a}^2 \boldsymbol{\nabla} \rho_1 - \boldsymbol{\nabla} \cdot \left(c_\mathrm{a}^2 \rho_1 \mathbb{I}\right), \tag{15.31}$$

and thus

$$\left(\mathbb{J}_m\right)_{ij} = c_\mathrm{a}^2 \rho_1 \delta_{ij} = p_1 \delta_{ij}. \tag{15.32}$$

15.3 Viscous damping of first-order acoustic waves

Hitherto, we have neglected viscous damping, but by maintaining non-zero viscosities in Eq. (15.1c), the first-order part of the Navier–Stokes equation (15.18b) is changed into

$$\rho_0 \partial_t \mathbf{v}_1 = -c_\mathrm{a}^2 \boldsymbol{\nabla} \rho_1 + \eta \nabla^2 \mathbf{v}_1 + \beta\eta\, \boldsymbol{\nabla}(\boldsymbol{\nabla} \cdot \mathbf{v}_1). \tag{15.33}$$

By taking the divergence of this equation and interchanging the order of differentiation in various terms, we obtain

$$\rho_0 \partial_t (\boldsymbol{\nabla} \cdot \mathbf{v}_1) = -c_\mathrm{a}^2 \nabla^2 \rho_1 + (1+\beta)\eta\, \nabla^2 (\boldsymbol{\nabla} \cdot \mathbf{v}_1). \tag{15.34}$$

From this a wave equation for ρ_1 is derived by use of the first-order continuity equation $\rho_0 \nabla \cdot \mathbf{v}_1 = -\partial_t \rho_1$,

$$-\partial_t^2 \rho_1 = -c_\mathrm{a}^2 \nabla^2 \rho_1 - \frac{(1+\beta)\eta}{\rho_0}\, \nabla^2 (\partial_t \rho_1). \tag{15.35}$$

We seek solutions with a harmonic time dependence,

$$\rho_1 = \rho_1(\mathbf{r})\, \mathrm{e}^{-\mathrm{i}\omega t}, \tag{15.36}$$

and arrive at a wave equation for the first-order density field ρ_1,

$$\omega^2 \rho_1 = -c_\mathrm{a}^2 \left[1 - \mathrm{i}\frac{(1+\beta)\eta\omega}{\rho_0 c_\mathrm{a}^2}\right] \nabla^2 \rho_1 = -\big[1 - \mathrm{i}\, 2\gamma\big] c_\mathrm{a}^2 \nabla^2 \rho_1. \tag{15.37}$$

Here, we have introduced the important acoustic damping factor γ defined by

$$\gamma(\omega) \equiv \frac{(1+\beta)\eta\,\omega}{2\rho_0 c_\mathrm{a}^2}. \tag{15.38}$$

For ultrasound waves used in acoustofluidics with $\omega \approx 10^7\ \mathrm{s}^{-1}$ in water the value of γ is quite small,

$$\gamma \approx \frac{2 \times 10^{-3}\ \mathrm{Pa\,s} \times 10^6\ \mathrm{s}^{-1}}{2 \times \left(10^3\ \mathrm{m\,s}^{-1}\right)^2 \times 10^3\ \mathrm{kg\,m}^{-3}} = 10^{-5} \ll 1. \tag{15.39}$$

The smallness of γ makes the approximation $[1 - \mathrm{i}\, 2\gamma] \approx [1 + \mathrm{i}\gamma]^{-2}$ quite accurate, so we can isolate $\nabla^2 \rho_1$ in Eq. (15.37) and obtain the final version of the wave equation for ρ_1,

Table 15.1 Typical parameter values for ultrasound in aqueous acoustofluidic systems.

frequency	f	1.0×10^6 Hz	wavelength	λ	1.5×10^{-3} m
angular frequency	ω	6.3×10^6 rad/s	wave number	k_0	4.2×10^3 m^{-1}
speed of sound	c_a	1.5×10^3 m/s	damping length	x_c	8.2×10^1 m
density	ρ_0	1.0×10^3 kg m^{-3}	damping coeff.	γ	3.0×10^{-6}

$$\nabla^2 \rho_1 = -\left[\frac{\omega}{c_\mathrm{a}}\,(1+\mathrm{i}\gamma)\right]^2 \rho_1 = -\left[k_0\,(1+\mathrm{i}\gamma)\right]^2 \rho_1 = -k^2\,\rho_1, \tag{15.40}$$

i.e. a Helmholtz equation for damped waves, defined in Eq. (E.22) in Appendix E, with a complex-valued wave number $k = k_0(1+\mathrm{i}\gamma)$, a real-valued wave number k_0 and the characteristic damping length x_c, see Eq. (E.17), given by

$$k_0 \equiv \frac{\omega}{c_\mathrm{a}} \tag{15.41a}$$

$$x_\mathrm{c} \equiv \frac{1}{\gamma k_0}. \tag{15.41b}$$

The wave equation for the first-order velocity field $\mathbf{v}_1$ is in general not simple to establish. However, in the special case of zero rotation, Eq. (1.27b) implies a simplification,

$$\nabla^2 \mathbf{v}_1 = \boldsymbol{\nabla}(\boldsymbol{\nabla}\cdot\mathbf{v}_1), \quad \text{for } \boldsymbol{\nabla}\times\mathbf{v}_1 \equiv \mathbf{0}. \tag{15.42}$$

This, together with the harmonic time dependence, Eq. (15.36), and the first-order harmonic continuity equation, $\boldsymbol{\nabla}\cdot\mathbf{v}_1 = -(\partial_t\rho_1)/\rho_0 = \mathrm{i}\omega\rho_1/\rho_0$, transforms the first-order Navier–Stokes equation (15.33) into

$$-\mathrm{i}\omega\rho_0\mathbf{v}_1 = -c_\mathrm{a}^2\boldsymbol{\nabla}\rho_1 + (1+\beta)\eta\,\boldsymbol{\nabla}\left[\mathrm{i}\frac{\omega}{\rho_0}\,\rho_1\right] = -c_\mathrm{a}^2(1+\mathrm{i}\gamma)^{-2}\boldsymbol{\nabla}\rho_1. \tag{15.43}$$

Here, $\mathbf{v}_1$ is proportional to $\boldsymbol{\nabla}\rho_1$, so by defining the potential ϕ_1 as

$$\phi_1(\mathbf{r},t) = -\mathrm{i}\frac{c_\mathrm{a}^2}{\omega\rho_0(1+\mathrm{i}\gamma)^2}\,\rho_1(\mathbf{r})\,\mathrm{e}^{-\mathrm{i}\omega t}, \tag{15.44}$$

we obtain the generalization of Eqs. (15.17a) and (15.17b) to the viscous case,

$$\mathbf{v}_1 = \boldsymbol{\nabla}\phi_1, \tag{15.45a}$$

$$\rho_1 = -\frac{\rho_0}{c_\mathrm{a}^2}\,(1+\mathrm{i}\gamma)^2\,\partial_t\phi_1 = -\frac{\rho_0 k^2}{\omega^2}\,\partial_t\phi_1. \tag{15.45b}$$

Since ϕ_1 is proportional to ρ_1 it is not surprising that ϕ_1 obeys the same wave equation as ρ_1. In Exercise 15.2 is shown how to derive this wave equation from Eqs. (15.45a) and (15.45b),

$$\nabla^2\phi_1 = \frac{(1+\mathrm{i}\gamma)^2}{c_\mathrm{a}^2}\,\partial_t^2\phi_1 = -\left[\frac{\omega}{c_\mathrm{a}}\,(1+\mathrm{i}\gamma)\right]^2\phi_1 = -k^2\,\phi_1. \tag{15.46}$$

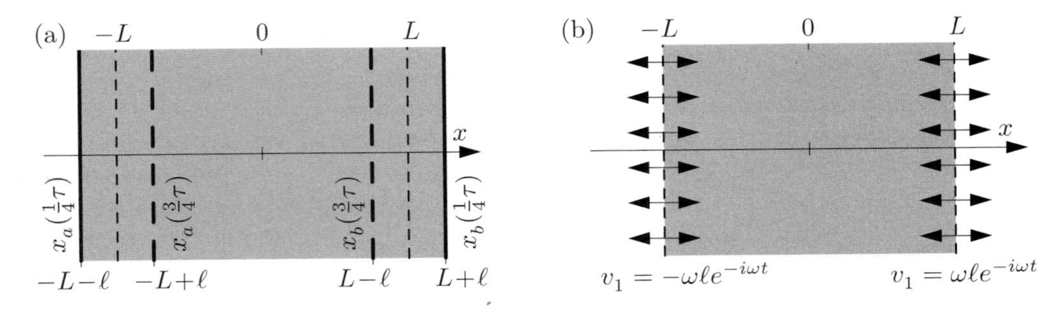

Fig. 15.1 (a) A liquid slab (dark gray) between two parallel, harmonically oscillating, planar walls at $x_a(t)$ and $x_b(t)$, respectively. At $t = 0$ the walls are at the equilibrium positions $x_a(0) = -L$ and $x_b(0) = L$. The amplitude of the oscillation is ℓ and the frequency is $f = 1/\tau$. (b) As the amplitude is minute, $\ell \ll L$, the wall positions are considered fixed, while the first-order velocity $v_1(t)$ at the walls is changing harmonically, $v_1(t) = \pm\omega\ell\, \mathrm{e}^{-\mathrm{i}\omega t}$.

15.4 Acoustic resonances

In acoustofluidics the acoustic fields are often imposed by placing the microfluidic chip on top of a piezoelectric ceramic element acting as an actuator. When the piezo-actuator is electrically biased with an AC voltage it vibrates and emits acoustic waves. As seen in Table 15.1 a vibration frequency of $f \approx 10^6$ Hz leads to ultrasound waves with a wavelength $\lambda \approx 1.5$ mm in water. When unloaded the amplitude of an ultrasound piezo-actuator is typically $\ell \approx 1$ nm.

As a simple example of acoustic actuation in microfluidics, we consider the 1D setup sketched in Fig. 15.1(a). Two planar walls both parallel to the yz-plane are placed at $x = -L$ and $x = L$, respectively. The space between is filled with a liquid, say water. The walls are forced to oscillate in antiphase at a frequency $f = 1/\tau \approx 1$ MHz and with an amplitude $\ell \approx 1$ nm. To simplify the calculation we neglect the actual tiny displacement of the walls and instead model the oscillation by the velocity boundary condition sketched in Fig. 15.1(a),

$$v_1(-L,t) = -\omega\ell\, \mathrm{e}^{-\mathrm{i}\omega t}, \qquad v_1(+L,t) = +\omega\ell\, \mathrm{e}^{-\mathrm{i}\omega t}. \tag{15.47}$$

We solve for the first-order velocity potential ϕ_1, and seek a solution to the wave equation (15.46) on the form of a superposition of a pair of counterpropagating plane waves with a complex wave number $k = k_0(1 + \mathrm{i}\gamma)$,

$$\phi_1(x,t) = \Big[A\mathrm{e}^{\mathrm{i}kx} + B\mathrm{e}^{-\mathrm{i}kx}\Big]\, \mathrm{e}^{-\mathrm{i}\omega t}. \tag{15.48}$$

The corresponding first-order velocity and density are

$$v_1(x,t) = \partial_x\phi_1(x,t) \quad = \mathrm{i}k\Big[A\mathrm{e}^{\mathrm{i}kx} - B\mathrm{e}^{-\mathrm{i}kx}\Big]\, \mathrm{e}^{-\mathrm{i}\omega t}, \tag{15.49a}$$

$$\rho_1(x,t) = -\frac{\rho_0 k^2}{\omega^2}\,\partial_t\phi_1 = \mathrm{i}\frac{\rho_0 k^2}{\omega}\,\Big[A\mathrm{e}^{\mathrm{i}kx} + B\mathrm{e}^{-\mathrm{i}kx}\Big]\, \mathrm{e}^{-\mathrm{i}\omega t}. \tag{15.49b}$$

The antisymmetric boundary condition on v_1 in Eq. (15.47) combined with Eq. (15.49a) leads to $A = B$, as well as the magnitude for the coefficients,

$$A = B = \frac{-\omega\ell}{2k\sin(kL)}, \tag{15.50}$$

which then leads to the following expressions for ϕ_1, v_1 and ρ_1,

$$\phi_1(x,t) = -\frac{\omega\ell}{k}\frac{\cos(kx)}{\sin(kL)}\,\mathrm{e}^{-\mathrm{i}\omega t} \approx -c_\mathrm{a}\ell\,\frac{\cos(k_0x) - \mathrm{i}\gamma k_0x\sin(k_0x)}{\sin(k_0L) + \mathrm{i}\gamma k_0L\cos(k_0L)}\,\mathrm{e}^{-\mathrm{i}\omega t}, \tag{15.51a}$$

$$v_1(x,t) = \omega\ell\,\frac{\sin(kx)}{\sin(kL)}\,\mathrm{e}^{-\mathrm{i}\omega t} \approx \omega\ell\,\frac{\sin(k_0x) + \mathrm{i}\gamma k_0x\cos(k_0x)}{\sin(k_0L) + \mathrm{i}\gamma k_0L\cos(k_0L)}\,\mathrm{e}^{-\mathrm{i}\omega t}, \tag{15.51b}$$

$$\rho_1(x,t) = -\mathrm{i}\rho_0k\ell\,\frac{\cos(kx)}{\sin(kL)}\,\mathrm{e}^{-\mathrm{i}\omega t} \approx -\mathrm{i}\rho_0k_0\ell\,\frac{\cos(k_0x) - \mathrm{i}\gamma k_0x\sin(k_0x)}{\sin(k_0L) + \mathrm{i}\gamma k_0L\cos(k_0L)}\,\mathrm{e}^{-\mathrm{i}\omega t}. \tag{15.51c}$$

We have utilized that $\gamma k_0L \ll 1$ to make Taylor expansions in k around k_0, and we note that when k_0L differs sufficiently from integer multiples of π, i.e. $\gamma \ll |k_0L - n\pi|$, then the imaginary parts of the denominators can be neglected. This corresponds to offresonance, and in this case the order of magnitude of the fields are given by their pre-factors,

$$|v_1(x,t)| \approx \omega\ell = \frac{\omega\ell}{c_\mathrm{a}}\,c_\mathrm{a}, \quad \text{(offresonance)}, \tag{15.52a}$$

$$|\rho_1(x,t)| \approx \rho_0k_0\ell = \frac{\omega\ell}{c_\mathrm{a}}\,\rho_0, \quad \text{(offresonance)}. \tag{15.52b}$$

The implicit perturbation parameter α can now be read off as the pre-factor $\omega\ell/c_\mathrm{a}$,

$$\alpha \equiv \frac{\omega\ell}{c_\mathrm{a}} \approx 10^{-6}, \tag{15.53}$$

where the value is calculated from the previously quoted parameter values, $\omega \approx 10^6$ rad/s, $\ell \approx 1$ nm and $c_\mathrm{a} \approx 10^3$ m/s.

More interesting perhaps is the acoustic resonances, where the acoustic fields acquire particularly large amplitudes, and where a large amount of energy is stored in the acoustic fields, see Fig. 15.2. Theoretically, the resonances are identified by the minima in the denominators of the fields in Eq. (15.51), i.e. for $\sin(k_0L) = 0$ or $k_0L = n\pi$, $n = 1, 2, 3, \ldots$,

$$k_0 = k_n \equiv n\,\frac{\pi}{L}, \quad n = 1, 2, 3, \ldots \quad \text{(resonance condition)}. \tag{15.54}$$

In practice, the resonance is achieved by tuning the frequency ω to one of the resonance frequencies ω_n,

$$\omega = \omega_n \equiv c_\mathrm{a}k_n = n\,\frac{\pi c_\mathrm{a}}{L}, \quad n = 1, 2, 3, \ldots \quad \text{(resonance frequency)}. \tag{15.55}$$

At resonance $\sin(k_0L) = 0$ and $\cos(k_0L) = 1$, so the acoustic fields become

$$\phi_1(x,t) \approx c_\mathrm{a}\ell\left[\frac{\mathrm{i}}{n\pi\gamma}\cos\left(n\pi\frac{x}{L}\right) + \frac{x}{L}\sin\left(n\pi\frac{x}{L}\right)\right]\mathrm{e}^{-\mathrm{i}\omega t}, \tag{15.56a}$$

$$v_1(x,t) \approx \omega\ell\left[\frac{-\mathrm{i}}{n\pi\gamma}\sin\left(n\pi\frac{x}{L}\right) + \frac{x}{L}\cos\left(n\pi\frac{x}{L}\right)\right]\mathrm{e}^{-\mathrm{i}\omega t}, \tag{15.56b}$$

$$\rho_1(x,t) \approx i\rho_0k_0\ell\left[\frac{\mathrm{i}}{n\pi\gamma}\cos\left(n\pi\frac{x}{L}\right) + \frac{x}{L}\sin\left(n\pi\frac{x}{L}\right)\right]\mathrm{e}^{-\mathrm{i}\omega t}. \tag{15.56c}$$

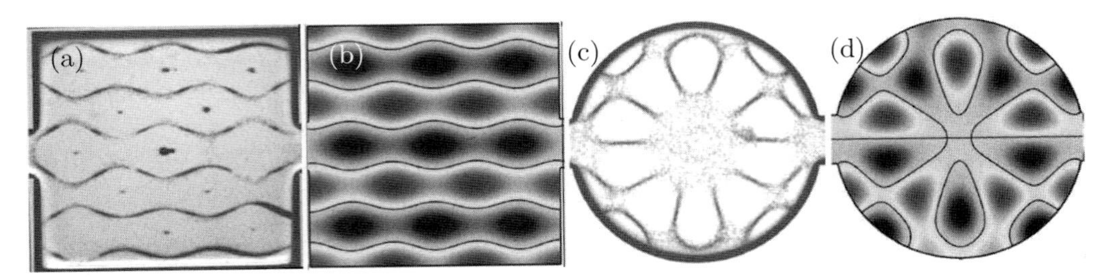

Fig. 15.2 (a) Picture of an acoustic resonance measured at $f = 2.17$ MHz in a 200 µm deep water-filled, square chamber of size 2 mm × 2 mm in a silicon chip with a Pyrex glass lid. The tracer particles (black spots) of diameter 10 µm collect at the pressure nodal lines. (b) Grayscale plot of the pressure field of the corresponding acoustic resonance found as a numerical solution of the Helmholtz equation using COMSOL. Pressure nodal lines are marked by black lines. (c) Picture of an acoustic resonance measured at $f = 2.42$ MHz in a 200 µm deep water-filled, circular chamber of diameter 2 mm; otherwise similar to panel (a). (d) Numerical simulation similar to panel (b), but adapted to the circular shape of the chamber in panel (c). Experimental pictures courtesy of S. Melker Hagsäter and numerical plots courtesy of Thomas Glasdam Jensen, DTU Nanotech.

From these expressions it follows that each of the fields acquires a resonant component with an amplitude that for $n = 3$ is a factor of $1/(n\pi\gamma) \approx 3 \times 10^4$ larger than the non-resonant component, e.g.

$$|v_1(x,t)| \approx \frac{1}{n\pi\gamma} \frac{\omega\ell}{c_\mathrm{a}} c_\mathrm{a}, \quad \text{(onresonance)}. \tag{15.57}$$

This is close to invalidating the entire perturbation approach since the perturbation factor α_res at resonance must be

$$\alpha_\mathrm{res} \equiv \frac{1}{n\pi\gamma} \frac{\omega\ell}{c_\mathrm{a}} \approx \frac{10^{-6}}{3 \times 10^{-5}} \approx 3 \times 10^{-2}. \tag{15.58}$$

In Exercise 15.3 we study the power dissipation at resonance due to viscosity in the liquid.

15.5 Acoustic waves in multilayer systems

Microfluidic systems often consist of layers of different materials, such as the one shown in Fig. 15.2 consisting of a silicon substrate, a water-filled chamber, a Pyrex glass lid and the surrounding air. It is therefore important in acoustofluidics to study the nature of wave propagation in such multilayer systems. For simplicity, we limit the discussion to the non-viscous 1D case, where undamped acoustic waves are propagating along in the x direction while interfaces between different media are planes parallel to the yz-plane and thus perpendicular to the direction of propagation. Consider an interface between medium a and medium b placed at $x = x_0$. The boundary conditions for the first-order acoustic fields at the interface are that the velocity and the pressure are continuous across the interface,

$$v_a(x_0,t) = v_b(x_0,t), \tag{15.59a}$$

$$p_a(x_0,t) = p_b(x_0,t), \tag{15.59b}$$

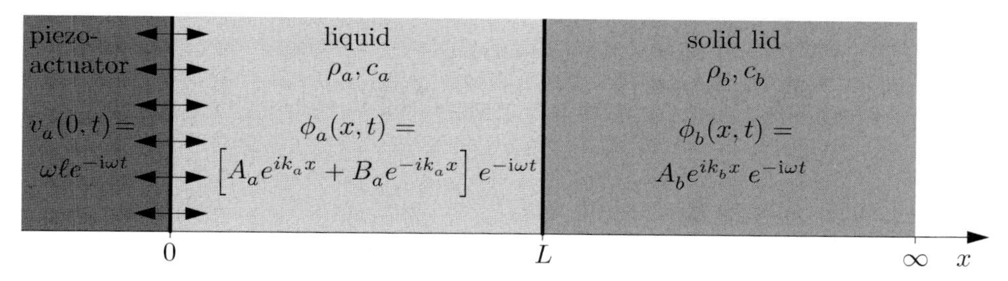

Fig. 15.3 Model of a three-layer system. A slab of liquid (light gray) of width L, density ρ_a and speed of sound c_a is sandwiched between a piezo-actuator (dark gray) oscillating with amplitude ℓ at angular frequency ω and an infinitely wide solid lid (gray) of density ρ_b and speed of sound c_b. The boundary condition at the piezo-actuator/liquid interface at $x = 0$ for the first-order velocity field in the liquid is $v_a(0,t) = \omega\ell\, \mathrm{e}^{-\mathrm{i}\omega t}$, while all sound waves are fully absorbed at infinity leaving only right-moving waves in the solid lid.

where the subscripts a and b refer to the two media. Introducing plane-wave velocity potentials ϕ_a and ϕ_b analogous to Eq. (15.48), we can reformulate the boundary conditions as

$$\partial_x \phi_a(x_0,t) = \partial_x \phi_b(x_0,t), \tag{15.60a}$$

$$\rho_a\, \phi_a(x_0,t) = \rho_b\, \phi_b(x_0,t). \tag{15.60b}$$

In the latter boundary condition we have used the fact that $p_j = \rho_j \partial_t \phi_j = -\mathrm{i}\omega\rho_j\phi_j$ and that due to continuity at the interface the frequency must be the same everywhere in the system. As we shall see shortly, these boundary conditions imply that the transmission and reflection of acoustic waves at the interface are governed by the so-called characteristic acoustic impedance Z_a and Z_b, which for a given material j is defined in terms of the density ρ_j and speed of sound c_j as

$$Z_j \equiv \rho_j\, c_j, \quad \text{for material } j. \tag{15.61}$$

As an explicit example of acoustofluidic multilayer systems we study the three-layer system sketched in Fig. 15.3. A semi-infinite piezo-actuator fills up the space $x < 0$, followed by a liquid slab in region a of width L for $0 < x < L$, and capped in region b by a semi-infinite solid lid for $x > L$. The piezo-actuator oscillates with a tiny nm-sized amplitude ℓ at angular frequency ω. The actual movement around $x = 0$ of the piezo-actuator/liquid interface is ignored and is only taken into account by the following velocity boundary condition for the liquid at $x = 0$,

$$v_a(0,t) = \omega\ell\, \mathrm{e}^{-\mathrm{i}\omega t}. \tag{15.62}$$

The first-order velocity potential ϕ_1 for the acoustic field takes the form

$$\phi_1(x,t) = \begin{cases} \phi_a(x,t) \;=\; \left[A_a \mathrm{e}^{\mathrm{i}k_a x} + B_a \mathrm{e}^{-\mathrm{i}k_a x}\right] \mathrm{e}^{-\mathrm{i}\omega t}, & \text{for } 0 < x < L, \\ \phi_b(x,t) \;=\; A_b \mathrm{e}^{\mathrm{i}k_b x}\, \mathrm{e}^{-\mathrm{i}\omega t}, & \text{for } x > L, \end{cases} \tag{15.63}$$

where total absorbtion of acoustic fields is assumed at $x = \infty$, thus leaving only right-moving waves in region b. The three unknown coefficients A_a, B_a and A_b are determined through the three boundary conditions Eqs. (15.60) and (15.62)

Table 15.2 Typical parameter values for 20 ℃ related to the characteristic acoustic impedances of materials commonly encountered in acoustofludics.

Material	Density ρ [kg m^{-3}]	Speed of sound c_a [m/s]	Impedance Z [kg/(m^2 s)]
Air	1.2×10^0	3.4×10^2	4.1×10^2
Water	1.0×10^3	1.5×10^3	1.5×10^6
PMMA polymer	1.2×10^3	1.6×10^3	1.9×10^6
Silicon	2.3×10^3	8.5×10^3	2.0×10^7
Pyrex glass	2.2×10^3	5.6×10^3	1.3×10^7

$$k_a \left[A_a e^{ik_a L} - B_a e^{-ik_a L}\right] = k_b\, A_b e^{ik_b L}, \tag{15.64a}$$

$$\rho_a \left[A_a e^{ik_a L} + B_a e^{-ik_a L}\right] = \rho_b\, A_b e^{ik_b L}, \tag{15.64b}$$

$$ik_a \left[A_a - B_a\right] = \omega\ell, \tag{15.64c}$$

and are found to be

$$A_a = c_a \ell\, \frac{\frac{1}{2}\,(1 + z_{ab})\, e^{-ik_a L}}{\sin(k_a L) + i z_{ab} \cos(k_a L)}, \tag{15.65a}$$

$$B_a = c_a \ell\, \frac{\frac{1}{2}\,(1 - z_{ab})\, e^{ik_a L}}{\sin(k_a L) + i z_{ab} \cos(k_a L)}, \tag{15.65b}$$

$$A_b = c_b \ell\, \frac{z_{ab}\, e^{-ik_b L}}{\sin(k_a L) + i z_{ab} \cos(k_a L)}, \tag{15.65c}$$

where we have introduced the impedance ratio z_{ab},

$$z_{ab} \equiv \frac{Z_a}{Z_b} = \frac{\rho_a c_a}{\rho_b c_b}. \tag{15.66}$$

We note that if the impedances are equal, i.e. $z_{ab} = 1$, which according to Table 15.2 is almost the case for a water/PMMA interface, then $A_a = A_b = -ic_a\ell$ and $B_a = 0$, and consequently the acoustic wave is fully transmitted through the interface without any reflection. Naturally, the wave still changes its propagation speed as it passes from the liquid to the solid. In Exercise 15.4 we study the two other special cases $z_{ab} \to 0$ and $z_{ab} \to \infty$.

The resonance behavior is determined by studying the velocity field v_a for the liquid in region a,

$$v_a(x,t) = \partial_x \phi_a(x,t) = i\omega\ell\, \frac{\frac{1}{2}\,(1 + z_{ab})\, e^{-ik_a(L-x)} - \frac{1}{2}\,(1 - z_{ab})\, e^{ik_a(L-x)}}{\sin(k_a L) + i z_{ab} \cos(k_a L)}\, e^{-i\omega t}. \tag{15.67}$$

From Table 15.2 we see that for water $Z_a \approx 10^6$ kg/(m^2 s), while for a hard solid like silicon or glass $Z_b \approx 10^7$ kg/(m^2 s). Hence, a typical value for the impedance ratio is $z_{ab} \approx 0.1$. The denominator in the expression for the velocity v_a is minimal when the $\sin(k_aL)$ term vanishes and leaves only the small $\mathrm{i}z_{ab}\cos(k_aL)$ term, which leads us to the resonance condition,

$$k_a \equiv n\,\frac{\pi}{L}, \quad n = 1, 2, 3, \ldots \quad \text{(resonance condition)}. \tag{15.68}$$

At resonance, $\sin(k_aL) = 0$ and $\cos(k_aL) = (-1)^n$, which reduces the expression for the velocity to

$$v_a(x,t) = \frac{\omega\ell}{z_{ab}}\Big[z_{ab}\cos(k_ax) + \mathrm{i}\sin(k_ax)\Big]\,\mathrm{e}^{-\mathrm{i}\omega t} \approx \mathrm{i}\,\frac{\omega\ell}{z_{ab}}\,\sin(k_ax)\,\mathrm{e}^{-\mathrm{i}\omega t}, \tag{15.69}$$

and the estimate for the magnitude of the onresonance velocity field becomes

$$|v_a(x,t)| \approx \frac{1}{z_{ab}}\,\frac{\omega\ell}{c_a}\,c_a, \quad \text{(onresonance, } z_{ab} \ll 1). \tag{15.70}$$

Likewise, as seen in Exercise 15.5, the estimate for the magnitude of the onresonance velocity field in the limit $z_{ab} \gg 1$, e.g. water capped by air, becomes

$$|v_a(x,t)| \approx z_{ab}\,\frac{\omega\ell}{c_a}\,c_a, \quad \text{(onresonance, } z_{ab} \gg 1). \tag{15.71}$$

In both cases the onresonance amplitude is smaller by one to three orders of magnitude compared to the large amplitude Eq. (15.57) in the monolayer system limited only by viscous damping.

15.6 Second-order acoustic fields

The linear theory of acoustics presented above offers no possibilities of achieving a DC drift velocity or an additional DC pressure gradient due to the presence of acoustic waves. The reason is that in a linear theory the harmonic drive $\cos(\omega t)$ enters all terms, which consequently all have a zero time average over a full oscillation period. However, the Navier–Stokes equation and the continuity equation are non-linear in the combined velocity and the density dependence, and expanding these equations to second order will introduce products of two first-order factors $\cos(\omega t)$, products that have the time average $\frac{1}{2}$ over a full period. Including the second-order terms in Eq. (15.17) leads to

$$p = p_0 + p_1 + p_2, \tag{15.72a}$$
$$\rho = \rho_0 + \rho_1 + \rho_2, \tag{15.72b}$$
$$\mathbf{v} = \mathbf{0} + \mathbf{v}_1 + \mathbf{v}_2. \tag{15.72c}$$

Combining Eqs. (15.1) and (15.72) we obtain the second-order equation of motion for the acoustic field,

$$p_2 = c_\mathrm{a}^2\rho_2 + \frac{1}{2}\left(\partial_\rho c^2\right)_0\rho_1^2 = c_\mathrm{a}^2\rho_2 + \frac{\gamma^*-1}{2}\,\frac{c_\mathrm{a}^2}{\rho_0}\,\rho_1^2, \tag{15.73a}$$

$$\partial_t\rho_2 = -\rho_0\boldsymbol{\nabla}\cdot\mathbf{v}_2 - \boldsymbol{\nabla}\cdot(\rho_1\mathbf{v}_1), \tag{15.73b}$$

$$\rho_0\partial_t\mathbf{v}_2 = -\boldsymbol{\nabla}p_2 - \rho_1\partial_t\mathbf{v}_1 - \rho_0(\mathbf{v}_1\cdot\boldsymbol{\nabla})\mathbf{v}_1 + \eta\nabla^2\mathbf{v}_2 + \beta\eta\boldsymbol{\nabla}(\boldsymbol{\nabla}\cdot\mathbf{v}_2), \tag{15.73c}$$

where, e.g. two terms in the last equation contain products of two first-order fields. The expression for the time average of such products in complex notation is given in Eq. (10.38) as $\langle A(t)B(t)\rangle = \frac{1}{2}\,\mathrm{Re}\big[A_0 B_0^*\big]$. If we consider the time average of Eq. (15.73c) we note that in steady state with a harmonic driving force, $\mathbf{v}_2$ must be periodic in the fundamental period, and consequently it can be decomposed in a Fourier series, perhaps containing a non-zero constant. However, after taking the time derivative the constant is eliminated, and furthermore, by a subsequent time average all the remaining (higher) harmonics are also eliminated, yielding $\rho_0\langle\partial_t\mathbf{v}_2\rangle = 0$. If we, furthermore, neglect the small contribution from the viscous terms we find

$$\boldsymbol{\nabla}\langle p_2\rangle = -\big\langle\rho_1\partial_t\mathbf{v}_1\big\rangle - \big\langle\rho_0(\mathbf{v}_1\!\cdot\!\boldsymbol{\nabla})\mathbf{v}_1\big\rangle = \rho_0\boldsymbol{\nabla}\Big\langle\frac{1}{2}\Big(\frac{1}{c_\mathrm{a}}\,\partial_t\phi_1\Big)^2 - \frac{1}{2}\big|\boldsymbol{\nabla}\phi_1\big|^2\Big\rangle, \tag{15.74}$$

where we have used the first-order expressions Eqs. (15.17a) and (15.17b) for $\mathbf{v}_1$ and ρ_1 in terms of ϕ_1, and $\langle p_2\rangle$ is obtained by straightforward integration of the gradient.

15.6.1 Acoustic radiation force

Let us consider the pressure deviation δp from the unperturbed static pressure,

$$\delta p \equiv p_1 + p_2. \tag{15.75}$$

In particular, we would like to calculate the resulting time-averaged acoustic radiation pressure force $\langle\mathbf{F}\rangle$ on some elastic body with surface $S(t)$ suspended in the fluid exposed to the acoustic field. Both the surface $S(t)$ and the volume $V(t)$ of the fluid surrounding the body depend on time, as the acoustic field in the fluid induces vibrations in the body. The equilibrium position of the surface is denoted S_0 so the surface deviation $S(t) - S_0$ gives at least rise to a first-order effect, so to second-order accuracy the acoustic radiation force becomes

$$\langle\mathbf{F}\rangle = \Big\langle\int_{S(t)}\mathrm{d}a\,(-\mathbf{n})\,\delta p\Big\rangle = \Big\langle\int_{S(t)}\mathrm{d}a\,(-\mathbf{n})\,p_1\Big\rangle + \Big\langle\int_{S_0}\mathrm{d}a\,(-\mathbf{n})\,p_2\Big\rangle, \tag{15.76}$$

where the surface integral notation is discussed in Section 2.2.2, and where it suffices to use the unperturbed surface S_0 for the contribution from the second-order pressure p_2.

The time average of the surface integral of the second-order pressure is straightforward to calculate, but the contribution from the first-order pressure demands greater care due to the time-dependent surface integral. One way of handling it is to utilize Reynolds' transport theorem that, as proven in Exercise 15.7, states the following for some time-dependent property $f(t)$ of the liquid with velocity field $\mathbf{v}$ in the volume $V(t)$:

$$\frac{\mathrm{d}}{\mathrm{d}t}\Big[\int_{V(t)}\mathrm{d}\mathbf{r}\,f(t)\Big] = \int_{V(t)}\mathrm{d}\mathbf{r}\,\partial_t f(t) \;+\; \int_{S(t)}\mathrm{d}a\,(\mathbf{n}\!\cdot\!\mathbf{v})\,f(t). \tag{15.77}$$

In steady state, as above with $\partial_t\mathbf{v}_2$, the time average of the time derivative of the integral is zero. In this special case, Reynolds' transport theorem therefore takes the form

$$\Big\langle\int_{V(t)}\mathrm{d}\mathbf{r}\,\partial_t f(t)\Big\rangle = -\Big\langle\int_{S(t)}\mathrm{d}a\,(\mathbf{n}\!\cdot\!\mathbf{v})\,f(t)\Big\rangle. \tag{15.78}$$

To get back to the p_1 term in Eq. (15.76), we first note that $-\boldsymbol{\nabla}p_1 = \partial_t(\rho_0\mathbf{v}_1)$ and then we apply Reynolds' transport theorem on $f(t) = \rho_0\mathbf{v}_1$,

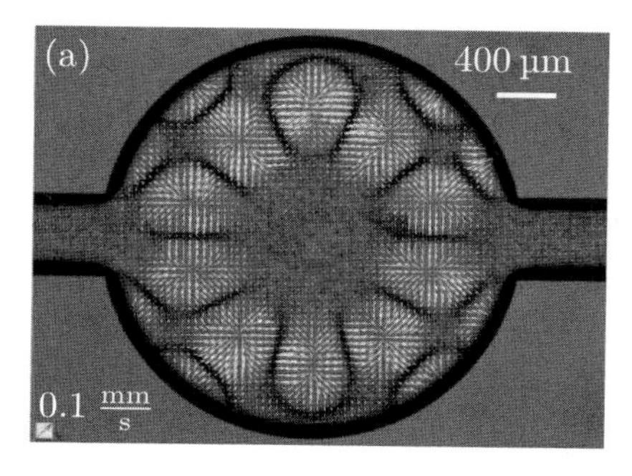

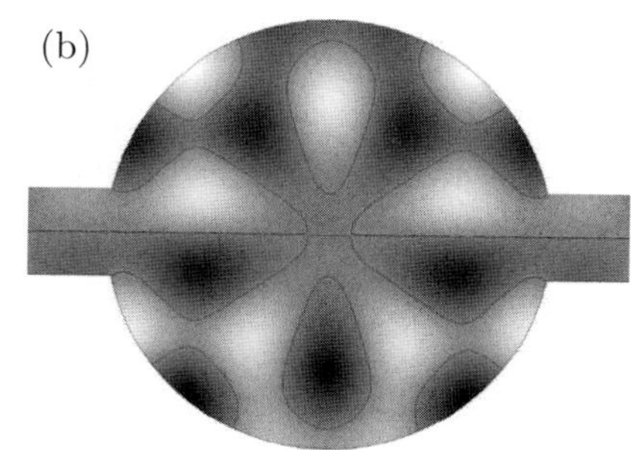

Fig. 15.4 (a) Top view of the acoustic radiation-force induced, initial velocities (white arrows) of $a = 5$ μm beads inside a 200 μm deep, glass-lid covered, water-filled, circular silicon microchamber just after turning on a $f = 2.417$ MHz piezo-actuator emitting ultrasound through the chamber from below. Also shown are the steady-state positions of the beads (bands of black spots) at the pressure nodes of the acoustic field. (b) Grayscale plot of the numerical solution to the Helmholtz equation (15.23) with $\omega/(2\pi) = 2.417$ MHz and Neumann boundary conditions. The pressure nodes are marked by thin black lines. Experimental pictures courtesy of S. Melker Hagsäter and numerical plots courtesy of Thomas Glasdam Jensen, DTU Nanotech.

$$\left\langle \int_{S(t)} da\,(-\mathbf{n})\,p_1 \right\rangle = \left\langle \int_{V(t)} d\mathbf{r}\,(-\boldsymbol{\nabla} p_1) \right\rangle = \left\langle \int_{V(t)} d\mathbf{r}\,\partial_t(\rho_0 \mathbf{v}_1) \right\rangle = \left\langle \int_{S(t)} da\,(\mathbf{n}\cdot\mathbf{v}_1)(\rho_0 \mathbf{v}_1) \right\rangle, \tag{15.79}$$

where Gauss' theorem is used in the first equality and Reynolds' transport theorem in the last. To second-order accuracy it is permissible to substitute $S(t)$ by S_0 in the above expression. The time-averaged acoustic radiation force $\langle \mathbf{F} \rangle$ is now given by substituting Eqs. (15.74) and (15.79) into Eq. (15.76),

$$\langle \mathbf{F} \rangle = -\rho_0 \left\langle \int_{S_0} da\,(\mathbf{n}\cdot\boldsymbol{\nabla}\phi_1)\boldsymbol{\nabla}\phi_1 \right\rangle - \frac{\rho_0}{2c_a^2} \left\langle \int_{S_0} da\,\mathbf{n}\big(\partial_t\phi_1\big)^2 \right\rangle + \frac{\rho_0}{2} \left\langle \int_{S_0} da\,\mathbf{n}\big|\boldsymbol{\nabla}\phi_1\big|^2 \right\rangle. \tag{15.80}$$

Note how all integrands are quadratic in derivatives of the first-order velocity potential ϕ_1, so that to second order it suffices to integrate over the constant equilibrium surface S_0.

With this theory it is simple, but often tedious, to calculate radiation forces on objects in acoustofluidics. One example of such a calculation is to consider a slab of some elastic material inserted parallel to and somewhere in between the two oscillating walls in Fig. 15.1.

For the more complicated case of a spherical particle of radius a in a standing-wave acoustic pressure field $p(x,t) = p_0\,\cos(kx)\mathrm{e}^{-\mathrm{i}\omega t}$, like the experimental result shown in Fig. 15.4, similar principles are used in the calculation. The time-averaged radiation force $\langle F_{\mathrm{sph}} \rangle$ acting on the particle is

$$\big\langle F_{\mathrm{sph}}(x) \big\rangle = \langle E_{\mathrm{ac}} \rangle\, 4\pi a^2\,(ka) \left[\frac{\frac{5}{3}\frac{\rho^*}{\rho} - \frac{2}{3}}{1 + 2\frac{\rho^*}{\rho}} - \frac{\left(\frac{k^*}{k}\right)^2}{3\frac{\rho^*}{\rho}} \right] \sin(2kx), \tag{15.81}$$

where unmarked and star-marked quantities refer to the liquid and the particle, respectively, and $\langle E_{\mathrm{ac}} \rangle$ is the average energy density of the incoming acoustic wave. It should be noted

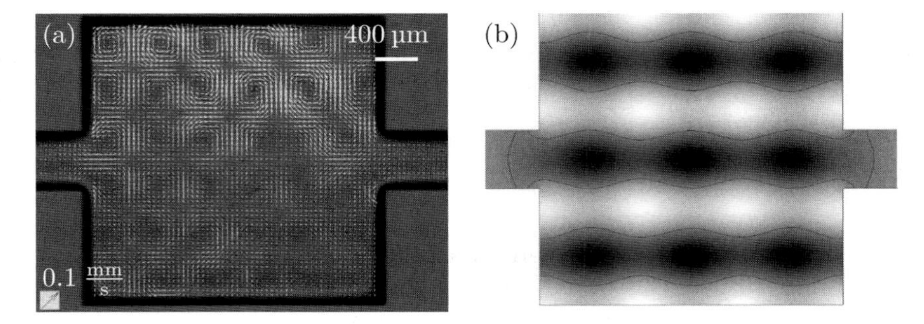

Fig. 15.5 (a) Top view of the steady-state acoustic streaming velocities (white arrows) of water inside a 200 μm deep, glass-lid covered, square silicon microchamber after turning on a $f = 2.17$ MHz piezo-actuator emitting ultrasound through the chamber from below. The liquid motion is detected by particle-image velocimetry on $a = 500$ nm beads suspended in the liquid. (b) Grayscale plot of the numerical solution to the Helmholtz equation (15.23) with $\omega/(2\pi) = 2.417$ MHz and Neumann boundary conditions. The pressure nodes are marked by thin black lines. Note how the streaming vortices in the experiment is period-doubled compared to the pressure field. Experimental pictures courtesy of S. Melker Hagsäter and numerical plots courtesy of Thomas Glasdam Jensen, DTU Nanotech.

that the sign of the pre-factor determines whether a particle is moved towards pressure nodes or pressure antinodes. Stiff particles tend to move to pressure nodes, while soft particles tend to move to pressure antinodes.

15.6.2 Acoustic streaming

If we assume that the time dependence of all first-order fields is harmonic, $e^{-i\omega t}$, then Eq. (15.73) has solutions for ρ_2 and $\mathbf{v}_2$ with non-zero time averages for $\langle\rho_2\rangle \neq 0$ and $\langle\mathbf{v}_2\rangle \neq 0$. In other words, the oscillating acoustic field imparts a time-independent component to the density and velocity field of the liquid. This phenomena is known as acoustic streaming, and an experimental example of this is shown in Fig. 15.5.

According to Exercise 10.7 the time average of a product is given by $\langle A(t)B(t)\rangle = \frac{1}{2}\,\mathrm{Re}\big[A_0 B_0^*\big]$, we find the time average of Eq. (15.73b) to be

$$\boldsymbol{\nabla}\cdot\langle\mathbf{v}_2\rangle = -\frac{1}{\rho_0}\boldsymbol{\nabla}\cdot\langle\rho_1\mathbf{v}_1\rangle, \tag{15.82}$$

while that of Eq. (15.73c), after taking the divergence and inserting Eq. (15.82), becomes

$$\nabla^2\langle\rho_2\rangle = -\frac{1}{2c_\mathrm{a}^2}\big(\partial_\rho c^2\big)_0\nabla^2\langle\rho_1^2\rangle - \frac{\rho_0}{c_\mathrm{a}^2}\boldsymbol{\nabla}\cdot\langle(\mathbf{v}_1\cdot\boldsymbol{\nabla})\mathbf{v}_1\rangle - \left[\frac{(1+\beta)\eta}{\rho_0 c_\mathrm{a}^2}\nabla^2 - \mathrm{i}\frac{\omega}{c_\mathrm{a}^2}\right]\boldsymbol{\nabla}\cdot\langle\rho_1\mathbf{v}_1\rangle. \tag{15.83}$$

To estimate the time-averaged speeds $\langle\mathbf{v}_2\rangle$ that one can hope to obtain by acoustic streaming, one can assume plane-wave solutions for the first-order fields. The calculations are straightforward, and using the usual parameters for water and an ultrasound frequency of the order 1 MHz, one finds $|\langle\mathbf{v}_2\rangle| \approx 10$ μm/s in agreement with the measurements presented in Fig. 15.5(a).

15.7 Further reading

A good starting point for further studies of acoustofluidics is the textbook by Lighthill (2003) followed by the classic papers by King (1934) and Yosioka and Kawasima (1955) on the acoustic radiation force on particles, and by Markham (1952) and Nyborg (1958) on acoustic streaming. A recent theory review paper is Hasegawa *et al.* (2000) while recent microfluidic applications can be found in Petersson *et al.* (2005), Sritharan *et al.* (2006), and Hagsäter *et al.* (2007).

15.8 Exercises

Exercise 15.1
The non-linear velocity term written as a gradient
Consider a potential flow $\mathbf{v} = \boldsymbol{\nabla}\phi$. Prove that the non-linear velocity term can be written as the gradient $(\mathbf{v}\cdot\boldsymbol{\nabla})\mathbf{v} = \frac{1}{2}\boldsymbol{\nabla}\big(|\boldsymbol{\nabla}\phi|^2\big)$ used in Eq. (15.5).

Exercise 15.2
The wave equation for the velocity potential including viscosity
Based on Eqs. (15.45a) and (15.45b), derive the wave equation for the first-order velocity potential ϕ_1 in the case of non-zero viscosity. Hint: employ the first-order continuity equation.

Exercise 15.3
The viscous power dissipation for acoustic resonances
Calculate the power dissipation $\partial_t W_{\mathrm{visc}}$ due to viscosity, see Eqs. (4.10) and (4.12), for the ultrasound resonance $k_0 L = n\pi$ in the liquid between two parallel planar walls vibrating in antiphase as sketched in Fig. 15.1.

Exercise 15.4
Acoustic impedance dependence in the three-layer system
Discuss the two limits $z_{ab} \to 0$ and $z_{ab} \to \infty$ of the acoustic impedance ratio in the three-layer system of Fig. 15.3 given the expressions Eq. (15.65).

Exercise 15.5
Acoustic resonance in the three-layer system
Calculate, in analogy with Eq. (15.70), the magnitude of the onresonance velocity field in the limit $z_{ab} \gg 1$, e.g. for water capped by air.

Exercise 15.6
First-order acoustic waves in 1D multilayer systems
Consider N layers parallel to the yz-plane of different materials. The jth layer lies between x_{j-1} and x_j. The velocity potential $\phi(x,t)$ is given by

$$\phi(x,t) = \phi_j(x,t) = \Big[A_j\,\mathrm{e}^{\mathrm{i}k_j x} + B_j\,\mathrm{e}^{-\mathrm{i}k_j x}\Big]\,\mathrm{e}^{-\mathrm{i}\omega t}, \quad j = 1,2,3,\ldots,N, \tag{15.84}$$

and the density and speed of sound for each layer is ρ_j and c_j, respectively.

(a) Determine the matrix T in the matrix equation

$$\begin{pmatrix} A_{j+1} \\ B_{j+1} \end{pmatrix} = T_{j+1,j}\begin{pmatrix} A_j \\ B_j \end{pmatrix}, \tag{15.85}$$

which expresses the coefficient pair (A_{j+1}, B_{j+1}) of layer $j+1$ in terms of the material parameters and the coefficient pair (A_j, B_j) of layer j.

(b) Establish a matrix equation for (A_N, B_N) in terms of (A_1, B_1). It is not necessary to multiply out products of matrices that might appear.

Exercise 15.7
Reynolds' transport theorem
Sketch a proof of Reynolds' transport theorem, Eq. (15.77). Hint: consider a finite but small time interval Δt and formulate the derivative as a difference quotient. Add and subtract a term with the volume evaluated at time t and the integrand at time $t+\Delta t$ and thereby form two difference quotients.

15.9 Solutions

Solution 15.1
The non-linear velocity term written as a gradient
Using the index notation we have $v_i = \partial_i\phi$. Therefore, the non-linear term can be rewritten as $\big[(\mathbf{v}\cdot\boldsymbol{\nabla})\mathbf{v}\big]_i = \big[(\partial_j\phi)\partial_j\big]\partial_i\phi = (\partial_j\phi)\partial_i(\partial_j\phi) = \partial_i\big[\frac{1}{2}(\partial_j\phi)^2\big] = \big[\frac{1}{2}\boldsymbol{\nabla}\big(|\boldsymbol{\nabla}\phi|^2\big)\big]_i$.

Solution 15.2
The wave equation for the velocity potential including viscosity
The first-order continuity equation reads $\boldsymbol{\nabla}\cdot\mathbf{v}_1 = -(\partial_t\rho_1)/\rho_0$. When substituting $\mathbf{v}_1$ on the left-hand side by Eq. (15.45a) it becomes $\nabla^2\phi_1 = -(\partial_t\rho_1)/\rho_0$. Now, insertion of Eq. (15.45b) in the new right-hand side, leads to $\nabla^2\phi_1 = \big[(1+\mathrm{i}\gamma)/c_\mathrm{a}\big]^2\partial_t^2\phi_1 = -\big[\omega(1+\mathrm{i}\gamma)/c_\mathrm{a}\big]^2\phi_1$.

Solution 15.3
The viscous power dissipation for acoustic resonances
At steady state the time-averaged power dissipated by viscosity in the liquid equals the mechanical power put into the liquid by the oscillating walls,

$$\partial_t W_\mathrm{visc} = \partial_t W_\mathrm{mech} = 2\langle v_1(L,t)p_1(L,t)\rangle_t = 2c_\mathrm{a}^2\langle v_1(L,t)\rho_1(L,t)\rangle_t, \tag{15.86}$$

where the factor of 2 is because there are two walls. Inserting the expressions for v_1 and ρ_1 from Eqs. (15.56b) and (15.56c) and using the rule, Eq. (10.38), for time averaging of complex functions, we find

$$\partial_t W_\mathrm{visc} = \frac{\rho_0 c_\mathrm{a}^2 k_0 \omega \ell^2}{2n\pi\gamma}. \tag{15.87}$$

For the usual parameters, $\omega = 10^6$ rad/s, $L = 1$ mm, $\rho_0 = 1400$ kg m^{-3}, and $\gamma = 10^{-6}$ we get an unrealistic large value for W_visc. The reason for this is the crudeness of the model neglecting dissipation in the piezo-actuator as well as in the walls and coupling losses between the actuator and the microfluidic channel.

Solution 15.4
Acoustic impedance dependence in the three-layer system
In the limit of $z_{ab} \to 0$, e.g. when the chamber in Fig. 15.3 is filled with a gas and the lid is a very hard material such as silicon, the resonance condition in Eq. (15.65) is $k_a L = n\pi$, which makes the denominator proportional to z_{ab}. At resonance $|A_a| \approx |B_a| \propto 1/z_{ab} \to \infty$, while $|A_b| \approx c_b\ell$, and the large-amplitude wave is confined to the chamber. Offresonance $|A_a| \approx |B_a| \approx c_a\ell$, while $|A_b| \propto z_{ab} \to 0$, and the small-amplitude wave is confined to the chamber.

In the opposite limit, $z_{ab} \to \infty$, the resonance condition is $k_a L = \big(n+\frac{1}{2}\big)\pi$ that due to the sine term makes the denominator of the order unity. At resonance $|A_a| \approx |B_a| \propto z_{ab} \to \infty$, while $|A_b| \approx c_b\ell$, and the large-amplitude wave is again confined to the chamber.

Offresonance $|A_a| \approx |B_a| \approx c_a \ell$, while $|A_b| \approx c_b \ell$, and the small-amplitude wave is not confined to the chamber.

Solution 15.5
Acoustic resonance in the three-layer system

In the limit $z_{ab} \gg 1$ the resonance condition for the velocity field $v_a(x,t)$ in Eq. (15.67) is $k_a L = \big(n+\frac{1}{2}\big)\pi$, which removes the large-valued $z_{ab}\cos(k_a L)$ term leaving behind only $\sin(k_a L) = (-1)^n$. In the denominator only terms proportional to z_{ab} are kept, and the onresonance velocity field becomes

$$v_a(x,t) \approx \mathrm{i}(-1)^n \omega \ell\, z_{ab}\, \cos\big[k_a(L-x)\big]\,\mathrm{e}^{-\mathrm{i}\omega t} = -\mathrm{i}(-1)^n \omega \ell\, z_{ab}\, \cos(k_a x)\,\mathrm{e}^{-\mathrm{i}\omega t}, \qquad (15.88)$$

and the expression in Eq. (15.71) for the onresonance amplitude has been verified.

Solution 15.6
First-order acoustic waves in 1D multilayer systems

(a) Consider the interface at $x = x_j$ between layer j and layer $j+1$. The two boundary conditions, Eq. (15.60) connecting the coefficient pairs (A_j, B_j) and (A_{j+1}, B_{j+1}), can be written as the following matrix equation,

$$\begin{pmatrix} \mathrm{e}^{\mathrm{i}k_{j+1}x_j} & -\mathrm{e}^{-\mathrm{i}k_{j+1}x_j} \\ \mathrm{e}^{\mathrm{i}k_{j+1}x_j} & \mathrm{e}^{-\mathrm{i}k_{j+1}x_j} \end{pmatrix} \begin{pmatrix} A_{j+1} \\ B_{j+1} \end{pmatrix} = \begin{pmatrix} \dfrac{k_j}{k_{j+1}}\mathrm{e}^{\mathrm{i}k_jx_j} & -\dfrac{k_j}{k_{j+1}}\mathrm{e}^{-\mathrm{i}k_jx_j} \\ \dfrac{\rho_j}{\rho_{j+1}}\mathrm{e}^{\mathrm{i}k_jx_j} & \dfrac{\rho_j}{\rho_{j+1}}\mathrm{e}^{-\mathrm{i}k_jx_j} \end{pmatrix} \begin{pmatrix} A_j \\ B_j \end{pmatrix}, \qquad (15.89)$$

where we have divided by k_{j+1} and ρ_{j+1} in row 1 and 2, respectively. The coefficient pair (A_{j+1}, B_{j+1}) can now be isolated by multiplying from left with the inverse coefficient matrix,

$$\begin{aligned} \begin{pmatrix} A_{j+1} \\ B_{j+1} \end{pmatrix} &= \frac{1}{2} \begin{pmatrix} \mathrm{e}^{-\mathrm{i}k_{j+1}x_j} & \mathrm{e}^{-\mathrm{i}k_{j+1}x_j} \\ -\mathrm{e}^{\mathrm{i}k_{j+1}x_j} & \mathrm{e}^{\mathrm{i}k_{j+1}x_j} \end{pmatrix} \begin{pmatrix} \dfrac{k_j}{k_{j+1}}\mathrm{e}^{\mathrm{i}k_jx_j} & -\dfrac{k_j}{k_{j+1}}\mathrm{e}^{-\mathrm{i}k_jx_j} \\ \dfrac{\rho_j}{\rho_{j+1}}\mathrm{e}^{\mathrm{i}k_jx_j} & \dfrac{\rho_j}{\rho_{j+1}}\mathrm{e}^{-\mathrm{i}k_jx_j} \end{pmatrix} \begin{pmatrix} A_j \\ B_j \end{pmatrix} \\ &= \frac{1}{2}\frac{k_j}{k_{j+1}} \begin{pmatrix} \big(z_{j,j+1}+1\big)\,\mathrm{e}^{\mathrm{i}(k_j-k_{j+1})x_j} & \big(z_{j,j+1}-1\big)\,\mathrm{e}^{-\mathrm{i}(k_j+k_{j+1})x_j} \\ \big(z_{j,j+1}-1\big)\,\mathrm{e}^{\mathrm{i}(k_j+k_{j+1})x_j} & \big(z_{j,j+1}+1\big)\,\mathrm{e}^{-\mathrm{i}(k_j-k_{j+1})x_j} \end{pmatrix} \begin{pmatrix} A_j \\ B_j \end{pmatrix}, \end{aligned} \qquad (15.90)$$

where, in analogy with Eq. (15.66), we have introduced the acoustic impedance ratio

$$z_{j,j+1} \equiv \frac{\rho_j c_j}{\rho_{j+1} c_{j+1}}. \qquad (15.91)$$

(b) By iterating the above result, interface by interface, it is possible to relate the coefficient pair (A_N, B_N) for layer N to the coefficient pair (A_1, B_1) for layer 1,

$$\begin{pmatrix} A_N \\ B_N \end{pmatrix} = T_{N,N-1}\, T_{N-1,N-2} \cdots T_{3,2}\, T_{2,1} \begin{pmatrix} A_1 \\ B_1 \end{pmatrix}. \qquad (15.92)$$

Solution 15.7
Reynolds' transport theorem
We write the time derivative as a difference quotient using a small but finite time interval Dt, and insert two compensating auxiliary terms,

$$\begin{aligned}\frac{\mathrm{d}}{\mathrm{d}t}\left[\int_{V(t)} \mathrm{d}\mathbf{r}\, f(t)\right] &\approx \frac{1}{\Delta t}\left[\int_{V(t+\Delta t)} \mathrm{d}\mathbf{r}\, f(t+\Delta t) - \int_{V(t)} \mathrm{d}\mathbf{r}\, f(t)\right] \\ &= \frac{1}{\Delta t}\left[\int_{V(t+\Delta t)} \mathrm{d}\mathbf{r}\, f(t+\Delta t) - \int_{V(t+\Delta t)} \mathrm{d}\mathbf{r}\, f(t)\right] \\ &\quad + \frac{1}{\Delta t}\left[\int_{V(t+\Delta t)} \mathrm{d}\mathbf{r}\, f(t) - \int_{V(t)} \mathrm{d}\mathbf{r}\, f(t)\right]. \end{aligned} \tag{15.93}$$

In the first term the volume is kept fixed at the same time $t + \Delta t$, so all variation is in the integrand. The pre-factor $1/\Delta t$ can therefore be taken inside the integral. In the second term, the integrand is kept fixed at time t, so all variation is in the integration volume. In the little time interval Δt only a thin shell near the surface $S(t)$ is changed. The volume of this thin shell is given by the surface area times the tiny displacement Δz that takes place perpendicular to any given part of the surface. This perpendicular displacement can be written as $\Delta z = \mathbf{n}\cdot(\mathbf{v}\Delta t)$, where $\mathbf{v}$ is the velocity of the surface at a given surface point and $\mathbf{n}$ the corresponding surface normal vector. So the difference in volume integral can symbolically be written as $\int_{V(t+\Delta t)}\mathrm{d}\mathbf{r} - \int_{V(t)}\mathrm{d}\mathbf{r} = \int_{S(t)}\mathrm{d}a\, (\mathbf{n}\cdot\mathbf{v})\, \Delta t$. We thus get

$$\begin{aligned}\frac{\mathrm{d}}{\mathrm{d}t}\left[\int_{V(t)} \mathrm{d}\mathbf{r}\, f(t)\right] &= \lim_{\Delta t\to 0}\left[\int_{V(t+\Delta t)} \mathrm{d}\mathbf{r}\, \frac{f(t+\Delta t) - f(t)}{\Delta t} + \frac{1}{\Delta t}\int_{S(t)} \mathrm{d}a\, (\mathbf{n}\cdot\mathbf{v})\, \Delta t\, f(t)\right] \\ &= \int_{V(t)} \mathrm{d}\mathbf{r}\, \partial_t f(t) + \int_{S(t)} \mathrm{d}a\, (\mathbf{n}\cdot\mathbf{v})\, f(t). \end{aligned} \tag{15.94}$$

The presence of a given species of molecules can be detected in a given solution by recording an absorption spectrum, as sketched in Fig. 16.1(c), where the transmitted intensity I is measured as a function of frequency f. Which specific molecular transitions that are recorded in absorption spectroscopy depends on the frequency range of the incoming radiation. In general, low-energetic infra-red light can induce vibrational or rotational transitions within the electronic ground state of the molecule, whereas the more energetic visible and ultraviolet radiation can induce electronic transitions. Thus, by varying the frequency of the incoming radiation it is possible to map out the various excitation energies inside a molecule. Whenever the frequency matches a transition energy the intensity I of the transmitted radiation falls below the background I_0. This dip in intensity is called an absorption line. As the absorption lines are unique for each species of molecule they can be used to detect the presence of the molecule in a given sample.

For absorption spectroscopy used in, e.g. biochemical sensing, one defines the transmittance $\mathcal{T}$ and the absorbance $\mathcal{A}$ in terms of the background intensity level I_0 and the transmitted intensity I as

$$\text{Transmittance} \quad \mathcal{T} \equiv \frac{I}{I_0}, \tag{16.22a}$$

$$\text{Absorbance} \quad \mathcal{A} \equiv -\log \mathcal{T} = \log \frac{I_0}{I}. \tag{16.22b}$$

The absorbance $\mathcal{A}$ depends in a simple way, known as Beer–Lambert's law, on the concentration c_m of the target molecule and the length ℓ of the path of the incident radiation in the medium. To derive Beer–Lambert's law we consider a medium of length ℓ in the direction x of propagation of the incident radiation and a slab of thickness Δx between x and $x + \Delta x$. The total cross-sectional area is denoted S, while the sum of the capture areas of the target molecules in the slab is denoted ΔS. For small concentrations the latter is proportional to the concentration c_m of the target molecules and the volume $S\Delta x$ of the slab under consideration,

$$\Delta S = \varepsilon c_\mathrm{m}\, S\Delta x, \tag{16.23}$$

where ε is a proportionality constant denoted the absorptivity. When passing through the slab, the radiation loses a small amount ΔI of its intensity $I(x)$ proportional to the relative caption area $\Delta S/S$,

$$\Delta I = -I(x)\frac{\Delta S}{S}. \tag{16.24}$$

Combining Eqs. (16.24) and (16.23) and taking the limit $\Delta x \to 0$ leads to the differential equation $\mathrm{d}I/\mathrm{d}x = -\varepsilon c_\mathrm{m}$, and finally Beer–Lambert's law

$$I(x) = I_0\, \mathrm{e}^{-\varepsilon c_\mathrm{m} x}, \tag{16.25a}$$

$$\mathcal{A} = \log\left[\frac{I_0}{I(\ell)}\right] = \varepsilon \ell c_\mathrm{m}. \tag{16.25b}$$

The conductivity-induced damping given in Eq. (16.17) is a specific example of Beer–Lambert's law. By noting that $I \propto |\mathbf{E}|^2$ we obtain

$$I(x) = I_0\, \mathrm{e}^{-2\gamma k_n x} \quad \text{(conductivity induced)}, \tag{16.26}$$

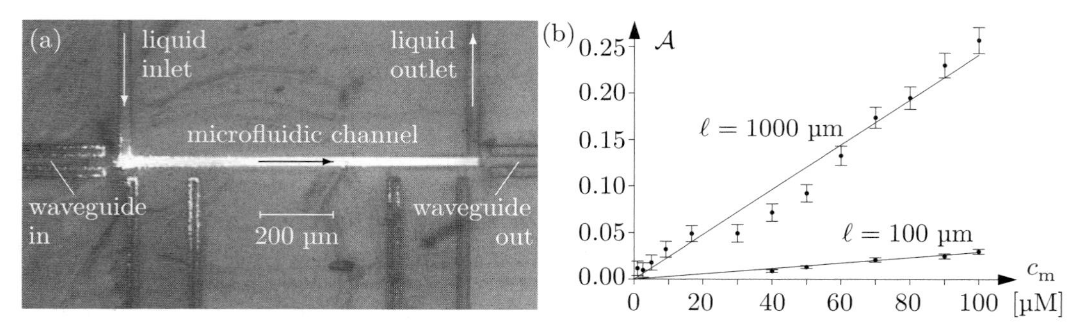

Fig. 16.2 Absorbance measurements and Beer–Lambert's law for a lab-on-a-chip system. (a) Picture of the chip containing a 30 µm wide U-shaped microfluidic channel, where the horizontal part of the U has the length $\ell = 1000$ µm and is aligned with waveguides and the ends. This design allows for onchip absorbance measurements. The waveguide and channel walls are made of 90 µm thick SU-8 polymer spun on top of a glass substrate. The lid is made of PMMA polymer. A solution of fluorescein is used to make the light path inside the microfluidic channel visible. (b) Absorbance measurements at 633 nm on the dye bromothymol blue having the absorptivity $\varepsilon = 2.65 \times 10^6$ M^{-1} m^{-1}. The linear fits illustrate Beer–Lambert's law for two different absorbance cells of lengths $\ell = 100$ µm and 1000 µm, respectively. Adapted from Mogensen *et al.*, Appl. Optics **42**, 4072 (2003), courtesy of Klaus Bo Mogensen and Jörg P. Kutter, DTU Nanotech.

and can thus identify the conductivity-induced absorptivity ε_σ as

$$\varepsilon_\sigma \equiv \frac{2\gamma k_n}{c_\mathrm{m}} = \frac{\sigma}{\epsilon c_0 c_\mathrm{m}}\, n. \tag{16.27}$$

For a mixture with N species of target molecules, which do not interact with each other, the total absorbance $\mathcal{A}$ is given by the sum of individual absorbances $\mathcal{A}_i$,

$$\mathcal{A} = \mathcal{A}_1 + \mathcal{A}_2 + \cdots + \mathcal{A}_N = \varepsilon_1 \ell c_1 + \varepsilon_2 \ell c_2 + \cdots + \varepsilon_N \ell c_N. \tag{16.28}$$

Beer–Lambert's law is only approximate, and in real experiments deviations from it are often encountered. One example is the influence of stray radiation, i.e. radiation that enters the detector as a result of scattering phenomena off the surfaces of prisms, lenses, *etc.*, and not directly from the sample. The observed absorbance $\mathcal{A}_\mathrm{obs}$ therefore involves the addition of the intensity I_s of this stray radiation,

$$\mathcal{A}_\mathrm{obs} = \log \frac{I_0 + I_\mathrm{s}}{I + I_\mathrm{s}}. \tag{16.29}$$

In Fig. 16.2 are shown two examples of Beer–Lambert's law applied to onchip absorption measurements by Mogensen *et al.*, Appl. Optics **42**, 4072 (2003). Two microfluidic channels with cross-sections 100 µm × 90 µm and of length $\ell = 1000$ µm and 100 µm were used as the absorbance cells. They were filled with weak concentrations c_m ranging up to 100 µM of the dye bromothymol blue having the absorptivity $\varepsilon = 2.65 \times 10^6$ M^{-1}m^{-1}. The absorbance measurements were performed at a wavelength of 633 nm, and they resulted in the two straight lines of absorbance versus concentration, thus confirming Beer–Lambert's law, Eq. (16.25b).

The sensitivity of the absorbance measurements is defined as the slope $\mathrm{d}\mathcal{A}/\mathrm{d}c_\mathrm{m} = \varepsilon\ell$, and as expected the measured sensitivity for the large path length $\ell = 1000$ µm is nearly one order of magnitude greater than that for $\ell = 100$ µm, namely $2.4 \times 10^{-3}\ \mathrm{M}^{-1}$ and $0.3 \times 10^{-3}\ \mathrm{M}^{-1}$, respectively. Due to noise problems in the experiment, the detection limit improved only by a factor of 2, from 30 µM to 15 µM, when changing the absorbance cell path length from $\ell = 100$ µm to 1000 µm, where under ideal conditions one would also expect an order of magnitude improvement.

Absorbance measurements is an important analysis tool in optofludics. However, it is clear that the drive towards miniaturization poses a problem for such measurements. As the path length ℓ decreases, the sensitivity decreases and the detection limits increases. One possible strategy for solving this fundamental problem, the introduction of photonic bandgap structures in optofluidics, is briefly discussed in Section 16.6.

16.3 Molecular fluorescence and phosphorescence

After an optical absorption process, as shown in the Jablonski[1] diagram of Fig. 16.1(b), the excited molecule returns to its ground state by one of several possible de-excitation processes.

The relaxation of excited molecules in liquid solutions is often dominated by the so-called non-radiative vibrational relaxation. It occurs when a molecule excited in some vibrational state collides with the surrounding solvent molecules and transfers its excess vibrational energy to the solvent molecules. This process, which leads to a tiny increase in the temperature of the medium, is so effective that the average lifetime of an excited vibrational state is only about 10^{-15} s. After vibrational relaxation the molecule is left in the lowest vibrational state of some excited electronic state, and from there the molecule can return either to its ground state by emitting a photon, or an excited vibrational level of a lower electronic state by other non-radiative relaxation processes. These processes are less efficient, so typically the average lifetime of the lowest vibrational state at room temperature is between 10^{-9} and 10^{-6} s. Note in the case of photon-emission that, due to internal energy losses, the wavelength λ_in of the absorbed photon may be shorter than the wavelength λ_out of the emitted photon.

Fluorescence is another relaxation process. It is a photon-emission process with many applications in chemical analysis. A given target molecule is first excited using an incident electromagnetic radiation at a specific frequency tuned to an electronic transition in the molecule. As mentioned above, for many molecules this radiative relaxation is dominated by non-radiative relaxation processes. However, if the latter are suppressed the emitted photon can be detected. In this case, the average lifetime for the excited state is of the order $10^{-9} - 10^{-6}$ s. The intensity I_f of the fluorescent radiation is proportional to the radiant intensity $I_0 - I$ absorbed by the system,

$$I_\mathrm{f} = K(I_0 - I). \tag{16.30}$$

If we let ε be the absorptivity of the fluorescing molecules and ℓ the length of the sample, we can rewrite Eq. (16.30) using Beer–Lambert's law, Eq. (16.25b), and obtain

$$I_\mathrm{f} = KI_0(1 - \mathrm{e}^{-\varepsilon\ell c}) \approx K\varepsilon\ell I_0\, c, \tag{16.31}$$

[1]The diagram is named after the Polish physicist Aleksander Jabłoński. According to my Polish students the name is pronounced something like "Ya-boing-ski".

where we have Taylor expanded the exponential function to first order. Thus, as I_f is proportional to both the incident radiation intensity I_0 and the concentration c of target molecules, it is a very sensitive method for analysis and biochemical sensing.

In micro- and optofluidics fluorescence is also a widely used method for tracing and imaging of liquid flows. Often, solutions of the fluorescent molecule fluorescein are used to map out flow properties of a given lab-on-a-chip system. Examples of this can be seen in Figs. 14.3 and 16.2(a).

In contrast to fluorescence, the phenomenon of phosphorescence is not so widely used in analytical chemistry, as substances with this property are relatively rare. Phosphorescence is due to occupation of the so-called excited triplet states, from which the molecule cannot easily relax to the normal singlet states due to angular momentum properties. The average lifetime of the excited triplet state is of the order $10^{-3} - 10^{3}$ s.

16.4 Onchip waveguides

As the active volumes in optofluidic devices by definition are small, it is often necessary to incorporate optical waveguides in the design. Thereby, the incident radiation for chemical analysis can be brought directly to the point where it is going to be used. One example is shown in Fig. 1.1, while another is presented in Fig. 16.2(a).

The waveguide relies on the concept of total internal reflection. Consider a waveguide made of a material with the index of refraction n_in, and let it be surrounded on the outside by a material with the index of refraction n_out. Snell's law states that for light inside the waveguide hitting the interface between the waveguide and the air with an angle of incidence θ_i will leave the interface with an angle of refraction θ_r, both angles with respect to the surface normal, given by

$$n_\mathrm{in} \sin\theta_i = n_\mathrm{out} \sin\theta_r. \tag{16.32}$$

In the case of $n_\mathrm{in} > n_\mathrm{out}$ there exists a critical angle of incidence θ_c for which $\theta_r = 90°$, so that the refracted light is just moving parallel to the surface of the waveguide, i.e. it is not really leaving the waveguide. From Snell's law it follows that

$$\theta_c = \arcsin\left(\frac{n_\mathrm{out}}{n_\mathrm{in}}\right). \tag{16.33}$$

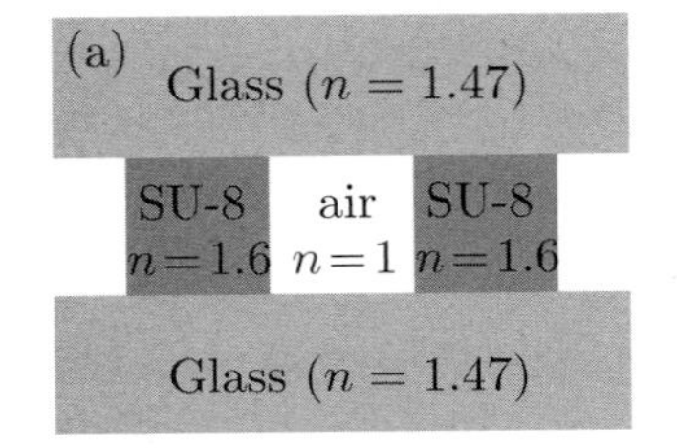

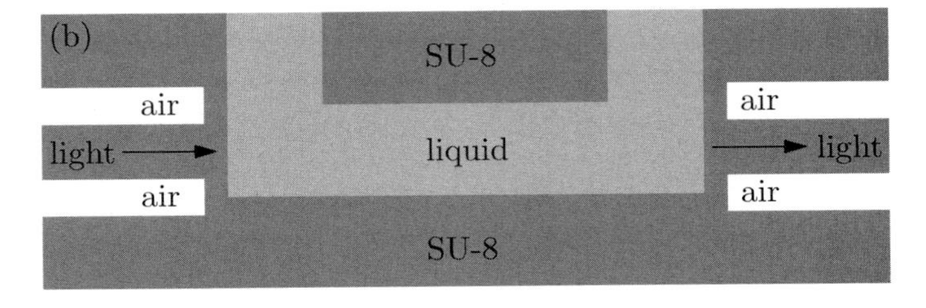

Fig. 16.3 (a) Cross-section of an onchip waveguide system with two waveguides made of the polymer SU-8 (dark gray). The sides are air or glass with lower indices of refraction. At other places on the chip the SU-8 layer constitutes the walls of microfluidic channels. (b) Top-view sketch of the onchip absorption setup from Fig. 16.2(a). The SU-8 polymer (dark gray) constitutes both waveguides (arrows) surrounded by air (white) and walls defining a microfluidic channel (light gray).

Snell's law implies that no angle of refraction can be found for light rays moving inside the waveguide and hitting the sidewalls with a small angle of incidence $\theta_i < \theta_c$. Thus, such light rays stay confined inside the waveguide by total reflection at the surface, and the critical angle is therefore also denoted the critical angle θ_c for total internal reflection.

Examples of onchip optofluidic systems with integrated waveguides are shown in Figs. 16.2 and 16.3.

16.5 Onchip laser sources

For many optofluidic applications it would be practical to integrate the light source in the chip. This can actually be done using, e.g. the fluidic dye laser shown in Fig. 16.4 and also in Fig. 4.3. In the following we sketch how such a laser works, i.e. how light amplification takes place.

Ultimately, the lasing action in a liquid dye laser is due to electronic transitions in the dye molecules and their influence on the index of refraction n of the solution. Let $\omega_0 = 2\pi(E_1 - E_0)/h$ denote the absorption frequency at which the dye molecule is excited from its ground state of energy E_0 to some excited state of energy E_1. Furthermore, let γ denote the decay rate, i.e. the probability per time unit that the excited molecule returns to its ground state. For a simple, classical model of the molecules in concentration c_{m}, it can then be shown, see Exercise 16.1, that an applied electric field $E\mathrm{e}^{-\mathrm{i}\omega t}$ will induce an electric

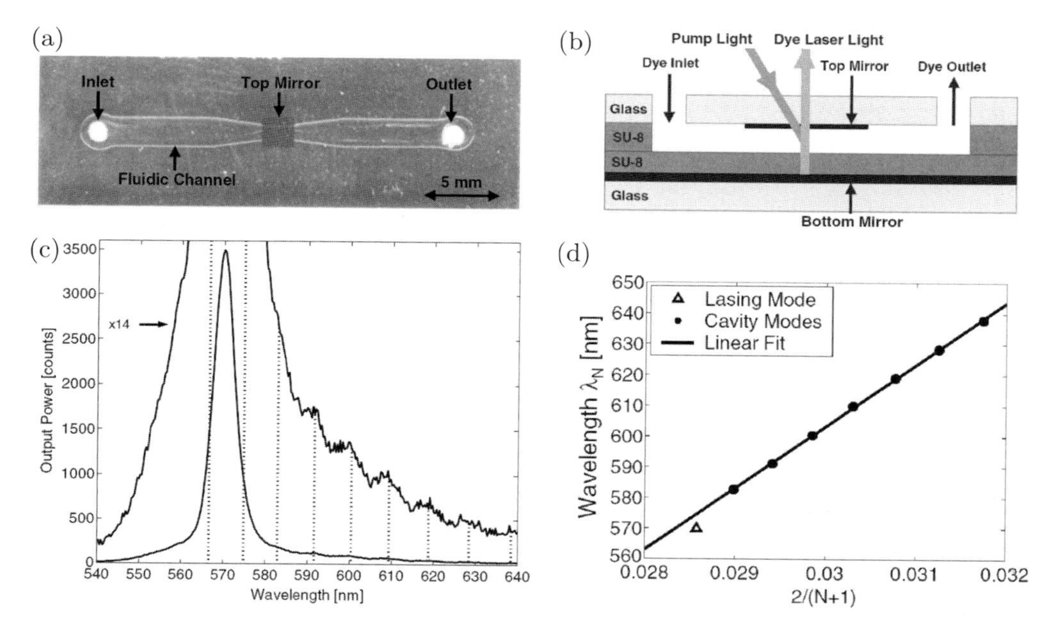

Fig. 16.4 (a) Top view picture of a microfluidic dye laser made in glass and the polymer SU-8. (b) Side view sketch of the laser design including the Cr/Au mirrors, SU-8 layers and the glass lid all assembled on a glass substrate. (c) Measurement of the output intensity showing both cavity modes ($\times 14$) and the strong lasing mode. (d) The measured wavelengths of the cavity modes as well as of the lasing mode. All figures are reproduced from B. Helbo, A. Kristensen and A. Menon, J. Micromech. Microeng. **13**, 307–311 (2003) by permission of IOP Publishing Ltd.

polarization P, defined in Eqs. (8.7) and (8.8), of magnitude

$$P(t) = P(\omega)\,\mathrm{e}^{-\mathrm{i}\omega t} \approx -\frac{\mathrm{e}^2 c_\mathrm{m}}{2m\epsilon_0\omega_0}\,\frac{1}{\omega-\omega_0+\mathrm{i}\gamma}\,\epsilon_0 E\,\mathrm{e}^{-\mathrm{i}\omega t}, \tag{16.34}$$

where m is the electron mass. We note that due to the decay rate γ the molecular polarization acquires an imaginary part. This implies that also the frequency-dependent index of refraction becomes complex, $n = n_\mathrm{re}+\mathrm{i}n_\mathrm{im}$, since by Eq. (16.10) we have $n(\omega) = \sqrt{\epsilon(\omega)/\epsilon_0} = \sqrt{1+P(\omega)/(\epsilon_0 E)}$. As shown in Exercise 16.2, the imaginary part of n becomes

$$n_\mathrm{im} = \frac{\mathrm{e}^2 c_\mathrm{m}}{4m\omega_0}\,\frac{\gamma}{(\omega-\omega_0)^2+\gamma^2}, \quad \text{(classical model)}, \tag{16.35}$$

and consequently a plane wave given by Eq. (16.8a) propagating in the x direction will be damped,

$$\mathbf{E}(\mathbf{r},t) = \mathbf{E}_0\,\mathrm{e}^{\mathrm{i}(nk_0x-\omega t)} = \mathbf{E}_0\,\mathrm{e}^{\mathrm{i}(n_\mathrm{re}k_0x-\omega t)}\,\mathrm{e}^{-n_\mathrm{im}k_0x}. \tag{16.36}$$

However, this conclusion is only true if n_im is positive. It turns out that in a more correct quantum-mechanical treatment of the absorption process the molecular concentration c_m must be replaced by the difference $c_\mathrm{m}^{(0)} - c_\mathrm{m}^{(1)}$ between the concentrations $c_\mathrm{m}^{(0)}$ and $c_\mathrm{m}^{(1)}$ of molecules in the ground state E_0 and the excited state E_1, respectively,

$$n_\mathrm{im} = \frac{\mathrm{e}^2\big(c_\mathrm{m}^{(0)} - c_\mathrm{m}^{(1)}\big)}{4m\omega_0}\,\frac{\gamma}{(\omega-\omega_0)^2+\gamma^2}, \quad \text{(quantum model)}. \tag{16.37}$$

With this more correct description it becomes clear that if it were possible to achieve population inversion, i.e. $c_\mathrm{m}^{(1)} > c_\mathrm{m}^{(0)}$, then the imaginary part n_im of the index of refraction would be come negative, and thus, according to Eq. (16.36), amplification of the light wave would result.

In thermal equilibrium, the ratio $c_\mathrm{m}^{(1)}/c_\mathrm{m}^{(0)}$ is given by

$$\frac{c_\mathrm{m}^{(1)}}{c_\mathrm{m}^{(0)}} = \mathrm{e}^{-(E_1-E_0)/k_\mathrm{B}T} = \mathrm{e}^{-hf/k_\mathrm{B}T}. \tag{16.38}$$

For optical frequencies $c_\mathrm{m}^{(1)}/c_\mathrm{m}^{(0)} \approx 10^{-35}$, a ratio so small that it is not possible to achieve inversion population by external pumping of the system. The simplest case where population inversion is possible is a three-level molecule $E_0 < E_1 < E_2$ with population concentrations $c_\mathrm{m}^{(0)}$, $c_\mathrm{m}^{(1)}$ and $c_\mathrm{m}^{(2)}$, respectively, such as sketched in Fig. 16.1(b).

Let Γ be the rate per unit time by which molecules are excited from E_0 to E_2, while γ_{ij} is the rate per unit time by which they relax from level E_j down to a lower-lying level E_i. In steady state, where $\mathrm{d}c_\mathrm{m}^{(i)}/\mathrm{d}t = 0$ for all levels, we have the following master equation relating the rates by which molecules enter and leave level i,

$$0 = \frac{\mathrm{d}c_\mathrm{m}^{(0)}}{\mathrm{d}t} = -\Gamma\,c_\mathrm{m}^{(0)} + \gamma_{01}\,c_\mathrm{m}^{(1)} + \gamma_{02}\,c_\mathrm{m}^{(2)}, \tag{16.39a}$$

$$0 = \frac{\mathrm{d}c_\mathrm{m}^{(1)}}{\mathrm{d}t} = -\gamma_{01}\,c_\mathrm{m}^{(1)} + \gamma_{12}\,c_\mathrm{m}^{(2)}, \tag{16.39b}$$

$$0 = \frac{\mathrm{d}c_\mathrm{m}^{(2)}}{\mathrm{d}t} = +\Gamma\,c_\mathrm{m}^{(0)} - (\gamma_{02}+\gamma_{12})\,c_\mathrm{m}^{(2)}. \tag{16.39c}$$

From Eq. (16.39)(b) and Eq. (16.39)(c) we find

$$c_\mathrm{m}^{(1)} = \frac{\gamma_{12}}{\gamma_{01}}\, c_\mathrm{m}^{(2)} = \frac{\gamma_{12}\Gamma}{\gamma_{01}(\gamma_{02}+\gamma_{12})}\, c_\mathrm{m}^{(0)}. \tag{16.40}$$

Population inversion $c_\mathrm{m}^{(1)} > c_\mathrm{m}^{(0)}$ can thus be achieved if the pumping rate Γ is above the following threshold,

$$\Gamma > \gamma_{01}\left[1+\frac{\gamma_{02}}{\gamma_{12}}\right]. \tag{16.41}$$

We are now in a position to state the conditions for sustaining laser oscillations in a liquid-filled microcavity. As an example, we take the simplest geometry, a Fabry–Perot cavity. This cavity consists of two plane-parallel mirrors placed in the yz-planes at $x=0$ and $x=\ell$. The mirrors are almost perfect in the sense that for light of intensity I incident on the mirror a fraction RI is reflected, where the coefficient $R\approx 1$ is denoted the reflectance. Such a cavity will sustain standing waves with the electric field of the form

$$E = E_0 \sin\big(k_n x\big)\, \mathrm{e}^{-\mathrm{i}\omega t}\mathrm{e}^{-n_\mathrm{im}k_0 x}, \tag{16.42a}$$

with the resonance condition on k_n given by

$$k_n = N\,\frac{\pi}{\ell}, \quad N=1,2,3,\dots. \tag{16.42b}$$

The corresponding cavity resonance wavelengths $\lambda_n^{(N)}$ and frequencies $\omega_\mathrm{c}^{(N)}$ are given by

$$\lambda_n^{(N)} = \frac{2\pi}{k_n(N)} = \frac{2}{N}\,\ell, \quad N=1,2,3,\dots, \tag{16.42c}$$

$$\omega_\mathrm{c}^{(N)} = 2\pi\,\frac{c}{\lambda_n^{(N)}} = N\pi\,\frac{c}{\ell} = N\,\frac{\pi}{n_\mathrm{re}}\,\frac{c_0}{\ell}, \quad N=1,2,3,\dots. \tag{16.42d}$$

Since intensity is proportional to the square of the electric field, we find that intensity amplification or gain can be characterized by a factor e^{Gx} given by

$$\mathrm{e}^{Gx} \equiv \left[\mathrm{e}^{-n_\mathrm{im}k_0 x}\right]^2 = \mathrm{e}^{-2n_\mathrm{im}k_0 x}, \tag{16.43}$$

so that the gain G itself through Eq. (16.37) is given by

$$G = \frac{\mathrm{e}^2\omega}{2m\omega_0 c}\,\frac{\gamma}{(\omega-\omega_0)^2+\gamma^2}\,\big(c_\mathrm{m}^{(1)} - c_\mathrm{m}^{(0)}\big). \tag{16.44}$$

The condition for having laser oscillations in the cavity is traditionally stated as the situation where the intensity loss during one full roundtrip, i.e. from $x=0$ to $x=\ell$ and back again, is compensated by the intensity gain induced by the inversion population. One roundtrip contains two reflections, one at each mirror, and a traversal length of 2ℓ. The threshold condition for lasing is therefore

$$R^2\mathrm{e}^{2G\ell} \equiv 1 \quad\Rightarrow\quad 2G\ell + \ln(R^2) = 0 \quad\Rightarrow\quad G = \frac{1-R^2}{2\ell}, \tag{16.45}$$

where we have used the Taylor expansion $\ln(1+x)\approx x$ with $x=R^2-1$, valid since $R\approx 1$. If other intensity losses are present in the cavity besides the tiny transmission through the mirrors these are conventionally collected in a term denoted the effective loss constant G_{loss}, so the final expression for the threshold G_{thres} of the gain to ensure lasing becomes

$$G_{\mathrm{thres}} = \frac{1-R^2}{2\ell} + G_{\mathrm{loss}}. \tag{16.46}$$

The oscillation frequency ω_{laser} of the laser turns out to lie in between the laser transition frequency ω_0 and the resonance frequency $\omega_{\mathrm{c}}^{(N)}$,

$$\omega_{\mathrm{laser}} = \frac{Q_0\omega_0 + Q_{\mathrm{c}}\omega_{\mathrm{c}}^{(N)}}{Q_0 + Q_{\mathrm{c}}}, \tag{16.47}$$

where the Q factors, related to the losses in the molecular oscillation ω_0 and the cavity oscillation $\omega_{\mathrm{c}}^{(N)}$, are defined as

$$Q_0 \equiv \frac{\omega_0}{2\gamma}, \tag{16.48a}$$

$$Q_{\mathrm{c}} \equiv \frac{\omega_{\mathrm{c}}^{(N)}W}{\mathcal{P}_{\mathrm{loss}}}, \tag{16.48b}$$

with W being the energy stored in the cavity and $\mathcal{P}_{\mathrm{loss}}$ the energy loss per unit time in the mirrors and the medium.

In the experimental results depicted in Fig. 16.4(c), we see two clear confirmations of the theory. As the wavelength of the excitation light source is swept through the interval from 540 nm to 630 nm a strong line appears centered around 570 nm, followed by much smaller but regular maxima. As shown in the insert the latter correspond to cavity resonances fulfilling condition Eq. (16.42c), whereas the strong line identified as the laser line in accordance with Eq. (16.47) differs from the cavity resonances.

It is possible to tune the frequency of the laser by changing the index of refraction in the cavity. For the microfluidic dye laser this is done by changing the concentration c_{m} of the dye in the microfluidic flow through the laser cavity.

16.6 Photonic bandgap structures in optofluidics

As mentioned at the end of Section 16.2, miniaturization poses a problem in optofluidics. The Beer–Lambert law predicts an exponentially damped intensity $I = I_0 \exp(-\varepsilon c_{\mathrm{m}}\ell)$ due to absorption, but as the optical path length ℓ decreases, so does the absorption, which eventually for small lengths leads to a loss of absorption signal. We will end this chapter on optofluidics by briefly mentioning a new development that addresses this problem. By introducing nanostructured dielectrics inside microfluidic channels it is possible to create photonic crystals that at specific propagation frequencies leads to the phenomenon of slow-light enhanced light–matter interaction. In the following, we take as an example the theoretical work in the group of N. Asger Mortensen at DTU Nanotech.

In Fig. 16.5(a) is shown an example of how a Beer–Lambert absorbance cell can be modified by inserting nanopatterned dielectric structures such as slabs, posts and a matrix with holes all forming regular lattices, the so-called photonic crystals. If we, for simplicity, assume that the electromagnetic damping in the dielectric structures are negligible in comparison to the (small) damping in the liquid, then the primary effect of the presence of the dielectric

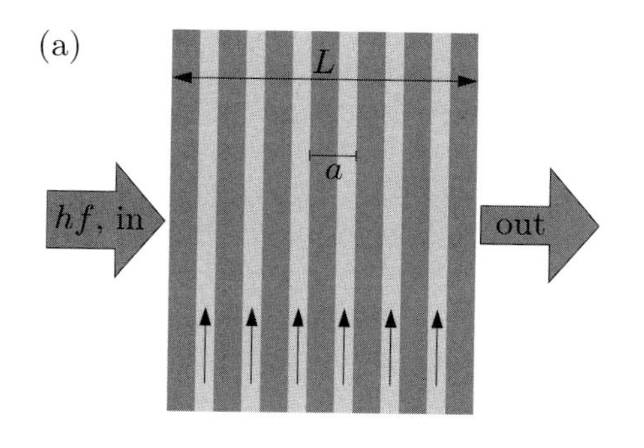

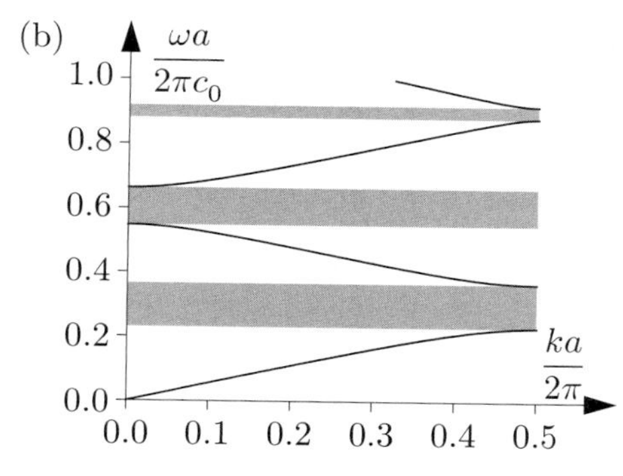

Fig. 16.5 (a) A Beer–Lambert absorbance cell of length L configured as a Bragg stack with alternating layers of parallel plates of some transparent dielectric (dark gray) and liquid-filled spacings (light gray) supporting a flow (black arrows). (b) Calculated dispersion relation $\omega(k)$ (full lines) for the electromagnetic modes in the liquid-filled Bragg stack. The widths of the wall and the liquid spacing is $0.2a$ and $0.8a$, while the respective indices of refraction are set to 3.0 and 1.33. The lowest three forbidden frequency intervals are marked by gray rectangles.

lattice structures is a change of the dispersion relation $\omega(k_n)$ of the electromagnetic field. This is illustrated in Fig. 16.5(b), where strong deviations from the usual linear dispersion, $\omega = c\,k_n$ given in Eq. (16.8d), are clearly seen.

Once a non-linear dispersion relation $\omega(k_n)$ is obtained the speed of light is strongly affected. This is because the speed of light c is given by the so-called group velocity v_{g}, which is obtained from the dispersion relation as

$$c \equiv v_{\mathrm{g}} \equiv \frac{\partial \omega}{\partial k_n}. \tag{16.49}$$

We see that in the case of a linear dispersion relation, this definition of the speed of light is in accordance with our previous definition. Note that for some specific frequencies the dispersion curves shown in Fig. 16.5(b) have very small slopes, which implies $v_{\mathrm{g}} \ll c_0$.

The implications of this on the Beer–Lambert law are quite dramatic. The conductivity-induced sensitivity $\varepsilon_\sigma c_{\mathrm{m}} = (\sigma/\epsilon c_0)\,n$, given in Eq. (16.27), changes in two ways. First, the index of refraction must be reinterpreted as $n = c_0/v_{\mathrm{g}}$, and secondly, a factor $f < 1$ that expresses the effective fraction of the electromagnetic field that exists in regions with the lossy liquid as opposed to the lossfree dielectric lattice. The resulting sensitivity is

$$\varepsilon_\sigma c_{\mathrm{m}} = \frac{\sigma}{\epsilon c_0}\,\frac{c_0}{v_{\mathrm{g}}} \times f = \frac{\sigma}{\epsilon c_0}\,n \times f\,\frac{c_0/n}{v_{\mathrm{g}}}. \tag{16.50}$$

As shown by Mortensen and Xiao it is possible by proper design of the photonic crystals to obtain drastic lowering of v_{g} while still maintaining an f factor close to unity. This technique of employing slow light opens up for continued downscaling of microfluidic biochemical sensing systems, because the hitherto expected loss of sensitivity due to length dependence in the Beer–Lambert law is countered by the enhancement factor $f(c_0/n)/v_{\mathrm{g}} \ll 1$ of Eq. (16.50).

16.7 Further reading

The three review papers by Verpoorte (2003), by Psaltis, Quake, and Yang, C. (2006) and by Monat, Domachuk, and Eggleton (2007) provide a very good overview of the emerging field of optofluidics, while Helbo, Kristensen, and Menon, A. (2003) and Balslev *et al.* (2006) are examples of experimental papers on waveguides and onchip lasers. Mortensen and Xiao (2007) is a recent example of theoretical optofluidics.

16.8 Exercises

Exercise 16.1
A simple classical model of molecular polarization
The frequency-dependent molecular polarization $P(\omega)$ in a simple classical model is given in Eq. (16.34). To derive this expression we consider an electron of charge e and mass m restricted to move classically in only 1D along the x axis in a molecule. When forced out of equilibrium, the electron will oscillate harmonically with the angular frequency ω_0 corresponding to the excitation frequency. Besides this harmonic restoring force, the electron is also influenced by the external electric field $E\mathrm{e}^{-\mathrm{i}\omega t}$ from the light wave by the force. Finally, the decay of an excited molecule back to its ground state, can classically be modelled as a velocity-dependent friction force $-2m\gamma\frac{\mathrm{d}x}{\mathrm{d}t}$.

(a) Write down Newton's second law applied to the molecular electron given the above mentioned classical model.

(b) Take as a trial solution the harmonically oscillating function $x(t) = x_0\mathrm{e}^{-\mathrm{i}\omega t}$, and determine the frequency-dependent amplitude $x_0(\omega)$.

(c) The damping is often minute, $\gamma \ll \omega_0$, so that the amplitude $x_0(\omega)$ is only significantly different from zero for $\omega \approx \omega_0$. Use this insight to simplify the above expression for $x_0(\omega)$

(d) If we assume that the polarization of the molecule is zero when the electron is in its equilibrium position, it follows from Eq. (8.7) that the dipole moment of the molecule is $p(t) = -ex(t)$, and thus by Eq. (8.8) that the polarization of the molecules in a concentration c_m is $P(t) = -ec_\mathrm{m}\, x(t)$. Use this to verify the correctness of Eq. (16.34) for the polarization $P(t) = P(\omega)\,\mathrm{e}^{-\mathrm{i}\omega t}$.

Exercise 16.2
Imaginary part of the index of refraction
Derive the expression given in Eq. (16.35) for the imaginary part n_im of the index of refraction from Eqs. (16.10) and (16.34).

Exercise 16.3
Number of photons from an ordinary 60 W light bulb
When the number of photons is very large the electromagnetic field is well described by the classical wave equation. Let us therefore estimate the number $\dot{N}$ of photons emitted per second by an ordinary 60 W light bulb, when it is given that only 10% of the power consumed by the bulb is converted into visible (yellow) light.

Exercise 16.4
Beer–Lambert's law and stray radiation
For a given medium with absorptivity ε, length ℓ and concentration c_0 the absorbance is found to be $\mathcal{A}_0$.

(a) Use Beer–Lambert's law, Eq. (16.25b), to derive an expression for the relative absorbance $\mathcal{A}/\mathcal{A}_0$ as a function of relative concentration c/c_0.

Now imagine that some stray radiation intensity I_s is present so that the observed absorbance $\mathcal{A}_{obs}$ no longer equals the actual absorbance $\mathcal{A}$ as seen in Eq. (16.29).

(b) Derive in the limit of small stray intensity $I_s/I_0, I_s/I_0 \ll 1$ an expression for $\mathcal{A}_{obs}$ in terms of $\mathcal{A}$.

Exercise 16.5
The radiation intensity of fluorescent emission
Verify the derivation of the expression Eq. (16.31) for the radiation intensity of fluorescent emission.

Exercise 16.6
Onchip waveguides of SU-8
The polymer SU-8 used for the waveguide shown in Fig. 16.2 has an index of refraction of 1.59. Calculate the critical angle for total internal reflection at an air interface.

16.9 Solutions

Solution 16.1
A simple classical model of molecular polarization
Given the oscillation frequency ω_0, the corresponding force in the harmonic oscillator is $-m\omega_0^2 x(t)$, and given the electric field, the electric force is $-e\,E\mathrm{e}^{-\mathrm{i}\omega t}$.

(a) With the explicit expressions for the forces, Newton's second law for the electron in the molecule becomes

$$m\frac{\mathrm{d}^2x}{\mathrm{d}t^2} = -m\omega_0^2 x - 2m\gamma\frac{\mathrm{d}x}{\mathrm{d}t} - eE\mathrm{e}^{\mathrm{i}\omega t}. \tag{16.51}$$

(b) Given the trial solution $x(t) = x_0\mathrm{e}^{-\mathrm{i}\omega t}$ Eq. (16.51) becomes

$$-m\omega^2 x_0\mathrm{e}^{-\mathrm{i}\omega t} = -m\omega_0^2 x_0\mathrm{e}^{-\mathrm{i}\omega t} + \mathrm{i}\,2m\gamma x_0\mathrm{e}^{-\mathrm{i}\omega t} - eE\mathrm{e}^{-\mathrm{i}\omega t}. \tag{16.52}$$

After division by $\mathrm{e}^{-\mathrm{i}\omega t}$, the amplitude is readily found to be

$$x_0(\omega) = \frac{eE}{m}\,\frac{1}{\omega^2 - \omega_0^2 + \mathrm{i}\,2\gamma\omega}. \tag{16.53}$$

(c) Given $\gamma \ll \omega_0$ we see from Eq. (16.53) that $x_0(\omega)$ only deviates significantly from zero for $\omega \approx \omega_0$. Hence, $\omega^2 - \omega_0^2 = (\omega+\omega_0)(\omega-\omega_0) \approx 2\omega_0\,(\omega-\omega_0)$, and $x_0(\omega)$ becomes

$$x_0(\omega) \approx \frac{eE}{2m\omega_0}\,\frac{1}{\omega - \omega_0 + \mathrm{i}\gamma}. \tag{16.54}$$

(d) From Eqs. (8.7) and (16.54) it follows that the time-dependent polarization $P(t)$ of the molecules in a concentration c_m is given by

$$P(t) = -ec_m\,x(t) = -\frac{e^2c_m}{2m\omega_0}\,\frac{1}{\omega-\omega_0+\mathrm{i}\gamma}\,E\,\mathrm{e}^{-\mathrm{i}\omega t}, \tag{16.55}$$

as stated in Eq. (16.34).

Solution 16.2
Imaginary part of the index of refraction
From Eq. (16.10) it follows directly that $n(\omega) = \sqrt{\epsilon(\omega)/\epsilon_0}$. Moreover, we see from Eq. (8.12) that $\epsilon\mathbf{E} = \epsilon_0\mathbf{E} + \mathbf{P}$ and thus, for this case with $\mathbf{E} = E\,\mathrm{e}^{-\mathrm{i}\omega t}\,\mathbf{e}_x$, that $\epsilon(\omega) = \epsilon_0 + P(\omega)/E$. We are therefore led to the expression $n(\omega) = \sqrt{1 + P(\omega)/(\epsilon_0 E)} \approx 1 + P(\omega)/(2\epsilon_0 E)$, where the Taylor expansion is valid for a weak solution of molecules. We then obtain

$$n_{\mathrm{im}} \approx \mathrm{Im}\left[\frac{P(\omega)}{2\epsilon_0 E}\right] = -\frac{\mathrm{e}^2 c_{\mathrm{m}}}{4m\epsilon_0\omega_0}\,\mathrm{Im}\left[\frac{1}{\omega - \omega_0 + \mathrm{i}\gamma}\right] = -\frac{\mathrm{e}^2 c_{\mathrm{m}}}{4m\epsilon_0\omega_0}\,\mathrm{Im}\left[\frac{\omega - \omega_0 - \mathrm{i}\gamma}{(\omega - \omega_0)^2 + \gamma^2}\right]$$
$$= \frac{\mathrm{e}^2 c_{\mathrm{m}}}{4m\epsilon_0\omega_0}\,\frac{\gamma}{(\omega - \omega_0)^2 + \gamma^2}. \tag{16.56}$$

Solution 16.3
Number of photons from an ordinary 60 W light bulb
The number of photons emitted per second is denoted $\dot{N}$, and the energy per photon is hf, so the emitted power can be written as $\mathcal{P} = \dot{N}\,hf$. On the other hand, it is given that $\mathcal{P} = 0.1 \times 60\ \mathrm{W} = 6\ \mathrm{W}$. Assuming that all emitted photons correspond to yellow light, we find from Table 16.1 that $f \approx 5 \times 10^{14}$ Hz. In conclusion

$$\dot{N} = \frac{\mathcal{P}}{hf} \approx \frac{6\ \mathrm{W}}{(5 \times 10^{14}\ \mathrm{Hz}) \times (6.63 \times 10^{-34}\ \mathrm{J\,s})} = 1.8 \times 10^{19}\ \mathrm{s}^{-1}. \tag{16.57}$$

Solution 16.4
Beer–Lambert's law and stray radiation
(**a**) From Beer–Lambert's law, Eq. (16.25b), we get directly

$$\mathcal{A} = \varepsilon\ell c = \varepsilon\ell c_0\frac{c}{c_0} = \frac{c}{c_0}\,\mathcal{A}_0. \tag{16.58}$$

(**b**) Using the Taylor expansion $\log(1+x) \approx x$ for $x \ll 1$ we get from Eq. (16.29)

$$\mathcal{A}_{\mathrm{obs}} = \log\frac{I_0 + I_{\mathrm{s}}}{I + I_{\mathrm{s}}} = \log\frac{I_0\left(1 + \frac{I_{\mathrm{s}}}{I_0}\right)}{I\left(1 + \frac{I_{\mathrm{s}}}{I}\right)} = \log\frac{I_0}{I} + \log\left(1 + \frac{I_{\mathrm{s}}}{I_0}\right) - \log\left(1 + \frac{I_{\mathrm{s}}}{I}\right)$$
$$\approx \log\frac{I_0}{I} + \frac{I_{\mathrm{s}}}{I_0} - \frac{I_{\mathrm{s}}}{I} = \log\frac{I_0}{I} + \frac{I_{\mathrm{s}}}{I_0}\left(1 - \frac{I_0}{I}\right) = \mathcal{A} + \left(1 - \mathrm{e}^{\mathcal{A}}\right)\frac{I_{\mathrm{s}}}{I_0}. \tag{16.59}$$

We see that the curve for the observed absorbance bends downward compared to the actual absorbance.

Solution 16.5
The radiation intensity of fluorescent emission
By combining Eq. (16.31) and Eq. (16.25b) we obtain

$$I_{\mathrm{f}} = K(I_0 - I) = KI_0(1 - \mathrm{e}^{-\varepsilon\ell c}) \approx KI_0\left(1 - [1 - \varepsilon\ell c]\right) = K\varepsilon\ell I_0\,c. \tag{16.60}$$

Solution 16.6
Onchip waveguides of SU-8
Inserting $n_{\mathrm{in}} = 1.59$ for SU-8 and $n_{\mathrm{out}} = 1$ for air in the expression for the critical angle Eq. (16.33) gives

$$\theta_c = \arcsin\left(\frac{1}{1.59}\right) = \arcsin(0.6289) = 38.9^\circ. \tag{16.61}$$

17 Nanofluidics

Nanofluidics is an emerging research field, which deals with fluid flow on the nanometer scale. It is being boosted by the ongoing development of nanotechnological tools and techniques, and it contains the potential for both basic research, e.g. an improved understanding of the limitations and ultimate breakdown of the classical continuum description of fluid dynamics for spatially confined liquids, as well as for technology, e.g. the development of devices for handling of single molecules in solution. In this chapter we shall study a few selected topics within nanofluidics.

17.1 Investigation of the no-slip boundary condition

Since its introduction in Eq. (3.1) we have in this book applied the no-slip boundary condition $\mathbf{v} = \mathbf{v}_\mathrm{wall}$ for liquids at the boundary of a solid wall moving with velocity $\mathbf{v}_\mathrm{wall}$. This boundary condition is well tested regarding liquid flow on the macroscale, but in the following we shall look into some of the experimental evidence for its validity in micro- and nanofluidics. Thanks to the development of new experimental techniques within the past decade, the no-slip hypothesis has been questioned in a number of experimental studies on the micro- and nanometer scale. While minor deviations from the no-slip boundary conditions have only negligible effects on macroscale liquid flow behavior, they could be of significance in micro- and nanofluidic systems.

We consider the infinite parallel-plate channel of Section 3.4.2 with stationary walls and a flow velocity parallel to the x axis, $\mathbf{v} = v_x(z)\,\mathbf{e}_x$. Already in the nineteenth century Navier discussed a more general boundary condition than no-slip $v_x(0) = 0$ at the bottom plate situated at $z = 0$. This so-called Navier boundary condition reads

$$v_x(0) = \lambda_\mathrm{s}\, \partial_z v_x(0), \quad \text{(wall at } z = 0\text{)}, \tag{17.1}$$

where λ_s is the slip length or Navier length defined as the distance behind the boundary where the tangent of the velocity field intersects the x axis. Note that for $\lambda_\mathrm{s} = 0$ we recover the usual no-slip boundary condition. A geometrical interpretation of the Navier boundary condition and the slip length λ_s is shown in Fig. 17.1(a). In the case of a general surface at rest with an outward pointing normal vector $\mathbf{n}$, see Fig. 2.1, the Navier boundary condition for the normal component $\mathbf{v}_\mathrm{n} = (\mathbf{n}\cdot\mathbf{v})\mathbf{n}$ and the tangential component $\mathbf{v}_\mathrm{t} = \mathbf{v} - (\mathbf{n}\cdot\mathbf{v})\mathbf{n}$ is

$$\mathbf{v}_\mathrm{n} = \mathbf{0}, \tag{17.2a}$$

$$\mathbf{v}_\mathrm{t} = -\lambda_\mathrm{s}(\mathbf{n}\cdot\boldsymbol{\nabla})\mathbf{v}_\mathrm{t}. \tag{17.2b}$$

Joseph and Tabeling (2005) have reported a careful measurement of the velocity profile of deionized water flowing in a 100 µm wide and 10 µm high microchannel with a transparent

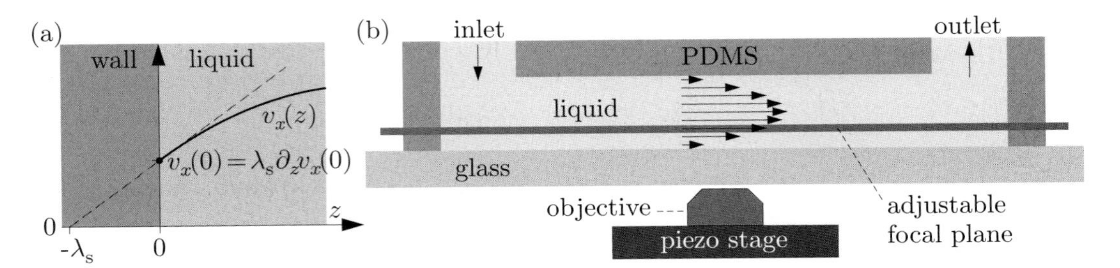

Fig. 17.1 (a) The Navier boundary condition $v_x(0) = \lambda_s\, \partial_z v_x(0)$, where the slip length λ_s is defined as the intersect of the slope of the velocity profile at the wall and the z axis. (b) Sketch of the experimental setup used by Joseph and Tabeling (2005) to measure the slip length in a microfluidic channel of height 10 µm. The velocity profile was scanned by adjusting the position of the focal plane with the piezo-stage carrying the microscope objective. For each position of the focal plane the velocity was measured using microparticle-image velocimetry.

glass bottom wall and the polymer PDMS for the other walls, see the sketch in Fig. 17.1(b). By using an objective with a large numerical aperture (NA = 1.3) they obtained a well-defined depth of field of 700 nm, and as the objective was mounted on a piezo-stage, they could move the focal plane in vertical steps of 50 nm up through the microchannel. The flow velocity was measured by particle-image velocimetry (PIV) on fluorescent tracer particles of radius $a = 50$ nm dissolved in the water in a volumetric concentration of 10^{-5}, i.e. small enough to give a good spatial resolution and large enough to suppress their Brownian motion. Each PIV recording involved a volume of size $25 \times 12 \times 0.5\ \mu\mathrm{m}^3$.

Some of their results are shown in Fig. 17.2. In panel (a) it is seen how well their data points for the velocity fall on top of a Poiseuille parabola. Deviations are seen close to the bottom wall situated at $z = 2$ µm, but these are explained by Debye-layer effects due to the different zeta-potentials of the tracer particles and the bottom wall. In panel (b) is seen a summary of several determinations of the slip length λ_s for different values of the shear rate $\partial_z v_x$ at the bottom wall. Joseph and Tabeling concluded this part of their measurements by stating the following slip length for water on glass:

$$\lambda_s = 50\ \mathrm{nm} \pm 50\ \mathrm{nm}, \quad \text{for water on glass.} \tag{17.3}$$

Their results do not invalidate the no-slip hypothesis, but on the other hand it is possible that a non-zero slip length less than 100 nm in fact does exist.

Let us analyze how a slip length $\lambda_s = 50$ nm would affect the hydraulic resistance $R_{\mathrm{hyd}}(\lambda_s)$ of microchannel in comparison with the no-slip resistance $R_{\mathrm{hyd}}(0)$. We consider a pressure-driven flow in an infinite parallel-plate channel of height h given a Navier boundary condition with a non-zero slip length λ_s on both bottom and top plates. Due to symmetry the problem is easier solved when placing the bottom plate at $\tilde{z} = -h/2$ and the top plate at $\tilde{z} = h/2$, instead of the usual $z = 0$ and $z = h$ positions. We denote the length, width, pressure drop and viscosity by L, w, Δp and η, respectively. As shown in Exercise 17.1 the velocity field $v_x(\tilde{z})$ becomes

$$v_x(\tilde{z}) = \frac{\Delta p}{2\eta L}\left[\left(1 + 4\frac{\lambda_s}{h}\right)\left(\frac{h}{2}\right)^2 - \tilde{z}^2\right], \tag{17.4}$$

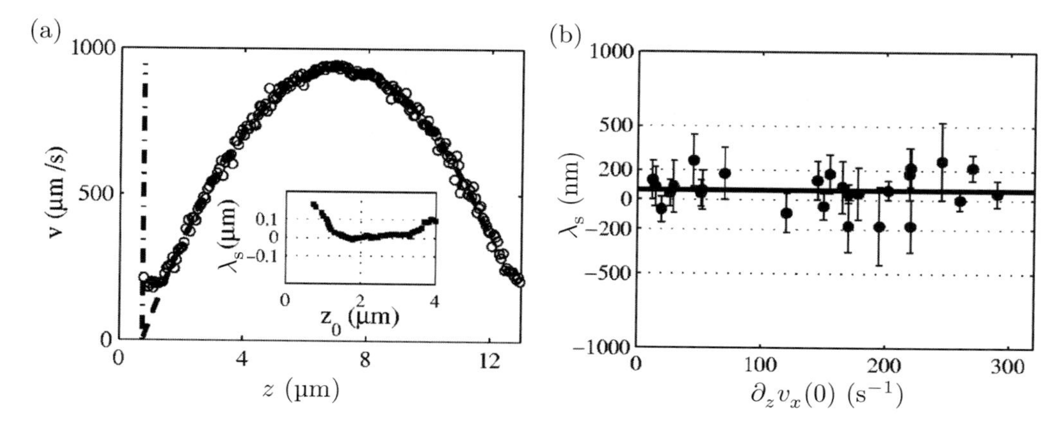

Fig. 17.2 Measurements of the velocity profile of deionized water in a 100 µm wide and 10 µm high microchannel. (a) Particle-image velocimetry measurements (open circles) of the velocity $v_x(z)$ as a function of the position z of the focal plane. The vertical dash-dotted line represents the position of the bottom glass plate of the microchannel. The dashed line, mostly covered by the data points, is the best fit to a parabola. The data points deviate from the parabola close to the bottom wall due to electrical effects stemming from the different zeta-potentials of the tracer particles and the wall, see Section 8.3. The inset shows the value obtained for the slip length λ_s as a function of left cutoff of the data points. Data points more than 1 µm away from the bottom wall leads to the same slip length. (b) The slip length λ_s measured for different values of the shear rate $\partial_z v_x(0)$ at the bottom wall. The overall result is $\lambda_\mathrm{s} = 50$ nm $\pm$ 50 nm. Figures reprinted by permission from P. Joseph and P. Tabeling, Phys. Rev. E **71**, 035303(R) (2005). Copyright (2005) by the American Physical Society.

from which we recover the usual no-slip solution when $\lambda_\mathrm{s} = 0$.

The explicit velocity field in Eq. (17.4) is easily integrated to yield the flow rate $Q(\lambda_\mathrm{s})$, from which the hydraulic resistance $R_\mathrm{hyd}(\lambda_\mathrm{s})$ is readily deduced, again see Exercise 17.1,

$$R_\mathrm{hyd}(\lambda_\mathrm{s}) = \frac{1}{1+6\frac{\lambda_\mathrm{s}}{h}}\,\frac{12\eta L}{h^3 w} = \frac{R_\mathrm{hyd}(0)}{1+6\frac{\lambda_\mathrm{s}}{h}}. \tag{17.5}$$

This expression reveals that if it were possible to increase the slip length to infinity, the hydraulic resistance, and thus viscous dissipation of energy, would vanish, which is indeed, an interesting perspective. However, even the small slip length $\lambda_\mathrm{s} = 50$ nm influences the hydraulic resistance significantly for channels of small height. For $h = 10$ µm we get $R_\mathrm{hyd}(\lambda_\mathrm{s}) = 0.97\,R_\mathrm{hyd}(0)$, a reduction of 3%, while for $h = 1$ µm we get $R_\mathrm{hyd}(\lambda_\mathrm{s}) = 0.77\,R_\mathrm{hyd}(0)$, a significant reduction of 23%.

In the literature are found reports of λ_s in the micrometer range, e.g. Tretheway and Meinhart (2002). Such extreme slip lengths, although desirable to achieve, are not very robust, probably because layers of gas forming between the liquid and the wall seem to be involved. It is fair to state that the last word has not yet been said about non-zero slip lengths for liquids flowing in micro- and nanofluidic systems.

17.2 Capillary filling of nanochannels

Using state-of-the-art nanotechnology it is possible to fabricate fluid channels having cross-sections of linear sizes in the nanometer range. In the literature, applications of such nanofluidic channels have been reported within studies of fundamental physical properties, Kameoka and Craighead (2001), van der Heyden, Stein, and Dekker (2005), Tas *et al.* (2004), as well as bio/chemical analysis, Bakajin (1998) and Reisner *et al.* (2007).

We use the work presented by Anders Kristensen's group at DTU Nanotech on capillary filling of nanochannels as an example of fluidics in flat, straight channels with rectangular cross-sections of heights less than 1 µm, see Persson *et al.* (2007). In Fig. 17.3 is shown a top view of the channel design and the principle of the fabrication method that leads to a control on the nanometer scale of the channel height. The nanofluidic properties are investigated by using the channel as a capillary pump, see Section 7.4.1. Pure (milli-Q) water or a 0.1 M NaCl electrolyte is loaded into the large micrometer-sized inlet channel, from where the liquid is sucked into the connected nanochannels by the capillary force. The advancement of the position $L(t)$ at time t of the front meniscus is recorded by a video camera attached to an optical microscope. The width of the channels is $w = 10$ µm, while the heights are less than 1 µm making the aspect ratio h/w less than 0.01. Neglecting the small aspect ratio correction, the expression Eq. (7.36) for the square $L^2(t)$ of the meniscus position becomes

$$L^2(t) = \frac{h\gamma\cos\theta}{3\eta}\,t \equiv a_\mathrm{p}\,t, \tag{17.6}$$

where $\cos\theta = \frac{1}{2}\big(\cos\theta_1 + \cos\theta_2\big)$ is the average between the cosine of the contact angle of the SiO_2 bottom wall and glass top lid, see Exercise 7.3, and where we have introduced the slope a_p for the expected linear dependence of L^2 versus time t for a Poiseuille flow profile. At 25 ℃ the experimental parameter values are $\eta = 0.89$ mPa s, $\gamma = 73$ mJ/m^3, and $\cos\theta = 0.96$, which results in the following dependence of channel height h for the expected slope a_p, based on Poiseuille flow,

$$a_\mathrm{p} = h \times 26 \text{ m/s}. \tag{17.7}$$

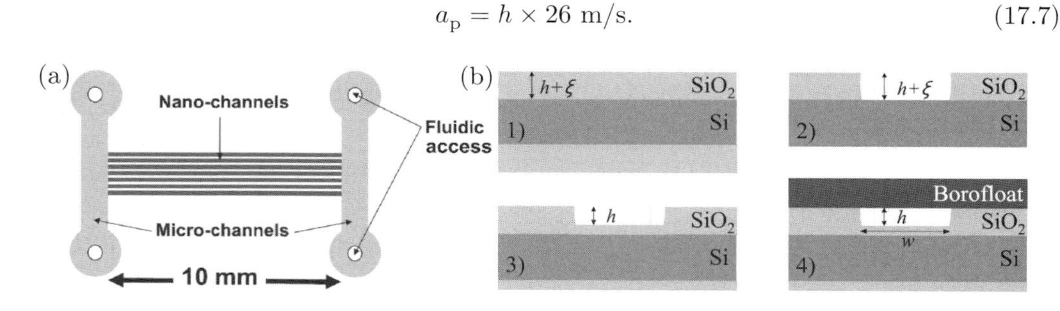

Fig. 17.3 (a) Top view of a chip containing seven nanochannels (dark gray) of length $L = 10$ mm, width $w = 10$ µm, and height h ranging from 14 to 300 nm. The nannochannels connect two microchannels (light gray) furbished with access ports through which liquid is introduced into the system. (b) The fundamental steps in the fabrication process: 1) First oxidation. 2) Wet isotropic BHF etch through an etch mask resulting in slightly sloped sidewalls. 3) Second oxidation, which is faster inside the channel region, where the oxide layer is thinner, than outside. 4) Bonding of glass lid. Adapted from Persson *et al.* (2007) courtesy of Anders Kristensen, DTU Nanotech.

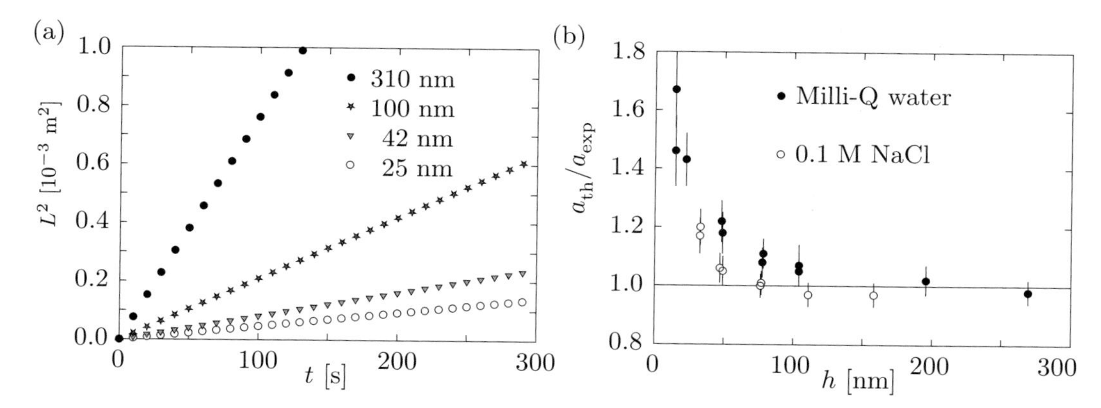

Fig. 17.4 (a) Measurements of the linear relationship $L^2 = a\,t$ between the square of the meniscus position L and time t for channel heights h = 25, 42, 100, and 310 nm. (b) The ratio $a_\mathrm{p}/a_\mathrm{exp}$ of the theoretical Poiseuille-flow slope a_p and the experimental slope a_exp for the capillary filling $L^2(t) = a\,t$ for the nanochannel. The filling liquids are Milli-Q water and an 0.1 M NaCl aqueous solution. Every data point is an averaged value from between 20 and 50 measurements. The channel width is w = 10 µm. Adapted from Persson *et al.* (2007) courtesy of Fredrik Persson, DTU Nanotech.

In Fig. 17.4 are shown some of the experimental results. Panel (a) is a scanning electron microscope picture of an actual nanochannel of height h = 75 nm and width w = 10 µm, while panel (b) contains graphs of $L(t)^2$ for four small channel heights h, as well as a plot of the ratio $a_\mathrm{exp}/a_\mathrm{p}$ as a function of channel height h between experimentally measured and theoretically expected slopes a_exp and a_p, respectively. The data exhibits two very clear features: First, the square L^2 of the meniscus position depends indeed linearly on time t, and secondly, while the measured slopes a_exp of the obtained straight lines agree well with the theoretically expected slope a_p for large channel heights, a significant and systematically increasing deviation is observed as the channel heights are decreased below approximately 100 nm.

Traditionally, the deviation of the measured slope from the slope expected from Poiseuille flow is expressed as the ratio a_p over a_exp. This emphasizes the observation of a relative increase in the resistance against capillary flow. As $a_\mathrm{p} \propto 1/\eta$ this increase can be summarized by introducing either a phenomenological effective viscosity η_eff, the hydraulic resistance ratio $R^\mathrm{exp}_\mathrm{hyd}/R^\mathrm{p}_\mathrm{hyd}$, or the flow rate ratio $Q_\mathrm{p}/Q_\mathrm{exp}$, where the indices "exp" and "p" refer quantities measured for the actual flow or expected for a pure Poiseuille flow, respectively, in the same nanochannel with a given Young–Laplace pressure drop Δp_surf,

$$\frac{a_\mathrm{p}}{a_\mathrm{exp}} = \frac{\eta_\mathrm{eff}}{\eta} = \frac{R^\mathrm{exp}_\mathrm{hyd}}{R^\mathrm{p}_\mathrm{hyd}} = \frac{Q_\mathrm{p}}{Q_\mathrm{exp}}. \tag{17.8}$$

If the experimental flow is a pure Poiseuille flow we expect a height-independent ratio $a_\mathrm{p}/a_\mathrm{exp} = 1$, but clearly the ratio is not constant so we must look for additional contributions to the flow. In the following, we first investigate the possible influence from electro-osmosis and then from a non-zero slip length.

Both pure water and 0.1 M NaCl are electrolytes, so as described in Section 8.3, the introduction of these liquids into the nanochannels leads to the formation of Debye-layers of thickness λ_D in the liquid adjacent to the channel walls. It follows from Eq. (8.26) that the Debye lengths for the two electrolytes in question are

$$\lambda_\mathrm{D}(\mathrm{NaCl},\ 10^{-1}\ \mathrm{M}) = 1\ \mathrm{nm} \ll h, \tag{17.9a}$$

$$\lambda_\mathrm{D}(\mathrm{water},\ 10^{-7}\ \mathrm{M}) = 1\ \mathrm{\mu m} > h. \tag{17.9b}$$

This, of course, just shows that the two electrolytes were carefully chosen to obtain the two extreme cases of Debye lengths either much smaller or much bigger than the channel heights. For simplicity, we apply the Debye–Hückel approximation in the analysis.

Even though no electrodes with applied voltages are attached in the capillary-filling experiments, electro-osmosis might come into play due to the charge convection currents $I_\mathrm{eo}^\mathrm{conv}$ and $I_\mathrm{p}^\mathrm{conv}$, introduced and analyzed in Section 9.2. As the position $L(t)$ of the meniscus propagate into the nanochannel driven by the Young–Laplace pressure Δp_surf, the Poiseuille-like flow sets up a non-zero charge convection current, $I_\mathrm{p}^\mathrm{conv} \neq 0$, given by Eq. (9.33). However, unless opposed by countercurrents, such a convection current would result in a gradual charging of the nanochannel, which in the long run would lead to unrealistic charging energies. Consequently, in a lowest-order approximation the total electric current I through a given cross-section situated at $x \ll L(t)$ must vanish,

$$I = I_\mathrm{p}^\mathrm{conv} + I_\mathrm{eo}^\mathrm{conv} + I_\mathrm{eo}^\mathrm{cond} \equiv 0. \tag{17.10}$$

The counterflowing electro-osmotic current can be established through a small charging of the region near the meniscus, which leads to the existence of an electric field E that drives the EO flow. The magnitude of E is found by combining Eq. (17.10) with the explicit expressions for the conduction and convection currents given by Eqs. (9.22) and (9.33),

$$\big[1 + \alpha\, g(s_\mathrm{o})\big]\, wh(\sigma_\mathrm{ion}^{+} + \sigma_\mathrm{ion}^{-})E = f(s_\mathrm{o})\, wh \frac{\epsilon\zeta}{\eta}\frac{\Delta p}{L}, \tag{17.11}$$

where

$$s_\mathrm{o} \equiv \frac{h}{2\lambda_\mathrm{D}}, \tag{17.12a}$$

$$\alpha \equiv \frac{\epsilon^2\zeta^2}{2\lambda_\mathrm{D}^2(\sigma_\mathrm{ion}^{+} + \sigma_\mathrm{ion}^{-})\eta}, \tag{17.12b}$$

$$f(s_\mathrm{o}) \equiv \left[1 - \frac{1}{s_\mathrm{o}}\tanh(s_\mathrm{o})\right], \tag{17.12c}$$

$$g(s_\mathrm{o}) \equiv \frac{1}{s_\mathrm{o}}\tanh(s_\mathrm{o}) - \mathrm{sech}^2(s_\mathrm{o}). \tag{17.12d}$$

These expressions lead to a determination of the magnitude v_eo of the EO flow necessary to guarantee a zero electric current

$$v_\mathrm{eo} \equiv \frac{\epsilon\zeta}{\eta}E = \frac{\alpha f(s_\mathrm{o})}{1 + \alpha g(s_\mathrm{o})}\,\frac{2\lambda_\mathrm{D}^2}{\eta}\,\frac{\Delta p}{L}. \tag{17.13}$$

Finally, by subtracting the counterflowing EO flow rate $Q_\mathrm{eo} = wh\, v_\mathrm{eo}\, f(s_\mathrm{o})$, given by Eq. (9.16), from the advancing Poiseuille flow $Q_\mathrm{p} = wh^3\Delta p_\mathrm{surf}/(12\eta L)$, given by Eq. (7.34), we obtain an estimate of the total flow rate Q,

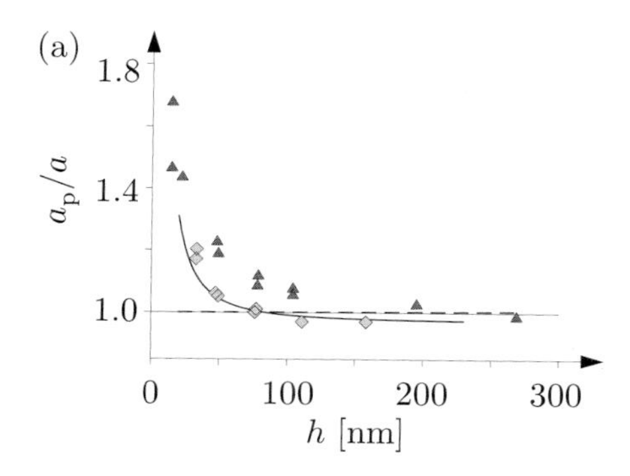

Fig. 17.5 (a) Calculated slope ratio $a_p/a = Q_p/Q$ as a function of channel height h obtained from Eq. (17.15) for 10^{-7} M water (dashed line) using $\zeta = 100$ mV and for 0.1 M NaCl (full line) using $\zeta = 250$ mV and a downshift of the base line with a factor of 0.97. The data from Fig. 17.4(b) are also shown: NaCl (light gray diamonds) and water (dark gray triangles).

$$\begin{aligned} Q = Q_p - Q_{eo} &= \frac{wh^3}{12\eta}\frac{\Delta p}{L} - \frac{\alpha f^2(s_o)}{1+\alpha g(s_o)}\frac{2wh\lambda_D^2}{\eta}\frac{\Delta p}{L} \\ &= Q_p\left[1 - \frac{6\alpha f^2(s_o)}{s_o^2\left[1+\alpha\, g(s_o)\right]}\right]. \end{aligned} \tag{17.14}$$

Returning to Eq. (17.8), we can now use Eq. (17.14) to derive an expression for the expected deviation of the slope a from pure Poiseuille flow slope a_p,

$$\frac{a_p}{a} = \frac{Q_p}{Q} = \left[1 - \frac{6\alpha f^2(s_o)}{s_o^2\left[1+\alpha\, g(s_o)\right]}\right]^{-1}. \tag{17.15}$$

Let us first focus on the NaCl electrolyte with a Debye length $\lambda_D = 1$ nm much smaller than the channel height h. Since 30 nm $< h <$ 300 nm the variable s_o lies in the range $15 < s_o < 150$, and the slope ratio $a_p/a = Q_p/Q$ from Eq. (17.15) is approximately given by

$$\frac{a_p}{a} = \frac{Q_p}{Q} \approx \left[1 - \frac{6\alpha}{s_o^2}\right]^{-1}, \quad \text{for } s_o \gg 1. \tag{17.16}$$

It is clearly seen that the theory predicts an increase in the deviation of the expected slope from unity. In Fig. 17.5 is shown a theoretical fit based on the full equation (17.15) with parameters corresponding to 0.1 M NaCl to the data. The agreement is fair, but actually the theory outlined here cannot explain the observed phenomena. This becomes clear when turning to the case of 10^{-7} M water, where the large Debye length $\lambda_D = 1$ µm is larger than the channel height h. The variable s_o now lies in the range $0.01 < s_o < 0.15$, and in this limit a Taylor expansion of a_p/a yields

$$\frac{a_p}{a} = \frac{Q_p}{Q} \approx 1 + \frac{2}{3}\,\alpha\, s_o^2, \quad \text{for } s_o \ll 1. \tag{17.17}$$

This behavior of the slope ratio is completely wrong, since it is decreasing from values above unity towards unity, as the channel height is decreased. Inserting the parameter values for pure water we obtain $Q_\mathrm{p}/Q \approx 1 + 0.27\, s_\mathrm{o}^2$, which increases from 1.000 at $s_\mathrm{o} = 0.01$ to 1.006 at $s_\mathrm{o} = 0.15$.

17.3 Squeeze flow in nanoimprint lithography

To fabricate nanofluidic channels one may choose to use the so-called nanoimprint lithography (NIL) process. Since this process itself utilizes nanofluidics, namely squeeze flow in liquid films with thicknesses of the order 100 nm, we will briefly study NIL in this section. NIL is a nanopatterning method, which combines nanometer-scale resolution with high throughput fabrication. A hard stamp containing the nanopattern is pressed into a thin polymer film deposited on a hard substrate. At sufficiently high temperature (above the glass-transition temperature T_g) the polymer melts and becomes a viscous liquid. Upon imprinting, as sketched in Fig. 17.6, the polymer therefore flows away from the regions beneath the protrusions of the stamp and into the cavities of the stamp. This flow process can be characterized as a nanofluidic squeeze flow in the polymer film, and in the following we analyze it assuming a simple model of the system.

We model the region below a given protrusion in the stamp as a parallel-plate system similar to the Couette flow system of Fig. 3.3, but now, as sketched in Fig. 17.7, moving the top plate downward antiparallel to the vertical z axis instead of parallel to the horizontal x axis. In squeeze flow, contrary to Couette flow, the height h of the liquid film between the bottom and top plates becomes a function $h(t)$ of time. We shall calculate how long it takes to squeeze the liquid film from some initial thickness h_0 to some final thickness h_f. This time is important as it sets the time scale for the NIL fabrication process. The stamp has the length $2L$ in the x direction and the width w in the y direction, so the polymer film fills the region $-L < x < L$ and $0 < y < w$.

To facilitate the calculation we make the following simplifying assumptions. The given, applied imprinting force F_imp on the stamp is constant in time, and for a sufficiently thin film the squeeze flow is so slow that in analogy with the capillary pump flow in Section 7.4.1 it can be regarded as quasi-steady. The validity of this fundamental assumption will be

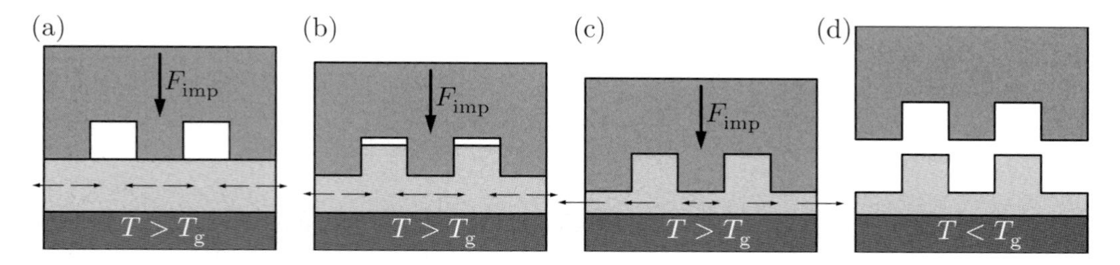

Fig. 17.6 The basic principle of nanoimprint lithography (NIL). (a) Initial phase: under the action of a constant force F_imp, the hard stamp is pressed into the soft, thin polymer film, which is deposited on a hard substrate and heated to above its glass-transition temperature T_g. (b) Intermediate phase: by the imprinting force F_imp the polymer film is slowly squeezed away from the regions beneath the protrusions of the stamp into the cavities or to the sides of the stamp. (c) Final phase: After complete filling of the cavities the polymer can only flow towards the sides of the stamp. (d) Demolding phase: the temperature is decreased below T_g, the polymer solidifies and the stamp is removed leaving behind the nanopattern from the stamp in the polymer film. Functional devices can be fabricated by further processing of the film.

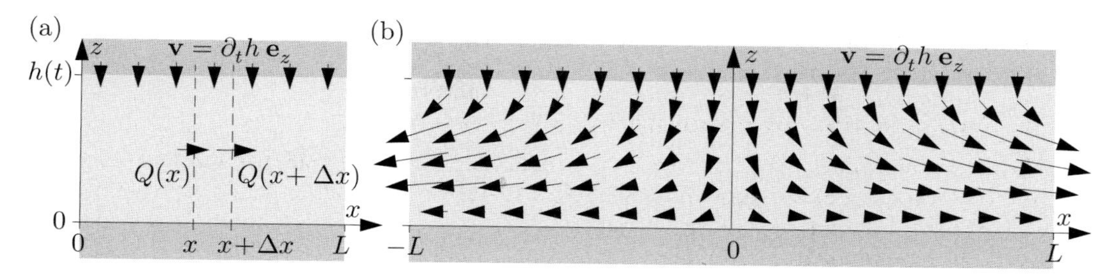

Fig. 17.7 Squeeze flow between parallel plates. (a) Conservation of mass for the slab of length Δx, width w (in the y direction), and height $h(t)$ leads to $Q(x+\Delta x) = Q(x) + \partial_t h\, w \Delta x$. (b) The velocity field Eq. (17.24), where it should be noted how the vertical component v_x increases as a function of horizontal distance x to the center of the channel at $x = 0$.

checked at the end of the calculation. If we, furthermore, consider $w \gg h(t)$ we can safely neglect edge effects from the edges at $y = 0$ and $y = w$, so at any given instant in time, where the film thickness is $h(t)$ and the speed of the stamp is $\partial_t h \equiv u_0$, the velocity field $\mathbf{v}$ must fulfill

$$\mathbf{v} = v_x(x,z)\, \mathbf{e}_x + v_z(x,z)\, \mathbf{e}_z, \text{ (quasi-steady 2D)}, \tag{17.18a}$$

$$\mathbf{v}(x,0) = \mathbf{0}, \qquad \text{(no-slip at fixed bottom plate)}, \tag{17.18b}$$

$$\mathbf{v}\big(x,h(t)\big) = \partial_t h\, \mathbf{e}_z \equiv -u_0 \mathbf{e}_z, \qquad \text{(no-slip at moving top plate)}. \tag{17.18c}$$

The rheological properties of polymers are complex, but we shall nevertheless assume that the polymer film can be described adequately as an incompressible liquid with a constant viscosity η. This assumption can be justified given the very low Reynolds number and the constant temperature in the squeeze nanoflow. Finally, given the large imprinting pressure F_imp that typically is applied in NIL processes we neglect the influence of gravity. The governing equations for the squeeze nanoflow are thus the Stokes equation and the continuity equation,

$$\eta\big(\partial_x^2 + \partial_z^2\big) v_x(x,z) = \partial_x p(x,z), \tag{17.19a}$$

$$\eta\big(\partial_x^2 + \partial_z^2\big) v_z(x,z) = \partial_z p(x,z), \tag{17.19b}$$

$$\partial_x v_x + \partial_z v_z = 0. \tag{17.19c}$$

Given the quasi-steady flow we expect a Poiseuille-like flow profile along the x direction. The flow rate $Q(x)$ at a specific cross-section x can be determined by using the continuity equation on the slab of length Δx, width w, and height $h(t)$ as sketched in Fig. 17.7(a). The outflow to the right is given by the sum of the inflow from the left and from the top (denoting $\partial_t h \equiv -u_0$) as

$$Q(x+\Delta x) = Q(x) + u_0 w\, \Delta x, \quad \text{or} \quad \partial_x Q(x) = u_0 w. \tag{17.20}$$

By symmetry, the flow rate must be zero at the center plane $x = 0$ of the channel, so the position-dependent flow rate is therefore found to be

$$Q(x) = u_0 w\, x. \tag{17.21}$$

Combining this with the assumption of a local Poiseuille flow (to be checked by the end of the calculation) between parallel plates of length Δx, leads to the following form of the pressure gradient along the x direction,

$$\partial_x p = \lim_{\Delta x \to 0} \frac{\Delta p}{\Delta x} = -\lim_{\Delta x \to 0} \frac{12\eta \Delta x Q(x)}{wh^3 \Delta x} = -\frac{12\eta u_0}{h^3}\, x. \tag{17.22}$$

The x component of the Stokes equation (17.19a) becomes

$$\eta(\partial_x^2 + \partial_z^2) v_x = \partial_x p = -\frac{12\eta u_0}{h^3}\, x, \tag{17.23}$$

with the following solution that satisfies the boundary conditions,

$$v_x(x,z) = \frac{6u_0}{h^3}\, z(h-z)\, x. \tag{17.24a}$$

From this expression for $v_x(x,z)$ and the continuity equation $\partial_z v_z = -\partial_x v_x$, Eq. (17.19c), the z-component of the velocity is easily found by integration to be

$$v_z(x,z) = \frac{u_0}{h^3}\, z^2(2z-3h). \tag{17.24b}$$

We see that in agreement first-order perturbation theory, Eq. (14.6), the velocity in the z direction is smaller than the velocity in the x direction by a factor of $z/x \approx h(t)/L$.

Given the two velocity components we can calculate the streamlines of the flow by following the procedure outlined in Eq. (14.24). Each streamline of the form $\big(x(z), z)\big)$ obeys the differential equation

$$\frac{\mathrm{d}z}{\mathrm{d}x} = \frac{v_z(x,z)}{v_x(x,z)} = \frac{z(2z-3h)}{6x(h-z)}. \tag{17.25}$$

By separation of the variables, see Exercise 17.5, we find the following explicit expression for the streamline through the point (x_0, h),

$$\mathbf{r}_{\{x_0,h\}}(z) = \begin{pmatrix} \dfrac{x_0 h^3}{z^2(3h-2z)} \\ z \end{pmatrix}. \tag{17.26}$$

A collection of streamlines is plotted in Fig. 17.8.

The non-trivial z-component of the velocity field implies a z-dependence in the pressure given by

$$\partial_z p = \eta(\partial_x^2 + \partial_z^2) v_z = \frac{6\eta u_0}{h^3}\,(z-h), \tag{17.27}$$

which together with $\partial_x p$ in Eq. (17.22) leads to the following pressure fulfilling the boundary condition $p(L,0) = p^*$,

$$p(x,z) = \frac{6\eta u_0}{h^3}\Big[\big(L^2 - x^2\big) - z\big(h - \tfrac{1}{2}z\big) + p^*\Big]. \tag{17.28}$$

We now have the solution for the velocity field $\mathbf{v}(x,z)$ and the pressure $p(x,z)$, which will allow us to calculate the imprinting time. However, before doing so we check if the

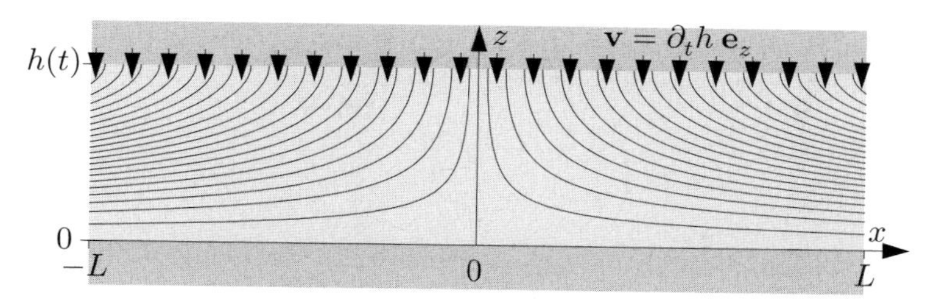

Fig. 17.8 The streamlines for squeeze flow between parallel plates, see Eq. (17.26) and Exercise 17.5. The downward velocity $\partial_t h$ of the top plate fulfills the criterion $\rho h|\partial_t h|/\eta \ll 1$ for quasi-steady motion.

solution in fact is in accordance with the assumptions. Most importantly, we have neglected the explicit time dependence $\rho\partial_t \mathbf{v}$ as well as the non-linear inertia $\rho(\mathbf{v}\cdot\boldsymbol{\nabla})\mathbf{v}$, so we now seek to explicitly state the criterion for this approximation to be valid, i.e. when these two terms are small compared to the only velocity-dependent term left in the equation, namely $\eta\nabla^2\mathbf{v}$.

The solutions for v_x and v_z are simple polynomials in x and z, so it is easy to obtain the following orders of magnitude estimates: $v_x \approx u_0 L/h$, $v_z \approx u_0$, $\partial_x \approx 1/L$, and $\partial_z \approx 1/h$, which for the inertial terms leads to $\rho(\mathbf{v}\cdot\boldsymbol{\nabla})v_i \approx (\rho u_0/h)v_i$, where $i = x, z$, while the viscous terms become $\eta\nabla^2 v_i \approx (\eta/h^2)v_i$. Since the time dependence in the solution only appears through $h(t)$ we can estimate the order of magnitude of the acceleration term as $\rho\partial_t v_i = \rho(\partial_t h)\partial_h v_i \approx (\rho u_0/h)v_i$, i.e. the same as the inertial term. The criterion for the validity of the approximative solution is that the Reynolds number-like quantity $\rho h u_0/\eta$ is much smaller than unity,

$$\frac{|\rho\partial_t v_i|}{|\eta\nabla^2 v_i|}, \frac{|\rho(\mathbf{v}\cdot\boldsymbol{\nabla})v_i|}{|\eta\nabla^2 v_i|} \approx \frac{(\rho u_0/h)v_i}{(\eta/h^2)v_i} = \frac{\rho h u_0}{\eta} = \frac{\rho h|\partial_t h|}{\eta} = \frac{\rho|\partial_t(h^2)|}{2\eta} \ll 1. \tag{17.29}$$

The slower the motion of the stamp the better is the approximative solution.

The imprinting time τ_{imp} is defined as the time it takes to squeeze the polymer film from a given initial thickness $h(0) = h_0$ to a given final thickness $h(\tau_{\mathrm{imp}}) = h_{\mathrm{f}}$. It is calculated by considering the average imprinting pressure $p_{\mathrm{imp}} = \langle p(x,z) - p^*\rangle$ derived from Eq. (17.28). If we disregard the small contribution from the z-dependent part of the pressure, which is suppressed by a factor of h^2/L^2 we obtain

$$p_{\mathrm{imp}} = \frac{1}{2L}\int_{-L}^{L} \mathrm{d}x\, \Big[p(x,0) - p^*\Big] = \frac{4\eta u_0 L^2}{h^3} = -\frac{4\eta L^2 \partial_t h}{h^3}. \tag{17.30}$$

The imprinting pressure can be expressed in terms of the constant imprinting force acting on the area $(2L)w$ of the top plate as, $p_{\mathrm{imp}} = F_{\mathrm{imp}}/(2Lw)$, and upon separation of the variables t and h in Eq. (17.30) we find

$$\tau_{\mathrm{imp}} = \int_0^{\tau_{\mathrm{imp}}} \mathrm{d}t = -\frac{4\eta L^2}{p_{\mathrm{imp}}}\int_{h_0}^{h_{\mathrm{f}}} \frac{\mathrm{d}h}{h^3} = \frac{2\eta L^2}{p_{\mathrm{imp}}}\left[\frac{1}{h_{\mathrm{f}}^2} - \frac{1}{h_0^2}\right] = \frac{\eta(2L)^3 w}{2F_{\mathrm{imp}}}\left[\frac{1}{h_{\mathrm{f}}^2} - \frac{1}{h_0^2}\right]. \tag{17.31}$$

In the literature, this expression is known as the Stefan equation.

From the Stefan equation also follows the explicit expression for the time dependence of the thickness $h(t)$ of the polymer film during squeeze flow,

$$h(t) = \left(\frac{p_{\mathrm{imp}}t}{2\eta L^2} + \frac{1}{h_0^2}\right)^{-\frac{1}{2}} \approx \sqrt{\frac{2\eta L^2}{p_{\mathrm{imp}}t}}. \tag{17.32}$$

With this expression at hand, we can check if the criterion Eq. (17.29) is fulfilled. By squaring and differentiating Eq. (17.32) we find after elimination of t by h that

$$\frac{\mathrm{d}h^2}{\mathrm{d}t} = \frac{p_{\mathrm{imp}}h^4}{2\eta L^2}. \tag{17.33}$$

Inserting appropriate parameter values for NIL by hot embossing in the polymer PMMA, for which $p_{\mathrm{imp}} \approx 10^6$ Pa, $\eta \approx 10^5$ Pa s, $\rho \approx 2 \times 10^3$ kg m^{-3}, and taking the typical length scales $h \approx 10^{-6}$ m and $L \approx 10^{-5}$ m we find

$$\frac{\rho|\partial_t(h^2)|}{2\eta} \approx 10^{-15} \ll 1. \tag{17.34}$$

One can safely conclude that the solution is accurate. For the same parameters, as studied in Exercise 17.6 the Stefan equation leads to imprinting times of the order

$$\tau_{\mathrm{imp}} \approx 10^2\ \mathrm{s}, \quad (1\ \mu\mathrm{m\ thick\ PMMA\ film\ at\ } 10^6\ \mathrm{Pa}). \tag{17.35}$$

17.4 Nanofluidics and molecular dynamics

In Section 1.3.2 the continuum description was introduced by applying the concept of averaging over molecular quantities in mesoscopic volumes. It is therefore of importance to verify and study the applicability of this fundamental description by direct simulation of liquids on the molecular level. One such method is the widely used molecular dynamics (MD) method. This method is well suited for simulating liquids in volumes of linear size less than 100 nm and for short time intervals less than 5 ns. Although severely restricted to this small space-time domain, the MD method nevertheless allows for first-principles calculations of liquid properties and behavior, and it might also provide important insight under conditions, where the continuum description fails, e.g. due to very strong spatial confinements or very high shear stresses.

The typical MD simulation comprises three main steps: (*i*) Setup of the geometry and the initial conditions, (*ii*) specification of the intermolecular interaction potential, and (*iii*) time integration of the molecular equation of motion.

In step (*i*) a calculational grid is defined to match the geometry of the given problem. A set of molecules i is introduced with positions $\mathbf{r}_i$ on randomly selected grid points, and each molecule is assigned with a random energy ε_i in accordance with the Maxwell distribution $f(\varepsilon_i) \propto \exp(-\varepsilon_i/k_{\mathrm{B}}T)$. In step (*ii*) a pair-interaction potential $V(r_{ij})$ is introduced for a pair of molecules positioned at $\mathbf{r}_i$ and $\mathbf{r}_j$ and with $r_{ij} \equiv |\mathbf{r}_i - \mathbf{r}_j|$. Often, the pair-potential is some variant of the Lennard-Jones potential V_{LJ} introduced in Section 1.3.1 and Exercise 1.2,

$$V_{\mathrm{LJ}}(r_{ij}) = 4\varepsilon\left[\left(\frac{\sigma}{r_{ij}}\right)^{12} - \left(\frac{\sigma}{r_{ij}}\right)^{6}\right]. \tag{17.36}$$

Finally, in step (*iii*) a time integration is performed of the molecular equation of motion, which typically is simply Newton's second law. For molecules of mass m moving in contact

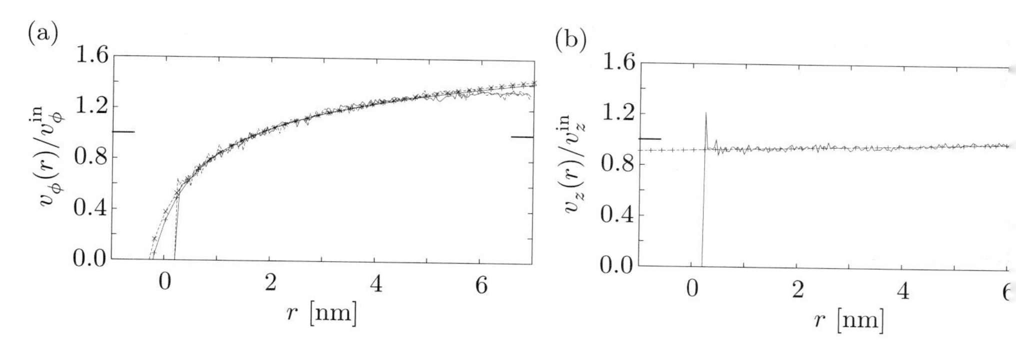

Fig. 17.9 Molecular dynamics calculation of the slip length for water flowing past a c nanotube of diameter 2.5 nm. (a) The azimuthal velocity component $v_\phi(r)$ normalized b inflow value as a function of the distance r from the surface of the carbon nanotube. The slip l is found to be $\lambda_s(v_\phi) = 0.40$ nm. (b) The axial velocity component $v_z(r)$ normalized by the i value as a function of the distance r from the surface of the carbon nanotube. The slip len found to be $\lambda_s(v_z) = 88$ nm. Figures reprinted by permission from J. H. Walther *et al.*, Phys. E **69**, 062201 (2004). Copyright (2004) by the American Physical Society.

with a dissipative thermal reservoir having the equilibration time τ the equation of m could be of the form

$$m\,\frac{\mathrm{d}^2\mathbf{r}_i}{\mathrm{d}t^2} = \sum_{j\neq i} \partial_{\mathbf{r}_i} V(r_{ij}) - \frac{m}{\tau}\,\frac{\mathrm{d}\mathbf{r}_i}{\mathrm{d}t}. \qquad (1$$

With modern computers it is possible to compute the behavior of up to 10^6 mole in the previously mentioned small space-time domain. Using advanced time integr methods and allowing for suitable equilibration time period, typically 10 to 100 τ, ave of the molecular properties can be obtained. The calculation domain is divided into a nu of bins α each situated around a position $\mathbf{x}_\alpha$, and, e.g. the Eulerian velocity $\mathbf{v}(\mathbf{x}_\alpha)$ is obta as the time average of the velocity of the molecules of mass m_i inside the bin,

$$\mathbf{v}(\mathbf{x}_\alpha) \equiv \frac{\left\langle \sum_{i\in\alpha} m_i \frac{\mathrm{d}\mathbf{r}_i}{\mathrm{d}t} \right\rangle_t}{\left\langle \sum_{i\in\alpha} m_i \right\rangle_t}. \qquad (1$$

The field of molecular dynamics is highly developed and many tricks and methods been introduced over the years. The above introduction is very rudimentary and the re interested in the MD method is referred to textbook by Allen and Tildesley (1994) o review paper by Koplik and Banavar (1995).

We end this short section on molecular dynamics by a brief presentation of the re obtained by the group of Koumoutsakos at ETH Zürich, see Walther *et al.* (2004) regar calculations of the slip length λ_s introduced in Section 17.1. In the center of a flat p lelepiped of size $L_x = 16.4$ nm, $L_y = 16.4$ nm, and $L_z = 2.1$ nm, a carbon nanotub diameter 2.5 nm was placed parallel to the z axes. The nanotube consisting of 640 at was surrounded by 1.8×10^4 water molecules. To ensure that the average velocity is n larger than the statistical spread, the flow speed was set to be of the order 100 m/s < 0.1 Although this value is much higher than the typical flow speed in microfluidics, it is

ce either non-linear or acoustic effects, so the obtained results can be

was chosen to have a non-trivial angle of 17° with respect to the and thereby there was both an azimuthal component v_ϕ and an axial velocity. These two velocity components were sampled statistically tational domain into 6 azimuthal bins and 200 radial bins, while d using 1600 radial bins. The result of the calculation is shown in n that the calculated slip length for the strongly curved azimuthal ry short, while it is much longer for the axial velocity component,

$$\lambda_s(v_\phi) = 0.40 \text{ nm}, \tag{17.39a}$$

$$\lambda_s(v_z) = 88 \text{ nm}. \tag{17.39b}$$

rbon the ngth flow h is Rev.

that the slip length is related to the presence of a stagnation nce to the particular geometry. It is interesting to note that the d for the water-carbon interface along the flat axial direction is imental result 50 nm ± 50 nm measured on the flat water-glass tion 17.1. The question about the value of the slip length remains as well as theoretically. It is one of the exciting research topics in

tion

37)

for a non-zero slip length in a flat channel

n flow in an infinite parallel-plate channel of height h given a Navier a non-zero slip length λ_s on both bottom and top plates.

sion for the velocity profile $v_x(z)$. Hint: due to symmetry the prob- en placing the bottom plate at $\tilde{z} = -h/2$ and the top plate at

ession for the hydraulic resistance $R_{\text{hyd}}(\lambda_s)$ based on the velocity xpress the result as $R_{\text{hyd}}(\lambda_s) = f(\lambda_s)\, R_{\text{hyd}}(0)$.

al results in Fig. 17.2(b) points toward a slip length $\lambda_s = 50$ nm. hange in the hydraulic resistance when changing λ_s from 0 nm to rochannels with height $h = 10$ µm and 1 µm.

les ion ges ber ned

38)

ave ler the

for a non-zero slip length in a circular channel

en flow in a straight channel of length L with a circular cross-section vier boundary condition with a non-zero slip length λ_s at the channel

essions for the slip length dependent velocity profile $v_x(z)$ and for $R_{\text{hyd}}(\lambda_s)$.

elative change in the hydraulic resistance when changing λ_s from two values of the radius $a = 5$ µm and 0.5 µm.

lts ng al- of ns ch a. ot

Exercise 17.3
Brownian motion of tracer particles
In the PIV measurements of the slip length tracer particles of radius $a = 50$ nm were employed.

(a) Calculate the diffusion constant D of the tracer particles in deionized water at room temperature.

(b) The PIV measurements involved a recording region of height $\Delta z = 500$ nm and a time interval $\tau = 20$ ms between each pair of pictures on which the PIV analysis is based. Discuss the likelihood of tracer particles blurring the PIV signal by diffusing in or out of the recording region in the time interval τ between the two pictures in a pair of PIV pictures.

Exercise 17.4
The Debye length of pure water and 0.1 M NaCl
The pH value of pure water is 7, which by definition means that the proton concentration is $c = 10^{-7}$ M. Calculate the Debye length λ_D for pure water and for 0.1 M NaCl used in the experiments of the capillary filling of nanochannels, see Section 17.2.

Exercise 17.5
Streamlines of the quasi-steady squeeze flow
Calculate the streamlines for the squeeze flow in the quasi-steady limit given by the velocity field $\big(v_x(x,z), v_z(x,z)\big)$ in Eq. (17.24). Hint: separate the variables in the defining differential equation $\mathrm{d}x/v_x = \mathrm{d}z/v_z$ and find x as a function of z.

Exercise 17.6
Nanoimprint lithography time
Consider a nanoimprint stamp covering the area 1 mm × 1 mm consisting of a number of parallel rectangular cavities of length $L_\mathrm{c} = 30$ µm in the x direction, width $w = 1$ mm in the y direction, and height $h_\mathrm{c} = 200$ nm in the z direction, and separated by protrusions of length $2L = 20$ µm in the x direction, see Fig. 17.6. The incompressible polymer film has a viscosity of $\eta = 10^5$ Pa s and an initial thickness of $h_0 = 300$ nm. The nanoimprint pressure is constant and given by $p_\mathrm{imp} = 5$ MPa.

(a) Calculate the final film thickness h_f defined as the thickness underneath the protrusions exactly at the moment where the cavities are filled with polymer. Hint: use the fact that the polymer can be treated as an incompressible liquid.

(b) Use the Stefan equation to estimate the time τ_imp it takes to squeeze the polymer film from the initial thickness h_0 to the final thickness h_f.

(c) Estimate the ratio ᚠ between the imprint speed $\partial_t h$ just after and just before the final film thickness has been reached[1]. Hint: although denoted the final thickness, the polymer film continues to be squeezed after the cavities are filled at the thickness h_f, however, now the polymer has to flow all the way to the edges of the stamp.

17.6 Solutions

Solution 17.1
Hydraulic resistance for a non-zero slip length in a flat channel
The starting point is the Navier–Stokes equation (3.20a).

[1] Here, at the end of the book there is a shortage of latin and greek letters for naming the variables, hence this use of the first letter ᚠ (fehu) in the futhark runic alphabet of my ancestors, the vikings.

(a) With the Navier boundary condition applied at $\tilde{z} = \pm h/2$, the solution to the Navier–Stokes equation has the symmetric form

$$v_x(\tilde{z}) = \frac{\Delta p}{2\eta L}\left(z_0^2 - \tilde{z}^2\right). \tag{17.40}$$

Application of the Navier boundary condition $\partial_{\tilde{z}} v_x = \lambda_s v_x$ at $\tilde{z} = -h/2$ leads to the condition $\lambda_s h = z_0^2 - (h/2)^2$ or

$$z_0 = \pm\frac{h}{2}\sqrt{1 + 4\frac{\lambda_s}{h}}, \tag{17.41}$$

which upon insertion into Eq. (17.40) yields Eq. (17.4).

(b) The flow rate is found by integration of Eq. (17.40),

$$Q = 2\frac{\Delta p\, w}{2\eta L}\int_0^{\frac{h}{2}} \mathrm{d}\tilde{z}\left(z_0^2 - \tilde{z}^2\right) = \frac{\Delta p\, w}{\eta L}\left[z_0^2\frac{h}{2} - \frac{1}{3}\left(\frac{h}{2}\right)^3\right] = \frac{\Delta p\, wh^3}{12\eta L}\left[1 + 6\frac{\lambda_s}{h}\right]. \tag{17.42}$$

Since $R_{\mathrm{hyd}} = \Delta p/Q$ we find the hydraulic resistance to be given by Eq. (17.5).

(c) For $h = 10$ µm we find $R_{\mathrm{hyd}}(\lambda_s) = R_{\mathrm{hyd}}(0)/(1+6\times 0.05/10) = 0.97 R_{\mathrm{hyd}}(0)$. Similarly, for $h = 1$ µm we find $R_{\mathrm{hyd}}(\lambda_s) = R_{\mathrm{hyd}}(0)/(1 + 6 \times 0.05/1) = 0.77\, R_{\mathrm{hyd}}(0)$.

Solution 17.2
Hydraulic resistance for a non-zero slip length in a circular channel
The starting point is the usual Poiseuille flow in a channel with circular cross-section presented in Section 3.4.4.

(a) Since only the boundary condition has changed, the velocity field as a form similar to Eq. (3.42a)

$$v_x(r) = \frac{\Delta p}{4\eta L}\left(r_0^2 - r^2\right), \tag{17.43}$$

where the radius a has been substituted with an unknown constant r_0 to be determined. We note that the Navier boundary condition Eq. (17.2b) becomes $v_x(a)\,\mathbf{e}_x = -\lambda_s\left[\partial_r v_x(a)\right]\mathbf{e}_x$, which leads to the condition

$$r_0 = a\sqrt{1 + 2\frac{\lambda_s}{a}}, \tag{17.44}$$

which upon insertion into Eq. (17.43) yields

$$v_x(r) = \frac{\Delta p}{4\eta L}\left[\left(1 + 2\frac{\lambda_s}{a}\right)a^2 - r^2\right]. \tag{17.45}$$

The flow rate is found as $Q = 2\pi\int_0^a \mathrm{d}r\, r v_x(r)$, which straightforwardly leads to the hydraulic resistance,

$$R_{\mathrm{hyd}}(\lambda_s) = \frac{Q(\lambda_s)}{\Delta p} = \frac{\pi a^4 \Delta p}{8\eta L}\frac{1}{1 + 4\frac{\lambda_s}{a}} = \frac{R_{\mathrm{hyd}}(0)}{1 + 4\frac{\lambda_s}{a}}. \tag{17.46}$$

(b) For $a = 5$ µm we find $R_{\mathrm{hyd}}(\lambda_s) = R_{\mathrm{hyd}}(0)/(1+4\times 0.05/5) = 0.96\, R_{\mathrm{hyd}}(0)$. Similarly, for $a = 0.5$ µm we find $R_{\mathrm{hyd}}(\lambda_s) = R_{\mathrm{hyd}}(0)/(1 + 4 \times 0.05/0.5) = 0.71\, R_{\mathrm{hyd}}(0)$. We note that when introducing a slip length λ_s, the hydraulic resistance decreases relatively more for a circular channel compared to a flat channel of height $h = 2a$. This is expected, since the lack of side walls in the flat channel implies no lowering of the resistance at the side regions of the channel upon introducing a non-zero slip length.

B

Dimensionless numbers

In microfluidics, several physical phenomena are considered simultaneously. It is therefore useful to characterize a given system by dimensionless numbers in the form of ratios between forces, time scales, length scales, or other relevant physical quantities of the involved phenomena. Below is given a list of some of the many dimensionless numbers encountered in microfluidics. Excerpt from *CRC Handbook of Chemistry and Physics.*

Table B.1 Dimensionless numbers in microfluidics.

Name	Symbol	Definition	Significance
Bond	*Bo*	$\frac{L^2 g}{\gamma}\,(\rho' - \rho)$	$\frac{\text{gravity}}{\text{surface tension}}$
Boussinesq	*B*	$\frac{v^2}{g\ell}$	$\frac{\text{inertia}}{\text{gravity}}$
Brinkman	*Br*	$\frac{\eta v^2}{\kappa \Delta T}$	$\frac{\text{viscous heat}}{\text{conducted heat}}$
Capillary	*Ca*	$\frac{\eta v}{\gamma}$	$\frac{\text{viscous force}}{\text{surface tension}}$
Crispation	*Cr*	$\frac{\eta D}{\gamma \ell}$	$\frac{\text{diffusion}}{\text{surface tension}}$
Dean	*D*	$\frac{(2a)^{\frac{3}{2}} v}{\nu \sqrt{2R}}$	$\frac{\text{transverse flow}}{\text{longitudinal flow}}$
Deborah	*D*	$\frac{\tau_{\text{relax}}}{\tau_{\text{obs}}}$	$\frac{\text{relaxation time}}{\text{observation time}}$
Eckert	*E*	$\frac{v^2}{c_{\text{p}} \Delta T}$	$\frac{\text{kinetic energy}}{\text{thermal energy}}$
Euler	*Eu*	$\frac{\Delta p}{\rho v^2}$	$\frac{\text{viscous pressure drop}}{\text{dynamic pressure}}$
Froude	*Fr*	$\frac{v^2}{g\ell}$	$\frac{\text{inertia}}{\text{gravity}}$

Name	Symbol	Definition	Significance
Knudsen	Kn	$\frac{\lambda}{\ell}$	$\frac{\text{mean free path}}{\text{length}}$
Lewis	Le	$\frac{D_{\text{th}}}{D}$	$\frac{\text{thermal conduction}}{\text{molecular diffusion}}$
Mach	M	$\frac{v}{c_{\text{a}}}$	$\frac{\text{speed}}{\text{speed of sound}}$
Péclet	$P\acute{e}$	$\frac{u\ell}{D}$	$\frac{\text{convection}}{\text{diffusion}}$
Poiseuille	Po	$\frac{a^2 \Delta p}{\eta \ell v}$	$\frac{\text{pressure force}}{\text{viscous force}}$
Prandtl	Pr	$\frac{\nu}{D_{\text{th}}}$	$\frac{\text{momentum diffusion}}{\text{heat diffusion}}$
Reynolds	Re	$\frac{\ell v}{\nu}$	$\frac{\text{inertial force}}{\text{vicous force}}$
Schmidt	Sc	$\frac{\nu}{D}$	$\frac{\text{momentum diffusion}}{\text{molecular diffusion}}$
Stokes	N_{St}	$\frac{\eta v}{\rho g \ell^2}$	$\frac{\text{viscous force}}{\text{gravitational force}}$
Strouhal	St	$\frac{\ell}{\tau v}$	$\frac{\text{flow time scale}}{\text{unsteady time scale}}$
Weber	We	$\frac{\rho \ell v^2}{\gamma}$	$\frac{\text{inertial force}}{\text{surface tension}}$
Weissenberg	Wi	$\dot{\gamma}\tau$	shear rate × relax. time
Womersley	α	$\frac{\ell}{\sqrt{\nu\tau}}$	$\frac{\text{length scale}}{\text{momentum diff. length}}$

C

Curvilinear co-ordinates

In this appendix we present the explicit co-ordinate representations of the equation of motion in Cartesian, cylindrical polar, and spherical polar co-ordinates. The choice of co-ordinates for solving a given problem is often dictated by the symmetry of the boundary conditions.

C.1 Cartesian co-ordinates

In Cartesian co-ordinates x, y, and z the position vector $\mathbf{r}$ of a point is given by

$$\mathbf{r} = x\mathbf{e}_x + y\mathbf{e}_y + z\mathbf{e}_z, \tag{C.1}$$

where the vectors $\mathbf{e}_x$, $\mathbf{e}_y$, and $\mathbf{e}_z$ are independent of the co-ordinates (x, y, z) and form an orthonormal basis, i.e. $\mathbf{e}_i \cdot \mathbf{e}_j = \delta_{ij}$ for $i, j = x, y, z$.

Any scalar function S depends on the three co-ordinates as

$$S = S(\mathbf{r}) = S(x, y, z), \tag{C.2}$$

while any vector function $\mathbf{V}(\mathbf{r})$ takes the form

$$\mathbf{V} = \mathbf{V}(\mathbf{r}) = \mathbf{e}_x V_x(x, y, z) + \mathbf{e}_y V_y(x, y, z) + \mathbf{e}_z V_z(x, y, z). \tag{C.3}$$

The differential operator $\boldsymbol{\nabla}$, denoted nabla, is defined as

$$\boldsymbol{\nabla} \equiv \mathbf{e}_x \partial_x + \mathbf{e}_y \partial_y + \mathbf{e}_z \partial_z. \tag{C.4}$$

Note that the differential operators ∂_i are written to the right of the basis vectors. While not important in Cartesian co-ordinates it is crucial when working with curvilinear co-ordinates. Once nabla has been introduced we can proceed and write down a number of important derivatives of scalar and vector functions.

C.1.1 Single derivatives

For a scalar function S only one quantity can be formed using the nabla operator, namely the gradient $\boldsymbol{\nabla} S$, which is a vector,

$$\boldsymbol{\nabla} S = \mathbf{e}_x(\partial_x S) + \mathbf{e}_y(\partial_y S) + \mathbf{e}_z(\partial_z S). \tag{C.5}$$

For a vector function$\mathbf{V}$ three quantities can be formed using the nabla operator. They are relatively simple to derive as the three basis vectors are independent of the co-ordinates. First, the divergence $\boldsymbol{\nabla}\cdot\mathbf{V}$ of a vector is a scalar,

$$\nabla\cdot\mathbf{V} = \partial_x V_x + \partial_y V_y + \partial_z V_z, \tag{C.6}$$

secondly, the rotation $\nabla \times \mathbf{V}$ of a vector is a vector,

$$\nabla \times \mathbf{V} = \mathbf{e}_x(\partial_y V_z - \partial_z V_y) + \mathbf{e}_y(\partial_z V_x - \partial_x V_z) + \mathbf{e}_z(\partial_x V_y - \partial_y V_x), \tag{C.7}$$

and thirdly, the gradient $\nabla\mathbf{V}$ of a vector is a tensor

$$\begin{aligned}\nabla\mathbf{V} = &+ (\partial_x V_x)\mathbf{e}_x\mathbf{e}_x + (\partial_x V_y)\mathbf{e}_x\mathbf{e}_y + (\partial_x V_z)\mathbf{e}_x\mathbf{e}_z \\ &+ (\partial_y V_x)\mathbf{e}_y\mathbf{e}_x + (\partial_y V_y)\mathbf{e}_y\mathbf{e}_y + (\partial_y V_z)\mathbf{e}_y\mathbf{e}_z \\ &+ (\partial_z V_x)\mathbf{e}_z\mathbf{e}_x + (\partial_z V_y)\mathbf{e}_z\mathbf{e}_y + (\partial_z V_z)\mathbf{e}_z\mathbf{e}_z.\end{aligned} \tag{C.8}$$

C.1.2 Double derivatives

The well-known Laplacian operator ∇^2 acting on a scalar function S is a scalar,

$$\nabla^2 S = \partial_x^2 S + \partial_y^2 S + \partial_z^2 S, \tag{C.9}$$

and the Laplacian acting on a vector function $\mathbf{V}$ is likewise a vector,

$$\nabla^2\mathbf{V} = \mathbf{e}_x\nabla^2 V_x + \mathbf{e}_y\nabla^2 V_y + \mathbf{e}_z\nabla^2 V_z. \tag{C.10}$$

C.1.3 Integrals

When integrating over the entire 3D space the integral takes the following form in Cartesian co-ordinates

$$\int_{\text{all}} \mathrm{d}\mathbf{r}\, f(\mathbf{r}) = \int_{-\infty}^{\infty} \mathrm{d}x \int_{-\infty}^{\infty} \mathrm{d}y \int_{-\infty}^{\infty} \mathrm{d}z\, f(x, y, z). \tag{C.11}$$

C.2 Cylindrical polar co-ordinates

As sketched in Fig. C.1, choosing the z axis as the cylinder axis (though often we let $\mathbf{e}_x$ be the cylinder axis), the cylindrical polar co-ordinates r, ϕ, and z are related to the Cartesian co-ordinates x, y, and z by

$$x = r\,\cos\phi, \tag{C.12a}$$

$$y = r\,\sin\phi, \tag{C.12b}$$

$$z = z, \tag{C.12c}$$

defined in the intervals $0 \leq r < \infty$, $0 \leq \phi \leq 2\pi$, and $-\infty < z < \infty$. The associated basis vectors $\mathbf{e}_r$, $\mathbf{e}_\theta$, and $\mathbf{e}_\phi$ are given by the derivatives of the Cartesian position vector $\mathbf{r}$ of Eq. (C.1) with respect to the cylindrical polar co-ordinates as

$$\mathbf{e}_r \equiv \partial_r\mathbf{r} = +\cos\phi\,\mathbf{e}_x + \sin\phi\,\mathbf{e}_y, \tag{C.13a}$$

$$\mathbf{e}_\phi \equiv \frac{1}{r}\,\partial_\phi\mathbf{r} = -\sin\phi\,\mathbf{e}_x + \cos\phi\,\mathbf{e}_y, \tag{C.13b}$$

$$\mathbf{e}_z \equiv \partial_z\mathbf{r} = \mathbf{e}_z. \tag{C.13c}$$

Note that the two basis vectors $\mathbf{e}_r$ and $\mathbf{e}_\phi$ depend on the co-ordinate ϕ, but all three vectors, nevertheless, form an orthonormal basis, i.e. $\mathbf{e}_i\cdot\mathbf{e}_j = \delta_{ij}$ for $i, j = r, \phi, z$.

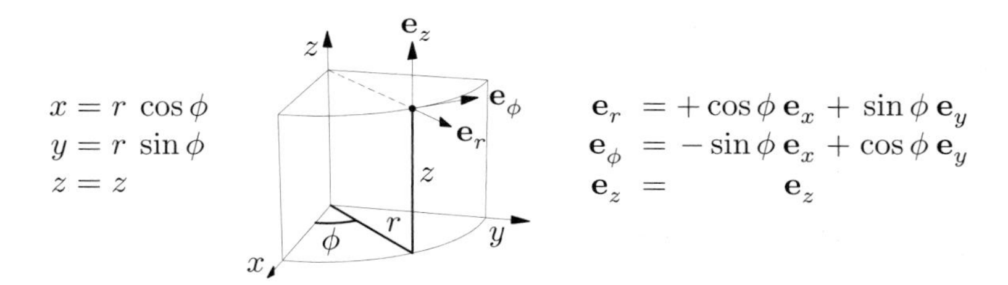

Fig. C.1 The cylindrical polar co-ordinates (r, ϕ, z) and the associated orthonormal basis vectors $\mathbf{e}_r$, $\mathbf{e}_\phi$, and $\mathbf{e}_z$.

Any scalar function S depends on the three co-ordinates as

$$S = S(\mathbf{r}) = S(r, \phi, z), \tag{C.14}$$

while any vector function $\mathbf{V}(\mathbf{r})$ takes the form

$$\mathbf{V} = \mathbf{V}(\mathbf{r}) = \mathbf{e}_r V_r(r, \phi, z) + \mathbf{e}_\phi V_\phi(r, \phi, z) + \mathbf{e}_z V_z(r, \phi, z). \tag{C.15}$$

Using the chain rule of differentiation and Eq. (C.13) the nabla operator of Eq. (C.4) can be transformed to cylindrical polar co-ordinates,

$$\boldsymbol{\nabla} \equiv \mathbf{e}_r \partial_r + \mathbf{e}_\phi \frac{1}{r}\, \partial_\phi + \mathbf{e}_z \partial_z. \tag{C.16}$$

In contrast to Cartesian co-ordinates, as noted above, the basis vectors of the cylindrical polar co-ordinates depend on the co-ordinates; in fact $\mathbf{e}_r$ and $\mathbf{e}_\phi$ both depend on ϕ leaving us with two non-vanishing derivatives of the basis vectors,

$$\partial_\phi \mathbf{e}_r = \mathbf{e}_\phi, \quad \text{and} \quad \partial_\phi \mathbf{e}_\phi = -\mathbf{e}_r. \tag{C.17}$$

Based on Eqs. (C.16) and (C.17) we can calculate the various derivatives in cylindrical polar co-ordinates.

C.2.1 Single derivatives

For a scalar function S only one quantity can be formed using the nabla operator, namely the gradient $\boldsymbol{\nabla} S$, which is a vector,

$$\boldsymbol{\nabla} S = \mathbf{e}_r (\partial_r S) + \mathbf{e}_\phi \frac{1}{r} (\partial_\phi S) + \mathbf{e}_z (\partial_z S). \tag{C.18}$$

For a vector function$\mathbf{V}$ three quantities can be formed using the nabla operator. Due to the dependence of the basis vectors on the co-ordinate ϕ, the expressions in cylindrical polar co-ordinates for these three quantities are slightly more complicated than those of the Cartesian co-ordinates. First, the divergence $\boldsymbol{\nabla}\cdot\mathbf{V}$ of a vector yielding a scalar,

$$\boldsymbol{\nabla}\cdot\mathbf{V} = \frac{1}{r}\, \partial_r (r V_r) + \frac{1}{r}\, \partial_\phi V_\phi + \partial_z V_z, \tag{C.19}$$

secondly, the rotation $\boldsymbol{\nabla} \times \mathbf{V}$ of a vector yielding a vector,

$$\boldsymbol{\nabla} \times \mathbf{V} = \mathbf{e}_r \Big(\frac{1}{r}\, \partial_\phi V_z - \partial_z V_\phi\Big) + \mathbf{e}_\phi \,\frac{1}{r}\, \Big(\partial_z V_r - \partial_r V_z\Big) + \mathbf{e}_z \,\frac{1}{r}\, \Big(\partial_r \big[r V_\phi\big] - \partial_\phi V_r\Big), \tag{C.20}$$

and thirdly, the gradient $\boldsymbol{\nabla}\mathbf{V}$ of a vector is a tensor, which, however, we will not present here.

C.2.2 Double derivatives

In cylindrical polar co-ordinates the Laplacian operator ∇^2 acting on a scalar function S is the following scalar,

$$\nabla^2 S = \frac{1}{r}\,\partial_r\Big(r\partial_r S\Big) + \frac{1}{r^2}\partial_\phi^2 S + \partial_z^2 S. \tag{C.21}$$

Taking the ϕ dependence of the basis vectors into account, we obtain the vector resulting from applying the Laplacian on a vector function $\mathbf{V}$,

$$\nabla^2\mathbf{V} = \mathbf{e}_r\Big(\nabla^2 V_r - \frac{2}{r^2}\,\partial_\phi V_\phi - \frac{1}{r^2}V_r\Big) + \mathbf{e}_\phi\Big(\nabla^2 V_\phi + \frac{2}{r^2}\,\partial_\phi V_r - \frac{1}{r^2}V_\phi\Big) + \mathbf{e}_z\Big(\nabla^2 V_z\Big), \tag{C.22}$$

where the Laplacian acting on the scalar components V_i, $i = r, \phi, z$ is calculated from Eq. (C.21).

C.2.3 Integrals

When integrating over the entire 3D space the integral takes the following form in cylindrical polar co-ordinates

$$\int_{\text{all}} \mathrm{d}\mathbf{r}\, f(\mathbf{r}) = \int_0^\infty \mathrm{d}r \int_0^{2\pi} \mathrm{d}\phi \int_{-\infty}^{\infty} \mathrm{d}z\; r\, f(r, \phi, z). \tag{C.23}$$

C.3 Spherical polar co-ordinates

Choosing $\mathbf{e}_z$ as the polar axis (though we often choose $\mathbf{e}_x$), the spherical polar co-ordinates r, θ, and ϕ are related to the Cartesian co-ordinates x, y, and z by

$$x = r\,\sin\theta\,\cos\phi, \tag{C.24a}$$

$$y = r\,\sin\theta\,\sin\phi, \tag{C.24b}$$

$$z = r\,\cos\theta, \tag{C.24c}$$

as sketched in Fig. C.2. They are defined in the intervals $0 \le r < \infty$, $0 \le \theta \le \pi$, and $0 \le \phi \le 2\pi$. The associated basis vectors $\mathbf{e}_r$, $\mathbf{e}_\theta$, and $\mathbf{e}_\phi$ are given by the derivatives of the Cartesian position vector $\mathbf{r}$ of Eq. (C.1) with respect to the spherical polar co-ordinates as

$$\mathbf{e}_r \equiv \partial_r\mathbf{r} = +\sin\theta\,\cos\phi\,\mathbf{e}_x + \sin\theta\,\sin\phi\,\mathbf{e}_y + \cos\theta\,\mathbf{e}_z, \tag{C.25a}$$

$$\mathbf{e}_\theta \equiv \frac{1}{r}\,\partial_\theta\mathbf{r} = +\cos\theta\,\cos\phi\,\mathbf{e}_x + \cos\theta\,\cos\phi\,\mathbf{e}_y - \sin\theta\,\mathbf{e}_z, \tag{C.25b}$$

$$\mathbf{e}_\phi \equiv \frac{1}{r\sin\theta}\,\partial_\phi\mathbf{r} = -\sin\phi\,\mathbf{e}_x + \cos\phi\,\mathbf{e}_y. \tag{C.25c}$$

Note that all three basis vectors depend on the co-ordinates θ and ϕ, but that they, nevertheless, form an orthonormal basis, i.e. $\mathbf{e}_i\!\cdot\!\mathbf{e}_j = \delta_{ij}$ for $i, j = r, \theta, \phi$.

Any scalar function S depends on the three co-ordinates as

$$S = S(\mathbf{r}) = S(r, \theta, \phi), \tag{C.26}$$

while any vector function $\mathbf{V}(\mathbf{r})$ takes the form

$$\mathbf{V} = \mathbf{V}(\mathbf{r}) = \mathbf{e}_r V_r(r, \theta, \phi) + \mathbf{e}_\theta V_\theta(r, \theta, \phi) + \mathbf{e}_\phi V_\phi(r, \theta, \phi). \tag{C.27}$$

$$x = r\,\sin\theta\,\cos\phi$$
$$y = r\,\sin\theta\,\sin\phi$$
$$z = r\,\cos\theta$$

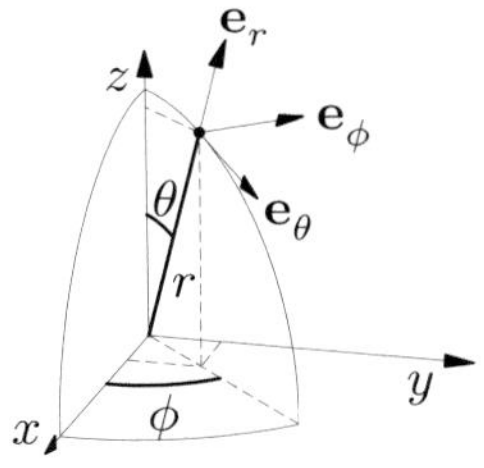

$$\begin{aligned}
\mathbf{e}_r &= \sin\theta\,\cos\phi\,\mathbf{e}_x + \sin\theta\,\sin\phi\,\mathbf{e}_y + \cos\theta\,\mathbf{e}_z\\
\mathbf{e}_\theta &= \cos\theta\,\cos\phi\,\mathbf{e}_x + \cos\theta\,\sin\phi\,\mathbf{e}_y - \sin\theta\,\mathbf{e}_z\\
\mathbf{e}_\phi &= -\sin\phi\,\mathbf{e}_x + \cos\phi\,\mathbf{e}_y
\end{aligned}$$

Fig. C.2 The spherical polar co-ordinates (r, θ, ϕ) and the associated orthonormal basis vectors $\mathbf{e}_r$, $\mathbf{e}_\theta$, and $\mathbf{e}_\phi$.

Using the chain rule of differentiation and Eq. (C.25) the nabla operator of Eq. (C.4) can be transformed to spherical polar co-ordinates,

$$\boldsymbol{\nabla} \equiv \mathbf{e}_r\partial_r + \mathbf{e}_\theta\frac{1}{r}\,\partial_\theta + \mathbf{e}_\phi\frac{1}{r\sin\theta}\,\partial_\phi. \tag{C.28}$$

In the case of spherical polar co-ordinates the dependence of the basis vectors on the co-ordinates θ and ϕ leads to five non-vanishing derivatives of the basis vectors,

$$\partial_\theta\mathbf{e}_r = +\mathbf{e}_\theta, \qquad \partial_\phi\mathbf{e}_r = +\sin\theta\,\mathbf{e}_\phi, \tag{C.29a}$$

$$\partial_\theta\mathbf{e}_\theta = -\mathbf{e}_r, \qquad \partial_\phi\mathbf{e}_\theta = +\cos\theta\,\mathbf{e}_\phi, \tag{C.29b}$$

$$\partial_\phi\mathbf{e}_\phi = -\sin\theta\,\mathbf{e}_r - \cos\theta\,\mathbf{e}_\theta. \tag{C.29c}$$

Based on Eqs. (C.28) and (C.29) we can calculate the various derivatives in spherical polar co-ordinates.

C.3.1 Single derivatives

For a scalar function S only one quantity can be formed using the nabla operator, namely the gradient $\boldsymbol{\nabla}S$, which is a vector,

$$\boldsymbol{\nabla}S = \mathbf{e}_r\partial_r S + \mathbf{e}_\theta\frac{1}{r}\,\partial_\theta S + \mathbf{e}_\phi\frac{1}{r\sin\theta}\,\partial_\phi S. \tag{C.30}$$

For a vector function $\mathbf{V}$ three quantities can be formed using the nabla operator. Due to the dependence of the basis vectors on the co-ordinate ϕ, the expressions in spherical polar co-ordinates for these three quantities are slightly more complicated than those of the Cartesian co-ordinates. First, the divergence $\boldsymbol{\nabla}\cdot\mathbf{V}$ of a vector yielding a scalar,

$$\boldsymbol{\nabla}\cdot\mathbf{V} = \frac{1}{r^2}\,\partial_r\Big(r^2V_r\Big) + \frac{1}{r\sin\theta}\,\partial_\theta\Big(\sin\theta\,V_\theta\Big) + \frac{1}{r\sin\theta}\,\partial_\phi V_\phi, \tag{C.31}$$

secondly, the rotation $\boldsymbol{\nabla}\times\mathbf{V}$ of a vector yielding a vector,

$$\boldsymbol{\nabla}\times\mathbf{V} = \mathbf{e}_r\frac{1}{r\sin\theta}\Big(\partial_\theta\big[\sin\theta V_\phi\big] - \partial_\phi V_\theta\Big) + \mathbf{e}_\theta\frac{1}{r}\Big(\frac{1}{\sin\theta}\partial_\phi V_r - \partial_r\big[rV_\phi\big]\Big) + \mathbf{e}_\phi\frac{1}{r}\Big(\partial_r\big[rV_\theta\big] - \partial_\theta V_r\Big), \tag{C.32}$$

and thirdly, the gradient $\boldsymbol{\nabla}\mathbf{V}$ of a vector is a tensor, which, however, we will not present here.

C.3.2 Double derivatives

In spherical polar co-ordinates the Laplacian operator ∇^2 acting on a scalar function S is the following scalar,

$$\nabla^2 S = \frac{1}{r^2}\,\partial_r\Big(r^2\partial_r S\Big) + \frac{1}{r^2\sin\theta}\,\partial_\theta\Big(\sin\theta\,\partial_\theta S\Big) + \frac{1}{r^2\sin^2\theta}\,\partial_\phi^2 S. \tag{C.33}$$

Taking the angular dependence of the basis vectors into account, we obtain the vector resulting from applying the Laplacian on a vector function $\mathbf{V}$,

$$\begin{aligned}\nabla^2\mathbf{V} = \mathbf{e}_r\Big(\nabla^2 V_r &- \frac{2}{r^2\sin^2\theta}\partial_\theta\big[\sin\theta\,V_\theta\big] - \frac{2}{r^2\sin\theta}\,\partial_\phi V_\phi - \frac{2}{r^2}\,V_r\Big)\\
+\,\mathbf{e}_\theta\Big(\nabla^2 V_\theta &- \frac{2\cos\theta}{r^2\sin^2\theta}\,\partial_\phi V_\phi + \frac{2}{r^2}\,\partial_\theta V_r - \frac{1}{r^2\sin^2\theta}\,V_\theta\Big)\\
+\,\mathbf{e}_\phi\Big(\nabla^2 V_\phi &+ \frac{2}{r^2\sin\theta}\,\partial_\phi V_r + \frac{2\cos\theta}{r^2\sin^2\theta}\,\partial_\phi V_\theta - \frac{1}{r^2\sin^2\theta}\,V_\phi\Big),\end{aligned} \tag{C.34}$$

where the Laplacian acting on the scalar components V_r, V_θ, and V_ϕ is calculated from Eq. (C.33).

C.3.3 Integrals

When integrating over the entire 3D space the integral takes the following form in spherical polar co-ordinates

$$\int_{\text{all}} \mathrm{d}\mathbf{r}\; f(\mathbf{r}) = \int_0^\infty \mathrm{d}r \int_0^\pi \mathrm{d}\theta \int_0^\infty \mathrm{d}z\; r^2\sin\theta\, f(r,\theta,\phi) = \int_0^\infty \mathrm{d}r \int_{-1}^1 \mathrm{d}(\cos\theta) \int_0^\infty \mathrm{d}z\; r^2\, f(r,\cos\theta,\phi). \tag{C.35}$$

D

The chemical potential

In this appendix we present an outline of the theory of the chemical potential of a weak solution of molecules B in a solvent A. More details can be found in the textbooks by Feynman (1972), Landau and Lifshitz (1982), and Kittel and Kroemer (2000). For brevity, we use the symbol τ for the absolute temperature, $\tau \equiv k_\mathrm{B}T$, in this appendix.

D.1 The partition function and the free energy

In thermodynamics the Helmholtz free energy F of any given subsystem, say a single molecule in solution, is given by the energy, temperature and entropy as $F = E - \tau S$. In statistical physics it can also be expressed in terms of the partition function Z_1 once all the possible energy states ε_n of the subsystem is known,

$$Z_1 = \sum_n \mathrm{e}^{\varepsilon_n/\tau}, \tag{D.1}$$

$$F_1 = -\tau \log Z_1. \tag{D.2}$$

Since the exponential of a sum equals the product of each exponential, it might be thought that the partition function Z_N of a collection of N identical non-interacting molecules (an ideal gas) is given by the product of N single-molecule partition functions Z_1. However, in Nature identical molecules are indistinguishable, thus configurations obtained by interchanging molecules between given quantum states are really the same configuation. Since N elements can be interchanged in $N!$ ways, this effect is accounted for by division with $N!$,

$$Z_N = \frac{1}{N!}\,(Z_1)^N. \tag{D.3}$$

The free energy of the non-interacting N-particle system can therefore be written as

$$F = -\tau \log Z_N = -N\tau \log Z_1 + \tau \log N! \approx -N\tau \log Z_1 + N\tau \log(N/\mathrm{e}), \tag{D.4}$$

where we in the last equality have used Stirling's formula valid for $N \gg 1$, $\log N! \approx N \log(N/\mathrm{e})$, and where e is the base of the natural logarithm.

To study solutions it is advantageous to change from the Helmholtz free energy, $F = F(V, \tau, N)$, which is a function of volume, temperature and particle number, to the Gibbs free energy,

$$G = G(p, \tau, N) \equiv F + pV, \tag{D.5}$$

which is a function of pressure, temperature and particle number. At constant pressure and temperature the Gibbs energy is proportional to the particle number, and it must therefore have the functional form

$$G(p,\tau,N) = N\mu(p,\tau), \tag{D.6}$$

where $\mu(p,\tau)$ is some function independent of N. This function can be interpreted as the energy of the last added particle,

$$\mu(p,\tau) = \left(\frac{\partial G}{\partial N}\right)_{p,\tau}, \tag{D.7}$$

and it is denoted the chemical potential of the molecules.

D.2 The chemical potential of a solution

Consider N_B molecules B in solution among the solvent molecules N_A. We consider only weak solutions $1 \ll N_B \ll N_A$ and define the concentration c as $c = N_B/N_A$.

Before adding the solute, the Gibbs free energy is given by

$$G_0 = N_A\mu_0(p,\tau). \tag{D.8}$$

In a weak solution the solute molecules B can be considered as non-interacting particles. The free energy Eq. (D.4) thus implies that the Gibbs energy of the solution must take the form

$$G = N_A\mu_0(p,\tau) + N_B\gamma(p,\tau,N_A) + N_B\tau\log(N_B/\mathrm{e}), \tag{D.9}$$

where $\gamma(p,\tau,N_A)$ is some function to be determined. However, it is clear that for a weak solution the correction term in Eq. (D.9) to $G_0 = N_A\mu_0(p,\tau)$ can only depend on the solvent molecule number N_A through the concentration $c = N_B/N_A$. This, combined with the functional form of Eq. (D.9) leads to

$$G = N_A\mu_0(p,\tau) + N_B\tau\Big[\psi(p,\tau) + \log(N_B/N_A)\Big], \tag{D.10}$$

where $\psi(p,\tau)$ is some function independent of the molecule numbers.

The chemical potentials μ_A and μ_B of the solvent and the solute, respectively, are now easily found by differentiation with respect to the molecule numbers N_A and N_B,

$$\mu_A = \frac{\partial G}{\partial N_A} = \mu_0(p,\tau) - \tau\frac{N_B}{N_A} = \mu_0(p,\tau) - \tau c, \tag{D.11}$$

$$\mu_B = \frac{\partial G}{\partial N_B} = \tau\psi(p,\tau) + \tau + \tau\log\left(\frac{N_B}{N_A}\right) \equiv \mu_0^B(p,\tau) + \tau\log\left(\frac{c}{c_0}\right). \tag{D.12}$$

E

The wave equation

In this appendix we give a brief presentation of some of the basic features of the standard wave equation in 1D and 3D and discuss the plane-wave solutions.

The standard wave equation for some field $\phi(x,t)$ depending on time t and only one space co-ordinate x can be written as

$$c^2\partial_x^2\phi(x,t) = \partial_t^2\phi(x,t). \tag{E.1}$$

For any doubly differentiable function $f(\tilde{x})$ of a variable $\tilde{x}$, we see by direct substitution that $\phi_1(x,t) = f(x-ct)$ is a solution to Eq. (E.1). Moreover, by shifting to a co-ordinate system $\tilde{x} = x - ct$ moving to the right with speed c, we see that in this moving co-ordinate system the solution has the constant shape $f(\tilde{x})$. Likewise, $\phi_2(x,t) = f(x+ct)$ is seen to be a solution with the shape f this time moving with the speed c to the left. The standard terminology is that

$$c \text{ is the phase velocity of the wave.} \tag{E.2}$$

Since the wave equation is linear we can write a general solution by superimposing a number of functions f_n,

$$\phi(x,t) = \sum_n \Big[A_n\, f_n(x-ct) + B_n\, f_n(x+ct)\Big]. \tag{E.3}$$

A special and very useful class of solutions is the plane waves with wavelength λ and frequency f or period $\tau = 1/f$, e.g.

$$\phi_+(x,t) = \cos\left[2\pi\left(\frac{x}{\lambda} - \frac{t}{\tau}\right)\right] = \cos(kx - \omega t), \tag{E.4a}$$

$$\phi_-(x,t) = \cos\left[2\pi\left(\frac{x}{\lambda} + \frac{t}{\tau}\right)\right] = \cos(kx + \omega t), \tag{E.4b}$$

where ϕ_+ and ϕ_- moves to the right and to the left, respectively, and where we have introduced the wave number k and the angular frequency ω by

$$k \equiv \frac{2\pi}{\lambda}, \tag{E.5a}$$

$$\omega \equiv \frac{2\pi}{\tau} = 2\pi f. \tag{E.5b}$$

The phase velocity c can be related to λ, f, k and ω by

$$c = \frac{\lambda}{\tau} = \lambda f = \frac{\omega}{k}. \tag{E.6}$$

The linearity of the wave equation makes it possible to benefit from the use of a complex number representation $\phi(x,t)$ of the real physical field ϕ_{phys}, since after finding a complex solution, the corresponding physical solution can be found by extracting the real part,

$$\phi_{\text{phys}}(x,t) \equiv \text{Re}\big[\phi(x,t)\big]. \tag{E.7}$$

For the special case of plane waves with complex amplitude $A = |A| \exp(\mathrm{i}\phi_A)$ we get the following expression using complex notation,

$$\phi^{(+)}_{\text{phys}}(x,t) = \text{Re}\Big[A\,\mathrm{e}^{\mathrm{i}(kx-\omega t)}\Big] = |A|\,\text{Re}\Big[A\,\mathrm{e}^{\mathrm{i}(kx-\omega t+\phi_A)}\Big], \tag{E.8a}$$

$$\phi^{(-)}_{\text{phys}}(x,t) = \text{Re}\Big[A\,\mathrm{e}^{\mathrm{i}(-kx-\omega t)}\Big] = |A|\,\text{Re}\Big[A\,\mathrm{e}^{\mathrm{i}(-kx-\omega t+\phi_A)}\Big]. \tag{E.8b}$$

Note that for $A = 1$ we have a cosine function, while for $A = -\mathrm{i}$ we have a sine function. Clearly, the imaginary part of the amplitude controls the phase of the plane wave. The complex exponential function and the trigonometric functions are related by

$$\mathrm{e}^{\mathrm{i}u} = \cos u + \mathrm{i}\sin u, \qquad \mathrm{e}^{-\mathrm{i}u} = \cos u - \mathrm{i}\sin u, \tag{E.9a}$$

$$\cos u = \frac{1}{2}\Big[\mathrm{e}^{\mathrm{i}u} + \mathrm{e}^{-\mathrm{i}u}\Big], \qquad \sin u = \frac{1}{2\mathrm{i}}\Big[\mathrm{e}^{\mathrm{i}u} - \mathrm{e}^{-\mathrm{i}u}\Big]. \tag{E.9b}$$

With plane waves as basis functions the general solution Eq. (E.3) can be written as a complex Fourier series

$$\phi(x,t) = \sum_k \Big[A_k\,\mathrm{e}^{\mathrm{i}(kx-\omega t)} + B_k\,\mathrm{e}^{\mathrm{i}(-kx-\omega t)}\Big] = \int_{-\infty}^{\infty} \mathrm{d}k\,\Big[A(k)\,\mathrm{e}^{\mathrm{i}(kx-\omega t)} + B(k)\,\mathrm{e}^{\mathrm{i}(-kx-\omega t)}\Big], \tag{E.10}$$

where the latter equality is valid in the limit of continuous wave numbers k. A single plane wave, as in Eq. (E.8a), has a constant amplitude A in all space. Such a wave is completely delocalized. Localized waves, also known as wave packets, can be constructed by adding waves with nearly the same wave number k_0. Consider the wave packet

$$\phi(x,t) = \int_{-\infty}^{\infty} \mathrm{d}k\,A(k)\,\mathrm{e}^{\mathrm{i}(kx-\omega t)}, \tag{E.11}$$

where the amplitude function $A(k)$ only differs from zero in the neighborhood of some fixed wave number k_0, e.g. like a Gaussian function $A(k) = \exp\big[-\ell^2(k-k_0)^2\big]$. The resulting wavepacket is more localized in space. Now, let us assume that the angular frequency ω has some dependence of the wave number k, maybe linear as in Eq. (E.6). We Taylor expand it around $k = k_0$, since $A(k)$ is only non-zero for $k \approx k_0$,

$$\omega = \omega(k) \approx \omega_0 + (\partial_k\omega_0)\,(k - k_0). \tag{E.12}$$

This is inserted into Eq. (E.11) and results in the following form of the wave packet,

$$\begin{aligned}\phi(x,t) &= \int_{-\infty}^{\infty} \mathrm{d}k\, A(k)\, \mathrm{e}^{\mathrm{i}\left(kx-\left[\omega_0+(\partial_k\omega_0)\,(k-k_0)\right]t\right)} \\ &= \left[\int_{-\infty}^{\infty} \mathrm{d}k\, A(k)\, \mathrm{e}^{\mathrm{i}(k-k_0)\left[x-(\partial_k\omega_0)t\right]}\right] \mathrm{e}^{\mathrm{i}(k_0x-\omega_0 t)} \\ &= F\big[x-(\partial_k\omega_0)t\big]\, \mathrm{e}^{\mathrm{i}(k_0x-\omega_0 t)}. \end{aligned} \tag{E.13}$$

The last equality is obtained by noting the particular x and t dependence of the k-integral. From its argument the envelope function F is seen to describe a fixed shape moving to the right with the speed $\partial_k\omega_0$. This speed is denoted the group velocity v_g,

$$v_\mathrm{g} \equiv \partial_k\omega_0. \tag{E.14}$$

For a linear dispersion relation, where $\omega = ck$, we obtain the group velocity $\partial_k\omega_0 = c$, and reach the conclusion that in this specific case, the group velocity and the phase velocity are equal.

So far we have treated k as a real number. However, the consequences of an imaginary part in the wave number are significant. Let us assume that the imaginary part of the wave number has the relative magnitude γ,

$$k = k_0\,(1+\mathrm{i}\gamma), \tag{E.15}$$

where both k_0 and γ are real. When inserting this wave number into the plane-wave expression (E.8a) we get

$$\phi(x,t) = A\mathrm{e}^{\mathrm{i}(kx-\omega t)} = A\mathrm{e}^{\mathrm{i}[k_0(1+\mathrm{i}\gamma)x-\omega t]} = A\mathrm{e}^{\mathrm{i}(k_0x-\omega t)}\,\mathrm{e}^{-\gamma k_0 x}. \tag{E.16}$$

It can be concluded that for $\gamma > 0$ the resulting wave is exponentially damped, while for $\gamma < 0$ it is exponentially growing. Such imaginary wave numbers appear in systems with dissipation or amplification, and thus have a significant physical interpretation. The characteristic damping length x_c is defined by the condition $\mathrm{e}^{-\gamma k_0 x_\mathrm{c}} = \mathrm{e}^{-1}$ and is thus determined by

$$x_\mathrm{c} \equiv \frac{1}{\gamma k_0}. \tag{E.17}$$

Finally, we briefly turn to the 3D case, where the wave equation takes the form

$$c^2\,\nabla^2\phi(\mathbf{r},t) = \partial_t^2\,\phi(\mathbf{r},t). \tag{E.18}$$

Plane waves are now characterized with a wavevector $\mathbf{k} = k\,\mathbf{e}_k$, where k remains the wave number and $\mathbf{e}_k$ is the propagation direction,

$$\phi(\mathbf{r},t) = A\,\mathrm{e}^{\mathrm{i}(\mathbf{k}\cdot\mathbf{r}-\omega t)} = A\,\mathrm{e}^{\mathrm{i}[k(\mathbf{e}_k\cdot\mathbf{r})-\omega t]}. \tag{E.19}$$

In the case of a product wavefunction of the form

$$\phi(\mathbf{r},t) = f(\mathbf{r})\,\mathrm{e}^{-\mathrm{i}\omega t}, \tag{E.20}$$

the wave equation becomes the Helmholtz equation

$$\nabla^2 f(\mathbf{r}) = -\frac{\omega^2}{c^2} f(\mathbf{r}) = -k^2 f(\mathbf{r}). \tag{E.21}$$

With a complex wavevector $k = k_0(1+\mathrm{i}\gamma)$ as in Eq. (E.15), we obtain the Helmholtz equation for damped waves,

$$\nabla^2 f(\mathbf{r}) = -k^2 f(\mathbf{r}) = -\left[k_0(1+\mathrm{i}\gamma)\right]^2 f(\mathbf{r}). \tag{E.22}$$

F
Numerical simulations

In most of the book we have studied analytical solutions to the Navier–Stokes equation and the convection-diffusion equation. Although these solutions are very important, they are in fact also very special: only in a few and highly symmetric cases is it possible to find analytical solutions. In the vast majority of cases we are forced to perform numerical simulations to get the solutions to the complicated set of partial differential equations that appears in theoretical microfluidics.

There are many ways to find the solutions to partial differential equations numerically. Most methods rely on some sort of discretization and transformation of the continuous equations into a matrix problem, which then can be solved by one of the many existing matrix solvers. The computer code to handle the problem can be written from scratch using general purpose software like Fortran, C, MATLAB or Mathematica, or by employing a specialized software for solving partial differential equations like ANSYS, Coventer, CFDACE, or COMSOL Multiphysics.

In this chapter we discuss COMSOL Multiphysics to illustrate how to solve problems in microfluidics numerically. COMSOL is designed to solve a wide range of partial differential equations comprising most of the equations appearing in physics and chemistry such as the Navier–Stokes equation, the Poisson equation, the Schrödinger equation, the convection-diffusion equation, the Maxwell equations, and problems combining these equations. COMSOL allows the unexperienced user to quickly get started and solve fairly complex microfluidics problems, while it still remains a powerful tool for the experienced user.

The basic numerical method used in COMSOL is the so-called finite-element method (FEM). While this method perhaps is not the first choice for fluidic problems exhibiting turbulence, it is very good for the low Reynolds number problems encountered in microfluidics. In the following, we shall take a brief look at the idea of FEM, and then move on to a short introduction to COMSOL including some examples. Once started on COMSOL the user can, with the help of the online user guides and user manuals, launch into a self-study and progress on his own.

A good account of the theory of FEM for microfluidic system is the M.Sc. thesis *Computational fluid dynamics in microfluidic systems* by Laurits Højgaard Olesen (DTU Nanotech, July 2003) available at the web-site www.nanotech.dtu.dk/microfluidics.

F.1 The finite-element method (FEM)

Consider a vector field $\mathbf{v}(\mathbf{r}, t)$ for which we want to solve a partial differential equation in some domain Ω,

$$\mathcal{D}\mathbf{v}(\mathbf{r}, t) = \mathbf{f}(\mathbf{r}, t), \quad \text{for } \mathbf{r} \in \Omega, \tag{F.1}$$

where $\mathcal{D}$ is some differential operator and where $\mathbf{f}(\mathbf{r},t)$ is the source or forcing term. The solution must typically respect some boundary conditions of the Dirichlet or the Neumann type (or a combination of the two),

$$v_i(\mathbf{r},t) = a(\mathbf{r}), \quad \text{for } \mathbf{r} \in \partial\Omega, \quad \text{(Diriclet boundary condition)}, \tag{F.2a}$$

$$(\mathbf{n}\cdot\boldsymbol{\nabla})v_i(\mathbf{r},t) = b(\mathbf{r}), \quad \text{for } \mathbf{r} \in \partial\Omega, \quad \text{(Neumann boundary condition)}, \tag{F.2b}$$

where $\mathbf{n}$ as usual is the outward-pointing normal vector of the surface. Solutions to the problem in the form of Eqs. (F.1) and (F.2) are called strong solutions.

F.1.1 Discretization using finite elements

Normally, it is not possible to obtain strong solutions for the problem at hand. Instead, one usually discretizes the otherwise continuous problem and obtains so-called weak solutions; weak in the sense of approximate.

A typical discretization of a 2D computational domain Ω is shown in Fig. F.1(a) in the form of a mesh containing a finite number of finite-sized triangular elements. These elements are the origin of the name FEM. Each element in the 2D mesh consists of a number of straight edges, and a number of corners denoted nodes.

It is now possible to introduce a finite set of N basis functions ϕ_j. As sketched in Fig. F.1(b) the jth basis function is only non-zero in the neighboring elements containing the jth node $\mathbf{r}_j$. At $\mathbf{r}_j$ itself the basis function becomes unity, $\phi_j(\mathbf{r}_j) \equiv 1$, while it decays continuously to zero on the edges connecting the neighboring nodes. Using these basis functions the solutions to the differential equation are sought on the following discrete form

$$v_i(\mathbf{r}) \approx \sum_{j=1}^{N} v_j^{(i)}\, \phi_j(\mathbf{r}), \tag{F.3}$$

where the coefficients $v_j^{(i)}$ are to be determined.

There are many choices for the specific form of the basis functions ϕ_j. Two commonly used forms are the linear and the quadratic interpolating functions shown in Fig. F.2.

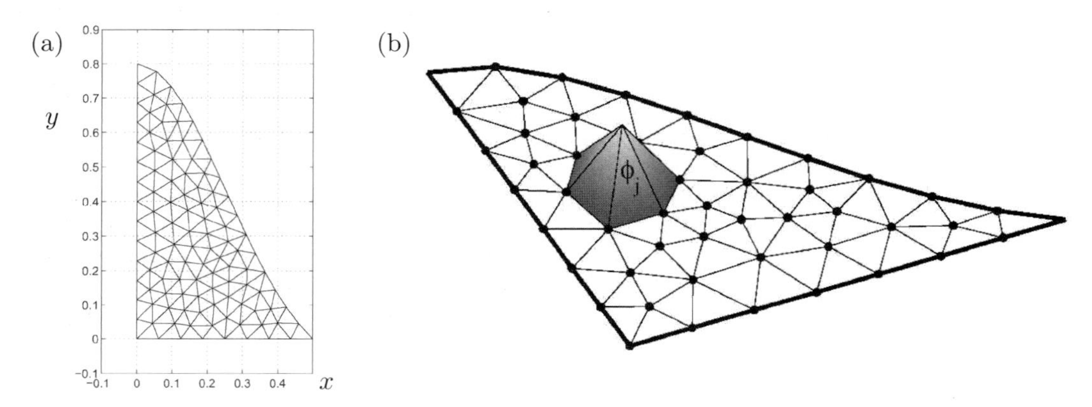

Fig. F.1 (a) The FEM mesh for half of the cross-section of the Gaussian-shaped microfluidic channel of Fig. 3.4(b). (b) The corresponding FEM domain showing one of the linearly interpolating basis function ϕ_j. Courtesy of Laurits Højgaard Olesen, DTU Nanotech.

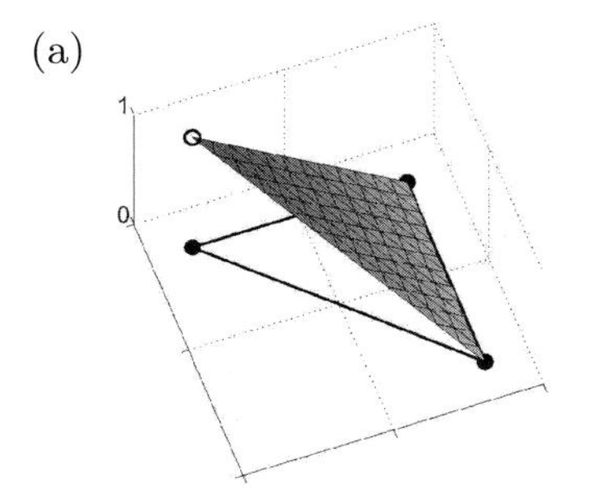

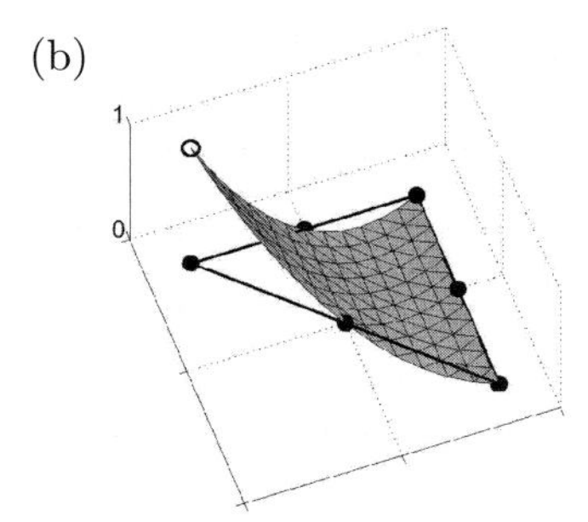

Fig. F.2 (a) A basis function corresponding to linear interpolation within an element. (b) A basis function corresponding to quadratic interpolation within an element. Courtesy of Laurits Højgaard Olesen, DTU Nanotech.

F.1.2 Weak solutions

In the theory of partial differential equations it has been shown that there are no conditions that ensure the existence of a unique solution to the problem Eqs. (F.1) and (F.2). However, such an existence and uniqueness theorem does in fact exist for the so-called weak solution to the problem. A weak solution is defined in the following by the introduction of the inner product in function space.

Consider the set $\mathcal{F}_\Omega$ of "well-behaved", real functions on Ω. The inner product $\langle u, v\rangle$ for any pair of functions $u, v \in \mathcal{F}_\Omega$ is defined by

$$\langle u, v\rangle \equiv \int_\Omega \mathrm{d}\mathbf{r}\; u(\mathbf{r})v(\mathbf{r}). \tag{F.4}$$

If we let $\mathbf{v}$ be a strong solution to Eq. (F.1), then $\mathcal{D}\mathbf{v} - \mathbf{f} = 0$. It is therefore natural to define the defect $d(\mathbf{w})$ for any function $\mathbf{w}$ as

$$d(\mathbf{w}) \equiv \mathcal{D}\mathbf{w} - \mathbf{f}. \tag{F.5}$$

A weak solution $\mathbf{w}$ to Eq. (F.1) is defined as a function with a defect $d(\mathbf{w})$ that has a vanishing inner product with any function v within the function space $\mathcal{F}_\Omega$,

$$\langle v_i, [d(\mathbf{w})]_i\rangle \equiv 0, \quad \text{for all } v_i \in \mathcal{F}_\Omega. \tag{F.6}$$

Clearly, a strong solution is also a weak solution, but the converse is not necessarily true.

The introduction of weak solutions relaxes the constraints that we must put on the functions belonging to $\mathcal{F}_\Omega$. The differential operators $\mathcal{D}$ in all of the partial differential equations we shall work with contain only up to second-order derivatives. For strong solutions this implies that the functions should have at least continuous first derivatives to avoid infinite second derivatives. However, for weak solutions it suffices to deal with continuous functions that are only piecewise differentiable. This can be seen by partial integration of the inner product of a function v with the term $\nabla^2 w$ contained in $\mathcal{D}w$ of Eq. (F.5) as follows,

$$\int_\Omega \mathrm{d}\mathbf{r}\; v\nabla^2 w = \int_{\partial\Omega} \mathrm{d}a\; v(\mathbf{n}\cdot\boldsymbol{\nabla})w - \int_\Omega \mathrm{d}\mathbf{r}\; (\boldsymbol{\nabla} v)\cdot(\boldsymbol{\nabla} w). \tag{F.7}$$

The surface integral is handled by invoking the boundary conditions Eq. (F.2) while the volume integral is well defined as long as v and w are just piecewise differentiable.

F.1.3 The Galerkin method

In the Galerkin method a weak solution $w_i(\mathbf{r}) = \sum_{j=1}^{N} w_j^{(i)}\,\phi_j(\mathbf{r})$ written in terms of the basis functions is obtained by demanding that its defect $d(w_i)$ has a zero inner product with all basis functions,

$$\langle \phi_k,\, \mathcal{D}w_i - f_i\rangle = 0, \quad k = 1, 2, \ldots, N. \tag{F.8}$$

This condition can be written as a finite matrix equation,

$$\sum_{j=1}^{N} \langle \phi_k,\, \mathcal{D}\phi_j\rangle\, w_j^{(i)} = \langle \phi_k,\, f_i\rangle. \tag{F.9}$$

The matrix problem of the Galerkin method is not simple to tackle numerically. To obtain an accurate solution it is often necessary to employ a fine mesh containing many elements. The $N \times N$ matrix of Eq. (F.9) thus becomes very large. However, since each basis function is non-zero in only one element, the matrix is sparse. This sparsity is utilized fully when implementing good computer codes for the finite-element method. The sparsity leads to a significant reduction in memory requirements since only the non-zero matrix elements need to be stored together with an index of where they are stored. Moreover, the sparsity implies a huge reduction in the number of arithmetic operations needed to solve the problem. For a full matrix problem this number is proportional to N^3 for standard Gauss elimination, but by using direct banded matrix schemes, or iterative methods like conjugate gradient methods or multigrid methods, it can be reduced to becoming proportional to N^2 or even N.

F.1.4 The Navier–Stokes equation in FEM

Using the Galerkin finite-element method the Navier–Stokes equation

$$\rho\Big[\partial_t \mathbf{v} + (\mathbf{v}\cdot\boldsymbol{\nabla})\mathbf{v}\Big] = -\boldsymbol{\nabla}p + \eta\nabla^2\mathbf{v} + \mathbf{f} \tag{F.10}$$

is rewritten in the spirit of Eq. (F.9) as

$$\rho\Big[\langle\phi_j, \partial_t\mathbf{v}\rangle + \langle\phi_j, (\mathbf{v}\cdot\boldsymbol{\nabla})\mathbf{v}\rangle\Big] + \eta\langle\boldsymbol{\nabla}\phi_j, \boldsymbol{\nabla}\mathbf{v}\rangle - \langle\boldsymbol{\nabla}\phi_j, p\rangle = \langle\boldsymbol{\nabla}\phi_j, \mathbf{f}\rangle - \int_{\partial\Omega} \mathrm{d}a\Big[\mathbf{n}p - \eta(\mathbf{n}\cdot\boldsymbol{\nabla})\mathbf{v}\Big]. \tag{F.11}$$

Here, partial integrations have been performed to eliminate the Laplace operator ∇^2 and the pressure gradient $\boldsymbol{\nabla}p$. The Navier–Stokes equation is as usual supplemented by the continuity equation $\boldsymbol{\nabla}\cdot\mathbf{v} = 0$ for incompressible fluids,

$$\langle\phi_j, \boldsymbol{\nabla}\cdot\mathbf{v}\rangle = 0. \tag{F.12}$$

When expanding the velocity components v_x and v_y as well as the pressure p in terms of the basis functions ϕ_j we obtain the following system of matrix equations

$$\rho\big[\mathbf{M}\dot{\mathbf{v}}_x + \mathbf{C}\mathbf{v}_x\big] + \eta\mathbf{K}\mathbf{v}_x - \mathbf{Q}_x^T\mathbf{p} = \mathbf{f}_x, \tag{F.13a}$$

$$\rho\big[\mathbf{M}\dot{\mathbf{v}}_y + \mathbf{C}\mathbf{v}_y\big] + \eta\mathbf{K}\mathbf{v}_y - \mathbf{Q}_y^T\mathbf{p} = \mathbf{f}_y, \tag{F.13b}$$

$$\mathbf{Q}_x\mathbf{v}_x + \mathbf{Q}_y\mathbf{v}_y = \mathbf{0}, \tag{F.13c}$$

where the column vectors $\mathbf{v}_x$, $\mathbf{v}_y$, and $\mathbf{p}$ contain the expansion coefficients for the velocity and pressure fields. The other matrices appearing are the mass matrix

$$\mathbf{M}_{jk} = \langle \phi_j, \phi_k \rangle, \tag{F.14}$$

the stiffness matrix

$$\mathbf{K}_{jk} = \langle \boldsymbol{\nabla}\phi_j, \boldsymbol{\nabla}\phi_k \rangle, \tag{F.15}$$

the convection matrix

$$\mathbf{C}_{jk} = \Big\langle \phi_j, \Big[\sum_m \phi_m v_x^{(m)}\Big]\partial_x\phi_k + \Big[\sum_m \phi_m v_y^{(m)}\Big]\partial_y\phi_k \Big\rangle, \tag{F.16}$$

and the divergence matrices

$$\mathbf{Q}_{x,jk} = \langle \phi_j, \partial_x\phi_k \rangle, \tag{F.17a}$$

$$\mathbf{Q}_{y,jk} = \langle \phi_j, \partial_y\phi_k \rangle. \tag{F.17b}$$

Finally, the force vectors $\mathbf{f}_x$ and $\mathbf{f}_y$ on the right-hand side include both the body force and the boundary integral from the partial integrations,

$$\mathbf{f}_{x,j} = \langle \phi_j, f_x \rangle - \int_{\partial\Omega} \mathrm{d}a\; \phi_k \big[n_x p - \eta(\mathbf{n}\cdot\boldsymbol{\nabla})v_x\big], \tag{F.18a}$$

$$\mathbf{f}_{y,j} = \langle \phi_j, f_y \rangle - \int_{\partial\Omega} \mathrm{d}a\; \phi_k \big[n_y p - \eta(\mathbf{n}\cdot\boldsymbol{\nabla})v_y\big]. \tag{F.18b}$$

F.2 The level set method and motion of interfaces

The level of numerical complexity increases when studying multiphase flow containing moving interfaces. When, e.g. applying the Young–Laplace pressure drop interface condition Eq. (7.9), the interface at which it is to be applied changes position in time. As a result, the geometries of the calculational domains under consideration are subject to dynamical equations of motion.

For a small oscillatory motion it suffices to make a parametric co-ordinate representation such as the one introduced in Section 14.1 and illustrated in Fig. 14.1. The parametric co-ordinates (ξ, ζ) are fixed in time, while the physical co-ordinates $\big(x(\xi,\zeta,t), z(\xi,\zeta,t)\big)$ depend on time. As a simple example of time-dependent co-ordinates in a single-phase system, let us assume that the top wall Eq. (14.2) in a liquid-filled shape-perturbed channel change its position $h(x,t)$ as a function of time due to some given external forces,

$$h(x,t) \equiv \Big[1 + \alpha\,\lambda\big(\tfrac{x}{L}\big)\cos(\omega t)\Big]\,h_0. \tag{F.19}$$

This results in the following time-dependent co-ordinate transformation

$$x(\xi,\zeta,t) = \xi\, L, \tag{F.20a}$$

$$z(\xi,\zeta,t) = \big[1 + \alpha\lambda(\xi)\cos(\omega t)\big]\,\zeta\, h_0, \tag{F.20b}$$

The coupling to the single-phase fluid domain is then given through the no-slip boundary condition Eq. (3.1) on the top wall,

$$\mathbf{v} = \mathbf{v}_{\rm wall} = \frac{\mathrm{d}}{\mathrm{d}t}\mathbf{r}(\xi,\zeta,t)\Big|_{(\xi,\zeta)=(\xi,1)}. \tag{F.21}$$

Let us move on to study the more complicated case of an interface between two phases, which moves over large distances, such as the meniscus in the capillary pump of Section 7.4. In this case, the above mentioned methods needs to be improved, e.g. by frequent remeshing of the parametric co-ordinate domains when too large deformations appear as a result of the interface motion. One of the methods able in an efficient way to follow the motion and evolution of an interface is the so-called level set method, see Sethian (1999), Osher and Fedkiw (2003), and Smereka and Sethian (2003).

In the level set method for a two-phase flow, say involving a liquid and a gas, an auxiliary field, the level set function $\phi(\mathbf{r},t)$ with values between zero and unity, is introduced to keep track of which respective parts of the computational domain the two fluids are occupying. The function takes the value unity where liquid is present, the value zero where gas is present, and intermediate values in the transition regions between the two phases. The definition of the level set function builds on the signed distance function $d(\mathbf{r})$ between a position $\mathbf{r}$ and the interface denoted ℓ at time t,

$$d(\mathbf{r},t) = \mathrm{sgn}\times\min_{\mathbf{r}_i\in\ell}\big(|\mathbf{r}-\mathbf{r}_i(t)|\big), \tag{F.22}$$

where $\mathbf{r}_i$ are the co-ordinates defining the interface ℓ, and where the sign function sgn is $+1$ on one side of the interface and -1 on the other. Finally, to avoid discontinuities in the problem, the interface transition is smeared out in a region of width ϵ. At the beginning of the simulation at $t=0$ the level set function thus takes the initial form,

$$\phi(\mathbf{r},0) = \begin{cases} 0, & \text{for} \quad d(\mathbf{r},0) < -\epsilon, \\ \frac{1}{2}+\frac{d(\mathbf{r},0)}{2\epsilon}+\frac{1}{2\pi}\sin\left(\pi\frac{d(\mathbf{r},0)}{\epsilon}\right), & \text{for} \quad -\epsilon < d(\mathbf{r},0) < \epsilon, \\ 1, & \text{for} \quad d(\mathbf{r},0) > \epsilon. \end{cases} \tag{F.23}$$

The basic time evolution of the interface is then given by the standard expression for convection in the velocity field $\mathbf{v}$,

$$\partial_t\phi + \boldsymbol{\nabla}\cdot(\mathbf{v}\phi) = 0, \tag{F.24}$$

with Eq. (F.23) as the initial condition. The advantage of the level set method is that it can handle multiple connected domains, such as seen during bubble formation or bubble breakup. Moreover, it is very easy to represent, e.g. the density and the viscosity of the entire domain as single functions interpolating between the values ρ_1, η_1 for the liquid and ρ_2 and η_2 for the gas,

$$\rho(\mathbf{r},t) = (\rho_1-\rho_2)\,\phi(\mathbf{r},t)+\rho_2, \tag{F.25a}$$
$$\eta(\mathbf{r},t) = (\eta_1-\eta_2)\,\phi(\mathbf{r},t)+\eta_2. \tag{F.25b}$$

It is more tricky to add the effects of surface tension. It is necessary to compute the mean curvature κ to be able to determine the Young–Laplace pressure drop, Eq. (7.8). This is done using the divergence of the interface normal vector $\mathbf{n}$, which is given by the normalized gradient of the level set function,

$$\kappa = -\boldsymbol{\nabla}\cdot\mathbf{n} = -\boldsymbol{\nabla}\cdot\left(\frac{\boldsymbol{\nabla}\phi}{|\boldsymbol{\nabla}\phi|}\right). \tag{F.26}$$

A serious problem with the level set method relates to lack of stability of the interface region, if it is governed by Eq. (F.24). As a function of time the sigmoid-shaped interface will broaden and eventually disappear. One solution to this problem was given by Olsson and Kreiss (2005). They introduced an artificial dynamics of the level set function, which results in stable numerics. First, an artificial diffusion was introduced by a term $\epsilon\nabla^2\phi$, and second, an artificial flux $\phi(1-\phi)\boldsymbol{\nabla}\phi/|\boldsymbol{\nabla}\phi|$ was introduced, a flux that, due to the bell-shaped gradient of ϕ, counters the broadening of the transition and squeezes it together. With these two source terms the time-evolution Eq. (F.24) for ϕ can be written in the conservative form

$$\partial_t\phi + \boldsymbol{\nabla}\cdot\left(\mathbf{v}\phi - \epsilon\boldsymbol{\nabla}\phi + \phi(1-\phi)\,\frac{\boldsymbol{\nabla}\phi}{|\boldsymbol{\nabla}\phi|}\right) = 0. \tag{F.27}$$

We notice that this governing equation for ϕ has the standard form of a diffusion-convection equation, which is easy to implement numerically.

Finally, ϕ couples to the Navier–Stokes equation not only through the velocity $\mathbf{v}$ appearing in Eq. (F.27), but also through the Young–Laplace pressure drop, as seen in Eq. (7.9). As the interface transition has been broadened the Young–Laplace pressure drop does not act on an infinitely thin interface, but rather on the broadened interface region defined by the bell-shaped gradient $\boldsymbol{\nabla}\phi$. The Navier–Stokes equation therefore takes the form

$$\rho\big[\partial_t v_i + (v_j\partial_j)v_i\big] = \partial_j\sigma_{ij} + \gamma\,\kappa\,|\boldsymbol{\nabla}\phi|\,n_i, \tag{F.28}$$

where γ is the surface tension and n_i is the ith component of the interface normal defined in Eq. (F.26). When simultaneously solving Eqs. (F.27) and (F.28) we can thus simulate a moving interface.

Bibliography

Ajdari, A. (2000). Pumping liquids using asymmetric electrode arrays. *Phys. Rev. E* **61**, R45–R48.

Ajdari, A. (2001). Transverse electrokinetic and microfluidic effects in micro-patterned channels: lubrication analysis for slab geometries. *Phys. Rev. E* **65**, 016301 1–9.

Allen, M. and Tildesley, D. (1994). *Computer simulations of liquids.* Clarendon Press, Oxford.

Aris, R. (1956). On the dispersion of a solute in a fluid flowing through a tube. *Roy. Soc.* **A235**, 67–77.

Atkins, P. W. (1994). *Physical chemistry, 5th edn.* Oxford University Press, Oxford.

Bakajin, O., Duke, T., Chou, C., Chan, S., Austin, R. and Cox, E. (1998). Electrohydrodynamic stretching of DNA in confined environments. *Phys. Rev. Lett.* **80**, 2737–2740.

Balslev, S., Jorgensen, A. M., Bilenberg, B., Mogensen, K. B., Snakenborg, D., Geschke, O., Kutter, J. P. and Kristensen, A. (2006). Lab-on-a-chip with integrated optical transducers. *Lab Chip* **6**, 213–217.

Batchelor, G. K. (2000). *An introduction to fluid dynamics.* Cambridge University Press, Cambridge.

Bazant, M. Z., Thornton, K. and Ajdari, A. (2004). Diffuse-charge dynamics in electrochemical systems Lab-on-a-chip with integrated optical transducers. *Phys. Rev. E* **70**, 021506 1–24.

Bear, J. (1972). *Dynamics of fluids in porous media.* American Elsevier Publishing Company, New York.

Berthier, J. and Silberzan, P. (2006). *Microfluidics for biotechnology.* Artech House, Boston.

Bird, R. B., Armstrong, R. C. and Hassager, O. (1987). *Dynamics of polymeric liquids, Volume 1.* John Wiley & Sons, Inc., New York.

Bird, R. B., Stewart, W. E. and Lightfoot, E. N. (2002). *Transport phenomena.* John Wiley & Sons, Inc., New York.

Brask, A., Goranović, G. and Bruus, H. (2003). Theoretical analysis of the low-voltage cascade electro-osmotic pump. *Sens. Actuators B* **92**, 127–132.

Brask, A., Kutter, J. P. and Bruus, H. (2005). Long-term stable electroosmotic pump with ion exchange membranes. *Lab chip* **5**, 730–738.

Brown, A. B. D., Smith, C. G. and Rennie, A.R. (2000). Pumping of water with ac electric fields applied to asymmetric pairs of microelectrodes. *Phys. Rev. E* **63**, 016305 1–8.

Bruus, H. and Flensberg, K. (2004). *Many-body quantum theory in condensed matter physics.* Oxford University Press, Oxford.

Cahill, B. P., Heyderman, L. J., Gobrecht, J. and Stemmer, A. (2004). Electro-osmotic streaming on application of traveling-wave electric fields. *Phys. Rev. E* **70**, 036305 1–14.

Chandrasekhar, S. (1961). *Hydrodynamic and hydromagnetic stability.* Oxford University Press, Oxford.

Chio, H., Jensen, M. J., Wang, X., Bruus, H. Attinger, D. (2006). Transient pressure drops of gas bubbles passing through liquid-filled microchannel contractions: an experimental study. *J. Micromech. Microeng.* **16**, 143–149.

Cussler, E. L. (1997). *Diffusion, mass transfer in fluid systems.* Cambridge University Press, Cambridge.

de Gennes, P.-G., Brochard-Wyart, F. and Quéré, D. (2003). *Capillarity and wetting phenomena: Drops, bubbles, pearls, waves.* Springer, New York.

Dullien, F. A. L. (1979). Porous media, fluid transport and pore structure. Academic Press Inc., New York.

El-Ali, J., Perch-Nielsen, I. R., Poulsen, C. R., Bang, D. D., Telleman, P. and Wolff, A. (2004). Simulation and experimental validation of a SU-8 based PCR thermocycler chip with integrated heaters and temperature sensor. *Sens. Actuators A* **110**, 3–10.

Faber, T. E. (1995). *Fluid dynamics for physicists.* Cambridge University Press, Cambridge.

Feynman, R. P. (1972). *Statistical mechanics: a set of lecture notes.* Addison-Wesley Publishing Company, Inc., New York.

Garstecki, P., Fuerstman, M. J. and Whitesides, G. M. (2005). Nonlinear dynamics of a flow-focusing bubble generator: An inverted dripping faucet. *Phys. Rev. Lett.* **94**, 234502 1–4.

Garstecki, P., Stone, H. A. and Whitesides, G. M. (2005). Mechanism for flowrate controlled breakup in confined geometries: A route to monodisperse emulsions. *Phys. Rev. Lett.* **94**, 164501 1–4.

Geschke, O., Klank, H. and Telleman, P. (eds.) (2004). *Microsystem engineering of lab-on-a-chip devices.* Wiley-VCH Verlag, Weinheim.

Gijs, M. A. M. (2004). Magnetic bead handling on-chip: new opportunities for analytical applications. *Microfluid Nanofluid* **1**, 22–40.

Hagsäter, S. M., Glasdam Jensen, T., Bruus, H. and Kutter, J. P. (2007). Acoustic resonancesin microfluidic chips: full-image micro-PIV experiments and numerical simulations. *Lab Chip*, **7**, 1336–1344.

Happel, J. and Brenner, H. (1983). *Low Reynolds number hydrodynamics.* Kluwer Boston Inc., Hingham.

Hasegawa, T., Kido, T., Iizuka, T. and Matsuoka, C. (2000). A general theory of Rayleigh and Langevin radiation pressures. *J. Acoust. Soc. Jpn. E* **21**, 145–152.

Helbo, B., Kristensen, A. and Menon, A. (2003) A micro-cavity fluidic dye laser. *J. Micromech. Microeng.* **13**, 307–311.

Jackson, J. D. (1975). *Classical electrodynamics.* John Wiley & Sons, Inc., New York.

Jensen, M. J., Goranović, G. and Bruus, H. (2004). The clogging pressure of bubbles in hydrophilic microchannel contractions. *J. Micromech. Microeng.* **14**, 876–883.

Jensen, M. J., Stone, H. A. and Bruus, H. (2006). A numerical study of two-phase Stokes flow in an axisymmetric flow-focusing device. *Phys. Fluids* **18**, 077103 1–10.
Jones, T. B. (1995). *Electromechanics of particles.* Cambridge University Press, Cambridge.
Joseph, P. and Tabeling, P. (2005). Direct measurement of the apparent slip length. *Phys. Rev. E* **71**, 035303(R) 1–4.
Kameoka, J. and Craighead, H. (2001). Nanofabricated refractive index sensor based on photon tunneling in nanofluidic channel. *Sens. Actuators B* **77**, 632–637.
Karniadakis, G. E. and Beskok, A. (2002). *Micro flows: fundamentals and simulation.* Springer-Verlag, New York.
Kilic, M. S., Bazant, M. Z. and Ajdari, A. (2007). Steric effects in the dynamics of electrolytes at large applied voltages. I & II. *Phys. Rev. E* **75**, 021502 1–16, *ibid.* 021503 1–11.
Kim, J.-W., Utada, A. S., Fernández-Nieves, A., Hu, Z. and Weitz, D. A. (2007). Fabrication of monodisperse gel shells and functional microgels in microfluidic devices. *Angew. Chem. Int. Ed.* **46**, 1819–1822.
King, L. V. (1934). On the acoustic radiation pressure on spheres. *Proc. R. Soc. (London)* **A147**, 212–240.
Kirby, B. J. and Hasselbrink, E. F. (2004). Zeta-potential of microfluidic substrates: 1, Theory, experimental techniques, and effects on separations. *Electrophoresis* **25**, 187–202.
Kittel, C. and Kroemer, H. (1982). *Thermal physics, 2nd edn.* W. H. Freeman and Company, New York.
Koplik, J. and Banavar, J. (1995). Continuum deductions from molecular hydrodynamics. *Ann. Rev. Fluid Mech.* **27**, 257–292.
Lamb, H. (1997). *Hydrodynamics.* Cambridge University Press, Cambridge.
Landau, L. D. and Lifshitz, E. M. (1982). *Statistical physics part 1, 3rd edn.* Course of theoretical physics vol. 5. Pergamon Press, Oxford.
Landau, L. D. and Lifshitz, E. M. (1993). *Fluid mechanics, 2nd edn.* Course of theoretical physics vol. 6. Pergamon Press, Oxford.
Landau, L. D., Lifshitz, E. M. and Pitaevskiĭ, L. P. (1984). *Electrodynamics of continuous media, 2nd edn.* Course of theoretical physics vol. 8. Pergamon Press, Oxford.
Laser, D. J. and Santiago, J. G. (2004). A review of micropumps *J. Micromech. Microeng.* **14**, R35–R64.
Lighthill, J. (2003). *Waves in fluids.* Cambridge University Press, Cambridge.
Lund-Olesen, T., Bruus, H. and Hansen, M. F. (2007). Quantitative characterization of magnetic separators: comparison of systems with and without integrated microfluidic mixers. *Biomed. Microdevices* **9**, 195–205.
Markham, J. J. (1952). Second-order acoustic fields: streaming with viscosity and relaxation. *Phys. Rev.* **86**, 497–502.
Ménétrier-Deremble, L. and Tabeling, P. (2006). Droplet breakup in microfluidic junctions of arbitrary angles. *Phys. Rev. E* **74**, 035303(R) 1–4.
Merzbacher, E. (1998). *Quantum mechanics, third edn.* John Wiley & Sons, Inc., New York.

Mikkelsen, C. and Bruus, H. (2005). Microfluidic capturing-dynamics of paramagnetic bead suspensions. *Lab Chip* **5**, 1293–1297.

Monat, C., Domachuk, P. and Eggleton, B. J. (2007). Integrated optofluidics: A new river of light. *Nature Photon.* **1**, 106–114.

Morrison, F. A. (2001). *Understanding rheology.* Oxford University Press, Oxford.

Mortensen, N. A., Okkels, F. and Bruus H. (2005). Reexamination of Hagen–Poiseuille flow: Shape dependence of the hydraulic resistance in microchannels. *Phys. Rev. E* **71**, 057301.

Mortensen, N. A., Olesen, L. H., Belmon, L. and Bruus H. (2005). Electrohydrodynamics of binary electrolytes driven by modulated surface potentials. *Phys. Rev. E* **71**, 056306.

Mortensen, N. A. and Xiao, S. (2007). Slow-light enhancement of Beer–Lambert–Bouguer absorption. *Appl. Phys. Lett.* **90**, 141108 1–3.

Nyborg, W. L. (1958). Acoustic streaming near a boundary. *J. Acoust. Soc. Am.* **30**, 329–339.

Olesen, L. H. (2003). *Computational fluid dynamics in microfluidic systems.* M.Sc. thesis, www.mic.dtu.dk/microfluidics, Technical University of Denmark.

Olesen, L. H., Bruus, H. and Ajdari, A. (2006). AC electrokinetic micropumps: The effect of geometrical confinement, Faradaic current injection, and nonlinear surface capacitance *Phys. Rev. E* **73**, 056313 1–16.

Olsson, E. and Kreiss, G. (2005). A conservative level set method for two phase flow. *J. Comp. Phys.* **210**, 225–246.

Osher, S. and Fedkiw, R. (2003). *Level set methods and dynamics implicit surfaces.* Springer-Verlag, New York.

Pamme, N. (2006). Magnetism and Microfluidics. *Lab Chip* **6**, 24–38.

Persson, F., Thamdrup, L. H., Mikkelsen, M. B. L., Jarlgaard, S.E., Skafte-Pedersen, P., Bruus, H. and Kristensen, A. (2007). Double thermal oxidation scheme for fabrication of SiO_2 nanochannels. *Nanotechnology* **18**, 245301 1-4.

Petersson, F., Nilsson, A., Holm, C., Jönsson, H. and Laurell, T. (2005). Continuous separation of lipid particles from erythrocytes by means of laminar flow and acoustic standing wave forces. *Lab Chip* **5**, 20–22.

Probstein, R. F. (1994). *Physicochemical hydrodynamics.* John Wiley & Sons, Inc., New York.

Psaltis, D., Quake, S. R. and Yang, C. (2006). Developing optofluidic technology through the fusion of microfluidics and optics. *Nature* **442**, 381–386.

Ramos, A., Morgan, H., Green, N. G., Gonzalez, A. and Castellanos, A. (2005). Pumping of liquids with travelling-wave electro-osmosis. *J. Appl. Phys.* **97**, 084906 1–8.

Reisner, W., Beech, J. P., Larsen, N. B., Flyvbjerg, H., Kristensen, A., and Tegenfeldt, J. O. (2007). Nanoconfinement-enhanced conformational response of single DNA molecules to changes in ionic environment. *Phys. Rev. Lett.* **99**, 058302.

Santiago, J. G., Wereley, S. T., Meinhart, C. D., Beebe, D. J. and Adrian, R. J. (1998). A particle image velocimetry system for microfluidics. *Exp. Fluids* **25**, 316–319.

Schift, H. and Kristensen, A. (2007). Nanoimprint lithography. Part A, Chap. 8 in *Springer Handbook of Nanotechnology. 2nd edn.* (ed. Bhushan, B.) Springer, New York.

Sethian, J. A. (1999). *Level set methods and fast marching methods.* Cambridge University Press, Cambridge.